Lecture Notes in Computer Science 1201

Edited by G. Goos, J. Hartmanis and J. van Leeuwen

Advisory Board: W. Brauer   D. Gries   J. Stoer

Springer
*Berlin
Heidelberg
New York
Barcelona
Budapest
Hong Kong
London
Milan
Paris
Santa Clara
Singapore
Tokyo*

Oded Maler   (Ed.)

Hybrid and Real-Time Systems

International Workshop, HART'97
Grenoble, France, March 26-28, 1997
Proceedings

 Springer

Series Editors

Gerhard Goos, Karlsruhe University, Germany

Juris Hartmanis, Cornell University, NY, USA

Jan van Leeuwen, Utrecht University, The Netherlands

Volume Editor

Oded Maler
CNRS-VERIMAG, Centre Equation
2, av. de Vignate, F-38610 Gières, France
E-mail: oded.maler@imag.fr

Cataloging-in-Publication data applied for

Die Deutsche Bibliothek - CIP-Einheitsaufnahme

Hybrid and real time systems : international workshop ; proceedings / HART, '97,
Grenoble, France, March 26 - 28, 1997. Oded Maler (ed.). - Berlin ; Heidelberg ; New
York ; Barcelona ; Budapest ; Hong Kong ; London ; Milan ; Paris ; Santa Clara ;
Singapore ; Tokyo : Springer, 1997
 (Lecture notes in computer science ; 1201)
 ISBN 3-540-62600-X

NE: HART <1997, Grenoble>; Maler, Oded [Hrsg.]; GT

CR Subject Classification (1991): C.1.m, C.3, D.2.1, D.4.7,F.3.1, F.1-2

ISSN 0302-9743
ISBN 3-540-62600-X Springer-Verlag Berlin Heidelberg New York

Typesetting: Camera-ready by author
SPIN 10549527   06/3142 – 5 4 3 2 1 0   Printed on acid-free paper

Preface

This volume contains the proceedings of HART'97, the international workshop on *Hybrid And Real-Time systems*, held in Grenoble, France, March 26–28, 1997. This workshop, sponsored by the European Community research grant HYBRID EC-US-043, was preceded by several formal and informal meetings, at Ithaca, New York (1991, 1994, 1996), Lyngby, Denmark (1992), Boston, Massachusetts (1993), Grenoble, France (1995) and New Brunswick, New Jersey (1995). The proceedings of some of these workshops can be found in volumes 736, 999, and 1066 of the Springer-Verlag Lecture Notes in Computer Science series.

The focus of the workshop was on mathematical methods for the rigorous and systematic design and analysis of hybrid systems. A hybrid system consists of digital devices that interact with analog environments. Driven by rapid advances in digital controller technology, hybrid systems are objects of investigation of increasing relevance and importance. The emerging area of hybrid systems research lies at the crossroads of computer science and control theory: computer science contributes expertise on the digital aspects of a hybrid system, and control theory contributes expertise on the analog aspects. Since the two research communities speak largely different languages, and employ largely different methods, it was the purpose of the workshop to bring together researchers from both computer science and control theory.

The three-day workshop featured 2 tutorial surveys, 5 invited presentations and 31 contributed papers (22 regular and 9 short), which were refereed and selected by the program committee before the workshop. I would like to thank the program committee members, R. Alur (Bell Labs), Z. Artstein (Weizmann), E. Asarin (Moscow), T. Henzinger (Berkeley), B. Krogh (CMU), Y. Lakhnech (Kiel), K. Larsen (Aalborg), N. Lynch (MIT), S. Nadjm-Tehrani (Linkoping), A. Pnueli (Weizmann), J. van Schuppen (CWI), E. Sontag (Rutgers), P. Varaiya (Berkeley) and F. Vaandrager (Nijmegen) for the effort they invested in the refereeing process. The help of the sub-referees J. Andersen, A. Bouajjani, M. Broucke, A. Deshpande, K. Havelund, K. Kristoffersen, J. Malec, T. Padron-McCarthy, C. Petersohn, A. Puri, A. van der Schaft, A. Skou, J. Springintveld, J. Stroemberg, and L. Urbina is gratefully acknowledged.

We are grateful to all invitees and contributors for making the workshop a success. In addition, I would like to thank my colleagues Sergio Yovine and Ahmed Bouajjani for their help, Chantal Costes for taking care of all the organizational details, and Joseph Sifakis and the rest of VERIMAG members for creating and maintaining an environment in which all this could happen.

Grenoble, France
December 1996

Oded Maler

Table of Contents

Verifying Liveness Properties
of Reactive Systems
(A Tutorial)

Amir Pnueli

Department of Computer Science, Weizmann Institute, Rehovot 76100, Israel
amir@wisdom.weizmann.ac.il

Abstract. In this tutorial, we will survey methods for the formal verification of liveness properties of reactive systems. A typical liveness property is the property of response, claiming that every p is eventually followed by q.

Most proofs of liveness properties rely on the construction of a well-founded ranking function. This is a function of the state which is guaranteed not to increase on every step of the computation and now and then must decrease, until the goal q is achieved. The fact that it is well-founded means that it cannot decrease infinitely many times.

We will consider a prefix of the following list of topics:

1. Verifying response properties of sequential programs, where an integer-valued ranking function is adequate.
2. Verifying responsiveness of reactive systems under the assumption of justice (weak fairness).
3. Verification diagrams for succinct presentation of reponse proofs.
4. Verifying responsiveness of reactive systems under general fairness (weak and strong): (a) A recursive rule, (b) Using progress measures.
5. Verifying reactivity (general progress) properties.
6. Liveness properties of clocked transition systems.
7. Approaches to verification of the non-Zeno'ness of a system.
8. Approaches to the verification of stability of hybrid systems.

The Lyapunov Method
(A Tutorial)

Zvi Artstein

Department of Mathematics, Weizmann Institute, Rehovot 76100, Israel
MTARTS@weizmann.weizmann.ac.il

Abstract. It is more than a century that the Lyapunov method is employed in the analysis of dynamical systems, primarily for purposes of stability and stabilization. In this tutorial we will survey the ideas underlying the method, and the scope of available techniques and applications. Of particular interest will be for us the tradeoff between the directness of the method, and its generality. Variants like vector Lyapunov functions, stability preserving mappings, invariance principles, etc., will be presented. The role of converse results will be discussed. Examples of Lyapunov functions will be displayed, with a comment on how such functions can be detected. The need to employ different Lyapunov forms in different structures will be demonstrated.

The role Lyapunov functions play in planning and control will also be discussed. Here a quantifier is added to the prescription. We shall notice the additional interplay with the policy variable.

Relating High-Level and Low-Level
Action Descriptions
in a Logic of Actions and Change

Erik Sandewall[1]

Department of Computer and Information Science
Linköping University, S-58183 Linköping, Sweden

Abstract. We address the problem of formally proving high-level effect descriptions of actions from low-level operational definitions. The low-level operational description has a hybrid character, involving both continuous and discrete change; the effect description is wholly or predominantly discrete. Both descriptions are expressed in a logic of actions and change, with extensions for characterizing continuous change, discontinuities, the distinction between true and estimated values of state variables, and the distinction between success and failure of an action. Both descriptions also require the use of nonmonotonicity in the logic in question. The transition from operational definition to effect description furthermore involves the creation of a closure with respect to the set of possible ways that the action can fail.
We outline, by means of an example, how these issues can be addressed as an extension of existing results on logics of actions and change.

1 Background and topic

The area of *cognitive robotics* is a branch of artificial intelligence (AI), and attempts to develop formal methods for the design of robots that reason about their own actions. The capability to reason about continuously changing properties of the world and about the properties and imperfections of sensors and actuators is important for this goal. However, other issues have so far received much more attention in this research.

It is generally recognized, in AI, that one has a need for high-level descriptions of actions which describe the effects of a robot action in terms of its preconditions and postconditions. These high-level descriptions are required for the purpose of planning sequences of actions which will achieve desired goals, and for the modification of those sequences (replanning). It is generally recognized, as well, that one must separate the main action description, which characterizes its normal conditions and effects, from the additional information describing various exceptional cases.

Finally, it is generally recognized that non-monotonic logics are needed in order to obtain that the action descriptions be concise and well structured.

We observe, however, that the *effect descriptions* of actions which are used e.g. for planning are rarely directly executable, if ever; they must be complemented by low-level *operational definitions* which specify how the action is actually performed in terms of continuous change, sensors, actuators, and control algorithms. In view of that observation, the present article addresses the following problem: *how can high-level effect descriptions of actions be formally proved from their low-level operational definitions.*

The present paper will describe our solution to this problem in terms of an example. Other, concurrent articles describe it in more general terms. The following concrete example will be used. Consider the move operation for a robot with the following properties: If the robot is told to move forward, it will gradually increase its velocity until it reaches 1 m/s, and then proceed at constant speed. The acceleration is a constant 0.5 m/s^2 until it reaches maximum speed. If the robot is told to stop, it stops instantly. There are also two cases where it will stop under the command of its motoric system, regardless of higher-level control: if its proximity sensor indicates an obstacle right in front of it, and if the emergency stop system which has been installed for safety reasons in the robot hall broadcasts a command to stop. The emergency stop is caused when any one of several alarm buttons is pushed; the stop request is then sent over an IR channel.

The relatively easy part of the problem is to arrange that the effect description for the normal case of the action follows from the operational definition. The hard part is to arrange that the auxiliary information which characterizes the failure of the action (for example because of power failure, obstacle in the way, etc) is represented in the context of the operational definition and the effect description, and that the latter kind of auxiliary information can be derived from the former one.

Although the action in the chosen example is an elementary one, we shall also reference related work showing how composite and higher-level actions can be defined in terms of elementary ones, in such a fashion that the same verification technique applies to them as well.

2 Formalism

Our approach will be to represent both the low-level and the high-level description of the action in terms of a first-order predicate calculus (FOPC) with several sorts, where time, represented as the real numbers, is one of those sorts. This is the traditional approach in AI and in this branch of research. Two primary predicates are used, H for *Holds* and D for *Do*, and they are defined as follows. $H(t, p)$ says that the "propositional fluent" or "propositional feature" p holds at time t. In other words, p is reified and $H(t, p)$ is the same as $p(t)$ in the case where p is atomic. (One reason for the reification is that some additional, auxiliary predicates also take such fluents or features as an argument). $D([s, t], a)$ says that the action a is performed over the time interval $[s, t]$.

A "feature" is then what is called a "state variable" in engineering; the word "variable" is avoided here because it has another connotation in logic. Non-propositional features are also admitted, using the notation $H(t, f : v)$ where $f : v$

is the proposition saying that the feature f has the value v.

We proceed now to describe which extensions to these common-sense notions are needed for the present purpose.

2.1 Separate Holds predicates for actual and estimated feature values

The operational definition for the robot shall not be expressed directly in terms of quantities in the world itself, but in terms of entities that are known to the robot, in particular through its sensors. Thus, rather than just having one feature (state variable, fluent) e.g. for the robot's position, one needs to distinguish between the actual position and the position as estimated by the robot itself. Statements about the relationship between those two quantities express our knowledge of the accuracy of the position sensor; the operational definition of the move action will use the estimated value; and the effect description of the same action will use the actual, physical position. The characterization of actuators is analogous to the one for sensors. These different kinds of information are then combined when effect descriptions are derived from operational action definitions.

Although in principle it would be possible to introduce two distinct state variables whenever needed, we prefer the more systematic approach proposed by Drakengren (in our research group) of complementing the traditional *Holds* predicate with a second predicate for the estimated value of a feature. Thus, we distinguish two predicates: a predicate H which specifies what is true in the world, and a predicate B which has the same argument structure as H and which expresses what value the agent (robot) *believes* is assigned to each feature. For example,

$$H(25.05, pos(r) : (45.2, 63.8))$$

will express that at time 25.05, the current value of the feature $pos(r)$ which is intended as the current position of the robot r in Cartesian room coordinates is $(45.2, 63.8)$. Similarly,

$$B(25.05, pos(r) : (45.4, 63.5))$$

will express that the estimated value for the same time is $(45.4, 63.5)$.

Since some feature values may be unknown, B generally allows nil as a feature value for the undefined value. For the moment, we do not allow for a set of alternative values in B, but this would be a natural generalization for the future.

Notice that B expresses the "raw" belief that the ego obtains from its sensors. It is possible that its reasoning capability will be used to question and revise some of those values, but that is not reflected in B as presently interpreted.

In this way we can describe how a sensor behaves relative to the world (B depends on H) and how an actuator behaves (H depends on B but also on context information).

2.2 Features for the current example

The following are the features (state variables) that are needed for the chosen simple application. For navigation, it is first of all necessary to specify the robot's attitude, that is, the direction that it is "looking".

attid [r] shall be either of N, S, E, or W, indicating where r is "looking" and with the obvious interpretations. North is in the direction of the y axis, and east is in the direction of the x axis.

pos[r] are the present coordinates of the robot r in the floorplane, represented as a complex number (x, y). Addition, subtraction, and multiplication of complex numbers have their standard definitions.

hdiag[r] is the half diagonal of the robot r, viewed as a rectangular object, in its own coordinate system. That is, it is a pair (x, y) where $2x$ is the width of the object and $2y$ is its length.

obstac[r] is true if the robot r is just runnig into an obstacle. In the context of B, this is detected by its obstacle detection sensor.

emergency is true if an emergency button is being pushed.

The arguments of these feature functions will sometimes be enclosed by round parentheses, sometimes with square brackets, according to the following convention. *attid*(r) is a *feature* which may appear in an argument of H ; *attid*[r] is the *value* of *attid*(r) in the current context. Thus,

$$\mathsf{H}(t, attid(r) : v) \leftrightarrow attid[r] = v$$

in a context where t is the current time.

Of these feature-valued functions, *attid*[r] and *hdiag*[x] for all mobile objects x are assumed to be always correctly known by the robot. This is expressed by

$$\mathsf{B}(t, attid(r) : v) \leftrightarrow \mathsf{H}(t, attid(r) : v)$$

and (illustrating at the same time how we would deal with the case where there are several kinds of rectangular objects, only some of which are mobile):

$$\forall x [\mathsf{Isa}(x, \mathsf{mobile}) \rightarrow (\mathsf{B}(t, hdiag(x) : v) \leftrightarrow \mathsf{H}(t, hdiag(x) : v))]$$

with implicit universal quantification over free variables, as usual. For *pos* we retain the possibility of a discrepancy between held and believed values.

2.3 Failible actions

For characterizing movement actions, it is necessary to have appropriate notation for dealing with the applicability, success, and failure of actions. We introduce several variants of the D ("Do") operator, namely one variant when the action succeeds and another one when it fails. The following are the details of the notation. In all cases, s and t are timepoints (usually s for starting time and t for termination time) and a is an action.

G(s, a): the action a is invoked ("go") at time s

A(s, a): the action a is applicable at time s

$D_s([s,t],a)$: the action a is executed successfully over the time interval $[s,t]$: it starts at time s and terminates successfully at time t.

$D_f([s,t],a)$: the action a is executed but fails over the time interval $[s,t]$: it starts at time s and terminates with failure at time t.

$D_c([s,t],a)$: the action a is being executed; the execution started at time s and has not been terminated before time t. (It may terminate at t or later).

It is also convenient to have the following abbreviation:

$D_n([s,t],a)$, defined as $D_c([s,t],a) \wedge \neg D_s([s,t],a)$.

For both D_s and D_f, s is the time when the action was invoked, and t is the exact time when it completes with success or failure.

A set of axioms stating the properties and relationships of these predicates is included in the full version of the paper. (See the WWW reference at the end of this paper). The following is an example of such an axiom. If an action is being executed, then it must have been invoked:

$$D_c([s,t],a) \rightarrow G(s,a)$$

As another example of the notation, here is the formula stating that a condition φ guarantees that an action always succeeds:

$$H(s,\varphi) \wedge G(s,a) \rightarrow \exists t[D_s([s,t],a)]$$

Ordinary action laws specify the action's effects when it succeeds. They are therefore written as usual and with D_s on the antecedent side: if preconditions apply and the action is performed successfully, then the postconditions result.

The full paper also contains an action schema for the description of actions in a particular special case, namely those actions whose effects are characterized by a precondition, a prevail condition, and a postcondition, and where the postcondition is at the same time the termination condition for the action. The prevail condition must be satisfied throughout the execution of the action; if it is violated then the action fails. The sections that follow below show how the formalism can be used for more general classes of actions.

3 Effect descriptions for actions

The effect descriptions shall be expressed in terms of what goes on in the world itself, and not in terms of internal quantities or beliefs in the robot. They are to consist of two parts: a *success description* which specifies the effects of the action if it succeeds, and a *failure description* which specifies the various ways that the action can fail.

The success description shall have the general form

$$D_s([s,t],a) \rightarrow (\text{results})$$

The applicability conditions need not be stated here, since they are implied by D_s according to the general axioms relating the above predicates.

One may consider two ways of writing the failure descriptions. A *closed* failure description has the general form

$$D_f([s, t], a) \rightarrow \neg A(s, a) \lor \delta_1 \lor \delta_2 \lor \ldots \lor \delta_k$$

where each of the δ_i characterizes one way or group of ways that the action may go wrong. An *open* failure description, on the other hand, is a set of axioms each of which has the form

$$\delta_i \rightarrow D_f([s, t], a)$$

The difference between these alternatives is not simply a question whether additional causes of error are to be edited into the single axiom for the closed description, or added as an additional axiom in the open description. In particular, the open description makes it possible to have cases of failure which are *implied* by underlying or more general circumstances, so that it is not necessary to enumerate all error reasons explicitly.

For these reasons, open failure descriptions would seem to be much more in agreement with the tradition in A.I. We shall argue, however, that for many purposes it is necessary to have a closed failure description on the effect level (that is, the high-level description of the action), since only so will one obtain a firm knowledge of when the action is guaranteed to work correctly. For the operational definition, on the other hand, it is quite appropriate to use an open failure description. The derivation of a closed failure description for the high level from an open failure description on the lower level is therefore a part of the problem being addressed here.

One of the axioms that were referenced in section 2 states that if an action is invoked then it either succeeds or fails. Using that axiom, it follows that either the success description or the failure description must apply whenever an action has been invoked. Also, if one can prove that all of the δ_i in a closed failure description must be false and that the applicability condition holds, then one is entitled to conclude that the action succeeds if invoked.

In our particular example, we obtain the following effect descriptions. The success description is

$$D_s([s, t], \text{GOTO}(r, p)) \rightarrow H(t, pos(r) : p)$$

saying that if r goes to p successfully, then at the end of the action period it is at p. Under the assumptions outlined above, the failure description is

$$D_f([s, t], \text{GOTO}(r, p)) \rightarrow \neg A(s, a) \lor H(t, obstac(r)) \lor H(t, emergency)$$

The problem now is to show how these effect descriptions can be derived from an operational description.

4 Operational definition of the move action

We shall introduce the first set of formulae for the operational definition of a simple move action in the framework that has just been defined. We do it for the elementary action GOTO that just moves ahead to a given position or to a given object.

4.1 Auxiliary constructs

The basic features for describing the robot have already been introduced. For navigation, it is convenient to use the robot's coordinate system for many navigation tasks. The following auxiliary definitions do that.

$xc(z)$ and $yc(z)$ extract the x and y coordinates, respectively, from a complex number.

$mc(z, r)$ transforms the coordinate $z = (x, y)$ in floorplan coordinates to the coordinate system of the robot (or other mobile object) r, where y is straight ahead in the direction of the attitude of r, and x is towards the right. This function is defined as

$$mc(z, r) = (z - pos[r]) * cdir(attid[r])$$

where the auxiliary function $cdir$ is defined as

$$\{N \mapsto (1, 0), S \mapsto (-1, 0), E \mapsto (0, 1), W \mapsto (0, -1)\}$$

Notice that these are mathematical functions; they are not features and are not part of the temporal interpretations. These mathematical functions are of course assumed known to the robot, in the sense that they can be used in the definitions of the operational action definitions.

4.2 Applicability of the action

We proceed to using the functions that have so been introduced, and begin with the applicability condition of the action GOTO. It can be defined as follows.

$$A(s, \text{GOTO}(r, p)) \leftrightarrow$$

$$B(s, |xc(mc(p, r))| \leq xc(hdiag[r]) \wedge$$

$$yc(mc(p, r)) > 0 \wedge$$

$$\partial pos[r] = \partial^2 pos[r] = (0, 0))$$

taking certain obvious liberties in the second argument of B. This definition means that the operation of going to the position p in an elementary way is applicable iff p is ahead of the robot r in its own forward direction, with a displacement towards the left or the right which is no more than the half-width of the robot. In other words, some point of the robot will touch p if the action is successful, but it will not necessarily be the centerline of the robot.

Notice that the applicability is defined in terms of what the robot believes, not in terms of what is actually the case. One may think of the action's operational definition as a software subroutine which is invoked in the way specified by the G predicate. This subroutine will first relate the parameters that are sent to it, to its present estimates of the state of the world, and if it considers that the requested action is impossible in the present situation, it will return immediately and without having engaged any actuator. This immediate return is modelled by the negation of A, and is assumed to take zero time. If the initial tests are passed, then the action is considered applicable in the sense of A, it takes non-zero time, and still it may either succeed or fail.

In this framework, it is evident that applicability must be *defined* in terms of "believed" rather than "held" values of features/fluents. Still, given that one has some knowledge about the relationship between believed and held values for each sensor (that is, about the properties of that sensor), one can of course expect to *derive* a condition for applicability in terms of "held" or actual values.

In the present, relatively simple approach, we assume that the robot has one specific belief about the position, and do not allow for the case that it only believes that the position is in a certain range.

4.3 Invocation and termination of the action

We now proceed to the invocation of GOTO. The general structure of the definitions is as follows: the proposition $G(GOTO(r, p))$ shall imply a certain momentary, discrete change in the world, which starts a chain of spontaneous change corresponding to the execution of the action, and finally there is (if necessary) another momentary, discrete change when the action terminates. The important thing is that that what goes on while the action is in progress is modelled using a technique for describing physical movements.

The following intuition may be useful. Suppose the robot were not a cart that is propelled by motors driving its wheels, but instead just a particle that slides without friction on a flat surface. What is needed then is to assign a velocity to it at the beginning of the action, allow it to run its course according to the laws of physics, and then reassign its velocity again when the destination has been reached.

The concept of reassignment is intuitively clear, particularly from a programming language background, but it requires special treatment in a logic of the present kind. Let us first show how it is used, and then proceed to its formal properties in a later section. Writing $H(t, f := v)$ for "the feature f adopts the value v at time t", the invocation of GOTO can be defined as follows:

$$\forall t[G(t, GOTO(r, p)) \rightarrow B(t, \partial^2 pos(r) := 0.5 * derdir(attid(r)))]$$

where the auxiliary function *derdir* is defined as

$$\{N \mapsto (0, 1), S \mapsto (0, -1), E \mapsto (1, 0), W \mapsto (-1, 0)\}.$$

(Unfortunately it does not come out the same as *cdir*). Here, the operator ∂ represents differentiation on both components of the position, ∂^2 is the second derivative, so it results in the acceleration viewed as a vector. No rotation is involved. Notice that we specify the robot's velocity in floorplan coordinates, and not relative to the direction where it is "looking". This is because we wish to treat the floorplan-coordinate position as the primary one, which is subject to the common routines about restriction of change, whereas robot coordinates must always be derived. (When the robot moves, then everything else in the room changes its position in the robot coordinates, so this does not go well with an assumption of more or less strict inertia or persistence).

Notice also that this definition specifies the new value (by the assignment operation) of *believed* acceleration. We use this here for indicating the value that is given to the actuator. Actually, it might have been preferable to have a third relation R besides H and B, reserving B for estimated feature values, and using R for reference values that the controller is requested to maintain. However, for the purpose of the present article it is sufficient to identify R with B, and to assume that the invocation of the action sets the appropriate value for the robot's acceleration for an instant, but for all subsequent times the believed value of the acceleration will be the estimated one.

We also stipulated an assumption that an initial acceleration of 0.5 will be held until the robot reaches the velocity of 1 m/s, and then the speed will become constant. We model this as change as something which happens without explicit intervention by the ego, and which therefore has the same status as a law of nature. It is expressed by

$$|\partial pos[r]| = 1 \wedge |\partial^2 pos[r]| \neq 0 \rightarrow |\partial^2 pos[r]| := 0$$

This applies of course for an idealized world; in a more realistic case one would have to compensate to hold velocity constant at the value 1.

Finally, we have to specify that the elementary action terminates with success if the robot reaches the target position:

$$B(t, yc(mc(p,r)) = 0) \wedge D_c([s,t], \text{GOTO}(r,p)) \rightarrow$$

$$D_s([s,t], \text{GOTO}(r,p)) \wedge B(t, \partial pos(r) := (0,0)) \wedge$$

$$B(t, \partial^2 pos(r) := (0,0))$$

4.4 Failure conditions for the action

We arrive now to the specifications of how the action can fail. In general, for each action there is a multitude of reasons why it may fail: various things can break, obstacles may come into the way, etc etc. In the AI literature this has been described by the term *qualification problem* (although the meaning of that term seems to have slipped in the last few years). It has been recognized since very long [3] that a nonmonotonic logic is needed for this: it is not possible to enumerate all failure reasons at once; some failures may be implied indirectly from various other and more general circumstances, and we wish to say that the action succeeds unless there is *some* reason why it fails.

The present logical system is no exception; the nonmonotonicity is introduced by preferring a model where an action continues over a model where it terminates, either by success or by failure.

To show how this works, consider the case where the ego terminates the action as failed if its proximity sensor *obstac(r)* senses something in the way although the destination object has not been reached. The corresponding proposition is as follows.

$$B(t, obstac(r)) \wedge D_n([s,t], \text{GOTO}(r,p)) \rightarrow$$

$$D_f([s,t], \text{GOTO}(r,p)) \wedge B(t, \partial pos(r) := (0,0)) \wedge$$

$$B(t, \partial^2 pos(r) := (0,0))$$

Notice how this rule simply implies a D_f expression for the action; other axioms may imply failure termination independently using similar consequents. The second condition for interrupting the action, namely that someone pushed a emergency button, can be added as follows:

$$B(t, emergency) \wedge D_n([s,t], \text{GOTO}(r,p)) \rightarrow$$

$$D_f([s,t], \text{GOTO}(r,p)) \wedge B(t, \partial pos(r) := (0,0)) \wedge$$

$$B(t, \partial^2 pos(r) := (0,0))$$

Referring back to the general axioms for the variants of the D operator, we see how the action, once invoked, is able to continue towards a success (when the successful termination condition is finally satisfied) unless there is a qualification that forces it to fail.

With this technique, additional reasons for failure of the action can be added in a modular fashion.

5 Nonmonotonicity

The first-order formalism using the *Holds* and *Do* predicates allows a considerable flexibility in expressing actions invoked by one or more agents, events which are caused directly or indirectly by the actions of agents, or which simply arise due to the laws of nature, extended duration and concurrency of actions and events, and so on. The present paper has scetched how this formalism can also deal with hybrid (mixed continuous-discrete) systems and with failure conditions of actions.

One very characteristic feature of the formalism that is obtained in this fashion is that, although it is based on classical FOPC, it must also incorporate some *nonmonotonic* technique in order that the intended conclusions shall be obtained for given scenarios. Monotonicity, in this context, means that if S is a set of axioms, $Th(S)$ is the set of conclusions that follow from S, and $S \subseteq S'$, it follows that $Th(S) \subseteq Th(S')$. In other words, previously obtained conclusions can be retained when more axioms are added. It is this traditional monotonicity property that is *not* upheld in logics of actions and change.

Technically speaking, nonmonotonicity is obtained by using *entailment methods* which restrict the set of models. If S is a set of axioms and $Mod(S)$ is the set of its classical models, then an entailment method is a way of obtaining a subset of $Mod(S)$ called the set of *selected models*. The corresponding set $T(S)$ of nonmonotonic conclusions is the set of formulae true in all selected models, so it is a superset of $Th(S)$. A number of entailment methods have been proposed and analyzed; most of them are based on the notion of imposing a preference relation on models, and of selecting models which are maximal with respect to the preference relation.

The traditional reason for introducing nonmonotoniciy in logics of actions and change was in order to capture intended closure properties with respect to the set of actions and to their effects. One wanted to assume that the actions that were explicitly stated in the scenario axioms were *the only relevant actions* in that scenario. One also wanted to assume that the effects of actions that were stated in an action law were *the only relevant effects* of the action described in the action law. Finally, it was recognized in general terms that qualification requires nonmonotonicity.

These rationale for nonmonotonicity continue to be valid, but the context of hybrid scenarios and actions for which a success/fail distinction has been defined add more to the picture. In particular, it allows us to bring a concrete formal treatment to the traditional intuitions about qualification.

Let us illustrate these aspects briefly through an example. Suppose one first describes a scenario through the following statements: at time 0 the robot r_4 is standing still at the origo, looking north, it is 2 units wide and 0.5 unit long, and there is no obstacle and noone pushes the emergency button. No action is performed. Furthermore, the effects of robots actions are as was specified above.

It is straight-forward to express these statements in the notation described here. The statement that no action is performed is expressed simply by not making any statement using the predicates D or G.

Under the assumptions mentioned above, one can conclude that at time 20, the robot is still in the position (0,0). Consider now the effect of adding the statement that at time 9, the robot r_4 starts to go towards position (0.5, 10), that is, one adds

$$G(9, \text{GOTO}(r_4, (0.5, 10)))$$

to the set of axioms. The logic should then be able to conclude that at time 20 the robot is in position (0,10), and that the GOTO action has been executed successfully over the time interval (9,20). More precisely, the given set of axioms after the addition of the invocation statement should have only one model, namely the one where the robot moves according to the given laws, and where

$$D_s([9, 20], \text{GOTO}(r_4, (0.5, 10)))$$

is entailed. Notice that the previous conclusion concerning the robot's position at time 20 is now invalid, and that another conclusions has taken its place. Notice also that the conclusion presently obtained depends crucially on the fact that no obstacle and no emergency need to be assumed during the interval (9,20).

Next, if one adds the additional statement

$$H(14, \textit{emergency})$$

which says that the emergency button was pushed at time 14, then the previous conclusion that the goto action succeeded over the interval $[9, 20]$ shall be replaced by the conclusion that it failed over the interval $[9, 14]$. The same holds, of course, if the statement

$$H(14, obstac(r_4))$$

is instead added as an axiom.

We see, therefore, that the characterization of this kind of hybrid system using a logic of action and change truly requires the logic *both* to characterize actions in terms of their operational definition, *and* to characterize the piecewise continuous but occasionally discontinuous behavior of positions, velocities, accelerations, etc. In this context, the change of value of a discrete-valued state variable is simply a special case of the discontinuity of a continuous-valued variable.

We described in some earlier papers [5, 4] how this type of hybrid change can be accomodated in a first-order logic. This approach is based on a few basic concepts:

- The basis of the approach is to embed differential calculus within a multi-sorted first-order logic with metric time as one of the sorts.
- Nonmonotonicity is dealt with by generalizing the traditional concept of minimizing or restricting change of discrete-valued fluents, to similarly minimizing or restricting discontinuities in piecewise continuous fluents [5, 4, 11].
- An additional key technique is to use the *occlusion* operator X that was introduced by us in [6] (it was later renamed "release" by Kartha and Lifschitz in [1]). We write $X(t, f)$ for saying that the fluent f is exempt from the restriction against discontinuities that applies elsewhere along the timeline. The possible range of values for occluded features is specified by action laws or other axioms.
- The reassignment operator := is used as an abbreviation: $H(t, f := v)$ is defined to mean $H(t, f : v) \wedge X(t, f)$, and similarly for B.

For the kind of robotic scenario that is discussed in the present article, one can use these principles – minimization or restriction of discontinuities, use of an occlusion operator, and viewing assignment as an abbreviation – both for the interpretation of the operational definitions of actions, and for describing the "physics" of the world where the robots operate.

One important feature of this approach is that it provides a unified way of interpreting the detailed behavior of the robot and the physical laws in the robot's environment. This means, for example, that the effect obtained by adding an axiom which explicitly says there is an obstacle at time 14, will also be obtained if one adds statements involving an additional robot r_2 whose movements are defined so that it gets in the way of r_4 at time 14. Thus, reasoning about qualifications or failure conditions is completely integrated into this nonmonotonic logic.

6 Relating the effect definition and the operational definition

We have now developed the effect definition and the operational definition side by side; let us denote them Σ_E and Σ_P (for program), respectively. The problem that we chose initially was how one can verify that Σ_E follows from Σ_P, using also the other relevant axioms (axioms for the D operators, definitions of how aggregated features depend on basic ones, definitions of sensors, etc).

At first sight, this appears to be straightforward, both in an informal sense and a formal sense. Informally, in any model where $D_s([s,t], \text{GOTO}(m))$ is true and all the axioms stated above are satisfied, the following must be the case:

1. Applicability holds (otherwise the action would have failed)
2. Invocation has been performed
3. The physical laws apply
4. No obstruction along the execution (because if so the action would have failed)
5. Condition for successful termination must be satisfied
6. Termination operations (resetting velocity to zero, etc) must have been performed.

From this, and also using the assumptions about sensor correctness, it should be straightforward to conclude the success description, and (with a bit more effort) the failure description of the action.

Similarly on the formal side: all the formulae are expressed in the same logical language; it is a first-order logic although a somewhat complicated one; its semantics is well defined. Isn't it just a matter of showing that the set of models of Σ_E contains the models for Σ_P?

This classical perspective has to be modified because of the need for non-monotonicity. The correctness criterion becomes, instead, that the set of *selected* models for Σ_E contains the selected models for Σ_P.

Entailment methods for actions which are described on the discrete level have been studied extensively, and relatively much is known about their range of applicability and other similar properties [7, 8, 9]. The effect-description level is simpler to deal with in the sense that one can (often) restrict it to only using discrete-valued state variables, but it also suffers from an additional complication: modelling the world in terms of qualitative change gives rise to apparent nondeterminism in addition to the one which may be genuine in the application. Entailment methods for hybrid systems involving mixed continuous and discrete change have been less studied, and no formal results about their range of correct applicability have been obtained yet. This is therefore a topic of continued research.

7 Composition of failible actions

The operative definitions of actions will usually be constructed by composition of simpler actions. It is straight-forward, in the approach described here, to introduce the customary operators for composition of actions, such as sequential composition, conditional choice of action, iteration of actions, successive attempts of several actions until one of them succeeds, and so forth. We have shown (see the WWW page mentioned at the end of the article) how this can be done, and that the required additional axioms are quite simple and straightforward. In fact, the verification criterion for low-level actions can be directly applied to high-level actions as well. This is because both kinds of actions as well as the composition operators have been defined in essentially the same logic.

Notice, by the way, that the operational definitions described here represent a kind of "programming language" for the actions; they describe the actions as seen from the executing robot, and not as seen from "outside". It is therefore quite natural to construct higher-level operations using composition operators that are familiar from programming languages.

8 Summary

We addressed the problem of formally proving high-level effect descriptions of actions from low-level operational definitions. Both descriptions are expressed in a logic of actions and change, with extensions for characterizing continuous change, discontinuities, the distinction between true and estimated values of state variables, and the distinction between success and failure of an action. Both descriptions also require the use of nonmonotonicity in the logic in question. The transition from operational definition to effect description furthermore involves the creation of a closure with respect to the set of possible ways that the action can fail.

We have outlined, by means of an example, how these issues can be addressed as an extension of existing results on logics of actions and change.

Additional material

The present paper is fairly brief due to the space constraints of the present workshop proceedings. The following WWW page contains links to additional articles on the same subject:

> http://www.ida.liu.se/~erisa/valrok/

and it will be maintained for the foreseeable future. A longer article has been published electronically by Linköping University Electronic Press, and will be maintained permanently at:

> http://www.ep.liu.se/ea/cis/1996/004/

It contains some of the material that has been mentioned above as missing in the present article (axioms omitted in subsection 2.3, axioms for action composition operators in section 7, etc).

References

1. G. Neelakantan Kartha and Vladimir Lifschitz. Actions with indirect effects (preliminary report). In *International Conference on Knowledge Representation and Reasoning*, pages 341–350, 1994.
2. Vladimir Lifschitz. Nested abnormality theories. *Artificial Intelligence*, 74:351–365, 1995.

3. John McCarthy and Patrick Hayes. Some philosophical problems from the viewpoint of artificial intelligence. In *Machine Intelligence, Vol. 4*, pages 463–502. Edinburgh University Press, 1969.
4. Tommy Persson and Lennart Staflin. A causation theory for a logic of continuous change. In *European Conference on Artificial Intelligence*, pages 497–502, 1990.
5. Erik Sandewall. Combining logic and differential equations for describing real-world systems. In *Proc. International Conference on Knowledge Representation, Toronto, Canada*, 1989.
6. Erik Sandewall. Filter preferential entailment for the logic of action in almost continuous worlds. In *International Joint Conference on Artificial Intelligence*, pages 894–899, 1989.
7. Erik Sandewall. The range of applicability of nonmonotonic logics for the inertia problem. In *International Joint Conference on Artificial Intelligence*, 1993.
8. Erik Sandewall. *Features and Fluents. The Representation of Knowledge about Dynamical Systems. Volume I*. Oxford University Press, 1994.
9. Erik Sandewall. Assessments of ramification methods that use static domain constraints. In *International Conference on Knowledge Representation and Reasoning*. Morgan Kaufmann, 1996.
10. Erik Sandewall and Ralph Rönnquist. A representation of action structures. In *National Conference on Artificial Intelligence*, pages 89–97, 1986.
11. Murray Shanahan. Representing continuous change in the event calculus. In *European Conference on Artificial Intelligence*, pages 598–603, 1990.
12. Yoav Shoham. *Reasoning about Change*. MIT Press, 1988.
13. Michael Thielscher. Computing ramifications by postprocessing. In *International Joint Conference on Artificial Intelligence*, 1995.

A New Algorithm for Discrete Timed Symbolic Model Checking

Jürgen Ruf and Thomas Kropf

University of Karlsruhe
Institute of Computer Design and Fault Tolerance (Prof. D. Schmid)
Kaiserstr. 12, 76128 Karlsruhe, Germany
Juergen.Ruf@informatik.uni-karlsruhe.de
Thomas.Kropf@informatik.uni-karlsruhe.de
http://goethe.ira.uka.de/hvg/cats

Abstract. When extending CTL with constructs for quantitative reasoning on time, either the linear algorithm complexity of standard model checking gets lost, efficient ROBDD techniques are no more applicable or the semantics becomes counter intuitive.

In this paper, we present a new approach to symbolic QCTL (Quantitative CTL) model checking. In contrast to previous approaches we use an intuitive QCTL semantics, provide an efficient model representation and the new algorithms require less iteration steps compared to translating the QCTL problem into CTL and using standard CTL model checking techniques. The approach is based on the observation, that we can identify two different state sets in temporal structures: main states, which constitute the reachability graph, and intermediate states, which are necessary to give a semantics to the timed state transitions. To represent intermediate states symbolically, we generalize the notion of characteristic functions. These functions are represented using multi-terminal BDDs (MTBDDs). The paper presents the new model checking algorithm as well as experimental results, showing the efficiency of the new approach.

1 Introduction

Formal verification is an important step in the system design process. Using model checking as verification technique, the validity of the specification has to be shown with regard to a model, representing the system to be verified. For special logics like CTL the complexity of the model checking algorithm is linear with regard to the formula length and the model size [5]. Techniques have been developed to efficiently represent state sets and transitions relations symbolically by using binary decision diagrams (ROBDDs, [14]). Proceeding this way systems with a large state space have become verifiable [9].

Standard CTL is not sufficient to cope with complex timing constraints. Thus various timed temporal logics have been presented to argue about timing. Logics based on real numbers model time as a continuous quantity[12]. The underlying structure contains clocks, which may be reset by a state transition. The value of the clocks may be used to determine possible state transitions. Specifications are given by a timed variant of CTL called TCTL. This approach is very expressive and allows a very accurate system modeling. However, the price to be paid is an exponential runtime with regard to the number of clocks and timing constraints, although symbolic model checking is possible in principle [16].

In many cases it is sufficient to use a discrete time model, e.g. for systems with an inherent clock cycle like traffic light controllers or communication protocols. Thus only natural numbers are assigned as transition times to state transitions. CTL is extended appropriately by allowing quantitative CTL operators, leading to languages like QCTL. Model checking is either based on a translation into a standard CTL model checking problem or by using dedicated representation techniques [15, 7]. The translation approach requires the mapping of timed structures to unit delay structures, leading to a state explosion if large transition times occur. This can be avoided when timed structures are used directly. In this case, however, the intuitive semantics of QCTL gets lost when composing temporal operators. Thus the resulting model checking algorithm is of limited use only (see section 4.1).

Hence it is desirable to find techniques for compactly representing timed structures by simultaneously keeping a clean QCTL semantics. This paper describes a solution to this problem. In addition to solving the described dilemma, to our knowledge for the first time multi-valued BDDs [2] are used to efficiently represent timed state transitions and state sets.

In section 2, we describe the fundamental data structures for symbolic model checking. Section 3 introduces characteristic functions and extended characteristic functions, which are used in the algorithms. In section 4, we introduce the quantitative computation tree logic (QCTL). We first show the state of the art and then we improve the semantics of QCTL. In section 5, the main QCTL model checking techniques are presented. We compare our approach to the state of the art with some experiments in section 6. Section 7 concludes the paper.

2 Binary Decision Diagrams

Symbolic model checking makes excessive use of Boolean functions. ROBDDs [14] are a compact and efficient representation of these functions. In contrast to ROBDDs, *Multi terminal binary decision diagrams* (MTBDDs [3]) represent *pseudo Boolean functions*. These functions map bit vectors to a finite set of elements.

Definition 2.1 (Pseudo Boolean Function)
Given that the set $\mathbb{B}$ contains the Boolean values $\{0, 1\}$, and a finite set A of elements, then each function $f : \mathbb{B}^n \to A$ is a pseudo Boolean function.

Like ROBDDs, MTBDDs may be represented by directed acyclic graphs where the leaf values contain elements of A.

There exist efficient routines to apply a terminal defined function to a MTBDD. These routines descend recursively the MTBDD graph starting at the root. If they reach a leaf, the terminal transformation is applied to the leaf value. There exist also an algorithm, recursively combining two MTBDDs. The terminal functions are defined over the set A.

3 Characteristic and Extended Characteristic Functions

Consider a set of elements S and a subset $T \subseteq S$. A way to represent this set is to use a function c_T mapping S to $\mathbb{B}$, where $c_T(x)$ holds if $x \in T$.

Set operations like union and intersection can be performed by Boolean operations on the corresponding characteristic functions [8].

Now assume that every element from S has an attribute. To use the concept of characteristic functions to express attributed elements, we extend these functions.

Definition 3.1 (Extended Characteristic Function)
Given a set S of elements, a set $T \subseteq S$ and a set A of attribute values. The extended characteristic function describing T is:

$$\tilde{c}_T : \quad S \quad \to A \cup \{\bot\}$$
$$\tilde{c}_T(s) := \begin{cases} t & \text{if } s \in T \text{ and } t \in A \text{ is the attribute value of } s \\ \bot & \text{if } s \notin T \end{cases}$$

The definition of set operations on extended characteristic functions is done in a two step manner. In the first step the operator is applied to the set elements with no regard of the element attributes. In the second step the operator is applied to the attribute values. The semantics of the second step is dependent on the attribute values.

4 QCTL - Quantitative Computation Tree Logic

Quantitative computation tree logic (QCTL) is an extension of CTL [4] by quantitative temporal operators. The logic has been introduced in [15, 7]. In QCTL each temporal operator of CTL carries a temporal scope as additional information, e.g. EF[5]f means there exists a path such that f holds in one of the next five states.

Although QCTL has the same expressiveness as CTL, specifications containing quantitative timing can be given in a more natural and succinct way. The semantics of QCTL may be given with regard to a standard unit-delay temporal structure. In fact, fragments of QCTL have been incorporated into the SMV model checker [10]. However, the virtues of QCTL are exhibited only when timed temporal structures are used for model checking. In this case it is possible to interpret the semantics of the QCTL operators directly, without translating them into the unit-delay model. Moreover timed temporal structures allow a more compact representation of the state space and the transition relation, especially if large delay times are used. Since the unit delay model needs additionally states, the number of variables encoding the states may increase.

4.1 State of the Art

To represent temporal structures Campos and Clarke suggested to use a timed transition relation, where delay times are encoded using additional Boolean values [15]. Another recent suggestion is to partition the transition relation, grouping together all transitions with the same delay in a separate ROBDD [7]. Unfortunately, if no unit-delay model is used, the QCTL semantics diverges from its CTL counterpart and is not what one would expect [6]. This hinders the use of nested QCTL expressions as shown in the following:

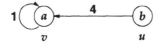

Fig. 1. A timed temporal structure

Example 4.1 *Given the timed temporal structure of figure 1. Consider the following three formulas and the sets of states where these formulas hold:*

$$EF[3]a \quad \cong \{v\}$$
$$EF[3]EF[3]a \cong \{v\}$$
$$EF[6]a \quad \cong \{u, v\}$$

The last two equations are not intuitive, because one expects that $EX[3]EX[3]$ *and* $EX[6]$ *yield in the same result.*

The reason for this effect is that these approaches consider only the existing states, but there are implicitly more states to be considered than are defined in the temporal structure. To avoid this counter intuitive semantics, we will redefine it.

4.2 Semantics of QCTL

In the following we present the semantics of QCTL as well as timed temporal structures.

Definition 4.1 (Timed Temporal Structure)
A timed temporal structure is a quintuple $\mathcal{M} = (AP, S, T, L_S, L_T)$.

- *AP is the set of atomic propositions*
- *S is the set of states*
- *T ⊆ S × S is the set of transitions with* $\forall s \in S. \exists s' \in S.(s, s') \in T$
- *$L_S : S \to \wp(AP)$ is the state labeling function*
- *$L_T : T \to \mathbb{N}_+$ assigns to each transition a delay time*

The standard CTL structure is extended by L_T, representing time delays at the edges. The delay times are non-zero natural numbers. The semantics of extended structures can be given in terms of single transition models by introducing new states (intermediate states) on timed edges carrying the same labels as the respective transition source state.

Next, we introduce derived temporal structures. These structures contain additionally states to represent a unit delay derivate of a timed temporal structure. This derived structure contains more information than a pure translation into CTL structure.

Definition 4.2 (Derived Temporal Structure)
A derived structure is a 6-tuple $\widetilde{M} = (AP, S_\prec, H, T, L_H, \tilde{L}_T)$.

- *AP is the set of atomic propositions*
- *$S_\prec$ is the set of states, with partial ordering $(\prec)$ on the states*
- *$H \subseteq S_\prec$ is the set of main states*
- *$T \subseteq H \times H$ is the set of transitions with $\forall s \in H. \exists s' \in H.(s, s') \in T$*
- *$L_H : H \to \wp(AP)$ is the main state labeling function*
- *$\tilde{L}_T : T \to \wp(S_\prec) \setminus \{\emptyset\}$ is the transition labeling function. The order $\prec$ is total on $\tilde{L}_T(t)$ for all $t \in T$*

For all $v \in S_\prec \setminus H$ there exists a $u \in H$ such that $u \prec v$.

The partial ordering ($\prec$ means comes earlier than) on the state space defines state chains for the transitions. These chains represent the delay time for the transition between two main states. States are only comparable if they are on the same transition. The least element of every partial ordering is the source main state of the transition. The main states of the derived model represent the states of the timed model. Only main states carry labels. The labels of the other states (intermediate states) are implicitly given by the predecessor main state. The transition labeling function $\tilde{L}_T$ maps transitions to the set of states lying on it.

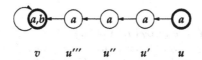

$$v \qquad u''' \qquad u'' \qquad u' \qquad u$$

Fig. 2. The derived temporal structure of figure 1

Example 4.2 *Given the example in figure 2. The derived temporal structure is formally defined by:*

$$\widetilde{M} = (\ \{a, b\}, \{u, u', u'', u''', v\}, \{u, v\}, \{(u, v), (v, v)\}, L_H, \tilde{L}_T)$$

The partial ordering of the states is: $u \prec u' \prec u'' \prec u'''$.
The state labeling function is: $L_H(u) = \{a\}, L_H(v) = \{a, b\}$.

The transition labeling function is: $\tilde{L}_T(u,v) = \{u, u', u'', u'''\}$,
$\tilde{L}_T(v,v) = \{v\}$, $\tilde{L}_T(u,u) = \emptyset$, $\tilde{L}_T(v,u) = \emptyset$
Note that $v \notin \tilde{L}_T(u,v)$ *but* $u \in \tilde{L}_T(u,v)$.

To define the new semantics of QCTL, we introduce the derived state labeling function which assigns labels to all states.

Definition 4.3 (Derived State Labeling Function)
Given the derived temporal structure $\widetilde{\mathcal{M}}$, *the derived state labeling function is:*

$$\tilde{L}_{S_\prec}(s): \quad S_\prec \quad \to AP$$
$$\tilde{L}_{S_\prec}(s) := \begin{cases} L_H(s) & \text{if } s \in H \\ L_H(u) & \text{if } u \in H \text{ and } u \prec s \end{cases}$$

Since the set of states on a transition is finite and since the ordering $(\prec)$ is total, there exists a minimum state (a main state) and a maximum state for each transition chain.

Definition 4.4 (Maximum State of a State Chain)
Given a transition $(u,v) \in T$ *of a derived structure* $\widetilde{\mathcal{M}}$, *the maximum state is*

$$\max_\prec: \quad H \times H \quad \to S_\prec$$
$$\max_\prec(u,v) := w \in \tilde{L}_T(u,v) \text{ where no } v' \in \tilde{L}_T(u,v) \text{ exists}$$
$$\text{such that } w \prec v'$$

The derived transition relation connects all states of S according to $\widetilde{\mathcal{M}}$ in a unit-delay manner.

Definition 4.5 (Derived Transition Relation)
Given the derived temporal structure $\widetilde{\mathcal{M}}$, *the derived transition relation* $\tilde{T} \subseteq S_\prec \times S_\prec$ *is:*

$$\tilde{T} = \left\{ (u,v) \,\middle|\, \begin{array}{l} (u,v) \in T \text{ and } \tilde{L}_T(u,v) = \{u\} \text{ or} \\ u \prec v \text{ and there exists no } w \in S_\prec \text{ such that } u \prec w \prec v \text{ or} \\ u \nprec v \text{ and there exists } w \in H \text{ such that } (w,v) \in T \text{ and} \\ u = \max_\prec(w,v) \end{array} \right\}$$

The first case connects two main states with no intermediate states. The second case connects the states on a transition. The third case connects the greatest intermediate state with the following main state.

Definition 4.6 (Derived Path)
A derived path $\tilde{p} := (s^0, s^1, \ldots)$ *of* $\widetilde{\mathcal{M}}$ *is a sequence of states with*
$\forall i \geq 0.(s^i, s^{i+1}) \in \tilde{T}$

Definition 4.7 (Semantics of QCTL)

Given the derived model $\widetilde{\mathcal{M}}$, a state s of the model and the QCTL formulas f and g. We write $s \models f$ for $\widetilde{\mathcal{M}}, s \models f$. The operators true, $\neg$, $\wedge$, EG and EU are defined analogously to the corresponding CTL operators.

$$s \models f \qquad :\Leftrightarrow f \in \tilde{L}_{S_{\prec}}(s) \text{ if } f \in AP$$

$$s \models EX[n]f \qquad :\Leftrightarrow \text{ there exists a derived path } \tilde{p} = (s^0, s^1, \ldots) \text{ in } \widetilde{\mathcal{M}} \text{ with } s^n \models f$$

$$s \models EG[n]f \qquad :\Leftrightarrow \text{ there exists a derived path } \tilde{p} = (s^0, s^1, \ldots) \text{ in } \widetilde{\mathcal{M}} \text{ with } \forall 0 \leq i \leq n.s^i \models f$$

$$s \models E(f\,U[n]\,g) :\Leftrightarrow \text{ there exists a derivedpath } \tilde{p} = (s^0, s^1, \ldots) \text{ in } \widetilde{\mathcal{M}} \text{ with } \exists 0 \leq i \leq n.(s^i \models g \text{ and} \forall 0 \leq j < i.(s^j \models f))$$

The validity and the satisfiability are defined analogously to the CTL semantics. The semantics of this QCTL variation is more intuitive, since we consider the intermediate states.

Example 4.3 *Revisit the example 4.1 and the derived temporal structure of figure 2. We now have the following*

$$\begin{aligned}
EF[3]a &\cong \{v, u', u'', u'''\} \\
EF[3]EF[3]a &\cong \{v, u, u', u'', u'''\} \\
EF[6]a &\cong \{v, u, u', u'', u'''\}
\end{aligned}$$

To show the flexibility of this semantics, we consider a CTL extension with intervals. An interval semantic of $EG[a, b]$ is:

$$s \models EG[a, b]f :\Leftrightarrow \text{ there exists a path } p = (s^0, s^1, \ldots) \text{ such that } \forall a \leq i \leq b.s^i \models f$$

This interval operator can easily be defined by the QCTL-formula:
$EG[a, b] := EX[a]EG[b - a]f$

5 QCTL Model Checking Algorithms

In the following we present a representation for the timed transition relation and for sets of states, which is more compact than the representation with unit delay transitions and which allows the explicit consideration of all intermediate states. Based on this representation we describe algorithms for QCTL model checking.

5.1 Representation of State Sets

Our approach is based on the observation, that main states and intermediate states can be treated differently:

main states have	intermediate states have
any number of predecessor states	exactly one predecessor state
at least one successor state	exactly one successor state
arbitrary labels	same labels as their predecessor main state

The idea is to represent states using extended characteristic functions. The characteristic function attributes transitions with state sets. In contrast to standard CTL model checking, we raise the characteristic function from the level of states to the level of transitions. The extended characteristic function maps transitions to the set of states lying on it:

$$\tilde{c} : H \times H \to \wp(S_\prec) \cup \{\bot\}$$
$$\tilde{c}(u, v) := \begin{cases} \wp(\tilde{L}_T(u, v)) & \text{if } (u, v) \in T \\ \bot & \text{if } (u, v) \notin T \end{cases}$$

The complete state space of a temporal structure is the union of all transition sets.

For a compact implementation, the set of states along a transition is represented by a bit vector. Every state on the transition is represented by one bit in the bit vector representation. The total ordering of states along a transition is reflected by the linear arrangement of the bits.

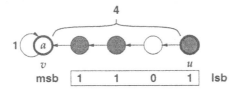

Fig. 3. The encoding of states with bit vectors

Example 5.1 *In figure 3 an example encoding of states by bit vectors is shown.*

Each transition is encoded by a pair of main states. Main states are encoded by Boolean types $\mathbb{B}^n$ which are the labels carried in the states. If we assume a maximal cardinality k of states on a transition then the transitions are given by a function $\tilde{c} : \mathbb{B}^n \times \mathbb{B}^n \to \mathbb{B}^k$.

In order to use this representation, we have to define some basic set operations for extended characteristic functions. As already explained in section 3, a union of two extended characteristic functions is done by first joining the set of transitions. In the second step, two identical transitions with different attributes (state sets) have to be reduced to one transition with one attribute. Hence, the union of two extended characteristic functions is defined as:

$$(\tilde{c}_1 \cup \tilde{c}_2)(u, v) := \begin{cases} \tilde{c}_1(u, v) & \text{if } \tilde{c}_2(u, v) = \bot \\ \tilde{c}_2(u, v) & \text{if } \tilde{c}_1(u, v) = \bot \\ \tilde{c}_1(u, v) \cup \tilde{c}_2(u, v) & \text{else} \end{cases}$$

The intersection of two extended characteristic functions can be computed analogously.

An adequate representation of QCTL state sets are extended characteristic functions encoded by multi terminal ROBDDs (MTBDDs). The transition encoding variables (a pair of labels) are stored in the internal nodes (variables) of the

MTBDD. The resulting attribute of a transition (the set of states belonging to a transition encoded by a bit vector) is stored in the leaves of the MTBDD.

5.2 Representation of Timed Structures

The encoding of the transition relation of a timed structure differs from the representation of sets of states as the intermediate states do not have to be represented. We only have to know which time delay a transition has. Thus, the corresponding extended characteristic function for the transition relation is of type $\tilde{c}_{T'} : H \times H \to \mathbb{N}$. It results from the transition relation (T) and the transition labeling function (L_T).

However, to achieve a homogenous representation, we map from transitions to states in $S_\prec$. Since we do not consider sets of states in the transition representation, we map to the maximum on a transition state of every transition.

This encoding of transition relations simplifies the model checking algorithm, as shown in the next section.

Analogous to the state set representation, we encode the maximum state by a bit vector. When using this representation in the algorithm, we interpret the most significant bit of the transition representation as the last state of the intermediate state chain. A non-existing transition is represented by a zero bit vector $(0\ldots0)$.

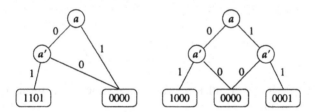

Fig. 4. The MTBDDs for the set of states and the transition relation of figure 3

Example 5.2 *Figure 4 shows the MTBDDs for the set of states and the transition relation of figure 3.*

5.3 Algorithms

We can modify the standard CTL model checking algorithms. Only the basic operators $\wedge$, $\neg$ and **EX** have to be considered for the adaption to the new data structure. The other operators can be implemented by these basic operators.

The construction of the set of states holding an atomic formula f is done in two steps. Firstly the main states labeled with f are computed. The second step adds all intermediate states on transitions starting at these main states. If $\tilde{c}_{S_\prec}$ represents the state space as an extended characteristic function, the function $\tilde{c}_f$ can be computed by:

$$\tilde{c}_f := \bigcup_{u,v \in H, f \in L_H(u)} \tilde{c}_{S_\prec}(u,v)$$

The result of the $\wedge$-operator can be computed by the intersection of the state sets of both operands $\tilde{c}_{f \wedge g} = \tilde{c}_f \cap \tilde{c}_g$. Since the computation of state sets is done bottom up, the sets of both operands have been already computed and are represented as extended characteristic functions.

The negation of a formula f is true in all states of $S_{\prec}$ without the states holding f. The following formula computes the set of states for the negation:

$$\tilde{c}_{\neg f}(u,v) = \begin{cases} \tilde{c}_{S_{\prec}}(u,v) \setminus \tilde{c}_f(u,v) & \text{if } \tilde{c}_f(u,v) \neq \bot \\ \bot & \text{otherwise} \end{cases}$$

The states holding $EX[1]f$ are direct predecessors of states holding f. In order to find the predecessors, it is necessary to distinguish between main and intermediate states. Because of the linear chains of intermediate states, it is easy to find their predecessors. We define a predecessor function for the states on a transition in dependence of the total ordering $\prec$:

$$pre : S_{\prec} \quad \rightarrow S_{\prec}$$
$$pre(v) := u \text{ with } u \prec v \text{ and there exists no } w \text{ such that } u \prec w \prec v$$

With this function, the set of predecessors of a state set A_f holding the formula f on one transition is defined by:

$$pre : \wp(S_{\prec}) \quad \rightarrow \wp(S_{\prec})$$
$$pre(A_f) := \{pre(u) | u \in A_f\}$$

We extend this function to all transitions in a relation. In terms of extended characteristic functions:

$$(pre(\tilde{c}_f))(u,v) := pre(\tilde{c}_f(u,v))$$

Applied to the state set representation with bit vectors, we obtain a simple right shift of all MTBDD leaf values.

There exists no predecessor of the least element (main state) on a transition. The predecessor states of main states are the greatest states of predecessor transitions. To extract these states out of the set of all states it is necessary to find the predecessor transitions and then extract the greatest states of them. Extracting the predecessors of the main states is done in three steps:

1. extracting transitions with main states holding f
2. extracting the labels of the main states holding f
3. finding the predecessor transitions of the main states in 2. and computing the maximum states on these transitions

To apply these actions to extended characteristic functions, we introduce two new operators:

Definition 5.1 (Main State Extraction)
Given an extended characteristic function $\tilde{c}$ representing a set of states. The transitions with main states can be computed by:

$$<>: (H \times H \to \wp(S_{\prec})) \to (H \times H \to \mathbb{B})$$
$$< \tilde{c} > (u, v) \quad := \begin{cases} 1 \ if \ u \in \tilde{c}(u, v) \\ 0 \ if \ u \notin \tilde{c}(u, v) \end{cases}$$

After applying this operator to the set of states holding f we obtain the characteristic function for the set of transitions, where in the starting main state f holds.

Definition 5.2 (Selection Operator)
The selection operator gets a source set and a selection set and returns elements of the source set dependent of the selection set. In terms of characteristic functions:

$$\tilde{\cdot}: (H \times H \to \wp(S_{\prec})) \times (H \to \mathbb{B}) \to (H \times H \to \wp(S_{\prec}))$$
$$\tilde{c}(w, u)\tilde{\cdot}c(u) \quad := \begin{cases} \bot & if \ \tilde{c}(w, u) = \bot \\ \emptyset & if \ \tilde{c}(w, u) \neq \bot \wedge c(u) = 0 \\ \tilde{c}(w, u) & if \ \tilde{c}(w, u) \neq \bot \wedge c(u) = 1 \end{cases}$$

Now we can formalize the three steps. Given the characteristic function $\tilde{c}_f$, which represents the set of states holding f.

1. We apply the new main state selection operator $c_1(u, v) := < \tilde{c}_f > (u, v)$. We now have a characteristic function mapping transitions to 0 or 1.
2. The extraction of the label of the starting main state is a projection:

$$proj : (H \times H \to \mathbb{B}) \to (H \to \mathbb{B})$$

Since the labels are represented by Boolean variables we can express this projection also by an existential quantification. Applied to the function c_1:

$$c_2(u) := \exists v. < \tilde{c}_f > (u, v)$$

The resulting characteristic function maps state labels to 1, if their corresponding state holds f.
3. In this step we have to extract the maximal states of predecessor transitions of main states holding f. Since the transition representation encodes the maximal states of all transitions, we have to select the required predecessor transitions. For this purpose we use the selection operator. Applied to the result of 2, we obtain:

$$\tilde{c}_3(w, u) = \tilde{c}_{T'}(w, u)\tilde{\cdot}c_2(u)$$

$\tilde{c}_{T'}$ is the encoding of the transition relation of the timed temporal structure. The extended characteristic function $\tilde{c}_3$ represents all direct predecessor states of main states.

Considering the intermediate states, we get the complete formula for the state set EX[1]f:

$$\tilde{c}_{EX[1]f}(w,u) = \tilde{c}_{T'}(w,u)\,\dot{\,}(\exists v. < \tilde{c}_f > (u,v)) \cup pre(\tilde{c}_f(u,v))$$

The set for EX[n] is obtained by executing the EX[1] operator n times. The EG[n] and the EU[n] operators use the standard fixed point iteration based on the new EX[1] operator. The terminal condition is extended by a time scope test, i.e. reaching a fixed point or exceeding the time scope terminates the algorithm. Detailed information on the realization of the new algorithms with MTBDDs can be obtained in [11].

6 Experimental Results

The overhead of the extended characteristic functions is justified mainly when dealing with timed structures with large delay times. The translation of timed structures into CTL structures causes additional intermediate states. These additional states blow up the ROBDD representation of the CTL structures. On the other hand, the number of MTBDD variables in the derived temporal structure representation is independent of the number of intermediate states.

For better testing, we have chosen a scalable model. We decided to check some properties of the *priority inheritance* protocol [1, 13]. For the modeling of this protocol, we used a time slice of 60 time units. All processes were modeled together in one timed temporal structure. The property we have verified is:

$$P = AG(process[max].state = try \rightarrow AF[70]process[max].state = active)$$

This property insures that the process with the highest priority will enter the critical section at latest 70 time steps after the trial to enter.

For the experiments[1] we used an ULTRAsparc station with 167 MHz. As model checker for the single transition model, we used standard CTL model checking with the SMV [10]. Hence we have to translate the timed models into single transition models. This transformation requires the explicit representation of the intermediate states. An upper bound for the number of nodes in the CTL structure is $O(|T|d_{\mathcal{M}})$, the product of the number of main states and the maximal delay time of the structure. The QCTL specification formulas also have to be transformed into CTL. This is done by explicit unrolling the fixpoint iteration. For example the formula EG[3]f becomes $f \wedge (EXf \wedge (EXf \wedge (EXf)))$. The growth of CTL formulas is heavily influenced by nested QCTL operators. E. g. the formula EG[t_1](EF[t_2]g) becomes

---

[1] The QCTL-Model Checker is a part of the C@S-System and is on-line testable at http://goethe.ira.uka.de/hvg/cats

$$\overbrace{\begin{array}{l} (g \lor \mathrm{EX}(\ldots(g \lor \mathrm{EX}g)\ldots))\land \\ \mathrm{EX}((g \lor \mathrm{EX}(\ldots(g \lor \mathrm{EX}g)\ldots))\land \\ \vdots \\ \mathrm{EX}((g \lor \mathrm{EX}(\ldots(g \lor \mathrm{EX}g)\ldots))\land \\ \mathrm{EX}((g \lor \mathrm{EX}(\ldots(g \lor \mathrm{EX}g)\ldots)))\ldots)) \end{array}}^{t_2+1} \left.\begin{array}{c} \\ \\ \\ \\ \\ \end{array}\right\} t_1+1$$

This leads to an coarse approximation of an upper bound of the formula length of $O(|f|(d_f+1)^{m_f})$ where $|f|$ denotes the length of the QCTL formula, d_f is the maximal time scope and m_f denotes the maximal nesting depth of timed QCTL operators. The growth of the structure and the formula leads to an increased number of iteration steps in the fixed point calculation as the number of iterations in CTL is linear in the structure size and the formula length [5]. We obtain for the number of iterations $O(|T|d_{\mathcal{M}}|f|(d_f+1)^{m_f})$. This exponential formula growth is avoidable, if the standard CTL model checking algorithms are adapted. The structures have to be expanded anyway. If we check a QCTL formula f with the new algorithms, we also have to consider the implicitly represented intermediate states. But the number of iterations caused by nested QCTL formulas is smaller, since we have to compute the set of states for each sub formula only once. This leads to $O(|T|d_{\mathcal{M}}|f|d_f)$ iteration steps. In both model checking algorithms every iteration step causes a constant number of ROBDD resp. MTBDD operations. The complexity of ROBDD and MTBDD-operations is the product of the sizes of the operands [?]. However all QCTL formulas of our experiments do not contain nested timed operators.

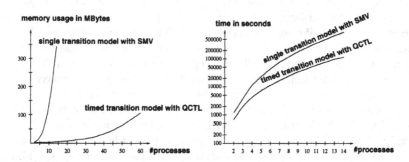

Fig. 5. The usage of memory and the runtimes for the verification of P

Figure 5 shows the usage of memory and the runtimes for the verification of the property P.

This specific example shows, that there exist problems which are better suited for QCTL model checking than for CTL model checking. In order to extract the parameters, when problems benefit from the QCTL representation, we studied both approaches with a large number of randomly generated models of different size. The QCTL formula checked against these models was: EX[20]q, with one of the atomic propositions q in the model. We have chosen the EX operator, since all other temporal operator are based on it. Moreover, the EX operator is a worst case test, since the other temporal operators in QCTL can terminate before reaching the time scope.

Figure 6 compares the memory usage and the run times of both model checkers. For one point in the plane we have constructed 20 different randomly generated temporal structures and have computed the average (for large delay times we have constructed 5 random temporal structures).

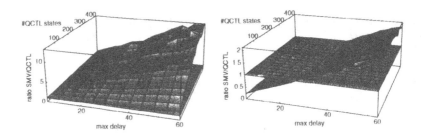

Fig. 6. Relative memory requirements and run times

7 Conclusions

Our aim was to find a representation of timed structures which requires less memory for the verification by correctly considering of all occurrences of intermediate states. We use an extension of characteristic functions which map transitions to sets of states. With these functions we have considered all intermediate states without explicitly representing them. As data structure for these characteristic functions we have used MTBDDs. Then we have extended the basic model checking routines to the MTBDD representation.

Our experiments showed that the representation of timed models is more compact than the single transition model. The representation of timed transition graphs with MTBDDs allow to verify bigger models as in the case of single transition representation. We also showed, that the basic routines for QCTL model checking work faster if they cross a specific value for the maximal time delay.

Acknowledgments

We are grateful to Hans Eveking, who made us aware of the counterintuitive semantics of QCTL which results if intermediate states are not considered [6].

References

1. S. Davari and L. Sha. Sources of unbounded priority inversion in real-time systems and a comparative study of possible solutions. In *Operating Systems Review*, pages 110–120. ACM, April 1992.

2. M. Fujita E. Clarke and X. Zhao. Applications of multi-terminal binary decision diagrams. Technical Report CMU-CS-95-160, School of Computer Science Carnegie Mellon University, Pittsburgh, PA 15213, April 1995.

3. E. Clarke, K.L. McMillian, X. Zhao, M. Fujita, and J.C.-Y. Yang. Spectral Transforms for large Boolean Functions with Application to Technologie Mapping. In *30th ACM/IEEE Design Automation Conference*, pages 54–60, Dallas, TX, June 1993.

4. E.Clarke, O. Grumberg, and D. Long. Verification Tools for Finite State Concurrent Systems. In J.W. de Bakker, W.-P. de Roever, and G. Rozenberg, editors, *A Decade of Concurrency-Reflections and Perspectives*, volume 803 of *Lecture Notes in Computer Science*, pages 124–175, Noordwijkerhout, Netherlands, June 1993. REX School/Symposium, Springer-Verlag.

5. E.M. Clarke, E.A. Emerson, and A.P. Sistla. Automatic Verification of Finite-State Concurrent Systems Using Temporal Logic Specifications. *ACM Transactions on Programming Languages and Systems*, 8(2):244–263, April 1986.

6. H. Eveking. private communication, March 1996.

7. J. Frößl, J. Gerlach, and T. Kropf. An Efficient Algorithm for Real-Time Model Checking. In *In Proccedings of the European Design and Test Conference*, pages 15–21, Paris, France, March 1996. IEEE Computer Society Press (Los Alamitos, California).

8. J. Lipson, editor. *Elements of Algebra and Algebraic Computing*. The Benjamin/Cummings Publishing Company, Inc., 1981.

9. J.R. Burch, E.M. Clarke, K.L. McMillan, D.L. Dill, and L.J. Hwang. Symbolic Model Checking: 10^{20} States and Beyond. In *Proceedings of the Fifth Annual IEEE Symposium on Logic in Computer Science*, pages 1–33, Washington, D.C., June 1990. IEEE Computer Society Press.

10. K.L. McMillan. The SMV system, symbolic model checking - an approach. Technical Report CMU-CS-92-131, Carnegie Mellon University, 1992.

11. T. Kropf and J. Ruf. Using MTBDDs for Discrete Timed Symbolic Model Checking. Technical Report of the SFB 358, August 1996.

12. R. Alur, C. Courcoubetics, and D.L. Dill. Model Checking for Real-Time Systems. In *Proceedings of the Fifth Annual IEEE Symposium on Logic in Computer Science*, pages 414–425, Washington, D.C., June 1990. IEEE Computer Society Press.

13. R. Rajkumar. *Task synchronisation in real-time systems*. PhD thesis, Carnegie Mellon University, 1989.

14. R.E. Bryant. Graph-Based Algorithms for Boolean Function Manipulation. *IEEE Transactions on Computers*, C-35(8):677–691, August 1986.

15. S.V. Campos and E. Clarke. Real-Time Symbolic Model Checking for Discrete Time Models. In T. Rus and C. Rattray, editors, *Theories and Experiences for Real-Time System Develpment*, AMAST Series in Computing. World Scientific Press, AMAST Series in Computing, May 1994.

16. T.A. Henzinger, X. Nicollin, J. Sifakis, and S. Yovine. Symbolic Model Checking for Real-Time Systems. In *7th. Symposium of Logics in Computer Science*, pages 394–406, Santa-Cruz, California, June 1992. IEEE Computer Scienty Press.

State Clock Logic:
A Decidable Real-Time Logic*

Jean-François Raskin and Pierre-Yves Schobbens
{jfr,pys}@info.fundp.ac.be

Computer Science Institute, University of Namur
Rue Grandgagnage 21, 5000 Namur, Belgium
Tel: +32 81 72 4990 Fax: +32 81 72 4967

Abstract. In this paper we define a real-time logic called SC logic. This logic is defined in the framework of State Clock automata, the state variant of the Event Clock automata of Alur et al [6]. Unlike timed automata [4], they are complementable and thus language inclusion becomes decidable. SC automata and SC logic are less expressive than timed automata and MITL but seem expressive enough in practice. A procedure to translate each SC formula into a SC automaton is presented. The main contribution of this paper is to complete the framework of this class of determinizable automata with a temporal logic and to introduce the notion of event clock in the domain of temporal logic.

1 Introduction

It is now widely recognized that the use of formal methods is very useful (and often necessary) for developing correct concurrent and reactive systems. This observation is still clearer when dealing with real-time and hybrid systems. One of the favorite formalisms to specify and verify concurrent systems are temporal logics. Temporal logics [15, 23] are modal logics that enable the expression of properties about a.o. the ordering of events in executions of concurrent programs [21]. These logics are used as tools for the verification of concurrent finite state programs [16, 11].

The properties that can be expressed in temporal logics are *qualitative* constraints about the ordering of events; *quantitative* timing constraints cannot be expressed. Logics that are able to express quantitative requirements are called *real-time logics* [9, 8]. Real-time logics have received a lot of attention from the research community [19, 8, 9, 2, 7]. Two main real-time logics with decidable model-checking problem have been defined [1] : MITL for the linear time framework and TCTL for the branching time framework[2].

---

* This work was supported by the Belgian National Fund for Scientific Research (FNRS) and by Belgacom.

[2] Note that the satisfiability problem is decidable for MITL but not for TCTL. Recently a real-time logic extending PTL with timed automata operators, called TATL [10], has been defined and a subset of this logic has a decidable model-checking problem

In the real-time framework, concurrent programs are modeled as timed automata [4], i.e. automata augmented with clocks and clock constraints. Unfortunately, the operational model of timed automata presents a drawback: timed automata are not closed under language complementation, unlike the untimed variety, and the language inclusion between two timed automata $L(A_1) \subseteq L(A_2)$ is undecidable. This problem essential a.o. for checking refinement, so in [6] Alur et al. think that timed automata may not be the best operational model for real-time systems. They propose a new type of timed automata, called *event clock (EC) automata* for which the inclusion problem is decidable.

In this paper, we propose a decidable linear real-time logic, based on the dense time semantics, which can express interesting properties of real-time systems. This logic, called *state clock (SC) logic*, has the nice property that for each formula of this logic, we can construct an *SC automaton* whose language contains exactly the models of the SC formula. The SC automata are a variant of EC automata and share the same nice properties, i.e. they are closed under all boolean operations; the language inclusion problem is thus decidable. The main contribution of this paper is to define a logic that completes the SC automata frameworks and to introduce the notion of *state clock* in the domain of temporal logic. To have some measure of the expressive power of states clocks in the domain of temporal logic we show that our logic can express a great part of the MITL definable properties. Furthermore all the most common real-time properties, e.g. bounded invariance, bounded response, time-out, ..., can be defined in SCL.

The rest of this paper is organized as follows. In section 2 we present SC automata, the state-based counterpart of EC automata. The SC logic is presented in section 3 and compared with MITL. Section 4 proposes a procedure that constructs a SC automaton that accepts exactly the models of a SC formula. Decision results and their complexity are also given.

2 SC Automata

As we already said, timed automata are finite state machines extended with the notion of time. With this extension, timed automata accepts timed words. In timed words [3], each symbol is paired with a real-valued time-stamp. In the *dense time* model, this time-stamp is a positive real number. In usual timed automata, the mechanism of clock resettings and clock constraints allow the specifier to define real-time requirements, for details and examples, see [4].

In [6], Alur et al. present a determinizable class of timed automata called *event clock automata*. This class of automata is closed under *union, intersection and complement*. Consequently the language inclusion problem is decidable for this class of automata. For event clock automata, the complement closure property is obtained by restricting the use of clocks: the clocks have a predefined association with symbols of the input alphabet. The *event-history clock* of the

---

[3] We will see later that timed automata can also accept timed sequences of states.

input symbol $a \in \Sigma$, noted x_a, is a history variable whose value is the time of the *last* occurrence of a relative to the current time. Symmetrically the *event-prophecy clock* of $a \in \Sigma$, noted y_a is a prophecy variable whose value is the time of the *next* occurrence of a relative to the current time.

Example 1. Let us consider the EC automaton of figure 1. This event-clock automaton contains 3 locations, l_0 is a start location. The constraint $x_a = 5$ decorating the edge starting from l_1 with the character **b** imposes that an **a** character must have been read exactly 5 time units before the edge is crossed. On the other hand the constraint $y_a < 2$ decorating the edge form l_1 to l_2 requires that each time this edge is crossed, an **a**-edge must be crossed within 2 time units.

Let us consider the execution of the automaton on the timed word $(a, 1.2), (b, 6.2), (c, 7), (b, 7.1), (b, 7.5), (b, 7.6), (a, 8.1), \ldots$ This is a possible prefix of a timed word accepted by the automaton.

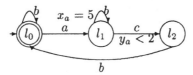

Fig. 1. Event-Clock automaton A_1.

As we can see in example 1, the values of the clocks are solely determined by the input word, not by the automaton.

We will now present SC automata. The main difference between an EC and a SC automaton is that in a SC automaton the locations are labeled with propositional symbols and constraints about history/prophecy variables. With this modification, SC automata generate timed sequences of states instead of timed words. Let us formally define the elements that compose a SC automaton :

Definition 1. A **State Clock (SC) automaton** $A = (\mathcal{P}, C_\mathcal{P}, L, L_0, E, \mathcal{L}, \Delta, \mathcal{F})$ consists of :

- a finite non-empty set of propositions $\mathcal{P}$;
- for each proposition of $\mathcal{P}$ there exists a history and a prophecy variable denoted respectively by x_p and y_p. This set of clocks is noted $C_\mathcal{P}$;
- a finite non-empty set L of locations;
- a non-empty set $L_0 \subseteq L$ of start locations;
- $E \subseteq L \times L$ a set of edges. Each edge is a couple (l, l') where l is the source location and l' is the target location;
- a proposition labelling function $\mathcal{L} : L \to 2^\mathcal{P}$ which labels each location $l \in L$ with the set of propositions that are true in that location;
- a constraint labelling function $\Delta : L \to 2^\mathbb{C}$, $\Delta(l)$ denotes the set of constraints that must be respected when staying in location l. The set $\mathbb{C}$ is the set of boolean combinations of constraints of the form $z \sim c$ where $z \in C_\mathcal{P}$, $c \in \mathbb{Q}^+$ and $\sim \in \{<, \leq, =, \geq, >\}$;

– a set $\mathcal{F}$ of sets $F_i \subseteq L$ of accepting locations.

Example 2. Let us illustrate the use of SC automata to model real-time reactive systems. This illustration is taken from the rail-road gate controller problem [18].

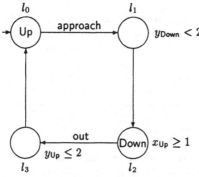

Fig. 2. State-Clock automata A_2.

In location l_0 the gate is Up [4]. When a train is approaching, the automaton evolves to the location l_1. This location is annotated by the constraint $y_{\mathsf{Down}} < 2$ imposing that the gate has to be down before 2 time units. The constraint $x_{\mathsf{Up}} \geq 1$ over the history variable associated with proposition Up, models the fact that the gate takes at least 1 time unit to go from state Up to state Down, i.e. to go down.

Although our SC automata are formally less expressive than timed automata, they are more declarative since they do not deal with the mechanics of clock resetting. Specifiers will find it easier and clearer to define a SC automaton to model their problem: in this sense SC automata are *practically* more expressive.

Furthermore, as we already said, our formalism presents the advantage of being closed under all boolean operations, thus the inclusion between languages of automata is decidable. This nice feature gives (theoretically) the opportunity of completely mechanizing the verification of e.g. refinement between an abstract description of a reactive module and a more operational one.

Alur et al. [6] have studied the expressive power of EC automata. Those automata are strictly less expressive than usual timed automata. Nevertheless they demonstrate that EC automata are expressive enough by showing that each *timed transition system* can be translated into an event-recording automata (a sub-class of EC automata). The SC automata share the same nice properties, and seem well-suited for modeling and verifying real-time systems.

a SC automaton defines a set of timed sequences of states. In such a sequence, each element is a state of the system (the set of propositions that are true) and an interval of time during which the system remains in that state. An *interval of*

---

[4] Let us note that the events (approach,out) are there for intuition but are not part of the automaton definition.

time is a non-empty convex subset of the real numbers, for example $I = (2, 5]$. In the following $l(I)$ denotes the left end point of I, $l(I) = 2$ in our example. Symmetrically, $r(I)$ denotes the right end point of I, $r(I) = 5$ in our example. Let us note that $I = (2, 5]$ is left open and right closed.

Definition 2. A **timed sequence of states** m is a mapping from N to couples (s_i, I_i) where s_i is a subset of $\mathcal{P}$ (the set of propositional symbols) and I_i is an interval of $\mathbb{R}^+$. Furthermore $I_0, I_1, I_2, \ldots$ is a sequence of intervals :

- the intervals I_i and I_{i+1} are adjacent, i.e. $\forall i \geq 0 \cdot r(I_i) = l(I_{i+1})$ and I_i is right open and I_{i+1} is left closed or I_i is right closed and I_{i+1} is left open.
- time progresses beyond any bounds : $\forall t \in \mathbb{R}^+ \cdot \exists i \geq 0: t \in I_i$.

Consequently, I is a partition of $\mathbb{R}^+$.

At each point of a timed sequence of states, each history and prophecy variable has a value. The value of the prophecy variable y_p at time t measures the time distance that separates t from the first subsequent state of the sequence were p is true. Symmetrically, the history variable x_p at time t measures the time distance that separates t from the last preceding state of the sequence where p was true.

Example 3. Let us consider the following timed sequence of states

$$m = (\{p\}, [0, 1.2]), (\{q\}, (1.2, 5)), (\{p\}, [5, 7)), \ldots$$

In this sequence, at time 1.5, the prophecy variable y_p is equal to 3.5; the proposition p is true at the 3rd state of the sequence and this state is entered at real-time 5.

If we consider the very similar timed sequence of state

$$m' = (\{p\}, [0, 1.2]), (\{q\}, (1.2, 5]), (\{p\}, (5, 7)), \ldots$$

were the second state is now right closed and the third one left open, it seems us natural to consider the same value (3.5) for the predicting variable y_p at time 1.5. We will thus give a definition to history/prophecy variable which is invariant to the fact that an interval is closed or open. Physically it is impossible to determine if an interval is open/closed. Therefore it seems to be a good option to use a formalism which is invariant to this aspect.

Here is the formal definition for the value of history/prophecy variables:

Definition 3 Clock valuation. The value of the clocks $x_p, y_p \in C_{\mathcal{P}}$ along an infinite timed sequence of states m at time $t \in I_i^m$, denoted by $\eta(m, i, t)$, are :

- **for the history variables:**

$$\eta(m, i, t)(x_p) = \begin{cases} 0 & \text{if } p \in s_i^m \\ t - t_j & \text{if } \exists j < i, \, p \in s_j^m \\ & \text{and } t_j = r(I_j^m) \text{ and } \forall k : j < k \leq i : p \notin s_k^m \\ \bot & \text{otherwise} \end{cases}$$

- **for the prophecy variable:**

$$\eta(m,i,t)(y_p) = \begin{cases} 0 & \text{if } p \in s_i^m \\ t_j - t & \text{if } \exists j > i,\ p \in s_j^m \text{ and } t_j = l(I_j^m) \\ & \text{and } \forall k : i \le k < j\colon p \notin s_k^m \\ \bot & \text{otherwise} \end{cases}$$

Let us note that a constraint $\bot \sim c$ always evaluates to false.

We have already mentioned that a SC automaton defines a set of timed sequences of states, called its language. A timed sequence of states m belongs to the language of a SC automaton if there is a run of this automaton on m. This is formally defined as follows:

Definition 4. A run of a SC automaton on an infinite timed sequence of states m is an infinite sequence $r = (l_0^r, I_0^r), (l_1^r, I_1^r), \ldots, (l_i^r, I_i^r), \ldots$ satisfying the following constraints:

- **Initiality**: $l_0 \in L_0$ and $0 \in I_0$. Initially, time is zero and the first location is an initial location;
- **Interval**: $I_0^r, I_1^r, \ldots, I_n^r, \ldots$ form an interval sequence.
- **Adequation**: $\forall t \in \mathbb{R}^+ : \mathcal{L}(l_i^r) = s_j^m$ where $t \in I_i^r$ and $t \in I_j^m$. The propositions that label the locations visited at time t is equal to the set of propositions that are true at time t in the infinite sequence of states m.
- **Consecution**: $\forall i > 0$, either $l_i^r = l_{i-1}^r$ (stuttering step), or $(l_{i-1}^r, l_i^r) \in E$;
- **Timing constraints**: $\forall i \ge 0 \cdot \forall t \in I_i^r : \eta(r,i,t) \models \Delta(l_i^r)$. The constraints about clocks, decorating locations, are respected.
- **Acceptance**: each $F_i \in \mathcal{F}$, there exists a location $l \in F_i$ which appears infinitely often along r.

Definition 5 Language. The timed sequence of states corresponding to an accepted run r of a SC automaton, is the sequence

$$(\mathcal{L}(l_0^r), I_0^r), (\mathcal{L}(l_1^r), I_1^r), \ldots, (\mathcal{L}(l_n^r), I_n^r), \ldots$$

The set of timed sequences of states corresponding to the accepted runs of a SC automaton A is called its language and is noted $L(A)$.

3 The Real-Time SC Logic

In this section we introduce the real-time logic associated with SC automata, called SC logic. This logic is a real-time extension of the temporal logic PTL (with past operators). The extension we propose is two indexed modal operators $\triangleright$ and $\triangleleft$ which are used to express real-time constraints. The semantics of those two operators are closely related to the notions of *prophecy* and *history* variables. $\triangleright_{\sim c} f$ allows the expression of constraints on the time of the next state where the formula f will be verified; symmetrically, $\triangleleft_{\sim c} f$ on the last state where f were verified. The modal operators $\triangleright$ and $\triangleleft$ generalize the semantics of history/prophecy variables as suggested in [6]. As we show later, the model checking and the satisfiability problems are decidable for our logic. In the linear framework, there are two other real-time logics which are decidable: MITL [5] and TATL [10].

We will now present formally the SC logic. Examples of specifications written in SC logic are given at the end of this section.

Definition 6 SC logic syntax. A formula of SC logic is composed of propositional symbols $p, p_1, p_2, .., q, ...$, usual boolean connectives $\vee$ and $\neg$, qualitative temporal operators : Until ($\mathcal{U}$) and Since ($\mathcal{S}$), real-time operators : state prophecy operator ($\triangleright$), state history operator ($\triangleleft$). A well-formed formula of SC logic satisfies the following syntactical rule:

$$f ::= p \mid f_1 \vee f_2 \mid \neg f \mid f_1 \mathcal{U} f_2 \mid f_1 \mathcal{S} f_2 \mid \triangleright_{\sim c} f \mid \triangleleft_{\sim c} f$$
Where $\sim \in \{<, \leq, =, \geq, >\}$, $p \in \mathcal{P}$ and f, f_1, f_2 are well formed formulae and c is a rational constant

The semantics that we will adopt for the until ($\mathcal{U}$) operator (an also for the since operator ($\mathcal{S}$)) is a bit different from its usual semantics as in MTL [8] or in MITL [5]. Our choice is similar to the choice made by Bouajjani et al [10] and is adopted for the same reasons. In general the semantics given to $f_1 \mathcal{U} f_2$ is the following: $f_1 \mathcal{U} f_2$ is verified in time t of a timed sequence of states m iff there exists some time t' ($t' \geq t$) where f_2 is verified and in all time $t'' \in [t, t')$, f_1 is verified. This semantics is *sensitive* to the fact that the interval $[t, t')$ is right open.

Example 4. With the semantics of $\mathcal{U}$ given above, the formula $p \mathcal{U} q$ is verified in the following model at time $t = 0$

$$m = (\{p\}, [0, 3)), (\{p, r\}, [3, 5)), (\{q\}, [5, 9)), \ldots$$

but not in the model m' similar to m but where the second interval is right closed.

In SC logic we will adopt a more intuitive semantics which is *invariant* to the fact that the interval $[t, t')$ is right open or closed. Futhermore as for the definition of history/prophecy variable, we define the semantics of the operators $\triangleright$, $\triangleleft$ in such a way that they are invariant to the fact that intervals are closed or open. To define the semantics of the SC logic elegantly, we define the notion of $^f Fine$ model, adapted from [1]. In the following definitions, we note $m \models f$ to express that m is a model of the SC formula f, $(m, i) \models f$ to express that f is verified in all points of interval I_i^m, $(m, i, t) \models f$ to express that f is verified at time $t \in I_i^m$ of m.

Definition 7 $^f Fine$ **model.** For a SC formula f, a timed state sequence $m = (s, I)$ is called $^f Fine$ iff for all $i \geq 0$, for all subformulae f_1 of f, for all $t, t' \in I_i^m$: $(m, i, t) \models f_1$ iff $(m, i, t') \models f_1$.

It is important to note that every model m can be refined in a $^f Fine$ model. It can be shown that a SC formula is satisfiable iff it has a $^f Fine$ model. The model m in the following definition, without loss of generality, is considered to be $^{f_1} Fine$ and $^{f_2} Fine$: [5]

Definition 8 SC logic semantics. An SC formula f holds in a timed sequence of states m at real-time $t \in \mathbb{R}^+$, where $t \in I_i^m$ iff (by recursion) :

- $(m, i, t) \models p$ iff $p \in s_i^m$, for $p \in \mathcal{P}$;
- $(m, i, t) \models f_1 \vee f_2$ iff $(m, i) \models f_1$ or $(m, i) \models f_2$;
- $(m, i, t) \models \neg f_1$ iff $(m, i) \not\models f_1$;
- $(m, i, t) \models f_1 \mathcal{U} f_2$ iff $\exists j \geq i: (m, j) \models f_2$ and $\forall k : i \leq k < j: (m, k) \models f_1$;
- $(m, i, t) \models f_1 \mathcal{S} f_2$ iff $\exists j \leq i: (m, j) \models f_2$ and $\forall k : j < k \leq i: (m, k) \models f_1$;
- $(m, i, t) \models \triangleright_{\sim c} f_1$ iff
  - $\exists j > i : (m, j) \models f_1$ and $\forall k : i \leq k < j : (m, k) \not\models f_1$ and $l(I_j^m) - t \sim c$
  - **Or** $(m, i) \models f_1$ and $0 \sim c$;
- $(m, i, t) \models \triangleleft_{\sim c} f_1$ iff
  - $\exists j < i : (m, j) \models f_1$ and $\forall k : j < k \leq i : (m, k) \not\models f_1$ and $t - r(I_j^m) \sim c$
  - **Or** $(m, i) \models f_1$ and $0 \sim c$;

A timed state sequence m is a model of a SC formula f, noted $m \models f$, iff $(m, 0, 0) \models f$[6].

As usual, we can define other temporal operators as syntactical abbreviations:

- **for the future:** $\top \equiv \neg f_1 \vee f_1$, $\Diamond f_1 \equiv \top \mathcal{U} f_1$, $\Box f_1 \equiv \neg \Diamond \neg f_1$, $f_1 \mathcal{U}_{\sim c} f_2 \equiv f_1 \mathcal{U} f_2 \wedge \triangleright_{\sim c} f_2$, $\Diamond_{\sim c} f_1 \equiv \top \mathcal{U}_{\sim c} f_1$, $\Box_{\sim c} f_1 \equiv \neg \Diamond_{\sim c} \neg f_1$, with $\sim \in \{<, \leq\}$;
- **for the past:** $\blacklozenge f_1 \equiv \top \mathcal{S} f_1$, $\blacksquare f_1 \equiv \neg \blacklozenge \neg f_1$, $f_1 \mathcal{S}_{\sim c} f_2 \equiv f_1 \mathcal{S} f_2 \wedge \triangleleft_{\sim c} f_2$, $\blacklozenge_{\sim c} f_1 \equiv \top \mathcal{S}_{\sim c} f_1$, $\blacksquare_{\sim c} f_1 \equiv \neg \blacklozenge_{\sim c} \neg f_1$, with $\sim \in \{<, \leq\}$;

Example 5 Some SC formulae. Here are some examples of SC formulae with their verbal meaning. The verbal interpretation characterizes the infinite timed state sequences m such that $m \models f$.

- $\Box q$: q is always true. Such a formula allows the specifier to assert invariance properties of a system.
- $\Box(p \rightarrow \Diamond_{\leq 5} q)$: a p position is always followed by a q position within 5 time units. Such a formula allows the specification of bounded response time.
- $\Box(p \rightarrow \triangleright_{=3} q)$: when a p position is encountered, the first following q position is at exactly 3 time units. Such a formula allows the assertion of exact response time (assuming no intervening request p).
- $\Diamond \Box q$: q will eventually hold permanently.
- $\Box(\triangleleft_{=3} q \rightarrow p)$. This formula asserts that if the last q position is exactly distant of 3 time units then p must be true now (time-out).

---

[5] Let us note that definition 7 and 8 are not cyclic since every timed sequence of state is propositions-Fine.

[6] anchored interpretation

- $\Box(q \rightarrow p\mathcal{S}_{\leq 3}r)$. When a q position is encountered then the last r position is distant at most of 3 time units and all intermediary positions were p positions.

However, there are properties that can not be expressed using SC logic:

Example 6. Every p state is followed by a q state exactly 5 time units later. Such a property can be expressed in MTL [8] as follows :$\Box(p \rightarrow \Diamond_{=5}q)$. This property is **not** expressed by the following SC formula :$\Box(p \rightarrow \triangleright_{=5} q)$ which is stronger since it requires that the **first** q is at exactly 5 time units.

As we can see, the SC logic is quite expressive. Most of the properties that are encountered when dealing with real-time systems, e.g. bounded response time, bounded invariance, time-out,..., can be easily and elegantly specified. In practice the use of the operator $\Diamond_{\sim c}$ can often be replaced by the stronger operator $\triangleright_{\sim c}$. The presence of the past operators is also a facility for the specifier.

To evaluate more deeply the expressive power of SCL we compare it to the logic MITL [5]. MITL is a linear real-time logic which is a subset of MTL [8]. In MITL the until operator, noted $\widetilde{\mathcal{U}}$ in the sequel, may be constrained by an interval I which is a convex subset of the positive real numbers. The decidability results for MITL are obtained by prohibiting the use of singular intervals. Here is our (open/closed invariant) formal semantics of the MITL $\widetilde{\mathcal{U}}_I$ operator:

Definition 9 $\widetilde{\mathcal{U}}_I$ Semantics. A formula $f_1\widetilde{\mathcal{U}}_I f_2$ [7] of MITL is satisfied in a model m at time t, noted $(m,t) \models_{MITL} f_1\widetilde{\mathcal{U}}_I f_2$ iff: $\exists t_1 \in t + I$ [8] s. t.:

- $\forall t' \geq t_1 : \exists t'' \in [t_1, t'] : (m, t'') \models_{MITL} f_2$
- $(m, t_1) \models_{MITL} f_1 \vee f_2$
- $\forall t' \in (t, t_1) : (m, t') \models_{MITL} f_1$

The semantics of the other logical operators and the syntactical abbreviations are defined as usual. We now show that the SC logic is a subset of MITL, more precisely of $MITL_{0,\infty}$ [5]. We give here the definition in MITL of the future temporal and real-time operators of SCL; the past operators can be defined symmetrically, the boolean connectives have the same semantics in the two logics.

(a1) $f_1\mathcal{U}f_2 \triangleq_{MITL} (f_1 \vee f_2) \wedge f_1\widetilde{\mathcal{U}}_{[0,\infty)} f_2$;

(a2) $\triangleright_{=c} f_1 \triangleq_{MITL} \widetilde{\Box}_{[0,c)}\neg f_1 \wedge \widetilde{\Diamond}_{[o,c]}f_1$;

(a3) $\triangleright_{\leq c} f_1 \triangleq_{MITL} \neg f_1\widetilde{\mathcal{U}}_{[0,c]}f_1$;

(a4) $\triangleright_{<c} f_1 \triangleq_{MITL} \neg f_1\widetilde{\mathcal{U}}_{[0,c)}f_1$;

---

[7] The unconstrained until operator is obtained by taking $I = [0, \infty)$.

[8] We have already underlined in the beginning of this section that we have taken a semantics for the $\mathcal{U}$ operator in SCL which is a bit different to the semantics given in MITL. To facilitate the comparison between the two logics we define the semantics of MITL in such a way that the until operator is invariant to the fact that the first interval is open or closed.

(a5) $\rhd_{\geq c} f_1 \triangleq_{MITL} \neg f_1 \wedge \neg f_1 \widetilde{\mathcal{U}}_{[c,\infty)} f_1$;

(a6) $\rhd_{>c} f_1 \triangleq_{MITL} \neg f_1 \wedge \neg f_1 \widetilde{\mathcal{U}}_{(c,\infty)} f_1$;

We now show which subset of MITL is expressible into SCL. All the constrained $\widetilde{\mathcal{U}}_I$ where $0 \in I$ and the unconstrained until can be expressed into SCL. In fact, if we consider that f_1 and f_2 are SCL expressible:

(b1) $f_1 \widetilde{\mathcal{U}} f_2 \triangleq_{SCL} \rhd_{=0} (f_1 \mathcal{U} f_2)$ [9]

(b2) $f_1 \widetilde{\mathcal{U}}_{[0,c)} f_2 \triangleq_{SCL} \rhd_{=0} (f_1 \mathcal{U} f_2) \wedge \rhd_{<c} f_2$;

(b3) $f_1 \widetilde{\mathcal{U}}_{[0,c]} f_2 \triangleq_{SCL} \rhd_{=0} (f_1 \mathcal{U} f_2) \wedge \rhd_{\leq c} f_2$;

Moreover a restricted form of $\widetilde{\mathcal{U}}_I$ with $l(I) \neq 0$ can be expressed into SCL:

(c1) $\neg f_2 \wedge (\neg f_2 \wedge f_1) \widetilde{\mathcal{U}}_{[a,b]} f_2 \triangleq_{SCL} \rhd_{=0} (f_1 \mathcal{U} f_2) \wedge \rhd_{\geq a} f_2 \wedge \rhd_{\leq b} f_2$

This formulae asserts that f_1 is verified until a *first* state where f_2 is verified is reached and this *first* state is distant of at least a time units and at most b time units. The other forms of intervals (open instead of closed or not right bounded) are trivially treated in the same way.

4 From SC Logic to SC Automata

The procedure we present here for the transformation of a SC formula into a SC automata is an adaptation of the usual tableau method [24] for the linear temporal logic PTL, see also [20]. The procedure for transforming a SC formula f into a SC automata A_f that accepts exactly runs on timed sequences of states that are models of f, i.e. $m \in L(A_f)$ iff $m \models f$, relies on a construction that uses the formulae of the closure of f. Let us first define formally the closure set of a SC formula :

Definition 10 Closure set. The closure set of a SC formula f, denoted $cl(f)$, is the smallest set of formulae satisfying the following requirements :

- $f \in cl(f)$;
- For every $f_1 \in cl(f)$ and f_2 a sub formula of f_1, $f_2 \in cl(f)$;
- For every $f_1 \in cl(f)$, $\neg f_1 \in cl(f)$. To keep $cl(f)$ finite, we identify $\neg\neg f_1$ with f_1 in any context;
- For every $f_1 \mathcal{U} f_2 \in cl(f)$, $\bigcirc(f_1 \mathcal{U} f_2) \in cl(f)$[10];
- For every $f_1 \mathcal{S} f_2 \in cl(f)$, $\bigodot(f_1 \mathcal{S} f_2) \in cl(f)$;

---

[9] The operator $\widetilde{\mathcal{U}}$ of MITL does not impose that f_1 must be true at time t of evaluation of $f_1 \widetilde{\mathcal{U}} f_2$.

[10] Let us note that neither $\bigcirc(f_1 \mathcal{U} f_2)$ nor $\bigodot(f_1 \mathcal{S} f_2)$ are SC logic formulae; they will be used to express connection requirements on the automaton $A_{\neg f}$.

4.1 Locations

The locations of the SC automaton A_f will be subsets of $cl(f)$. If a formula g belongs to a location l of A_f, the meaning is that when the automaton A_f is in location l in a run r then the formula g is verified in the states of the timed sequence of states corresponding to the locations l in the run r. Obviously, all possible subsets of the closure set are not candidate for representing a position in a model. For example, a subset of $cl(f)$ which contains both p and $\neg p$ can not be a candidate for a position in a model as this state is not satisfiable. To make the notion of candidate for a model position clearer, we define the notion of atom.

Definition 11 Atom. An atom over f is a subset $A \subseteq cl(f)$ satisfying the following requirements :

- RA_p : the atom A is propositionally consistent and complete. More formally :
  - For every $f_1 \in cl(f)$, $f_1 \in A$ iff $\neg f_1 \notin A$.
  - For every $f_1 \vee f_2 \in cl(f)$, $f_1 \vee f_2 \in A$ iff $f_1 \in A$ or $f_2 \in A$
- RA_{exp} : the atom A respects the semantics of the $\mathcal{U}$ and the $\mathcal{S}$ operators :
  - For every $f_1 \mathcal{U} f_2 \in cl(f)$, $f_1 \mathcal{U} f_2 \in A$ iff either :
    * $f_2 \in A$
    * $f_1, \bigcirc(f_1 \mathcal{U} f_2) \in A$
  - For every $f_1 \mathcal{S} f_2 \in cl(f)$, $f_1 \mathcal{S} f_2 \in A$ iff either :
    * $f_2 \in A$
    * $f_1, \odot(f_1 \mathcal{S} f_2) \in A$

The definition of atom represents a necessary condition for subsets of $cl(f)$ to be mutually satisfiable.

Definition 12 Locations of A_f. The set of locations L of the automaton $A_f = (\mathcal{P}, C, L, L_0, E, \mathcal{L}, \triangle, \mathcal{F})$ is the set of atoms and the subset of start locations L_0 is the subset of atoms $\{A \mid f \in A \wedge \forall(f_1 \mathcal{S} f_2) \in Cl(f) : \odot(f_1 \mathcal{S} f_2) \notin A\}$.

4.2 Transitions

The formula $\bigcirc(f_1 \mathcal{U} f_2)$ and $\odot(f_1 \mathcal{S} f_2)$ of $cl(f)$ are used to formulate the connection requirement of the formula automaton. In fact $\bigcirc(f_1 \mathcal{U} f_2)$ means that at the following location, the formula $f_1 \mathcal{U} f_2$ must be satisfied and $\odot(f_1 \mathcal{S} f_2)$ means that $f_1 \mathcal{S} f_2$ was satisfied at the previous location. We can now formulate more rigorously the connection requirement :

Definition 13 Connection requirement. In the formula automaton A_f of a SC formula f, the location $l \in L$ is connected by an edge to the location $l' \in L$, i.e. $(l, l') \in E$, iff the following requirements are satisfied :

- For every $\bigcirc(g) \in cl(f)$: $\bigcirc(g) \in l$ iff $g \in l'$
- For every $\odot(g) \in cl(f)$: $g \in l$ iff $\odot(g) \in l'$

4.3 Fatalities

The semantics of the formula $p\mathcal{U}q$ expresses that the formula p must stay true until a q state is reached. In this case, q is a fatality in the sense that in all models, a $p\mathcal{U}q$ state is always followed by some q state. The fulfillment of fatalities can be ensured by the mechanism of acceptance of Büchi automata and relies on the following theorem [20] :

Theorem 14. *Let m be a model of a SC formula f and $g \in cl(f)$, a formula of the form $p\mathcal{U}q$, promising a formula q. Then, m contains infinitely many positions $j \geq 0$ such that:$(m, j) \models \neg g$ or $(m, j) \models q$.*

We say that an execution of a formula automaton A_f fulfills the fatalities of a formula f iff for every formula $g \in cl(f)$ promising a formula q, the execution contains infinitely many $\neg g$ locations or q locations. To restrict the accepted execution of the formula automaton A_f to executions that fulfill the fatalities of f, we use the mechanism of accepting sets.

Definition 15 Accepting sets of a formula automaton. The set $\mathcal{F}$ of accepting sets of the formula automaton A_f of a SC formula f is defined as follow : for each formula $g \in cl(f)$, of the form $p\mathcal{U}q$, there exists a different accepting set $F_i \in \mathcal{F}$ that contains exactly the states A such that : $A \models \neg(p\mathcal{U}q)$ or $A \models q$;

4.4 Real-Time

The fulfillment of the real-time requirements can be obtained thanks to constraints over prophecy and history variables. Here we use a generalization of the definition given to state clocks. For each formula $\triangleleft_{\sim c} f_1$ (respectively $\triangleright_{\sim c} f_1$) $\in cl(f)$, we have an history (a prophecy) variable noted x_{f_1} (y_{f_1}). Those clocks are called *formula history* and *formula prophecy* variables. The value of those clocks (x_f, y_f) along a timed sequence of states m is obtained by trivially adapting the definition 3: replace $p \in s_i^m$ by $(m, i) \models f$ and $p \notin s_i^m$ by $(m, i) \not\models f$. We construct the constraints over these variables as follows :

Definition 16 Clock constraints. The clocks constraints $\triangle$ decorating the locations of the formula automaton A_f of a SC formula f are defined as follows :

- $\forall \triangleleft_{\sim c} f_1 \in cl(f)$:
  - if $\triangleleft_{\sim c} f_1 \in l$ then $y_{f_1} \sim c$ decorates l;
  - if $\triangleleft_{\sim c} f_1 \notin l$ then $\neg(y_{f_1} \sim c)$ decorates l.
- $\forall \triangleright_{\sim c} f_1 \in cl(f)$:
  - if $\triangleright_{\sim c} f_1 \in l$ then $x_{f_1} \sim c$ decorates l;
  - if $\triangleright_{\sim c} f_1 \notin l$ then $\neg(x_{f_1} \sim c)$ decorates l. Where $\sim \in \{<, \leq, =, \geq, >\}$

The state labeling function is trivially defined as follows :

Definition 17 Labeling function. $\mathcal{L}(l) = \{p \mid p \in \mathcal{P} \wedge p \in l\}$, where $l \in L$.

Theorem 18. *The timed language accepted by the formula automaton A_f is exactly the set of models of the SCL formula f, i.e. $L(A_f) = \{m \mid (m, 0, 0) \models f\}$.*

4.5 Region construction

The principles of the so-called region construction [3] which transforms a timed automaton into an untimed finite state machine, called the region automaton, can be applied to SC automata, see the full paper [22] for details. The following theorem is the basis for an algorithmic analysis of SC automata :

Theorem 19. *The number of locations in the region automaton $R(A)$ of a SC automata A is finite and $O(n \cdot 2^{m \cdot \log c \cdot m})$, where n is the number of locations in A, m is the number of clocks and c is the largest constant appearing in A.*

Theorem 20. *The language of $R(A)$ corresponds exactly to $Untimed(L(A))$. Consequently, the timed language of A is empty iff the language of $R(A)$ is empty.*

The theorem 18 and theorem 20 give us the possibility to decide the model-checking as well as the satisfiability/validity problems for SCL. In fact:

Theorem 21. *The model-checking problem, i.e. $L(A) \subseteq^? L(f)$[11], is decided by testing if $L(A) \cap L(A_{\neg f}) = \varnothing$. The validity of an SCL formula is decided by testing if $L(A_{\neg f}) = \varnothing$.*

4.6 Complexity results

The procedure that we propose for deciding SCL constructs first a SC automaton which is transformed into an untimed automaton, the region automaton, for checking emptiness. The following lemma and theorem characterize the size of the constructed automaton for a given SCL formula f:

Lemma 22. *The number of locations of the SC automaton A_f is $O(2^n)$ where n is the length of the formula f (the number of propositions, modal operators and logical connectives), the number of clocks is $O(n)$ and the largest constant appearing in the constraints in A_f is the largest constant c appearing in f.*

Theorem 23. *From lemma 22 and theorem 19 it can be established that the number of regions of the region automaton corresponding to the SC automaton A_f of a SCL formula f is $O(2^{n \cdot \log c \cdot n})$ where n is the length of the formula f and c is the largest constant appearing in f.*

It can be shown that the search of an accepting path in the region automaton can be performed without explicitly constructing the entire automaton. In fact, the complexity of the decision of SCL is in PSPACE.

Theorem 24. *The satisfiability/validity problem for SCL is decidable in* PSPACE.
Proof. *We have that $SCL \subset MILT_{0,\infty}$. The satisfiability/validity problem for $MILT_{0,\infty}$ is* PSPACE *[5], deciding SCL is thus at most in* PSPACE. *As our logic contains the until operator of PTL then SCL is at least* PSPACE *[12].*

---
[11] $L(f) = \{m \mid (m, 0, 0) \models f\}$.

5 Conclusion and Future Work

In this paper, we have presented a real-time logic based on the dense time semantics. The SC logic can be used to express most interesting properties of real-time reactive systems; the presence of past operators is often useful for specifications and dealt with simply. The satisfiability problem and the model-checking problem are decidable. The SC logic formulae can be transformed in an elegant way into SC automata. SC automata allow the declarative specification of real-time constraints, are closed under all boolean operations and are still expressive enough to model real-time systems.

The real-time operators that we have presented here are straightforward counterparts of the *history* and *prophecy* variables. The key for the determinization of SC automata is not the particular definition we have given (see Definition 3) to the prophecy/history variables but, as it is shown in [6] the property that *at each step during a computation, all clock values are determined solely by the input word*. So other types of clocks can be defined with their corresponding real-time operators.

The characteristic that each clock of a SC automaton is attached to a proposition can be exploited to conduct static analysis of SC automata. Partial evaluation techniques and abstract interpretation frameworks [13, 17, 14] will be studied in the context of the SC logic and SC automata.

References

1. R. Alur. *Techniques for Automatic Verification of Real-Time Systems*. PhD thesis, Stanford University, 1991.
2. R. Alur, C. Courcoubetis, and D. Dill. Model-checking for real-time systems. In *Proceedings of the 5th Symposium on Logic in Computer Science*, pages 414–425, Philadelphia, June 1990.
3. R. Alur, C. Courcoubetis, and D.L. Dill. Model-checking in dense real-time. *Information and Computation*, 104(1):2–34, 1993. Preliminary version appears in the Proc. of 5th LICS, 1990.
4. R. Alur and D.L. Dill. A theory of timed automata. *Theoretical Computer Science*, 126:183–235, 1994. Preliminary version appears in Proc. 17th ICALP, 1990, LNCS 443.
5. R. Alur, T. Feder, and T.A. Henzinger. The benefits of relaxing punctuality. *Journal of the ACM*, 43(1):116–146, 1996.
6. R. Alur, L. Fix, and T.A. Henzinger. A determinizable class of timed automata. In *Proceedings of the Sixth Conference on Computer-Aided Verification*, Lecture Notes in Computer Science 818, pages 1–13. Springer-Verlag, 1994.
7. R. Alur and T.A. Henzinger. Logics and models of real time: a survey. In J.W. de Bakker, K. Huizing, W.-P. de Roever, and G. Rozenberg, editors, *Real Time: Theory in Practice*, Lecture Notes in Computer Science 600, pages 74–106. Springer-Verlag, 1992.
8. R. Alur and T.A. Henzinger. Real-time logics: complexity and expressiveness. *Information and Computation*, 104(1):35–77, 1993. Preliminary version appears in the Proc. of 5th LICS, 1990.

9. R. Alur and T.A. Henzinger. A really temporal logic. *Journal of the ACM*, 41(1):181–204, 1994. Preliminary version appears in Proc. 30th FOCS, 1989.

10. A. Bouajjani and Y. Lakhnech. Temporal Logic + Timed Automata : Expressiveness and Decidability. In *Proc. Intern. Conf. on Concurrency Theory (CONCUR'95), Philadelphia, August 1995*. LNCS 962, 1995.

11. J.R. Burch, E.M. Clarke, K.L. McMillan, D.L. Dill, and L.J. Hwang. Symbolic model checking: 10^{20} states and beyond. In *Proceedings of the 5th Symposium on Logic in Computer Science*, pages 428–439, Philadelphia, June 1990.

12. E.M. Clarke, E.A. Emerson, and A.P. Sistla. Automatic verification of finite-state concurrent systems using temporal logic specifications. *ACM Transactions on Programming Languages and Systems*, 8(2):244–263, January 1986.

13. P. Cousot and R. Cousot. Abstract interpretation: A unified lattice model for static analysis of programs by construction or approximation of fixpoints. In *Conference Record of Fourth ACM Symposium on Programming Languages (POPL'77)*, pages 238–252, Los Angeles, California, January 1977.

14. Dennis René Dams. *Abstract Interpretation and Partition Refinement for Model Checking*. PhD thesis, Eindhoven University of Technology, P.O. Box 513, 5600 MB Eindhoven, The Netherlands, July 1996.

15. E.A. Emerson. *Handbook in Theoretical Computer Science, Formal Models and Semantics*, chapter Temporal and Modal Logic, pages 995–1072. Elsevier, 1990.

16. R. Gerth, D. Peled, M. Y. Vardi, and P. Wolper. Simple on-the-fly automatic verification of linear temporal logic. In *Proc. 15th Work. Protocol Specification, Testing, and Verification*, Warsaw, June 1995. North-Holland.

17. N. Halbwachs. Verification of Linear Hybrid Systems by Means of Convex Approximations. In *Proceedings of SAS'94, Lecture Notes in Computer Science 864*, pages 223–237, 1994.

18. C.L. Heitmeyer and N. Lynch. The generalized railroad crossing: A case study in formal verification of real-time systems. In *Proc. of the IEEE Real-Time Systems Symposium, San Juan, Puerto Rico*, December 7-9, 1994.

19. Ron Koymans. *Specifying message passing and time-critical systems with temporal logic*. LNCS 651, Springer-Verlag, 1992.

20. Z. Manna and A. Pnueli. *Temporal Verification of Reactive Systems : Safety*. Springer-Verlag, Berlin, January 1995.

21. A. Pnueli. The temporal logic of programs. In *Proc. 18th IEEE Symposium on Foundation of Computer Science*, pages 46–57, 1977.

22. J.-F. Raskin and P.-Y. Schobbens. State Clock Logic: a Decidable Real-Time Logic. Research Paper RP-10-96, Computer Science Department, FUNDP, Namur (Belgium), May 1996.

23. C. Stirling. Comparing linear and branching time temporal logics. In B. Banieqbal, H. Barringer, and A. Pnueli, editors, *Temporal Logic in Specification*, volume 398, pages 1–20. Lecture Notes in Computer Science, Springer-Verlag, 1987.

24. P. Wolper. The tableau method for temporal logic: An overview. *Logique et Analyse*, (110–111):119–136, 1985.

From Quantity to Quality*

Thomas A. Henzinger and Orna Kupferman

UC Berkeley, EECS Department, Berkeley, CA 94720-1770, U.S.A.
Email: {tah,orna}@eecs.berkeley.edu

Abstract. In temporal-logic model checking, we verify the correctness of a program with respect to a desired behavior by checking whether a structure that models the program satisfies a temporal-logic formula that specifies the behavior. The model-checking problem for the branching-time temporal logic CTL can be solved in linear running time, and model-checking tools for CTL are used successfully in industrial applications. The development of programs that must meet rigid real-time constraints has brought with it a need for real-time temporal logics that enable quantitative reference to time. Early research on real-time temporal logics uses the discrete domain of the integers to model time. Present research on real-time temporal logics focuses on continuous time and uses the dense domain of the reals to model time. There, model checking becomes significantly more complicated. For example, the model-checking problem for TCTL, a continuous-time extension of the logic CTL, is PSPACE-complete.

In this paper we suggest a reduction from TCTL model checking to CTL model checking. The contribution of such a reduction is twofold. Theoretically, while it has long been known that model-checking methods for untimed temporal logics can be extended quite easily to handle discrete time, it was not clear whether and how untimed methods can handle the reset quantifier of TCTL, which resets a real-valued clock. Practically, our reduction enables anyone who has a tool for CTL model checking to use it for TCTL model checking. The TCTL model-checking algorithm that follows from our reduction is in PSPACE, matching the known bound for this problem. In addition, it enjoys the wide distribution of CTL model-checking tools and the extensive and fruitful research efforts and heuristics that have been put into these tools.

1   Introduction

Temporal logics can describe a temporal ordering of events and have been adopted as a powerful tool for specifying and verifying concurrent programs [Pnu77, MP92]. In temporal-logic *model checking*, we verify that a program meets a desired behavior by checking that a mathematical model of the program satisfies a temporal-logic formula that specifies the behavior [CE81, QS81]. We distinguish between four levels of temporal reasoning. The verification methods induced by these levels differ in the interpretation given to time. The first level allows only qualitative reference to time. The classical method of CTL model checking belongs to this level. There, the program is modeled as a state-transition

---

* This research was supported in part by the ONR YIP award N00014-95-1-0520, by the NSF CAREER award CCR-9501708, by the NSF grant CCR-9504469, by the AFOSR contract F49620-93-1-0056, by the ARO MURI grant DAAH-04-96-1-0341, by the ARPA grant NAG2-892, and by the SRC contract 95-DC-324.036.

graph, and the correct behaviors of the program are specified in the qualitative branching temporal logic CTL, which allows temporal operators such as "always" and "eventually." For example, if *req* and *grant* are atomic propositions, then the CTL formula

$$\psi = AG(req \to AF\,grant)$$

asserts that in all computations of the program, every request is eventually granted. The model-checking problem for CTL can be solved in linear time. More precisely, given a state-transition graph K and a CTL formula ψ, we can determine whether K satisfies ψ in time $O(|K| \cdot |\psi|)$ [CES86].

The development of programs that must meet rigid *real-time* constraints has brought with it a need for *real-time temporal logics* that enable quantitative reference to time [EMSS90, AH92]. We consider here a real-time extension of CTL, which we call CTL+clocks. The syntax of CTL+clocks extends the syntax of CTL by allowing reference to a set of *clock variables*. Formulas of CTL+clocks can refer to the values of the clocks, and may contain a *reset quantifier* $c.\varphi$, which resets a clock c to the value 0. For example, if c is a clock, then the formula

$$\psi' = AG[c.req \to AF(grant \wedge (c \leq 2))]$$

strengthens the formula ψ above by putting an upper bound on the time that may elapse before a grant is given: if the clock c is started at the time of a request, then the value of c is at most 2 at the time of the subsequent grant. The exact meaning of ψ' depends on the formal interpretation of CTL+clocks. The formulas of CTL+clocks can be interpreted in three different ways, forming the following three levels of quantitative temporal reasoning.

Pioneering work on real-time temporal logics allowed very simple quantitative reference to time. In [EMSS90], Emerson et al. interpret CTL+clocks formulas[2] over state-transition graphs (see also [CCM+94]). Each transition in the graph advances the time by one time unit. Hence, this level of temporal reasoning uses the *discrete* domain of the *integers* to model time, and it uses quantitative reference to time only in the specifications. It is quite clear that, when interpreted over state-transition graphs, CTL+clocks formulas can be translated to equivalent CTL formulas, and the problem of CTL+clocks model checking can be reduced to the problem of CTL model checking.[3] For example, when interpreted over state-transition graphs, the formula ψ' above is equivalent to the CTL formula

$$\psi'' = AG(req \to (grant \vee AX(grant \vee AX\,grant))),$$

which states that every request is granted within two transitions. The main limitation of this level of temporal reasoning is that while discrete time suffices for modeling globally-clocked programs, continuous time is required for modeling the composition of independently-clocked programs.

This limitation has been removed in the third level of temporal reasoning, known as the *fictitious-clock* approach [HMP92, AH93, AH94]. At this level, transitions happen in continuous time, but are recorded by a global digital clock, in discrete time. Accordingly, time is viewed as a state variable that ranges over the domain of the integers. Some transitions in the state-transition graph are designated as *tick* transitions (i.e., transitions of the

---

[2] The logic used in [EMSS90] is RTCTL, which is a strict syntactic subset of CTL+clocks.

[3] We note that while the translation involves an exponential blow-up, using a more sophisticated approach, model checking can still be done in time linear in the original CTL+clocks formula [EMSS90].

global digital clock). Whenever a tick transition is taken, time is advanced by one time unit. Hence, any number of program transitions can be taken in one time unit. By introducing a new atomic proposition, "*tick*", we can translate CTL+clocks formulas, when interpreted over state-transition graphs with tick transitions, to CTL formulas.[4] For example, the CTL+clocks formula ψ' above is equivalent to the CTL formula

$$\psi''' = AG(req \rightarrow A((\neg tick)U(grant \vee (tick \wedge AXA((\neg tick)U grant))))).$$

Hence, CTL+clocks model checking can be reduced to CTL model checking on this level of temporal reasoning as well [AH92]. The main limitation of this level is its limited accuracy. For example, the formula ψ''' asserts only that in all computations of the program, every request is followed by a grant within more than one and less than three time units. Also, as in the previous level, we have to fix a time granularity, which may cause a blow-up in the state space.

Much present research on qualitative reasoning focuses on *dense* time and uses the domain of the *reals* to model time in both the state-transition graph and the specification. On this fourth level of temporal reasoning, we model the programs as *timed automata* [AD94], where real-valued clocks keep track of timing constraints. The fourth level constitutes the most expressive way of specifying real-time programs. With this semantics, the logic CTL+clocks is called TCTL [ACD93], and the formula ψ' asserts that in all computations of the program, every request is followed by a grant within at most two time units. *TCTL model checking* is the problem of determining whether a given TCTL formula is satisfied in a given timed automaton.

The introduction of dense time in the model makes quantitative reasoning more complicated. For example, while the satisfiability problem for CTL+clocks when interpreted over discrete-time models (levels two or three) is decidable, it is undecidable for TCTL. Indeed, algorithms that handle satisfiability or model checking of CTL, and which are applicable to verification methods induced by the first three levels of temporal reasoning, cannot be easily extended to handle TCTL. The reason is the dense time domain of TCTL, which induces state-transition graphs with infinitely many states. It was shown, however, in [AD94], that each timed automaton induces a finite quotient of the infinite state space, such that two equivalent states satisfy the same TCTL formulas. More precisely, the automaton partitions the infinite time domain of clock valuations into finitely many *regions*, each of which can be viewed as a set of clock constraints (e.g., $2 < clock_1 < 3$; $clock_2 = 1$). This finite quotient is used in [ACD93] in order to solve the model-checking problem for TCTL. Alur et al. also prove that the problem is PSPACE-complete. The importance of the model-checking problem has led to the development of several other model-checking algorithms for TCTL [HNSY94, LL95, SS95, HKV96], all trying to cope with the large state space that needs to be stored.

This space problem, known as the *state-explosion problem*, is the main computational limitation of all the verification methods induced by the four levels of temporal reasoning. It constitutes one of the most challenging issues in the area of computer-aided verification and is the subject of active research. Most of the efforts during the last two decades have focused on pure qualitative reasoning, yielding CTL model-checking tools (e.g., SMV, VIS, CADP) that can handle systems with large state spaces [McM93, CGL93, BHSV+96, FGK+96]. Model-checking algorithms for TCTL adopt some of the techniques used in the tools for

---

[4] Again, the translation involves an exponential blow-up that is unnecessary for model checking.

CTL model checking. Still, TCTL model-checking tools are less successful than CTL model-checking tools, both in their level of performance and in their distribution. The reason is not only the clear computational advantage of CTL, but also the broad attention that CTL model-checking tools have enjoyed.

In this paper we suggest a reduction from TCTL model checking to CTL model checking. The contribution of such a reduction is twofold. Theoretically, it completes the picture of the four levels of temporal reasoning. While it has long been known that the first three levels are inter-reducible, our reduction shows that, as far as model-checking is concerned, the fourth level can also be reduced to the first three levels. Practically, our reduction enables anyone who has a tool for CTL model checking to use it for TCTL model checking. The TCTL model-checking algorithm that follows from our reduction is in PSPACE, matching the known bound for this problem. In addition, it enjoys the extensive and fruitful research efforts that have been put into CTL model-checking tools.

The reduction is not complicated. Given a timed automaton $\mathcal{U}$ and TCTL formula ψ, we construct a state-transition graph $untime(\mathcal{U})$, of size exponential in the size of $\mathcal{U}$, and a CTL formula $untime(\psi)$, of length linear in the length of ψ, such that $\mathcal{U}$ satisfies ψ iff $untime(\mathcal{U})$ satisfies $untime(\psi)$. The graph $untime(\mathcal{U})$ is essentially the region graph used in [ACD93], augmented by a new atomic proposition and new transitions, which handle the reset quantifier $c.\varphi$. When we evaluate TCTL formulas, the reset quantifier causes the course of evaluation to "jump around in the graph." Having to evaluate a formula $c.\varphi$ in a state w, we actually evaluate the formula φ in another state, which differs from w only in that the value of the clock c is reset to 0. Such jumps are replaced in $untime(\mathcal{U})$ by new transitions. In $untime(\psi)$, path quantification guarantees that whenever we come to evaluate $c.\varphi$, the current value of c is 0, and thus no jump is required.

2 Definitions

2.1 Kripke Structures and Timed Structures

We model programs without real-time constraints by *Kripke structures*. A Kripke structure is a tuple $\mathcal{K} = \langle AP, W, R, w^0, \sigma \rangle$, where AP is a finite set of atomic propositions, W is a set of states, $R \subseteq W \times W$ is a total transition relation (every state has at least one successor), $w^0 \in W$ is an initial state, and $\sigma : W \to 2^{AP}$ maps to each state a set of atomic propositions true in the state. A *path* in $\mathcal{K}$ is an infinite sequence of states $w_0, w_1, w_2, \ldots$ such that for all $i \geq 0$, we have $\langle w_i, w_{i+1} \rangle \in R$.

We model real-time programs by *timed structures*. Timed structures extend traditional Kripke structures by labeling each transition with a nonnegative real number denoting its duration. Formally, a timed structure is a tuple $\mathcal{T} = \langle AP, W, R, w^0, \sigma \rangle$, where AP, W, w^0, and σ are as in a Kripke structure, and $R \subseteq W \times \mathbb{R} \times W$ (we denote by $\mathbb{R}$ the set of all nonnegative real numbers). A *path* in $\mathcal{T}$ is an infinite sequence of pairs $\langle w_0, \delta_0 \rangle, \langle w_1, \delta_1 \rangle, \ldots$, such that for all $i \geq 0$, we have $\langle w_i, \delta_i, w_{i+1} \rangle \in R$. A *timed word* is an infinite sequence $\tau \in (2^{AP} \times \mathbb{R})^\omega$. We sometimes refer to a timed word as a function $\tau : \mathbb{N} \to 2^{AP} \times \mathbb{R}$ and use $\tau_1(i)$ and $\tau_2(i)$ to refer to the i'th event and duration, respectively, in τ.

2.2 Timed Automata

We represent real-time programs by timed automata. We now define timed automata and the timed structures induced by them.

Given a set C of clocks, a *clock environment* $\mathcal{E} : C \to \mathbb{R}$ assigns to each clock a non-negative real value. Given a clock environment $\mathcal{E}$, a set $S \subseteq C$ of clocks and a nonnegative real $\delta \in \mathbb{R}$, we define $progress(\mathcal{E}, S, \delta)$ as the clock environment $\mathcal{E}'$ that resets all clocks in S and advances all clocks in $S \setminus C$ by δ; that is, for all $c \in C$, we have

$$\mathcal{E}'(c) = \begin{cases} 0 & \text{if } c \in S, \\ \mathcal{E}(c) + \delta & \text{if } c \notin S. \end{cases}$$

For two clock environments $\mathcal{E}$ and $\mathcal{E}'$, we say that $\mathcal{E} < \mathcal{E}'$ iff for every clock $c \in C$, we have $\mathcal{E}(c) < \mathcal{E}'(c)$. We use $\mathcal{E}^0$ to denote the clock environment that assigns 0 to all clocks. For a set C of clocks, a formula (also referred sometimes as *clock constraint*) in $guard(C)$ is one of the following:

- **true, false,** or $c \sim v$, where $c \in C$, $v \in \mathbb{N}$, and $\sim \in \{\geq, >, \leq, <\}$,
- $\theta_1 \vee \theta_2$ or $\theta_1 \wedge \theta_2$, where θ_1 and θ_2 are formulas in $guard(C)$.

A *timed automaton* is a tuple $\mathcal{U} = \langle AP, C, L, E, P, inv, l^0 \rangle$, where

- AP is a finite set of atomic propositions,
- C is a finite set of program clocks,
- L is a finite set of locations,
- $E : L \to 2^{guard(C) \times 2^C \times L}$ is a nondeterministic transition function,
- $P : L \to 2^{AP}$ assigns to each location a set of atomic propositions,
- $inv : L \to guard(C)$ assigns to each location an invariant,
- $l^0 \in L$ is an initial location.

A *position* of $\mathcal{U}$ is a pair $\langle l, \mathcal{E} \rangle \in L \times \mathbb{R}^C$; that is, a position describes a location and a clock environment. Given a position $\langle l, \mathcal{E} \rangle$ and a time delay $\delta \in \mathbb{R}$, we say that the position $\langle l', \mathcal{E}' \rangle$ is a δ-*successor* of $\langle l, \mathcal{E} \rangle$ iff there exists a triple $\langle \theta, S, l' \rangle \in E(l)$ such that the following three conditions hold:

1. $progress(\mathcal{E}, \emptyset, \delta) \models \theta$.
2. For every nonnegative real $\delta' < \delta$, we have $progress(\mathcal{E}, \emptyset, \delta') \models inv(l)$.
3. $\mathcal{E}' = progress(\mathcal{E}, S, \delta)$.

Timed automata run on timed words. A *run* r of a timed automaton on the timed word $\mathcal{U}$ is an infinite sequence of positions of $\mathcal{U}$. Thus, $r \in (L \times \mathbb{R}^C)^\omega$. We sometimes refer to a run as a function $r : \mathbb{N} \to L \times \mathbb{R}^C$. Given a timed word $\tau : \mathbb{N} \to 2^{AP} \times \mathbb{R}$, a run r of $\mathcal{U}$ on τ satisfies the following. For every $i \geq 0$, let $r(i) = \langle l_i, \mathcal{E}_i \rangle$. Then:

- $l_0 = l^0$ and $\mathcal{E}_0 = \mathcal{E}^0$.
- For every $i \geq 0$, we have $P(l_i) = \tau_1(i)$.
- For every $i \geq 0$, we have $\langle l_{i+1}, \mathcal{E}_{i+1} \rangle$ is a $\tau_2(i)$-successor of $\langle l_i, \mathcal{E}_i \rangle$.

We say that $\mathcal{U}$ *accepts* the timed word τ iff there exists a run of $\mathcal{U}$ on τ. The *language* of $\mathcal{U}$ is the set of all timed words that are accepted by $\mathcal{U}$. Each timed automaton induces a timed structure. Formally, the *timed structure of* $\mathcal{U}$ is

$$\mathcal{T}(\mathcal{U}) = \langle AP, L \times \mathbb{R}^C, R, \{\langle l^0, \mathcal{E}^0 \rangle\}, \sigma \rangle,$$

where R and σ are defined as follows:

- $R(\langle l, \mathcal{E} \rangle, \delta, \langle l', \mathcal{E}' \rangle)$ iff $\langle l', \mathcal{E}' \rangle$ is a δ-successor of $\langle l, \mathcal{E} \rangle$.
- For all states $\langle l, \mathcal{E} \rangle \in L \times \mathbb{R}^C$, we have $\sigma(\langle l, \mathcal{E} \rangle) = P(l)$.

Note that the state set of $\mathcal{T}(\mathcal{U})$ is infinite and that $\mathcal{T}(\mathcal{U})$ may have an infinite branching degree. It is easy to see that a timed word is accepted by $\mathcal{U}$ iff it is induced by a path, starting at $\langle l^0, \mathcal{E}^0 \rangle$, in $\mathcal{T}(\mathcal{U})$.

2.3 The Real-time Branching Temporal Logic TCTL

We specify properties of real-time programs using real-time temporal logics. We consider here TCTL, a real-time extension of the branching temporal logic CTL with clocks [ACD93]. Formulas of TCTL are defined with respect to the sets AP and $C_\mathcal{U}$ of the program's atomic propositions and clocks, respectively, and a set C_ψ of specification clocks. We consider TCTL formulas in positive normal form in which negation may apply to atomic propositions only. Given $AP, C_\mathcal{U}$, and C_ψ, a formula of TCTL is one of the following:

- **true, false**, p, or $\neg p$, where $p \in AP$.
- $c \sim v$, where $c \in C_\mathcal{U} \cup C_\psi$, $v \in \mathbb{N}$, and $\sim \in \{\geq, >, \leq, <\}$.
- $\varphi_1 \vee \varphi_2$, $\varphi_1 \wedge \varphi_2$, $E\varphi_1 U \varphi_2$, $A\varphi_1 U \varphi_2$, $E\varphi_1 \tilde{U} \varphi_2$, or $A\varphi_1 \tilde{U} \varphi_2$, where φ_1 and φ_2 are TCTL formulas.
- $c.\varphi$, where $c \in C_\psi$ and φ is a TCTL formula.

The temporal operator $\tilde{U}$ is dual to the U ("until") operator. For example, the formula $E\varphi_1 \tilde{U} \varphi_2$ is equivalent to the formula $\neg A(\neg\varphi_1)U(\neg\varphi_2)$. In addition, we use the usual $\rightarrow$ ("implies"), F ("eventually"), and G ("always") abbreviations. The reset quantifier $c.\varphi$ binds all free occurrences of c in φ. We denote by $free(\psi) \subseteq C_\mathcal{U} \cup C_\psi$ the set of clocks free in ψ (i.e., these that have a non-bound occurrence). We denote by $bound(\psi) \subseteq C_\psi$ the set of all clocks bound in ψ (i.e., clocks c for which $c.\varphi$ is a subformula of ψ). For example, the formula $\psi = (c_1 = 5) \vee c_2.AF(c_2 > 0)$ has $free(\psi) = \{c_1\}$ and $bound(\psi) = \{c_2\}$. By renaming clocks we can make sure that no occurrence of a clock in ψ is bound to more than one reset quantifier and that $bound(\psi)$ and $free(\psi)$ are disjoint. A TCTL formula ψ is *closed* iff $free(\psi) = \emptyset$. We denote by $cl(\psi)$ the set of all subformulas of ψ. It is easy to see that the size of $cl(\psi)$ is bounded by the length of ψ.

TCTL formulas are interpreted over pairs that consist of a state of a timed structure $\mathcal{T}(\mathcal{U})$, where $C_\mathcal{U}$ is the set of $\mathcal{U}$'s clocks, and a clock environment $\mathcal{E} : C_\mathcal{U} \cup C_\psi \rightarrow \mathbb{R}$. We use $w, \mathcal{E} \models \varphi$ to indicate that a formula φ holds at state w with clock environment $\mathcal{E}$ (with respect to the given timed structure $\mathcal{T}$). A formal definition of the relation $\models$ can be found in [HNSY94]. We will define later the semantics of TCTL formulas when interpreted over quotient graphs induced by timed structures.

For a timed structure $\mathcal{T}$ and a TCTL formula ψ, we say that $\mathcal{T} \models \psi$ iff $\langle w^0, \mathcal{E}^0 \rangle \models \psi$. The *model-checking problem* for TCTL is defined as follows: given a timed automaton $\mathcal{U}$ and a closed TCTL formula ψ, determine whether $\mathcal{T}(\mathcal{U}) \models \psi$ (denoted $\mathcal{U} \models \psi$).

2.4 Regions and Region Structures

Each set C of clocks induces infinitely many clock environments. Given a set C of clocks and a set G of clock constraints in $guard(C)$, we can partition the clock environments in

$\mathcal{R}^C$ into finitely many equivalence classes such that all clock environments in the same class are indistinguishable by clock constraints in G. It was proven in [AD94] that a sufficient condition for two environment clocks to be indistinguishable is agreement on the integral parts of all clocks values and agreement on the ordering of the fractional parts of all clock values. This leads to the following definition of *regions*. For $x \in \mathbb{R}$, let $\lfloor x \rfloor$ and $\langle x \rangle$ denote the integer and the fractional parts of x, respectively. Also, for each clock $c \in C$, let v_c be the largest integer v for which $x \sim v$ is a subformula of some clock constraint in G. We define an equivalent relation $\approx \subseteq \mathbb{R}^C \times \mathbb{R}^C$. For two clock environments $\mathcal{E}$ and $\mathcal{E}'$, we have $\mathcal{E} \approx \mathcal{E}'$ iff the following three conditions hold:

1. For all $c \in C$, either $\lfloor \mathcal{E}(c) \rfloor = \lfloor \mathcal{E}'(c) \rfloor$, or $\mathcal{E}(c) > v_c$ and $\mathcal{E}'(c) > v_c$.
2. For all $\{c, d\} \subseteq C$ with $\mathcal{E}(c) \leq v_c$ and $\mathcal{E}(d) \leq v_d$, we have $\langle \mathcal{E}(c) \rangle \leq \langle \mathcal{E}(d) \rangle$ iff $\langle \mathcal{E}'(c) \rangle \leq \langle \mathcal{E}'(d) \rangle$.
3. For all $c \in C$ with $\mathcal{E}(c) \leq v_c$, we have $\langle \mathcal{E}(c) \rangle = 0$ iff $\langle \mathcal{E}'(c) \rangle = 0$.

We now define a region to be an equivalence class of the relation $\approx$.

We denote the set of all regions induced by C and G by $\Upsilon(C, G)$. When C and G are clear or not important, we use only Υ. Let $rep : \Upsilon \to \mathbb{R}^C$ map each region to an arbitrary representative clock environment in it. We represent a region π by $rep(\pi)$. A clock environments $\mathcal{E}$ then belongs to π iff $\mathcal{E} \approx rep(\pi)$. We sometime represent a region also by a finite set of clock constraints (e.g., $[x = 1; 2 < z < 3]$). A clock environment $\mathcal{E}$ then belongs to π iff it satisfies all its clock constraints. Following the definition of regions, the constraints that represent π specify the integral part of all clocks, the order among the fractional parts, and whether they are equal to 0.

Lemma 1. [AD94] *The number of regions in $\Upsilon(C, G)$ is bounded by $|C|! \cdot 2^{|C|} \cdot \prod_{c \in C}(2v_c + 2)$.*

For a region π and a formula $\varphi \in G$, we say that π satisfies φ (denoted $\pi \models \varphi$) iff $rep(\pi)$ satisfies φ. Note that by the definition of regions, π satisfies φ iff all clock environments in π satisfy φ. Given a clock constraint $c \sim v$ in G, let $reg(c \sim v)$ be the union of all regions that satisfy $c \sim v$.

Each region has a unique successor region. Intuitively, the successor of a region π is obtained from π by letting time pass. The function $succ : \Upsilon \to \Upsilon$ maps a region π to its successor region. For a region π and a clock environment $\mathcal{E}$, we have that $\mathcal{E} \in succ(\pi)$ iff $rep(\pi) \not\approx \mathcal{E}$, $rep(\pi) < \mathcal{E}$, and for every clock environment $\mathcal{E}'$ with $rep(\pi) < \mathcal{E}' < \mathcal{E}$, we have $\mathcal{E}' \approx rep(\pi)$ or $\mathcal{E}' \approx \mathcal{E}$. So, for example, if π has a clock constraint $c = v$, for $v \neq v_c$, its successor has a clock constraint $v < c < v + 1$. If $v = v_c$, then the successor has a clock constraint $c > v$, reflecting the fact that once c is larger than v_c, we are no longer interested in how large it is.

For $\pi \in \Upsilon(C, G)$ and a set S of clocks, we denote by $\pi[S := 0]$ the region obtained from π by resetting all clocks in S. That is, $\pi[S := 0]$ contains the clock environment $progress(rep(\pi), S, 0)$. For a guard $\varphi \in G$ and $S \subseteq C$ we denote by $\varphi[S := 0]$ the guard obtained from φ by replacing with 0 all clocks in S. Also, let $\Upsilon[0]$ denote the region where all clocks in C are set to 0.

Consider a timed automaton $\mathcal{U} = \langle AP, C_{\mathcal{U}}, L, E, P, inv, l^0 \rangle$ and a TCTL formula ψ over $AP, C_{\mathcal{U}}$, and a set C_ψ of specification clocks. Let G be the set of clock constraints in $\mathcal{U}$ and ψ, and let $\Upsilon = \Upsilon(C_{\mathcal{U}} \cup C_\psi, G)$ be the set of regions induced by $\mathcal{U}$ and ψ. We define a *region position* of $\mathcal{U}$ as a pair $\langle l, \pi \rangle \in L \times \Upsilon$. When we say that $\mathcal{U}$ is in region position

$\langle l, \pi \rangle$, we mean that $\mathcal{U}$ is in location l and that its clock environment is in π. We say that a region position $\langle l, \pi \rangle$ is *admissible* iff π satisfies $inv(l)$. We know that the automaton $\mathcal{U}$ can be only in admissible region positions. Moreover, when $\mathcal{U}$ is in region position $\langle l, \pi \rangle$, we know what its possible next region positions are: the automaton $\mathcal{U}$ can either take an *edge transition* and move to another location, possibly resetting some clocks, or take a *time transition* and stay in l while the values of the clocks change and meet the successor region. This leads to the following definition of the *region structure* $\mathcal{R}(\mathcal{U}, \psi)$ induced by $\mathcal{U}$ and ψ.

We define $\mathcal{R}(\mathcal{U}, \psi) = \langle AP \cup \Upsilon, W, R, w^0, \sigma \rangle$ to be the following Kripke structure:

- The set $W \subseteq L \times \Upsilon$ of states consists of all the admissible positions of $\mathcal{U}$.
- $R(\langle l, \pi \rangle, \langle l', \pi' \rangle)$ iff one of the following two conditions holds:
  1. $l = l'$ and $\pi' = succ(\pi)$. These transitions correspond to time transitions in $\mathcal{U}$ and we call them *time transitions*.
  2. There exists a transition $\langle \theta, S, l' \rangle \in E(l)$ such that π satisfies θ and $\pi' = \pi[S := 0]$. These transitions correspond to edge transitions in $\mathcal{U}$ and we call them *edge transitions*.
- The initial state w^0 is $\langle l^0, \Upsilon[0] \rangle$. If w^0 is not admissible, the automaton $\mathcal{U}$ is not an interesting real-time program.
- For all states $\langle l, \pi \rangle \in W$, we have $\sigma(\langle l, \pi \rangle) = P(l) \cup \{\pi\}$.

We can interpret a TCTL formula ψ with respect to a state in the timed automaton $\mathcal{U}$ and a clock environment for the specification clocks by means of the region structure $\mathcal{R}(\mathcal{U}, \psi)$. We use $\langle l, \pi \rangle \models \varphi$ to indicate that a subformula φ of ψ holds at state $\langle l, \pi \rangle$ of $\mathcal{R}(\mathcal{U}, \psi)$. Note that as π refers to the values of the clocks in both $C_{\mathcal{U}}$ and C_{ψ}, we do not need a clock environment for the specification clocks. The relation $\models$ is defined inductively as follows:

- For all l and π, we have $\langle l, \pi \rangle \models$ **true** and $\langle l, \pi \rangle \not\models$ **false**.
- For $p \in AP$, we have $\langle l, \pi \rangle \models p$ iff $p \in \sigma(\langle l, \pi \rangle)$, and $\langle l, \pi \rangle \models \neg p$ iff $p \notin \sigma(\langle l, \pi \rangle)$.
- For $c \in C$, $v \in \mathbb{N}$, and $\sim \in \{\geq, >, \leq, <\}$, we have $\langle l, \pi \rangle \models c \sim v$ iff $\pi \models c \sim v$.
- $\langle l, \pi \rangle \models \varphi_1 \vee \varphi_2$ iff $\langle l, \pi \rangle \models \varphi_1$ or $\langle l, \pi \rangle \models \varphi_2$.
- $\langle l, \pi \rangle \models \varphi_1 \wedge \varphi_2$ iff $\langle l, \pi \rangle \models \varphi_1$ and $\langle l, \pi \rangle \models \varphi_2$.
- $\langle l, \pi \rangle \models E\varphi_1 U \varphi_2$ iff there exists a path $\langle l_0, \pi_0 \rangle, \langle l_1, \pi_1 \rangle \ldots$ in $\mathcal{R}(\mathcal{U}, \psi)$ with $\langle l_0, \pi_0 \rangle = \langle l, \pi \rangle$, and there exists $i \geq 0$ such that $\langle l_i, \pi_i \rangle \models \varphi_2$, and for all $0 \leq j < i$ we have $\langle l_j, \pi_j \rangle \models \varphi_1$.
- $\langle l, \pi \rangle \models A\varphi_1 U \varphi_2$ iff for every path $\langle l_0, \pi_0 \rangle, \langle l_1, \pi_1 \rangle \ldots$ in $\mathcal{R}(\mathcal{U}, \psi)$ with $\langle l_0, \pi_0 \rangle = \langle l, \pi \rangle$, there exists $i \geq 0$ such that $\langle l_i, \pi_i \rangle \models \varphi_2$ and for all $0 \leq j < i$ we have $\langle l_j, \pi_j \rangle \models \varphi_1$.
- $\langle l, \pi \rangle \models E\varphi_1 \tilde{U} \varphi_2$ iff there exists a path $\langle l_0, \pi_0 \rangle, \langle l_1, \pi_1 \rangle \ldots$ in $\mathcal{R}(\mathcal{U}, \psi)$ with $\langle l_0, \pi_0 \rangle = \langle l, \pi \rangle$ such that for every $i \geq 0$ for which $\langle l_i, \pi_i \rangle \not\models \varphi_2$, there exists $j < i$ such that $\langle l_j, \pi_j \rangle \models \varphi_1$.
- $\langle l, \pi \rangle \models A\varphi_1 \tilde{U} \varphi_2$ iff for every path $\langle l_1, \pi_1 \rangle, \langle l_2, \pi_2 \rangle \ldots$ in $\mathcal{R}(\mathcal{U}, \psi)$ with $\langle l_1, \pi_1 \rangle = \langle l, \pi \rangle$ and for every $i \geq 0$ for which $\langle l_i, \pi_i \rangle \not\models \varphi_2$, there exists $j < i$ such that $\langle l_j, \pi_j \rangle \models \varphi_1$.
- $\langle l, \pi \rangle \models c.\varphi$ iff $\langle l, \pi[c := 0] \rangle \models \varphi$.

We say that $\mathcal{R}(\mathcal{U}, \psi) \models \psi$ iff $w^0 \models \psi$. Several works on real-time temporal logics consider a more elaborated semantic for TCTL, where path quantification ranges only over paths for which time diverges [HNSY94]. As we discuss in Section 4, our algorithm can be easily extended to handle this semantics as well.

By the definition of regions, we have the following.

Theorem 2. [HNSY94] *For every timed automaton $\mathcal{U}$ and TCTL formula ψ, we have $\mathcal{U} \models \psi$ iff $\mathcal{R}(\mathcal{U}, \psi) \models \psi$.*

3   Reducing TCTL Model Checking to CTL Model Checking

3.1   Untiming the Program

Consider a timed automaton $\mathcal{U} = \langle AP, C_{\mathcal{U}}, L, E, P, inv, l^0 \rangle$ and a TCTL formula ψ over $AP, C_{\mathcal{U}}$, and a set C_ψ of specification clocks. Let $\mathcal{R}(\mathcal{U}, \psi)$ be the region structure induced by $\mathcal{U}$ and ψ. Many states in $\mathcal{R}(\mathcal{U}, \psi)$ are not reachable. Indeed, the transitions in $\mathcal{R}(\mathcal{U}, \psi)$ may reset some of the clocks in $C_{\mathcal{U}}$ but can never reset clocks in C_ψ. We need these unreachable states because ψ may contain reset quantifiers that cause the course of the evaluation of ψ to "jump" into these states. When we untime the program, we make these states reachable. Below we define the *Kripke structure* $\mathcal{K}(\mathcal{U}, \psi)$ induced by $\mathcal{U}$ and ψ. The structure $\mathcal{K}(\mathcal{U}, \psi)$ is very similar to $\mathcal{R}(\mathcal{U}, \psi)$. Each state w in $\mathcal{K}(\mathcal{U}, \psi)$ corresponds to two states $\langle w, \mathbf{T} \rangle$ and $\langle w, \mathbf{E} \rangle$ in $\mathcal{R}(\mathcal{U}, \psi)$. The copy of w annotated with $\mathbf{T}$ can be reached only by time transitions. The copy of w annotated with $\mathbf{E}$ can be reached only by edge transitions. Time transitions in $\mathcal{K}(\mathcal{U}, \psi)$ may reset an arbitrary subset of the clocks in C_ψ. Formally,

$$\mathcal{K}(\mathcal{U}, \psi) = \langle AP \cup \Upsilon \cup \{\mathbf{T}\}, W \times \{\mathbf{T}, \mathbf{E}\}, R, \langle w^0, \mathbf{E} \rangle, \varsigma \rangle,$$

where W, w^0, and are as in $\mathcal{R}(\mathcal{U}, \psi)$, and R and ς are defined as follows:

- $R(\langle l, \pi, b \rangle, \langle l', \pi', b' \rangle)$ iff one of the following two conditions holds:
    1. $b' = \mathbf{T}$, $l = l'$, and there exists a set $S_\psi \subseteq C_\psi$ of specification clocks such that $\pi' = succ(\pi)[S_\psi := 0]$. These transitions correspond to time transitions in $\mathcal{U}$ and we call them *time transitions*.
    2. $b' = \mathbf{E}$ and there exists a transition $\langle \theta, S, l' \rangle \in E(l)$ such that π satisfies θ and $\pi' = \pi[S := 0]$. These transitions correspond to edge transitions in $\mathcal{U}$ and we call them *edge transitions*.
- The set of atomic propositions that hold in each state is as the one in $\mathcal{R}(\mathcal{U}, \psi)$, only that we add the atomic proposition $\mathbf{T}$ to states in $W \times \{\mathbf{T}\}$. Thus, $\varsigma(\langle w, b \rangle)$ is $\sigma(w)$ if $b = \mathbf{E}$, and is $\sigma(w) \cup \{\mathbf{T}\}$ if $b = \mathbf{T}$.

Note that the specification clocks are changed only in time transitions, when each specification clock is either advanced by the same amount as the program clocks, or it is reset. The duplication of the states and the new atomic proposition $\mathbf{T}$ enable us to distinguish between states that are reached by a time transition and states that are reached by an edge transition. Following Lemma 1, the size of $\mathcal{K}(\mathcal{U}, \psi)$ is exponential in the size of $\mathcal{U}$ and the length of ψ.

3.2   Untiming the Specification

We define a function

$$untime : \text{TCTL formulas} \rightarrow \text{CTL formulas}$$

such that for every timed automaton $\mathcal{U}$ and TCTL formula ψ, we have $\mathcal{U} \models \psi$ iff $\mathcal{K}(\mathcal{U}, \psi) \models untime(\psi)$. We define the function $untime$ by means of a function

$$f : \text{TCTL formulas} \times \text{sets of specification clocks} \to \text{CTL formulas}.$$

For a TCTL formula ψ, the CTL formula $untime(\psi)$ is then the formula $f(\psi, \emptyset)$.

For a set $S \subseteq C$ of clocks, we use $S = 0$ as an abbreviation for $\bigwedge_{c \in S}(c = 0)$ and use $S > 0$ as an abbreviation for $\bigwedge_{c \in S}(c > 0)$. Note that when $S = \emptyset$, the formulas $S = 0$ and $S > 0$ evaluate to **true**. Consider a path ρ in $\mathcal{K}(\mathcal{U}, \psi)$. For sets S_1 and S_2 of clocks, we use $fair(S_1, S_2)$ to abbreviate a path formula stating that the clocks in S_1 are never reset and the clocks in S_2 are always reset along the path ρ. That is,

$$fair(S_1, S_2) = XG(\mathbf{T} \to (S_1 > 0)) \wedge G(S_2 = 0).$$

Given a TCTL formula φ and a set $S \subseteq C_\varphi$ of clocks, we define $f(\varphi, S)$ by induction on the structure of φ as follows (we first present the mapping into CTL* formulas and then translate them, in Section 3.4, to CTL).

- $f(\mathbf{true}, S) = \mathbf{true}$ and $f(\mathbf{false}, S) = \mathbf{false}$.
- For $p \in AP$, we have $f(p, S) = p$ and $f(\neg p, S) = \neg p$.
- For a clock constraint $c \sim v$, we have $f(c \sim v, S) = reg((c \sim v)[S := 0])$.
- $f(\varphi_1 \vee \varphi_2, S) = f(\varphi_1, S) \vee f(\varphi_2, S)$.
- $f(\varphi_1 \wedge \varphi_2, S) = f(\varphi_1, S) \wedge f(\varphi_2, S)$.
- $f(E\varphi_1 U\varphi_2, S) = E[fair(S, bound(E\varphi_1 U\varphi_2)) \wedge f(\varphi_1, \emptyset)Uf(\varphi_2, \emptyset)]$.
- $f(A\varphi_1 U\varphi_2, S) = A[fair(S, bound(A\varphi_1 U\varphi_2)) \to f(\varphi_1, \emptyset)Uf(\varphi_2, \emptyset)]$.
- $f(E\varphi_1 \tilde{U}\varphi_2, S) = E[fair(S, bound(E\varphi_1 \tilde{U}\varphi_2)) \wedge f(\varphi_1, \emptyset)\tilde{U}f(\varphi_2, \emptyset)]$.
- $f(A\varphi_1 \tilde{U}\varphi_2, S) = A[fair(S, bound(A\varphi_1 \tilde{U}\varphi_2)) \to f(\varphi_1, \emptyset)\tilde{U}f(\varphi_2, \emptyset)]$.
- $f(c.\varphi_1, S) = f(\varphi_1, S \cup \{c\})$.

Intuitively, the set S in $f(\varphi, S)$ contains all the free clocks in φ that should be reset and then never reset again once we come to evaluate φ. When we evaluate a formula φ, path quantification ranges only over paths in which the clocks in $bind(\varphi)$ are always reset. This restricted quantification is imposed by the second conjunct in the path formula $fair$. Consider a clock $c \in S$. We know that c enters S as a result of being a binding clock in a formula of the form $c.\varphi_1$. Hence, when c enters S is is not free and therefore, by the above rules, it is reset. We "release" the clock c and path quantification becomes restricted to paths in which c and the other clocks in S are never reset (this is imposed by the first conjunct in $fair$) and in which the clocks in $bind(\varphi_1)$ remain always reset.

If φ is a clock constraint, then releasing c is done by simply assigning 0 to c. If φ is of the form $E\varphi_1 U\varphi_2$, $A\varphi_1 U\varphi_2$, $E\varphi_1 \tilde{U}\varphi_2$, or $A\varphi_1 \tilde{U}\varphi_2$, then releasing c is done by updating the parameters of the formula $fair$. For example, the formula

$$\psi = AG[c.req \to AF(grant \wedge (c \leq 2))],$$

mentioned earlier in the introduction, has

$$untime(\psi) = A[Gc_0 \to G(req \to A[XG(T \to \neg c_0) \to F(grant \wedge c_{\leq 2})],$$

where c_0 and $c_{\leq 2}$ abbreviate $reg(c = 0)$ and $reg(c \leq 2)$, respectively.

It is easy to see that the length of $untime(\psi)$ is linear in the length of ψ (we ignore quadratic blow-up caused by specifying sets of clocks and sets of regions in $fair$; such a blow-up can be easily handled by new atomic propositions).

3.3 Correctness of the Reduction

We now prove the correctness of the reduction. Along the proof, we use $\langle l, \pi \rangle \models_{\mathcal{R}} \psi$ to indicate that the state $\langle l, \pi \rangle$ satisfies the TCTL formula ψ in the region structure $\mathcal{R}(\mathcal{U}, \psi)$ and we use $\langle l, \pi \rangle \models_{\mathcal{K}} \varphi$ to indicate that the states $\langle l, \pi, \mathbf{E} \rangle$ and $\langle l, \pi, \mathbf{T} \rangle$ satisfy the CTL formula φ in the Kripke structure $\mathcal{K}(\mathcal{U}, \psi)$. Note that as $\langle l, \pi, \mathbf{E} \rangle$ and $\langle l, \pi, \mathbf{T} \rangle$ have exactly the same future, and they differ only in the value of $\mathbf{T}$, they agree on satisfaction of CTL formulas that do not refer to the value of $\mathbf{T}$ in the present.

Theorem 3. *For every timed automaton $\mathcal{U}$, TCTL formula ψ, location l of $\mathcal{U}$, subformula φ of ψ, region $\pi \in \Upsilon$, and set $S \subseteq C_\psi$ of specification clocks such that $S \cap bound(\varphi) = \emptyset$, the following are equivalent:*

1. $\langle l, \pi[S := 0] \rangle \models_{\mathcal{R}} \varphi$.
2. $\langle l, \pi[S \cup bound(\varphi) := 0] \rangle \models_{\mathcal{K}} f(\varphi, S)$.

Proof. The proof proceeds by induction on the structure of φ.

- The cases where φ is of the form **true**, **false**, p, or $\neg p$ are immediate.

- For φ that is a clock constraint $c \sim v$, we have $f(c \sim v, S) = reg((c \sim v)[S := 0])$. Since the clocks in S are reset anyway, then $\langle l, \pi[S := 0] \rangle \models_{\mathcal{R}} (c \sim v)$ holds iff $\langle l, \pi[S := 0] \rangle \models_{\mathcal{R}} (c \sim v)[S := 0]$. Since the satisfaction of $c \sim v$ depends only on the value of c and is independent of l, then the latter, according to the definition of the mapping reg, holds iff $\langle l, \pi[S := 0] \rangle \models_{\mathcal{K}} reg((c \sim v)[S := 0])$; thus iff $\langle l, \pi[S := 0] \rangle \models_{\mathcal{K}} f(c \sim v, S)$. Finally, as $bound(\varphi) = \emptyset$, the latter holds iff $\langle l, \pi[S \cup bound(\varphi) := 0] \rangle \models_{\mathcal{K}} f(c \sim v, S)$, and we are done.

- For $\varphi = \varphi_1 \wedge \varphi_2$, we have that $f(\varphi, S) = f(\varphi_1, S) \wedge f(\varphi_2, S)$. By the semantics of TCTL, $\langle l, \pi[S := 0] \rangle \models_{\mathcal{R}} \varphi_1 \wedge \varphi_2$ iff $\langle l, \pi[S := 0] \rangle \models_{\mathcal{R}} \varphi_1$ and $\langle l, \pi[S := 0] \rangle \models_{\mathcal{R}} \varphi_2$. Since $S \cap bound(\varphi) = \emptyset$, then $S \cap bound(\varphi_1) = \emptyset$. Therefore, the induction hypothesis is applicable, and $\langle l, \pi[S := 0] \rangle \models_{\mathcal{R}} \varphi_1$ iff $\langle l, \pi[S \cup bound(\varphi_1) := 0] \rangle \models_{\mathcal{K}} f(\varphi_1, S)$.

Consider the set of clocks $bound(\varphi) \setminus bound(\varphi_1)$. By the syntax of TCTL, these clocks do not appear in φ_1, and therefore, as $S \cap bind(\varphi) = \emptyset$, they do not appear in in $f(\varphi_1, S)$ either. Hence $\langle l, \pi[S \cup bound(\varphi_1) := 0] \rangle \models_{\mathcal{K}} f(\varphi_1, S)$ iff $\langle l, \pi[S \cup bound(\varphi) := 0] \rangle \models_{\mathcal{K}} f(\varphi_1, S)$. Similarly, $\langle l, \pi[S \cup bound(\varphi_2) := 0] \rangle \models_{\mathcal{K}} f(\varphi_2, S)$ iff $\langle l, \pi[S \cup bound(\varphi) := 0] \rangle \models_{\mathcal{K}} f(\varphi_2, S)$. We thus have that $\langle l, \pi[S := 0] \rangle \models_{\mathcal{R}} \varphi_1 \wedge \varphi_2$ iff $\langle l, \pi[S \cup bound(\varphi) := 0] \rangle \models_{\mathcal{K}} f(\varphi_1, S)$ and $\langle l, \pi[S \cup bound(\varphi) := 0] \rangle \models_{\mathcal{K}} f(\varphi_2, S)$. This holds iff $\langle l, \pi[S \cup bound(\varphi) := 0] \rangle \models_{\mathcal{K}} f(\varphi_1, S) \wedge f(\varphi_2, S)$. This, by the definition of f, holds iff $\langle l, \pi[S \cup bound(\varphi) := 0] \rangle \models_{\mathcal{K}} f(\varphi_1 \wedge \varphi_2, S)$, and we are done.

The proof is similar for φ of the form $\varphi_1 \vee \varphi_2$.

- For $\varphi = E\varphi_1 U\varphi_2$, we have $f(E\varphi_1 U\varphi_2, S) = E[fair(S, bound(\varphi)) \wedge f(\varphi_1, \emptyset)U f(\varphi_2, \emptyset)]$. Consider a state $\langle l, \pi \rangle$ in $\mathcal{R}(\mathcal{U}, \psi)$. Assume first that $\langle l, \pi[S := 0] \rangle \models_{\mathcal{R}} \varphi$. Then, by the semantic of TCTL, there is a path $\langle l_0, \pi_0 \rangle, \langle l_1, \pi_1 \rangle \ldots$ in $\mathcal{R}(\mathcal{U}, \psi)$ with $\langle l_0, \pi_0 \rangle = \langle l, \pi[S := 0] \rangle$, and there exists $i \geq 0$ such that $\langle l_i, \pi_i \rangle \models_{\mathcal{R}} \varphi_2$, and for all $0 \leq j < i$ we have $\langle l_j, \pi_j \rangle \models_{\mathcal{R}} \varphi_1$. By the induction hypothesis (applied with $\varphi = \varphi_2$, $l = l_i, \pi = \pi_i$, and $S = \emptyset$), $\langle l_i, \pi_i \rangle \models_{\mathcal{R}} \varphi_2$ iff $\langle l_i, \pi_i[bound(\varphi_2) := 0] \rangle \models_{\mathcal{K}} f(\varphi_2, \emptyset)$. Similarly, for all $0 \leq j < i$, we have that $\langle l_j, \pi_j \rangle \models_{\mathcal{R}} \varphi_1$ iff $\langle l_j, \pi_j[bound(\varphi_1) := 0] \rangle \models_{\mathcal{K}} f(\varphi_1, \emptyset)$. As in the $\varphi_1 \wedge \varphi_2$ case, this implies, that $\langle l_i, \pi_i \rangle \models_{\mathcal{R}} \varphi_2$ iff $\langle l_i, \pi_i[bound(\varphi) := 0] \rangle \models_{\mathcal{K}} f(\varphi_2, \emptyset)$, and for all $0 \leq j < i$, we have that $\langle l_j, \pi_j \rangle \models \varphi_1$ iff $\langle l_j, \pi_j[bound(\varphi) := 0] \rangle \models_{\mathcal{K}} f(\varphi_1, \emptyset)$. Hence, the induction hypothesis is applicable as follows. Consider the sequence of regions $\eta = \eta_0, \eta_1, \ldots$ and the sequence of attributions $b_0, b_1, \ldots \in \{\mathbf{T}, \mathbf{E}\}^\omega$, where

- $\eta_0 = \pi[S \cup bound(\varphi) := 0]$ and $b_0 = \mathbf{E}$.
- For every $i \geq 0$, the region η_{i+1} and the attribution b_{i+1} are defined as follows.
  - If the transition in $\mathcal{R}(\mathcal{U}, \psi)$ from $\langle l_i, \pi_i \rangle$ to $\langle l_{i+1}, \pi_{i+1} \rangle$ is an edge transition, in which case there exists $S_\mathcal{U} \subseteq C_\mathcal{U}$ for which $\pi_{i+1} = \pi_i[S_\mathcal{U} := 0]$, then $\eta_{i+1} = \eta_i[S_\mathcal{U} := 0]$ and $b_{i+1} = \mathbf{E}$.
  - If the transition in $\mathcal{R}(\mathcal{U}, \psi)$ from $\langle l_i, \pi_i \rangle$ to $\langle l_{i+1}, \pi_{i+1} \rangle$ is a time transition, then $\eta_{i+1} = succ(\eta_i)[bound(\varphi) := 0]$ and $B_{i+1} = \mathbf{T}$.

Note that for every $k \geq 0$, we have $\eta_k = \pi_k[bound(\varphi) := 0]$. Thus, clearly, for all $k \geq 0$, we have that $\eta_k \models (bound(\varphi) = 0)$. Also, for all $k \geq 1$, since the only specification clocks that are reset along ρ are these in $bind(\varphi)$ and since $S \cap bind(\varphi) = \emptyset$, we have that $\eta_k \models \mathbf{T} \rightarrow (S > 0)$. Indeed, the value of a clock that is not reset in a time transition must become greater than 0. Since $bound(\varphi) \subseteq C_\psi$ then, by the definition of $\mathcal{K}(\mathcal{U}, \psi)$, the sequence $\rho = \langle l_0, \eta_0, b_0 \rangle, \langle l_1, \eta_1, b_1 \rangle, \ldots$ is a path in $\mathcal{K}(\mathcal{U}, \psi)$. As detailed above, by the induction hypothesis, we have that $\langle l_i, \eta_i \rangle \models_\mathcal{K} f(\varphi_2, \emptyset)$, and for all $0 \leq j < i$, we have that $\langle l_j, \eta_j \rangle \models f(\varphi_1, \emptyset)$. In addition, ρ satisfies $fair(S, bound(\varphi))$. Hence, $\langle l, \pi[S \cup bound(\varphi) := 0] \rangle \models_\mathcal{K} E[fair(S, bound(\varphi)) \wedge f(\varphi_1, \emptyset) U f(\varphi_2, \emptyset)]$, and we are done.

Assume now that $\langle l, \pi[S \cup bound(\varphi) := 0] \rangle \models_\mathcal{K} f(\varphi, S)$. Therefore, there exists a path $\langle l_0, \pi_0, b_0 \rangle, \langle l_1, \pi_1, b_1 \rangle, \ldots$ in $\mathcal{K}(\mathcal{U}, \psi)$ with $\langle l_0, \pi_0, b_0 \rangle = \langle l, \pi[bound(\varphi) \cup S := 0], \mathbf{E} \rangle$, such that the following hold. First, for all $k \geq 1$, we have that $\pi_k \models \mathbf{T} \rightarrow (S > 0)$. Second, for all $k \geq 0$, we have that $\pi_k \models (bound(\varphi) = 0)$. In addition, there exists $i \geq 0$ such that $\langle l_i, \pi_i, b_i \rangle \models f(\varphi_2, \emptyset)$, and for all $0 \leq j < i$, we have $\langle l_j, \pi_j, b_j \rangle \models f(\varphi_1, \emptyset)$. Consider the sequence of regions $\eta = \eta_0, \eta_1, \ldots$ defined as follows.

- $\eta_0 = \pi[S := 0]$.
- For every $i \geq 0$, the region η_{i+1} is defined as follows.
  - If $b_{i+1} = \mathbf{E}$, in which case there exists $S_\mathcal{U} \subseteq C_\mathcal{U}$ for which $\pi_{i+1} = \pi_i[S_\mathcal{U} := 0]$, then $\eta_{i+1} = \eta_i[S_\mathcal{U} := 0]$.
  - If $b_{i+1} = \mathbf{T}$, then $\eta_{i+1} = succ(\eta_i)$.

Note that for every $k \geq 0$, we have $\pi_k = \eta_k[bound(\varphi) := 0]$. Also, since no specification clocks are reset along η, it is guaranteed that the sequence $\langle l_0, \eta_0 \rangle, \langle l_1, \eta_1 \rangle, \ldots$ is a path in $\mathcal{R}(\mathcal{U}, \psi)$. As detailed above, by the induction hypothesis, we have that $\langle l_i, \eta_i \rangle \models_\mathcal{K} \varphi_2$, and for all $0 \leq j < i$, we have that $\langle l_j, \eta_j \rangle \models \varphi_1$. Hence, as $\langle l_0, \eta_0 \rangle = \langle l, \pi[S := 0] \rangle$, it follows that $\langle l, \pi[S := 0] \rangle \models_\mathcal{R} \varphi$, and we are done.

The proof is similar for φ of the form $A\varphi_1 U \varphi_2$, $E\varphi_1 \tilde{U} \varphi_2$, or $A\varphi_1 \tilde{U} \varphi_2$.

- For $\varphi = c.\varphi_1$, we have $f(\varphi, S) = f(\varphi_1, S \cup \{c\})$. By the semantics of TCTL, we have $\langle l, \pi[S := 0] \rangle \models_\mathcal{R} \varphi$ iff $\langle l, \pi[S \cup \{c\} := 0] \rangle \models_\mathcal{R} \varphi_1$. Since $c \notin bound(\varphi_1)$ and $bound(\varphi_1) \subseteq bound(\varphi)$, then $S \cap bound(\varphi) = \emptyset$ implies that $(S \cup \{c\}) \cap bound(\varphi_1) = \emptyset$. Therefore, by the induction hypothesis, $\langle l, \pi[S \cup \{c\} := 0] \rangle \models_\mathcal{R} \varphi_1$ iff $\langle l, \pi[S \cup \{c\} \cup bound(\varphi_1) := 0] \rangle \models_\mathcal{K} f(\varphi_1, S \cup \{c\})$. As $bound(\varphi) = \{c\} \cup bound(\varphi_1)$, this holds iff $\langle l, \pi[S \cup bound(\varphi) := 0] \rangle \models_\mathcal{K} f(\varphi_1, S \cup \{c\})$; thus, iff $\langle l, \pi[S \cup bound(\varphi) := 0] \rangle \models_\mathcal{K} f(\varphi, S)$, and we are done. $\blacksquare$

Theorems 2 and 3 imply the following.

Corollary 4. *For every timed automaton $\mathcal{U}$ and TCTL formula ψ, the following are equivalent:*

1. $\mathcal{U} \models \psi$.
2. $\mathcal{K}(\mathcal{U}, \psi) \models untime(\psi)$.

The transition from $\mathcal{U}$ to $\mathcal{K}(\mathcal{U}, \psi)$ involves an exponential blow-up, and the translation of ψ into $untime(\psi)$ involves only a linear blow-up. The model-checking problem for CTL can be solved in space that is polynomial in the specification and only poly-logarithmic in the Kripke structure [BVW94]. Corollary 4 then suggests a PSPACE model-checking algorithm for TCTL, matching the known lower bound [ACD93].

3.4 Moving from CTL* to CTL

The formula $fair(S_1, S_2)$ that we use in the definition of *fair* is not a CTL formula. Moreover, when we use the formula *fair*, it comes before a boolean connective ($\wedge$ or $\rightarrow$) that relates it to formulas of the form $\varphi_1 U \varphi_2$ or $\varphi_1 \tilde{U} \varphi_2$. Hence, as defined now, the function *untime* translates TCTL formulas to CTL* formulas. In this section we redefine $f(\varphi, S)$ for φ of the form $E\varphi_1 U \varphi_2, A\varphi_1 U \varphi_2, E\varphi_1 \tilde{U} \varphi_2$, or $A\varphi_1 \tilde{U} \varphi_2$, so that the resulting formula will be a CTL formula. Recall that

$$fair(S_1, S_2) = XG(\mathbf{T} \rightarrow (S_1 > 0)) \wedge G(S_2 = 0).$$

Thus, clearly, we could define

$$fair(S_1, S_2) = (S_2 = 0) \wedge [XG((\mathbf{T} \rightarrow (S_1 > 0)) \wedge (S_2 = 0))],$$

which has the form $\xi_1 \wedge XG\xi_2$, where $\xi_1 = (S_2 = 0)$ and $\xi_2 = (\mathbf{T} \rightarrow (S_1 > 0)) \wedge (S_2 = 0)$ are propositional formulas. It follows that we have to translate the following four CTL* formulas to CTL formulas:

1. $E(\xi_1 \wedge XG\xi_2 \wedge \varphi_1 U \varphi_2)$.
2. $A((\xi_1 \wedge XG\xi_2) \rightarrow \varphi_1 U \varphi_2)$.
3. $E(\xi_1 \wedge XG\xi_2 \wedge \varphi_1 \tilde{U} \varphi_2)$.
4. $A((\xi_1 \wedge XG\xi_2) \rightarrow \varphi_1 \tilde{U} \varphi_2)$.

We translate the four formulas to a fragment of CTL* in which the path formulas may contain two temporal operators connected by a boolean operator. Formulas of this fragment have equivalent formulas in CTL [KG96].

1. $E(\xi_1 \wedge XG\xi_2 \wedge \varphi_1 U \varphi_2) = \xi_1 \wedge [(\varphi_2 \wedge EXEG\xi_2) \vee (\varphi_1 \wedge EXE((G\xi_2) \wedge \varphi_1 U \varphi_2)]$.
2. $A((\xi_1 \wedge XG\xi_2) \rightarrow \varphi_1 U \varphi_2) = (\neg\xi_1) \vee \varphi_2 \vee (\varphi_1 \wedge AXA((G\xi_2) \rightarrow \varphi_1 U \varphi_2)) \vee AXAF\neg\xi_2$.
3. $E(\xi_1 \wedge XG\xi_2 \wedge \varphi_1 \tilde{U} \varphi_2) = \xi_1 \wedge [(\varphi_1 \wedge \varphi_2 \wedge EXEG\xi_2) \vee (\varphi_2 \wedge EXE((G\xi_2) \wedge \varphi_1 \tilde{U} \varphi_2))]$.
4. $A((\xi_1 \wedge XG\xi_2) \rightarrow \varphi_1 \tilde{U} \varphi_2) = (\neg\xi_1) \vee (\varphi_1 \wedge \varphi_2) \vee (\varphi_2 \wedge AXA((G\xi_2) \rightarrow \varphi_1 \tilde{U} \varphi_2)) \vee AXAF\neg\xi_2$.

Finally, as the formula $EXEG\xi_2$ is valid in $\mathcal{K}(\mathcal{U}, \psi)$, we replace it with **true** and replace its negation $AXAF\neg\xi_2$ with **false**. Accordingly, we now have.

1. $f(E\varphi_1 U \varphi_2, S) = \xi_1 \wedge [\varphi_2 \vee (\varphi_1 \wedge EXE((G\xi_2) \wedge \varphi_1 U \varphi_2))]$.
2. $f(A\varphi_1 U \varphi_2, S) = (\neg \xi_1) \vee \varphi_2 \vee (\varphi_1 \wedge AXA((G\xi_2) \rightarrow \varphi_1 U \varphi_2))$.
3. $f(E\varphi_1 \tilde{U} \varphi_2, S) = \xi_1 \wedge [(\varphi_1 \wedge \varphi_2) \vee (\varphi_2 \wedge EXE((G\xi_2) \wedge \varphi_1 \tilde{U} \varphi_2))]$.
4. $f(A\varphi_1 \tilde{U} \varphi_2, S) = (\neg \xi_1) \vee (\varphi_1 \wedge \varphi_2) \vee (\varphi_2 \wedge AXA((G\xi_2) \rightarrow \varphi_1 \tilde{U} \varphi_2))$.

This completes the translation of $untime(\psi)$ into CTL.

4 Discussion

In this paper we suggested a reduction from TCTL model checking to CTL model checking. Recall that the way we define the semantics for TCTL, we do not require path quantification to range only over paths for which time diverges. diverge. Since we can replace the divergance requirement by a fairness constraint on $\mathcal{K}(\mathcal{U}, \psi)$ (see [HKV96]), it is easy to extend our algorithm to handle a semantics in which path quantification ranges only over divergent paths. Then, TCTL model checking is reduced to Fair-CTL model checking. By [KV95], the latter can be solved with the same space complexity as CTL model checking. Hence, the PSPACE complexity is preserved.

Our reduction handles the reset quantifier of TCTL by augmenting the region graph induced by a timed automaton with new transitions and limiting path quantification in the formula. As such, our reduction can be easily adjusted to handle model checking of TCTL formulas when interpreted with respect to hybrid systems with finite bisimulations [Hen95].

The advantage of the algorithm that follows from our reduction is the existence of fine-tuned tools for CTL model checking. The algorithm can be optimized further by exploiting the special structure of $\mathcal{K}(\mathcal{U}, \psi)$. For example, the optimization suggested in [HKV96], which integrates states that differ only in their region element into a single state, can be used also here. It remains to be seen how the algorithm performs in practice.

References

[ACD93]   R. Alur, C. Courcoubetis, and D. Dill. Model-checking in dense real-time. *Information and Computation*, 104(1):2–34, May 1993.

[AD94]   R. Alur and D. Dill. A theory of timed automata. *Theoretical Computer Science*, 126(2):183–236, 1994.

[AH92]   R. Alur and T.A. Henzinger. Logics and models of real time: a survey. In J.W. de Bakker, K. Huizing, W.-P. de Roever, and G. Rozenberg, editors, *Real Time: Theory in Practice*, Lecture Notes in Computer Science 600, pages 74–106. 1992.

[AH93]   R. Alur and T. Henzinger. Real-time logics: Complexity and expressiveness. *Information and Computation*, 104(1):35–77, May 1993.

[AH94]   R. Alur and T.A. Henzinger. A really temporal logic. *Journal of the ACM*, 41(1):181–204, 1994.

[BHSV+96] R.K. Brayton, G.D. Hachtel, A. Sangiovanni-Vincentelli, F. Somenzi, A. Aziz, S.-T. Cheng, S. Edwards, S. Khatri, T. Kukimoto, A. Pardo, S. Qadeer, R.K. Ranjan, S. Sarwary, T.R. Shiple, G. Swamy, and T. Villa. VIS: a system for verification and synthesis. In *Computer Aided Verification, Proc. 8th Int. Workshop*, volume 1102 of *Lecture Notes in Computer Science*, pages 428–432. Springer-Verlag, 1996.

[BVW94]   O. Bernholtz, M.Y. Vardi, and P. Wolper. An automata-theoretic approach to branching-time model checking. In *Computer Aided Verification, Proc. 6th Int. Conference*, volume 818 of *Lecture Notes in Computer Science*, pages 142–155, 1994.

[CCM+94] S. Campos, E.M. Clarke, W. Marrero, M. Minea, and H. Hiraishi. Computing quantitative characteristics of finite-state real-time systems. In *Proceedings of the 15th Annual Real-time Systems Symposium*. IEEE Computer Society Press, 1994.

[CE81] E.M. Clarke and E.A. Emerson. Design and synthesis of synchronization skeletons using branching time temporal logic. In *Proc. Workshop on Logic of Programs*, volume 131 of *Lecture Notes in Computer Science*, pages 52–71. Springer-Verlag, 1981.

[CES86] E.M. Clarke, E.A. Emerson, and A.P. Sistla. Automatic verification of finite-state concurrent systems using temporal logic specifications. *ACM Transactions on Programming Languages and Systems*, 8(2):244–263, January 1986.

[CGL93] E.M. Clarke, O. Grumberg, and D. Long. Verification tools for finite-state concurrent systems. In J.W. de Bakker, W.-P. de Roever, and G. Rozenberg, editors, *Decade of Concurrency – Reflections and Perspectives (Proceedings of REX School)*, Lecture Notes in Computer Science, pages 124–175. Springer-Verlag, 1993.

[EMSS90] E.A. Emerson, A.K. Mok, A.P. Sistla, and J. Srinivasan. Quantitative temporal reasoning. In *Proc. 2nd Workshop on Computer Aided Verification*, volume 531 of *Lecture Notes in Computer Science*, pages 136–145. Springer-Verlag, 1990.

[FGK+96] J.-C. Fernandez, H. Garavel, A. Kerbrat, L. Mounier, R. Mateescu, and M. Sighireanu. CADP: a protocol validitation and verification toolbox. In *Computer Aided Verification, Proc. 8th Int. Workshop*, volume 1102 of *Lecture Notes in Computer Science*, pages 437–440. Springer-Verlag, 1996.

[Hen95] T.A. Henzinger. Hybrid automata with finite bisimulations. In Z. Fülöp and F. Gécseg, editors, *ICALP 95: Automata, Languages, and Programming*, Lecture Notes in Computer Science 944, pages 324–335. Springer-Verlag, 1995.

[HKV96] T.A. Henzinger, O. Kupferman, and M.Y. Vardi. A space-efficient on-the-fly algorithm for real-time model checking. In *Proc. 7th Conferance on Concurrency Theory*, Pisa, August 1996. Springer-Verlag.

[HMP92] T.A. Henzinger, Z. Manna, and A. Pnueli. What good are digital clocks? In W. Kuich, editor, *ICALP 92: Automata, Languages, and Programming*, Lecture Notes in Computer Science 623, pages 545–558. Springer-Verlag, 1992.

[HNSY94] T.A. Henzinger, X. Nicollin, J. Sifakis, and S. Yovine. Symbolic model checking for real-time systems. *Information and Computation*, 111:193–244, 1994.

[KG96] O. Kupferman and O. Grumberg. Buy one, get one free! *Journal of Logic and Computation*, 6(4), 1996.

[KV95] O. Kupferman and M.Y. Vardi. On the complexity of branching modular model checking. In *Proc. 6th Conferance on Concurrency Theory*, pages 408–422, Philadelphia, August 1995. Springer-Verlag.

[LL95] F. Laroussinie and K. G. Larsen. Compositional model checking of real time systems. In *Proc. 6th Conferance on Concurrency Theory*, pages 27–41, Philadelphia, August 1995. Springer-Verlag.

[McM93] K.L. McMillan. *Symbolic model checking*. Kluwer Academic Publishers, 1993.

[MP92] Z. Manna and A. Pnueli. *The Temporal Logic of Reactive and Concurrent Systems: Specification*. Springer-Verlag, Berlin, January 1992.

[Pnu77] A. Pnueli. The temporal logic of programs. In *Proc. 18th IEEE Symposium on Foundation of Computer Science*, pages 46–57, 1977.

[QS81] J.P. Queille and J. Sifakis. Specification and verification of concurrent systems in Cesar. In *Proc. 5th International Symp. on Programming*, volume 137, pages 337–351. Springer-Verlag, Lecture Notes in Computer Science, 1981.

[SS95] O.V. Sokolsky and S.A. Smolka. Local model checking for real-time systems. In *Computer Aided Verification, Proc. 7th Int. Workshop*, Lecture Notes in Computer Science 939, pages 211–224, Liege, July 1995.

Verifying Periodic Task-Control Systems

Vlad Rusu*

Abstract. This paper deals with the automated verification of a class of task-control systems with periods, durations, and scheduling specifications. Such systems are translated into Periodic Hybrid Automata for verification. We show that safety, liveness, and time-bounded properties are decidable for the considered systems.

Keywords: task-control systems, periodic hybrid automata, verification, decidability.

1 Introduction

The automated verification of real-time systems has made considerable progress in the last years, since verification methods based on *timed/hybrid automata* and *Timed Computational Tree Logic* [ACH+95] were developed. A main result is that the logic TCTL is decidable on timed automata [ACD90, HNSY94]: as a consequence, verification of TCTL properties is decidable for real-time formalisms that are translatable into timed automata [NSY92, JMO93, DOY94].

But timed automata cannot express all the aspects of real-time applications, like for instance a task *preempting* another task. Indeed, preemption needs the more powerful model of hybrid automata to be expressed; but the verification of TCTL properties is in general undecidable on hybrid automata [HKPV95]. However, in [RR96] we proved the decidability of TCTL for a class of task-control systems, under the restriction that tasks are not preempted too often, such that all of them eventually terminate in bounded time. In this paper, we prove decidability of TCTL for another class of systems, without the former restriction; in exchange, the tasks are constrained to be *periodic*.

To express the task control, we introduce in section 2 a simple language with its untimed semantics (finite automata) and timed semantics (Periodic Hybrid Automata, to model the durations and periods of tasks). Section 3 presents the decidability results, and we conclude in section 4.

2 A Language for Task-Control

We define a simple language to describe a set of tasks running and preempting each other to obtain a processor for execution. When a task has been preempted,

* Laboratoire d'Automatique de Nantes (CNRS N°823, École Centrale Nantes, Univ. Nantes) 1 rue de la Noë - BP92101 44321 Nantes Cedex 3, France. E-mail Vlad.Rusu@lan.ec-nantes.fr

64

its execution may be later resumed at the last preemption point. Consider for instance a computer system that iterates a task **A**, that can be preempted by tasks **D** or (exclusively) **B**; and task **B** can itself be preempted by **C**. This is modeled by the automaton of figure 2, next page (for the moment, ignore the formulæ around the automaton). Task **A** runs at location **A**; task **D** can preempt **A** and it runs (while **A** is preempted) at location **D,A_pre**. Similarly, **B** runs while **A** is preempted at location **B,A_pre**, and **C** runs while **A** and **B** are preempted, at location **C,A_pre,B_pre**. This will be expressed syntactically as **loop(A preempted by (D exclusive(B preempted by C)))**. The language syntax is :

$$T ::= Task \mid (T\ T) \mid \mathbf{loop}(T) \mid (T\ \mathbf{exclusive}\ T) \mid$$
$$(Task\ \mathbf{preempted\ by}\ T)$$
$$Task ::= \mathbf{A, B, C} \ldots$$

with the following restriction: *the relation $\prec$ on the tasks such that "A $\prec$ B iff B occurs in a A preempted by T construct"*, *is a strict order*. So, 'higher-level' tasks can preempt 'lower-level' ones, but not the contrary. This restriction will be used for proving the decidability results in section 3.

Untimed semantics: finite automata. The untimed semantics of a task-control program is given as a finite-state automaton, generated directly while the program is top-down parsed. For this, we define a rewriting system, whose root is a one-state automaton; the whole program labels this state (fig. 1(a)). The program is parsed top down, and when a grammar rule applies, the state is expanded into a state construction depending on the rule (fig 1(b)-(e)). The new states are connected to their neighbours in the following manner: all incoming (resp. outcoming) transitions of the replaced state, have their destination (resp. origin) in the initial (resp. final) replacing states.

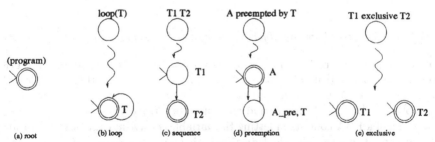

Figure 1 *Rewriting states;* $>\!\!\bigcirc$ = *initial,* $\circledcirc$ = *final state*

Timed semantics: Periodic Hybrid Automata. We now introduce in the model such quantitative features as the durations and the periods of tasks. More precisely, all tasks will be characterized by some *duration*, and tasks that can preempt other tasks will also be characterized by some *period* — time interval between two consecutive occurrences. To model this, we add to the finite automaton two sets of variables: the *duration variables* and the *period variables*. We obtain the model of Periodic Hybrid Automata.

The Periodic Hybrid Automaton (PHA) of a task-control program consists of:
— the finite automaton of the program;
— for each task A_i in the program, a *duration variable* a_i;
— for each task A_j that can preempt another task, a *period variable* α_j;
— for each location of the automaton:
  • an *invariant* $a_i \leq D_{A_i}$ if task A_i is active at that location; $D_{A_i} \in \mathbb{Q}^{>0}$ is the duration of task A_i (there is exactly one active task at each location);
  • for each duration variable a_i, a *differential law* $a_i' = 1$ if task A_i is active at that location, $a_i' = 0$ otherwise; and for each period variable α_j, a differential law $\alpha_j' = 1$ if task A_j is active, preempted, or if it can preempt a task which is active at that location, $\alpha_j' = 0$ otherwise;
— for each transition, a *guard* and a *reset*:
  • for transitions that correspond to task endings, the *guard* is $a_i = D_{A_i}$ if task A_i ends on that transition, and the *reset* applies to duration variable a_i;
  • for transitions that correspond to task preemptions, the *guard* is $\alpha_j = T_{A_j}$ if the transition is a preemption by task A_j ($T_{A_j} \in \mathbb{Q}^{>0}$ is called the period of task A_j). The *reset* applies to variable α_j, and to the period variables of the tasks that can preempt A_j. $\square$

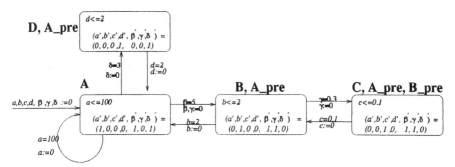

Figure 2 *Periodic Hybrid Automaton for example program.*

Figure 2 represents the PHA for program **loop(A preempted by (D exclusive(B preempted by C)))** in which tasks **A, B, C, D** have durations 100, 2, 0.1 and 2 time units, and **B, C, D** have periods 5, 0.3, and 3 time units. The duration variables are $\{a, b, c, d\}$, and period variables are $\{\beta, \gamma, \delta\}$.

A *run* of a PHA consists in letting the variables evolve at control states inside the invariants, and crossing the transitions when guards are true [ACH+95]. But unlike the previous, we shall define runs such that *period variables are frozen when they reach the period value*; and *transitions are fired as soon as possible*. These two features allow to express what happens when a task's period has been exceeded. Consider the example of figure 2 when **B** is running (at location **B, A_pre**), and suppose **B** preempted **A** more than 5 time units (**B** 's period), because **B** itself was preempted by several occurrences of **C**. When **B** completes, it should preempt **A** at once because its period is exceeded. This happens precisely because variable β was frozen at value 5 and because the transitions are urgent.

We now make more precise the semantics of PHA. Suppose the PHA has n duration variables and m period variables. A *state* is a triple $(l, \overline{a}, \overline{\alpha})$ where l is a location, $\overline{a}$ is a n-vector of real values for duration variables *that satisfy location l 's invariant*, and $\overline{\alpha} = (\alpha_j)_{j=\overline{1,m}}$ is vector of real values for period variables. For $t \geq 0$, $\overline{a} + t$ denotes the vector $(a_i + t)_{i=\overline{1,n}}$, and $\overline{\alpha} + t$ denotes the vector $(max\{\alpha_j + t, T_j\})_{j=\overline{1,m}}$ where T_j is the period of the task with period variable α_j. A *time step* of duration $t \geq 0$, denoted $(l, \overline{a}, \overline{\alpha}) \rightarrow_t (l, \overline{a'}, \overline{\alpha'})$, is defined by $\overline{a'} = \overline{a} + t$ and $\overline{\alpha'} = \overline{\alpha} + t$. A *transition step* $(l, \overline{a}, \overline{\alpha}) \rightarrow (l', \overline{a'}, \overline{\alpha'})$ consists in crossing a transition whose guard is enabled (i.e. is satisfied by $\overline{a}, \overline{\alpha}$), and resetting the values of variables according to the transition.

A *run* is a sequence of time and transition steps, such that transition steps are *urgent [BGK+96]: as soon as* a transition gets enabled, the next step consists in crossing that transition (if several transition are enabled simultaneously, one is chosen arbitrarily). The *duration of a run* is the sum of the durations of time steps, as the transition steps are instantaneous.

3 Verifying properties: symbolic analysis

The time-bounded (resp.time-unbounded) reachability properties of Periodic Hybrid Automata are of the form: *starting from an initial set of states ψ, and by remaining in an intermediary set of states ψ_1, is it possible to reach a final set of states ψ_2, by a run of duration at most $c \in \mathbb{Q}^{>0}$ time units (resp. of arbitrary duration)*. The time-bounded (resp. time-unbounded) liveness are similar except for the modality *possible* which is replaced by *inevitable*. For example, a time-unbounded liveness property for the automaton of figure 2 is: *starting from location A with all variables equal to 0, is it inevitable that task A terminates*.

These properties can be verified by *symbolic analysis*[HNSY94, ACH+95]: infinite sets of states are represented by *symbolic states* and infinite sets of runs are represented by *symbolic runs*. A symbolic state is defined by a location and a domain of values for variables identifiable with a polyhedron in $\mathbb{R}^n$; and a symbolic run is a sequence of symbolic states, simulating all the runs that start from a given set of states and cross a given sequence of transitions. Symbolic runs can be computed algorithmically [HPR94].

We now prove that symbolic analysis terminates on PHA: we first show that time-bounded reachability is decidable; then we show that time-unbounded reachability reduces to the time-bounded version; and finally we indicate that liveness properties reduce to the previous two.

Proposition 1. *Time-bounded reachability is decidable on PHA.*
Proof. Define the cyclic runs to be the runs that cross twice some transition. Then there exists a uniform low-bound $\epsilon > 0$ for the durations of cyclic runs, since these necessarily perform a complete execution of at least one task. Then, the bounded-duration runs can be decomposed into a bounded number of cyclic runs, that cross a bounded number of transitions, which are simulated by bounded-length symbolic runs [RR96]. □

Next, we will show that PHA satisfy the following *periodicity condition*: (PC) there exists $T > 0$ and a finite number N of states such that any run of the PHA meets a state among the above, in at most T time units. This property is important because if it holds, time-unbounded reachability reduces to the time-bounded version (in time $(N+1) \times T$ [RR96]), which is decidable (Prop.1).

Definition. A *remarkable point* of a task A is a set of values for variables of tasks greater or equal than A^1, at an instant when A is being preempted or is ending. A *total duration* of task A is a possible duration between the beginning and the end of A (including preemption time).

Proposition 2. For each task: (a) there exist a finite number of remarkable points; (b) there exist a finite number of total durations, including ∞.
Proof idea. By induction on the preemption order $\prec$ of tasks. The key point is our particular definition of runs (cf. previous page): period variables are frozen when they reach period value, and transitions are fired as soon as possible. $\square$

Proposition 3. The periodicity condition holds for PHA.
Proof sketch. We first show (c): there exists $T > 0$ and a finite set V of reals with the property: *for each task A, there exists a task B greater or equal than A, such that when A is active, a state will be reached in time at most T, where B is active and where values of variables for tasks greater or equal than A are in V.*

Let V be the finite union of all remarkable points, T_1 the greatest finite total duration, T_2 the sum of all period values, and $T = T_1 + T_2$. Consider a task A. When A is active then either 1. it eventually terminates or 2. it never terminates. In the first case, by Proposition 2(b), it will terminate in at most T_1 time units. This ending state is a remarkable point of A, and at that moment, by Proposition 2(a), the values of variables for tasks greater and equal than A are in V.

In the second case, there exists a task $B \succeq A$, that is preempted an infinity of times and does not terminate. We choose B minimal with the above property. Suppose first $B \succ A$; then B runs while all tasks A through B are preempted. By construction of the PHA, starting from a state when A is active, one *will* reach in at most T_2 time units, a state where B is running, all tasks A through B are preempted, and period variables of tasks A through B are frozen at period value[2]. Moreover, since task B will never terminate, all tasks A through B remain preempted forever, so the duration variables of those tasks remain forever unchanged. So, starting from a state with A active, one will reach in at most T_2 time units a state where B is active, and where values of variables of tasks A through B are in V *and will remain forever unchanged.*

Next, B itself is preempted an infinity of times and does not terminate; but all tasks that preempt B terminate in at most T_1 time units (by choosing B minimal and Proposition 2(b)). The preemptions of B constitute remarkable points of B, in finite number by Proposition 2(a). Then, one of these preemptions is repeated an infinity of times, at least once every T_1 time units, and at the preemption instant, the values of variables for tasks greater or equal than B are in V.

---

[1] with regards to the preemption order $\prec$ defined with the task-control language syntax
[2] this holds because of time amount T_2, the sum of all period values

So, starting from a state where A is active, one will reach in at most $T = T_1 + T_2$ time units a state where B is active and where variables of all tasks greater or equal than A are in V. This proves (c) when $B \succ A$. The case $A = B$ is similar (it reduces to the previous paragraph). Note that when $B \succ A$ we have more (+): the values of variables of tasks A through B *remain forever unchanged*. We can now obtain the Periodicity Condition (PC) by applying (c+) to a a smallest task with regards to $\prec$. $\qquad\square$

Using Proposition 3, time-unbounded reachability reduces to time-bounded reachability, which is decidable (Proposition 1). Also, time-unbounded liveness reduces to time-bounded liveness, which reduces to reachability [HNSY94].

4 Conclusion

We have presented a decidability result that allows the verification of TCTL properties for a class of task-control systems with durations, periods, and preemptions. The verification model is Periodic Hybrid Automata, a variant of hybrid automata. The proof lies on the fact that the infinite timed behaviour reduces to a finite number of 'remarkable points'.

The author wishes to thank Olivier Roux, Pablo Argon, and the anonymous reviewers of HART'97 for useful comments and constructive criticism.

References

ACD90. R. Alur, C. Courcoubetis, and D. Dill. Model-checking for real-time systems. In *Proc. IEEE 5th Symp. Logic in Computer Science, LNCS*, 1990.

ACH+95. R. Alur, C. Courcoubetis, N. Halbwachs, T. Henzinger, P. Ho, X. Nicollin, A. Olivero, J. Sifakis, and S. Yovine. The algorithmic analysis of hybrid systems. *Theoretical Computer Science B*, 137, January 1995.

BGK+96. J. Bengtsson, D. Griffioen, K. Kristoferssen, K.Larsen, F.Larsson, P.Petersson, and W.Yi. Verification of an audio protocol with bus collision using UPPAAL. In *Computer-Aided Verification, LNCS*, 1996.

DOY94. C. Daws, A. Olivero, and S. Yovine. Verifying ET-LOTOS programs with KRONOS. In *Proc. FORTE'94, LNCS*, 1994.

HKPV95. T.A. Henzinger, P.W. Kopke, A. Puri, and P. Varaiya. What's decidable about hybrid automata. In *STOCS'95, LNCS*, 1995.

HNSY94. T.A. Henzinger, X. Nicollin, J. Sifakis, and S. Yovine. Symbolic model-checking for real-time systems. *Information and Computation*, (111), 1994.

HPR94. N. Halbwachs, Y. E. Proy, and P. Raymond. Verification of linear hybrid systems by means of convex approximations. In *International Symposium on Static Analysis, LNCS*, 1994.

JMO93. M. Jourdan, F. Maraninchi, and A. Olivero. Verifying quantitative real-time properties of synchronous programs. In *Computer-Aided Verification, LNCS 697*, 1993.

NSY92. X. Nicollin, J. Sifakis, and S. Yovine. Compiling real-time specifications into extended automata. *IEEE Transactions on Software Engineering*, 18(9):794–804, 1992.

RR96. O. Roux and V. Rusu. Uniformity for the decidability of hybrid automata. In *Internat. Static Analysis Symposium, LNCS 1145*, pages 301–316, 1996.

A Case Study in Timed CSP: The Railroad Crossing Problem

Luming Lai and Phil Watson

Department of Computing, University of Bradford,Bradford BD7 1DP, UK

Abstract. We use timed CSP, which is an extension of the formal method CSP to problems with a real-time component, to tackle a benchmark problem for real-time systems, the railroad crossing problem.

1 The problem and associated assumptions

The railroad crossing problem was specified in Heitmeyer [5, 4] and the description was further clarified in [3].

> The system to be developed operates a gate at a railroad crossing. The railroad crossing I lies in a region of interest R, i.e. $I \subseteq R$. A set of trains travel through R on multiple tracks in both directions. A sensor system determines when each train enters and exits region R. To describe the system formally, we define a gate function $g(t) \in [0, 90]$ where $g(t) = 0$ means the gate is down and $g(t) = 90$ means the gate is up. We also define a set $\{\lambda_i\}$ of *occupancy intervals*, where each occupancy interval is a time interval during which one or more trains are in I. The ith occupancy interval is represented as $\lambda_i = [\tau_i, \nu_i]$ where τ_i is the time of the ith entry of a train into the crossing when no other train is in the crossing and ν_i is the first time since τ_i that no train is in the crossing (i.e. the train that entered at τ_i has exited as have any trains that entered the crossing after τ_i).

> Given two constants ξ_1 and ξ_2, $\xi_1 > 0$, $\xi_2 > 0$, the problem is to develop a system to operate the crossing gate that satisfies the following two properties.

Safety Property: $t \in \bigcup_i \lambda_i \implies g(t) = 0$ (The gate is down during all occupancy intervals.)

Utility Property: $t \notin \bigcup_i [\tau_i - \xi_1, \nu_i + \xi_2] \implies g(t) = 90$ (The gate is up when no train is in the crossing.)

The description, though seemingly comprehensive, contains a number of ambiguities and omissions. One defect in the specification is as follows. It is nowhere explicitly forbidden in the English statement of the problem, that the gate should be raised and almost immediately lowered. In reality there would be an interval γ_c sufficient for at least one car to cross between the gate reaching the fully raised position and a *down* signal being sent; otherwise the gate would not be raised at all. We call this the **Common Sense Property**.

It is interesting to reflect on an imaginary abstracted version of the English specification of the problem, in which the requirements remain the same but no reference is made to any real world system. Then we would have no reason to derive the Common Sense Property: only our knowledge of real railroad crossings drives us to do this.

Various constraints on the parameters of the problem are given in [4] or derived in [3]. These include:

1. $\epsilon_2 \geq \epsilon_1$. The time for the slowest train to reach I after entering R, ϵ_2, is greater than or equal to the time for the fastest train to reach I, ϵ_1
2. $\epsilon_1 \geq \gamma_d + \gamma_c + \gamma_u$. Here γ_d and γ_u are the times required to lower and raise the physical gate, respectively. If the property does not hold we can never open the gate after the first train has passed without losing Safety. Clearly this is not the intent of the English statement of the problem
3. $\xi_1 - \gamma_d \geq \epsilon_2 - \epsilon_1$. The speeds of the fastest and slowest trains on the track must not differ so greatly that safety and utility are impossible to guarantee together
4. $\xi_2 \geq \gamma_u$. The time taken to raise the gate, γ_u, is small enough that the Utility Property is not violated
5. $\epsilon_2 \geq \xi_1$. Otherwise we could send a *down* signal before a train is detected approaching. Clearly this would be unrealistic.

We make only one assumption which is not strictly justified by the English specification: that for an *up* signal to be sent, R should have been empty at some time since the last *up* signal was sent. This is not an unreasonable assumption.

2 The timed CSP specification of the railroad crossing

Space limitations preclude a full description of the features and use of Timed CSP [1]. We introduce some of its features as and when they occur.

GateController and Counter Processes. The main process for the railroad crossing in timed CSP consists of two component processes.

$$GateController \stackrel{\text{def}}{=} (Counter \,|[\,\{k, m\}\,]|\, Controller) \setminus \{k, m\}$$

Here $P \,|[\, A \,]|\, Q$ means that processes P and Q synchronise on the events in the set A.

Most of the interest lies in the *Controller* process. *Counter* keeps an implicit count (in the index of the current sub-process) of the number of trains currently in R and communicates with the *Controller* process in two ways: a k signal means that a train has arrived in an otherwise empty R; an m signal means that the last train in R has just left.

$$Counter \stackrel{\text{def}}{=} a \rightarrow k \rightarrow Counter_1$$
$$Counter_i \stackrel{\text{def}}{=} a \rightarrow Counter_{i+1} \sqcap d \rightarrow Counter_{i-1} \qquad \text{for } i \geq 1$$
$$Counter_0 \stackrel{\text{def}}{=} m \rightarrow Counter,$$

Here $\sqcap$ is non-deternministic choice, while a represents the arrival of a train in R and d the departure of a train from the crossing.

Controller Process. For brevity we define the following abbreviations, which are based on required properties of the specification. Let

$t_{s1} \stackrel{def}{=} \xi_1 - \epsilon_2 + \epsilon_1 - \gamma_d$. The length of the interval during which a *down* signal may be sent. Note that $t_{s1} \geq 0$ by assumption 3. above

$t_{u1} \stackrel{def}{=} \epsilon_2 - \xi_1$. The lower time bound for sending a *down* signal. We call t_{u1} the *inward utility point* and $t_{u1} \geq 0$ by 5. above

$t_{u2} \stackrel{def}{=} \xi_2 - \gamma_u$. We call t_{u2} the *outward utility point* and $t_{u2} \geq 0$ by 4. above.

$t_{c1} \stackrel{def}{=} \gamma_u + \gamma_c$. The time it takes to raise the gate and let one car through

$t_{s2} \stackrel{def}{=} \epsilon_1 - (\gamma_u + \gamma_c + \gamma_d)$. The upper time bound for sending a *down* signal to satisfy the Safety and Common Sense Properties; $t_{s2} \geq 0$ by assumption 2. in Section 1.

Thus far we have used (untimed) CSP. The subsequent specification will omit the fixed delay δ between CSP events for clarity. It is untidy (especially in the proofs) but not difficult to introduce δ to the following specification.

*The specification of **Controller**.* To define new processes which can (additionally) satisfy the Common Sense Property, we need the following abbreviations in addition to those above: The *Controller* is defined as follows:

$$Controller \stackrel{def}{=} k \xrightarrow{t_{u1}} down@t_1\{t_{s1}\} \to P$$

$$P \stackrel{def}{=} m \to (up@t_2\{t_{u2}\} \to k@t_3 \xrightarrow{max\{t_{c1}-t_3,\, t_{u1}\}} down@t_4\{min\{t_{s1},\, t_{s2}+t_3\}\} \to P$$
$$\square\ k@t_5\{t_{u2}\} \to up@t_6\{min\{t_{u2}-t_5,\, t_{s2}\}\} \xrightarrow{max\{t_{c1},\, t_{u1}-t6\}}$$
$$down@t_7\{min\{t_{s1},\, t_{s2}-t_6\}\} \to P).$$

Note the additional complexity entailed by the various ways in which the requirements for our Common Sense Property can conflict with those for Safety and Utility.

To prove that the above processes satisfy the Safety, Utility, and Common Sense Properties, we need the following lemmas.

Lemma 1 *Counter monitors the number of trains in R correctly and sends a signal "k" when the first train enters an empty R and a signal "m" when the last train leaves R.*

We define a number of predicates for structured specifications as follows.

$$a \underline{from}\ t\ \underline{until}\ t'\ (s, X) \stackrel{def}{=} a \notin \sigma(X \uparrow [t,\, min\{t',\, begin(s \uparrow [t, \infty) \mid a)\}])).$$

If a process satisfies this specification, then event a must become available at time t, and must remain available until either time t' or the time at which the next a is observed, whichever is smaller.

$$a \ \underline{at} \ I \ (s, \ X) \ \overset{\text{def}}{=} \ \exists t : I \bullet \langle (t, a) \rangle \ \underline{in} \ s.$$

Event a must be observed at some time during time interval I. Now, we can define the liveness property of an event a during time interval I as follows:

$$a \ \underline{live} \ [t, \ t'] \ (s, \ X) \ \overset{\text{def}}{=} \ a \ \underline{from} \ t \ \underline{until} \ t' \ \vee \ a \ \underline{at} \ [t, \ t'],$$

which means that either event a is available during time interval $[t, \ t']$, or it occurs during it. We also define a predicate as follows:

$$a \ \underline{precedes} \ (t, \ b) \ (s, \ X) \ \overset{\text{def}}{=} \ last(s \uparrow [0, \ t)) \ = \ a,$$

which means that a immediately precedes the b at time t.

Notice that, in the following lemmas, theorems and their proofs, we can only prove that our processes will behave as required to satisfy the Safety, Utility, and Common Sense Properties. To guarantee these properties, the cooperation of their environment is also required, such as the gate will not be jammed and the *up* and *down* signals will be received by the gate whenever they are sent.

Lemma 2 *When the first train enters the empty region R, Controller will be ready to send a "down" signal in time for the gate to be fully closed before the train enters the crossing, i.e.,*

Controller **sat**
$$k \ \underline{at} \ t \wedge \neg(m \ \underline{precedes} \ (t, k)) \ \Rightarrow \ down \ \underline{live} \ [t + max\{t_{c1}, t_{u1}\}, \ t + \epsilon_1 - \gamma_d]$$
$$\wedge \ k \ \underline{at} \ t \wedge m \ \underline{precedes} \ (t, k) \wedge up \ \underline{at} \ (t + t_6) \wedge t_6 \leq min\{t_{u2}, t_{s2}\} \ \Rightarrow$$
$$down \ \underline{live} \ [t + max\{t_{u1}, t_6 + t_{c1}\}, \ t + \epsilon_1 - \gamma_d], \tag{1}$$

and $max\{t_{u1}, t_6 + t_{c1}\} \leq \epsilon_1 - \gamma_d.$

Lemma 3 *When the first train enters an empty region R, Controller will wait until it reaches the inward utility point t_{u1} before sending a "down" signal, i.e.,*

$$\text{Controller} \quad \textbf{sat} \quad k \ \underline{at} \ t \ \Rightarrow \ \neg(down \ \underline{at} \ [t, \ t + t_{u1})) \tag{2}$$

Lemma 4 *When the last train leaves R, Controller will be ready to send an "up" signal before the outward utility point t_{u2} so that the gate will be fully up before the train passes the second utility point ξ_2, i.e.,*

$$\text{Controller} \ \textbf{sat} \ m \ \underline{at} \ t \ \Rightarrow \ up \ \underline{live} \ [t, \ t + t_{u2}].$$

Theorem 1 (The Common Sense Property) *Controller satisfies the Common Sense Property, i.e., after an "up" signal being sent, it will wait for at least t_{c1} time units before sending a "down" signal:*

$$\text{Controller} \ \textbf{sat} \ up \ \underline{at} \ t \ \Rightarrow \ \neg(down \ \underline{at} \ [t, \ t + t_{c1}]).$$

Theorem 2 (The Safety Property) *GateController satisfies the Safety Property, provided that the gate behaves properly, i.e.,*

$$t \in \bigcup_i [\tau_i, \nu_i] \Rightarrow g(t) = 0.$$

Theorem 3 (The Utility Property) *GateController satisfies the Utility Property, provided that the gate behaves properly, i.e.,*

$$t \notin \bigcup_i [\tau_i - \xi_1, \nu_i + \xi_2] \Rightarrow g(t) = 90.$$

3 Related Work

The railroad crossing problem is a real-time specification problem of great generality and surprising subtlety. It has been attempted by numerous authors with as many different techniques [5], though not until now with timed CSP, to the best of the authors' knowledge, although a simpler version of the system was specified in timed CSP in [3].

Due to space restrictions, it has not proved possible to present the proofs of the Lemmas and Theorems presented here. These appear in [7]. The use of the Timed CSP Proof System was found to be natural, simple and efficient. Comparison [3] with the proof systems available in other formal notations is a strong argument in favour of Timed CSP.

We have considered the case of a crossing over a single track, which may be easily extended to multiple tracks, and indeed other authors take the same approach. We assume that there is a sensor to detect when a train starts to enter R, and a sensor to detect when a train has completely left I. Other authors [8] attempting this problem have assumed the existence of additional sensors, but this is not justified by the English specification above.

Our general approach has been to regard our task as one of presenting the original English statement of the problem in a formal language and proving that any implementation of this specification has certain properties. In particular, we have made our additional assumptions explicit. This approach - regarding the provider of the English specification as a customer and the final CSP specification as the customer's desired product - seems natural but has not been universally adopted. Some authors [9] have narrowed the specification so far from the English statement of the problem that they have produced an implementation instead of a specification.

4 Conclusions

We have specified the benchmark railroad crossing problem in Timed CSP and noted some of the lessons learned about the problem and the formal method.

Timed CSP is an elegant formal method for real-time systems. The language is concise, expressive and natural; the proof system is easy to use. The existence of a dedicated proof system for timed CSP means that it scores over other timed process algebras which lack such a system.

The interplay of the conflicting requirements leads to a many-branching process. It is instructive to consider the ways in which more or less justified real-world assumptions can be used to control and reduce this branching.

The biggest benefit of doing the detailed proofs (see [7]) is that they have helped a great deal in the design of the system. Even during the proof, we often had to go back to correct the definitions of our processes. There are well-known difficulties in scaling up the application of formal methods to real-world problems, but in such cases the difficulties of producing a correct design without use of formal techniques are scaled up also. We believe that it would be very difficult to design a correct real-time system of any realistic size without the help of formal methods.

References

1. J. Davies. *Specification and Proof in Real-Time CSP*. D.Phil thesis, Computing Laboratory, Oxford University, published by Cambridge University Press, 1993.
2. J. Davies. Setting real-time CSP. Internal note, Computing Laboratory, Oxford University, 1994.
3. A.S. Evans, D.R.W. Holton, L. Lai and P. Watson. A comparison of formal real-time specification languages. In *Proceedings of the Northern Formal Methods Workshop*, Ilkley, UK, 1996.
4. C.L. Heitmeyer and N. Lynch. The generalized railroad crossing: a case study in formal verification of real-time systems. NRL Memorandum Report NRL/MR/5540-94-7619, Navy Research Laboratory, Washington DC, USA, 1994.
5. C.L. Heitmeyer and D. Mandrioli (eds.). *Formal Methods for Real-Time Computing*. John Wiley & Sons, 1996.
6. C.A.R. Hoare. *Communicating Sequential Processes*. Prentice Hall International, London, 1985.
7. L. Lai, P. Watson, A case study in Timed CSP: the Railroad Crossing Problem, Technical Report CS-01-97, Dept. of Computing, University of Bradford, January 1997.
8. I. Lee, H. Ben-Abdallah and J. Choi. A process algebraic method for the specification and analysis of real-time systems. In [5].
9. W.D. Young. Modelling and verification of a simple real-time railroad gate controller. In M.G. Hinchey and J.P. Bowen (eds.), *Applications of Formal Methods*, Prentice Hall, 1995.

Analysis of Slope-Parametric Hybrid Automata

Frédéric Boniol[1], Augusto Burgueño[1] *, Olivier Roux[2] **, and Vlad Rusu[2]

[1] ONERA-CERT, Département d'Informatique,
2 av. E. Belin BP4025, 31055 Toulouse Cedex 4, France.
[2] LAN (URA CNRS N. 823, Ecole Centrale de Nantes, Université de Nantes),
1 rue de la Noë - BP92101 44321 Nantes Cedex 3, France.

Abstract. This paper addresses the analysis of slope-parametric hybrid automata: finding conditions on the slopes of the automaton variables, for some safety property to be verified. The problem is shown decidable in some practical situations (e.g. finding the running speeds of tasks in a real time application, for all tasks to respect their deadlines). The resolution technique generalizes polyhedral-based symbolic analysis and it involves reasoning about polyhedra with parametric shapes.

Keywords: real-time system, hybrid automata, parametric polyhedra.

1 Introduction

The parametric analysis of a real-time system consists in computing some parameters of the system to insure its 'correct' behaviour. Among the system parameters, some important ones are the running speeds of the tasks that compose it. We define here Slope-Parametric Hybrid Automata to model real-time systems whose running speeds should be adjusted for meeting some safety requirement. Then, parametric analysis consists in finding conditions on the slopes of the automaton variables, for some given property to be true. This is a new approach, since (up to our knowledge) only delay-parametric analysis has been studied and implemented (cf. [AHV93] and the HyTech tool [HH94]).

We describe a procedure which performs parametric analysis on slope-parametric hybrid automata. The procedure does not terminate in general; however, in the case of *time-bounded reachability* properties and *uniformly low-bounded* slope-parametric hybrid automata, the procedure terminates and the obtained conditions are necessary and sufficient. We believe that these automata are powerful enough to model many real-time applications. This leads for instance to a (theoretical) means for finding the running speeds, for all tasks in an application to respect their deadlines. While costly in practice, we think that the main interest of the procedure lies in setting a first solution, and in extending the usual polyhedral calculus [Zie95] with computation on parametric-shaped polyhedra.

---

* Supported by Research Grant of the Spanish Ministry of Education. This research was carried out while the author was visiting the LAN.
** Corresponding author. Email: roux@lan.ec-nantes.fr. Institut Universitaire de France.

2 Slope-Parametric Hybrid Automata

Definition. Slope-Parametric Hybrid Automata are a generalization of Multi-rate Automata [ACH+95]. A multirate automaton is a hybrid automaton whose continuous evolution laws are defined by differential equations of the form $\dot{x} = c$ where x is a variable ranging over $\mathbb{R}$ and c is a constant in $\mathbb{R}$ (the slope). Thus, for a multirate automaton the slopes are constant. A slope-parametric hybrid automaton allows evolution laws of the form $\dot{x} = k$, where k is a parameter in $\mathbb{R}$: the slopes may be either constants or parameters. Invariants and transition conditions of slope-parametric hybrid automata are expressed in the usual way ($\bigwedge_i x_i \sim c_i$, where x_i is a variable, c_i is a constant in $\mathbb{R}$ and $\sim \in \{<, \leq, >, \geq\}$) and the only allowed updates in variables when crossing an edge are reinitializations to zero.

Example. The slope-parametric hybrid automaton of figure 1 has two vertices, L_1 and L_2, and two variables, x and y. For vertex L_1 the evolution law of x is parametric ($\dot{x} = k_1$) while the one for y is constant ($\dot{y} = 1$). Same case for vertex L_2 with a different parameter for x ($\dot{x} = k_2$). The guard of the edge from vertex L_1 to vertex L_2 is ($x \geq 1 \wedge x \leq 2 \wedge y \geq 3 \wedge y \leq 8$). Only one reinitialization takes place on the edge from L_1 to L_2 ($x := 0$). The invariant for both vertices is *true* (not shown in the figure).

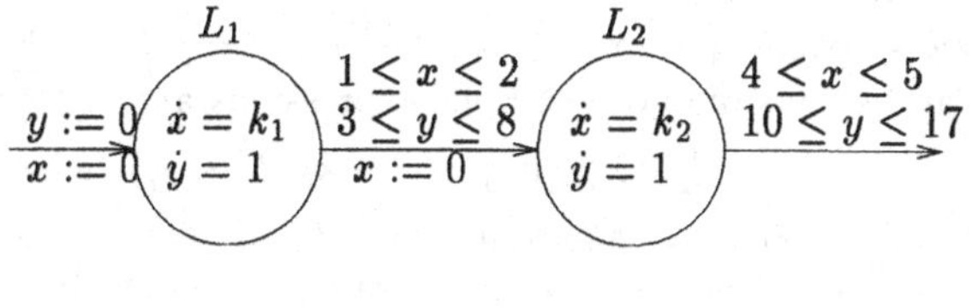

Figure 1. A slope-parametric hybrid automaton

Parametric analysis. The problem of parametric analysis is formulated as follows: given a *slope-parametric hybrid automaton*, a *temporal logic formula expressing a reachability property* and a *set of intervals in $\mathbb{R}$* (one for each parameter), find the relations among parameters (in the corresponding intervals) such that the formula is true. For instance, in the automaton of figure 1, find values for parameters $k_1 \in [-10, 10], k_2 \in [0, 5]$, for the TCTL formula $\varphi = [L_1 \wedge x = 0 \wedge y = 0] \wedge \exists true \mathcal{U}[L_2 \wedge x \geq 4 \wedge x \leq 5 \wedge y \geq 10 \wedge y \leq 17]$ to be true. This formula means that, starting from the initial set of states defined by vertex L_1 with $x = 0$ and $y = 0$, it is possible to reach the final set of states, defined by vertex L_2 with $x \in [4, 5]$ and $y \in [10, 17]$.

3 Computing parameter values

Parametric analysis comes to operating with parametric polyhedra. The latter are described by sets of linear equations and inequations whose coefficients can be either *constants* or *symbolic expressions on the parameters*; for instance, $x \geq 0 \wedge y \geq 0 \wedge k_1 \cdot y - x \geq 0 \wedge x - k_1 \cdot y \geq 0$, where k_1 is a parameter. The operations involved in parametric analysis are *extension*, *restriction* and *projection*, that incrementally generate conditions on the parameters, at each transition from a vertex to the next one. We first demonstrate their use on an example and then describe them in more detail.

Example. Consider the slope-parametric hybrid automaton in figure 1 and the problem of finding values for parameters $k_1 \in [-10, 10], k_2 \in [0, 5]$, for the formula $\varphi = [L_1 \wedge x = 0 \wedge y = 0] \wedge \exists true \mathcal{U} [L_2 \wedge x \geq 4 \wedge x \leq 5 \wedge y \geq 10 \wedge y \leq 17]$ to be true. For this, we *extend* the initial region $x = 0 \wedge y = 0$ at vertex L_1, in the parametric direction given by $\dot{x} = k_1, \dot{y} = 1$, and then *restrict* the values of parameter k_1 such that the previous extension intersects the guard $1 \leq x \leq 2$, $3 \leq y \leq 8$. Next, we *project* the obtained intersection on plane $x = 0$ (this corresponds to reinitializing x on the transition). Likewise, we continue by *extending* the obtained projection in the direction given by $\dot{x} = k_2, \dot{y} = 1$ at vertex L_2, and terminate by *restricting* the values of parameters k_1, k_2 such that the last extension intersects the final region $x \geq 4 \wedge x \leq 5 \wedge y \geq 10 \wedge y \leq 17$.
Extension and projection are operations on parametric polyhedra, and restriction means finding the values of parameters such that a given parametric polyhedron is non-void.

Extension means: given a parametric polyhedron $P = \bigwedge_{j=1}^{m} \left(\Sigma_{i=1}^{n} a_{i,j} \cdot x_i \succ b_j \right)$ where $a_{i,j}, b_j$ can be constants or symbolic expressions and $\succ \in \{>, \geq\}$, find the polyhedron: $\overrightarrow{P} = \exists t \geq 0. \left[\bigwedge_{j=1}^{m} \left(\Sigma_{i=1}^{n} a_{i,j} \cdot (x_i - k_i \cdot t) \succ b_j \right) \right]$ where each k_i is the slope for variable x_i (constant or parameter). This is *forward continuous simulation* [ACH+95] except that we consider parametric polyhedra and directions. This imposes to consider several cases for eliminating the $\exists$ quantifier in the equivalent expression $\overrightarrow{P} = \exists t \geq 0. \left[\bigwedge_{j=1}^{m} \left(\Sigma_{i=1}^{n} a_{i,j} \cdot x_i - t \cdot (\Sigma_{i=1}^{n} a_{i,j} \cdot k_i) \succ b_j \right) \right]$. The cases to consider are, for each sum-of-products $\Sigma_{i=1}^{n} a_{i,j} \cdot k_i$, the possibility that it is negative, positive or 0. As in [AHH93] we eliminate the existential quantifier by dropping the inequations that correspond to negative sums-of-products $(\Sigma_{i=1}^{n} a_{i,j} \cdot k_i < 0)$, keeping the inequations that correspond to positive or zero sums-of-products $(\Sigma_{i=1}^{n} a_{i,j} \cdot k_i \geq 0)$, and linearly combining pairs of inequations that correspond to one negative and one positive sum-of-products.
The point here is that the sums-of-products are symbolic, so in general we will not be able to tell at sight the sign of the expressions $\Sigma_{i=1}^{n} a_{i,j} \cdot k_i$. We must then consider all the possible cases for these signs, so the extended polyhedron $\overrightarrow{P}$ has at most 3^m different forms, following the possible signs of the m expressions $\Sigma_{i=1}^{n} a_{i,j} \cdot k_i$. For the example of figure 1, vertex L_1, the extension of the region $P = \{x = 0 \wedge y = 0\}$ gives three cases: 1) for $k_1 = 0, \overrightarrow{P_1} = \{x = 0 \wedge y \geq 0\}$,

2) for $k_1 > 0$, $\vec{P_2} = \{x \geq 0 \wedge y \geq 0 \wedge k_1 \cdot y - x \geq 0 \wedge x - k_1 \cdot y \geq 0\}$, and 3) for $k_1 < 0$, $\vec{P_3} = \{-x \geq 0 \wedge y \geq 0 \wedge -k_1 \cdot y + x \geq 0 \wedge -x + k_1 \cdot y \geq 0\}$. The possible signs of sums-of-products (like $k_1 > 0$, $k_1 = 0$, $k_1 < 0$) constitute relations on the parameters (slopes) that should be remembered when further propagating the extended polyhedron $\vec{P}$. Note also that the parametric slopes have become coefficients in the inequations of the extended polyhedron (i.e. $k_1 \cdot y - x \geq 0$).

Restriction is: given a parametric polyhedron $P = \bigwedge_{j=1}^{m} (\Sigma_{i=1}^{n} a_{i,j} \cdot x_i \succ b_j)$ (remember that $a_{i,j}, b_j$ are constants or symbolic expressions on the parameters) find the possible values of parameters such that P is non-empty. Here is a simple example of restriction (figure 1, vertex L_1): for the three parametric polyhedra obtained at previous step (one for each case $k_1 = 0, k_1 > 0$ and $k_1 < 0$) we compute their intersection with the guard $\{1 \leq x \leq 2 \wedge 3 \leq y \leq 8\}$ and find the values of k_1 such that the intersection is non-empty. For 1) $k_1 = 0$ it is empty: $\vec{P_1} \cap \{1 \leq x \leq 2 \wedge 3 \leq y \leq 8\} = \emptyset$. For 2) $k_1 > 0$, $\vec{P_2} \cap \{1 \leq x \leq 2 \wedge 3 \leq y \leq 8\}$ is non-empty iff $\frac{1}{8} \leq k_1 \leq \frac{2}{3}$. For 3) $k_1 < 0$ the intersection is always empty. The general case can be treated as follows: the polyhedron P is non-empty iff the expression $\exists x_1.\exists x_2 \ldots \exists x_n.P$ is 'true'. The formal elimination of all variables in the previous expression generates a symbolic condition on the parameters, *that precisely constitutes the condition for P to be non-empty.* Variables $x_1 \ldots x_n$ can be eliminated one by one by successively applying the Fourier-Motzkin elimination algorithm [Zie95] that we now describe.

The Fourier-Motzkin elimination algorithm. This algorithm computes, given a system of linear inequations $P = \bigwedge_{j=1}^{m} (\Sigma_{i=1}^{n} a_{i,j} x_i \succ b_j)$, the system obtained by eliminating a variable say x_k : $P \Updownarrow_k = \exists x_k. \bigwedge_{j=1}^{m} (\Sigma_{i=1}^{n} a_{i,j} x_i \succ b_j)$. The idea is to consider the possible signs of the coefficients $a_{k,j}$: as in the case of extension, eliminating variable x_k leads to at most 3^m possible forms of the result, depending on the signs of the m coefficients $a_{k,j}$. For a given combination of signs, denote $J_>$ (respectively, $J_=$, $J_<$) the subsets of indices of $\{1, \ldots, m\}$ such that $a_{k,j} > 0$ (respectively $= 0$, < 0). Then, $P \Updownarrow_k$ is obtained by keeping the inequations indexed by $J_=$, by eliminating the inequations indexed by $J_<$ and $J_>$ (i.e. keep only the inequations where x_k does not occur), and by linearly combining pairs of inequations (one indexed by some $j_> \in J_>$, the other indexed by some $j_< \in J_<$) to eliminate variable x_k. For all $j_> \in J_>$ and $j_< \in J_<$, generate the following linear combination: $a_{k,j_>} [\Sigma_{i=1}^{n} a_{i,j_<} x_i \succ b_{j_<}] - a_{k,j_<} [\Sigma_{i=1}^{n} a_{i,j_>} x_i \succ b_{j_>}]$ which is equivalent to $\Sigma_{i=1}^{n} [a_{k,j_>} a_{i,j_<} - a_{k,j_<} a_{i,j_>}] x_i \succ [a_{k,j_>} b_{j_<} - a_{k,j_<} b_{j_>}]$. In this last inequation, the coefficient of x_k is 0 so variable x_k has been eliminated.

Projection on the $x_k = 0$ plane means: given a parametric polyhedron $P = \bigwedge_{j=1}^{m} (\Sigma_{i=1}^{n} a_{i,j} \cdot x_i \succ b_j)$, find the parametric polyhedron $P|_{x_k=0}$ obtained by projecting P on the plane $x_k = 0$. Consider again the example of figure 1. In the previous step (restriction) we had obtained only one valid case ($k_1 > 0$) for which the parametric polyhedron is non-empty: $\vec{P_2} = \{x \geq 1 \wedge x \leq 2 \wedge y \geq 3 \wedge y \leq 8$

$\wedge k_1 \cdot y - x \geq 0 \wedge x - k_1 \cdot y \geq 0\}$ when $\frac{1}{8} \leq k_1 \leq \frac{2}{3}$. The projection of $\overrightarrow{P_2}$ on the plane $x = 0$ is $\overrightarrow{P_2} \mid_{x=0} = \{x \geq 0 \wedge x \leq 0 \wedge y \geq 3 \wedge y \leq 8 \wedge k_1 \cdot y \geq 1 \wedge k_1 \cdot y \leq 2\}$. In general, to obtain the projected parametric polyhedron we use the Fourier-Motzkin elimination algorithm and the identity $P\mid_{x_k=0} = \exists x_k.P \wedge (x_k = 0)$.

The operations at work. To obtain the conditions on the parameters for a reachability formula to be true, combine the three operations as follows. First, choose a vertex path that links a vertex from the initial region, to a vertex from the final region (as defined by the reachability formula). Then, iterate the three operations on that vertex path, to generate a *tree* whose nodes are pairs (vertex, parametric polyhedron), and whose edges are labeled by symbolic conditions on the parameters. Starting from the pair (initial vertex, polyhedron defined by the initial values of variables) as the root, apply the *extension* procedure to the initial polyhedron to generate all the possible extended polyhedra, and associate a node (successor of the root) to each one. The branch leading to a node is labeled with the condition under which the node's extended polyhedron was obtained. Likewise, for each new node, generate its successors by applying the *restriction* procedure, and label the new branches correspondingly. For the lastly obtained successors, continue with the *projection* procedure. Iterate the sequence of procedures in this order until the final vertex is reached, and terminate by a restriction to intersect the final region.

At this point, any sequence of branches from the root to a leaf defines a sufficient condition on the parameters, for the final region to be reachable from the initial one (it is the conjunction of the conditions on all branches). However, two problems arise. First, such a condition might be unsatisfiable. Second, there might be an infinity of vertex paths in the automaton, from a formula's initial vertex to a final one; this means that in order to generate all the conditions (necessary and sufficient) on the parameters for the reachability to hold, we might need to iterate the above operations on an infinite number of vertex paths. This is why we consider a restricted class of automata and formulas, for which we have to analyze only a finite number of vertex paths.

Uniformly Low-bounded Hybrid Automata, Time-bounded Reachability. A relaxed hybrid automaton is a rectangular hybrid automaton [HKPV95] obtained from a slope-parametric hybrid automaton by replacing each parameter by the corresponding interval. For instance, a relaxed hybrid automaton is obtained from the slope-parametric hybrid automaton of figure 1 by replacing '$= k_1$' with '$\in [-10, 10]$' and '$= k_2$' with '$\in [0, 5]$'. If the relaxed hybrid automaton is *uniformly low-bounded* (i.e. there is a uniform, strictly positive low-bound on the durations of cyclic runs) then any time-bounded reachability formula is decidable and furthermore *there are only a finite number of vertex paths by which reachability is possible* [RR96]. In consequence, we have to iterate the three above-mentioned operations only on a finite number of vertex paths, because if there exists a solution to parametric analysis, it will be found on some of these paths. Thus, we can find the necessary and sufficient conditions on the parameters such that the time-bounded reachability formula is true.

As an application, consider a real-time system composed of a number of tasks running at adjustable speeds. We consider the problem of finding the necessary and sufficient conditions on the speeds for all tasks to terminate within given deadlines. Such a real-time system can be modeled by a uniformly low-bounded hybrid automaton [RR96], and the property that a task terminates within a (constant) delay c can be expressed by the negation of a time-bounded reachability formula (expressing that a task cannot be continuously running for a time c). Thus, we have a theoretical means to solve the given parametric problem.

4 Conclusion

We have presented a new approach to the parametric analysis of hybrid automata, which focuses on computing slopes of variables (rather than delays) for some safety requirement to be respected. The method is an extension of the polyhedra-based symbolic analysis [ACH+95] of hybrid automata. The main operation is the Fourier-Motzkin's algorithm [Zie95] to deal with parametric polyhedra, which imposes to consider a large number of cases. We are currently studying several ways of improving its efficiency. This work is a first step towards the study of the more general case of hybrid automata with evolution laws of the form $\dot{x} + \dot{y} = k$.

Acknowledgments. We thank the anonymous referees for their useful comments on the previous version of this paper.

References

ACH+95. R. Alur, C. Courcoubetis, N. Halbwachs, T. A. Henzinger, P-H. Ho, X. Nicollin, A. Olivero, J. Sifakis, and S. Yovine. The algorithmic analysis of hybrid systems. *Theoretical Computer Science*, 138:3–34, 1995.

AHH93. R. Alur, T. A. Henzinger, and P-H. Ho. Automatic symbolic verification of embedded systems. In *Proc. 14th Annual Real-time Systems Symposium, RTSS'93*, pages 2–11. IEEE Computer Society Press, 1993.

AHV93. R. Alur, T. A. Henzinger, and M. Y. Vardi. Parametric real-time reasoning. In *Proc. of the 25th Annual ACM Symposium on Theory of Computing, STOC'93*, pages 592–601, 1993.

HH94. T. A. Henzinger and P.-H. Ho. HyTech: the cornell HYbrid TECHnology tool. In P. Antsaklis, W. Kohn, A. Nerode, and S. Sastry, editors, *Hybrid Systems II*, volume 999 of *Lecture Notes in Computer Science*, pages 265–294. Springer-Verlag, 1994.

HKPV95. T. A. Henzinger, P. W. Kopke, A. Puri, and P. Varaiya. What's decidable about hybrid automata? In *Proc. of the 27th Annual ACM Symposium on Theory of Computing, STOC'95*, pages 373–382, 1995.

RR96. O. Roux and V. Rusu. Uniformity for the decidability of hybrid automata. In *Internat. Static Analysis Symposium, LNCS 1145*, pages 301–316, 1996.

Zie95. G. M. Ziegler. *Lectures on Polytopes*, volume 152 of *Graduate Texts in Mathematics*. Springer-Verlag, 1995.

Comparing Timed C/E Systems with Timed Automata*
(Abstract)

R. Huuck[1], Y. Lakhnech[1], L. Urbina[1],
Institut für Informatik und Praktische Mathematik der Christian-Albrechts-Universität zu Kiel,
Preußerstr. 1-9, D-24105 Kiel, Germany.
S. Engell[2], S. Kowalewski[2], J. Preußig[2]
Lehrstuhl für Anlagensteuerungstechnik, Fachbereich Chemietechnik, Universität
Dortmund. D-44221 Dortmund, Germany.

[1] Email: {rhu, yl, lu}@informatik.uni-kiel.de
[2] {engell, stefan. joerg}@ast.chemietechnik.uni-dortmund.de

Abstract. We investigate the relationship between timed c/e systems and timed automata. We provide an effective function that associates to each timed c/e system an "equivalent" timed automaton. Equivalence has to be understood as describing the same set of trajectories. A benefit from providing such a function is that analysis tools developed for timed automata can now be applied to analyze timed c/e systems. We also provide an effective function that translates each timed automaton into an equivalent timed c/e system.

1 Introduction

Condition/event systems (CESs) were introduced in [6] as a modeling paradigm for discrete event systems (DESs). In this paper we will focus on the timing extension *timed condition/event systems* (TCESs) [4]. Our main result is that TCESs [4] and timed automata (TAs) [2] are expressively equivalent. In fact, we provide elementary transformations from one model to the other. The motivation of this work is two fold. TCESs are used in the community of control theory for describing real-time systems. On the other hand, the framework of TAs has become a standard model, at least in the community of computer scientists. Therefore, it is necessary to study the relationship between these models. The second motivation is of practical nature. TAs have been intensively studied from different point of views including the automata theoretic view [2], the verification view [1, 5, 3], and the logic view [7, 3]. This led to a variety of tools that can analyze real-time systems modeled by TAs. Therefore, a benefit from providing an elementary function transforming a TCES into a TA is that analysis tools developed for TAs can now also be applied to analyze real-time systems modeled by TCESs.

In Section 2 we introduce CESs and TCESs. Then, in Section 3 we define the model of TAs we are considering. Section 4 presents the translations from one model to the other and Section 5 contains some concluding remarks.

* This work has been supported by the German Research Council (DFG) in the special program KONDISK (analysis and synthesis of technical systems with continuous-discrete dynamics) under the grants Ro 1122/2 and En 152/19.

2   C/E Systems and Timed C/E Systems

Here, we introduce *condition/event systems* [6] and their real–time version [4]. We consider two different sorts of signals, *condition signals* and *event signals*. In this context, the word condition is used to refer to states of the system. Given a nonempty finite set U of *condition symbols* a function $s_u : [0, \infty) \to U$ is called *condition signal* over U, if it is right–continuous and finite–variable. The set of all condition signals over U is denoted by $C(U)$. We call U a *condition alphabet*. Given a finite set V of *event symbols* and a *null symbol* $0 \notin V$. A function $s_v : [0, \infty) \to V^0 = V \cup \{0\}$ is called *event signal* over V, if $s_v(0) = 0$ and if in any bounded interval I in $[0, \infty)$, there are finitely many points $t \in I$ such that $s_v(t) \neq 0$. An event symbol $v \in V$ is called a *proper event symbol*. The set of all event signals over V is denoted by $E(V)$. We refer to V as an *event alphabet*.

A *condition/event system* (CES) is characterized by a tuple (U, V, Y, Z, S), where U and Y are the input, respectively, output condition alphabets, V and Z are the input, respectively, output event alphabets, and $S : C(U) \times E(V) \to 2^{C(Y) \times E(Z)}$ is the *system behavior function*. S can be identified with a relation $S_R \subseteq \text{beh}(U, V, Y, Z) = C(U) \times E(V) \times C(Y) \times E(Z)$. In the sequel we tacitly make use of this observation. We also often represent a CES only by its behavior function since it is easy to infer the other components from this one. Moreover, if any of the alphabets U, Y is a singleton or any of the alphabets V, Z is empty, we omit it in the description of the considered CES.

C/e systems with discrete state realizations are CESs that can be described operationally by so–called discrete c/e systems. A *discrete c/e system* (DCES) is given by a tuple $D = (U, V, Q, Y, Z, f, g, h, q_0)$, where

- U, V, Y, and Z are as above, Q is a finite set of *states*,
- $f : Q \times U \times V^0 \to 2^Q \setminus \{\emptyset\}$ is the *state transition function*,
- $g : Q \times U \to Y$ is the *condition output function*,
- $h : Q \times Q \times V^0 \to Z^0$ is the *event output function*, and
- $q_0 \in Q$ is the *initial state*.

A DCES is called *well–behaved*, if it satisfies *1.)* Stuttering: $\forall q \in Q. \forall u \in U. q \in f(q, u, 0)$, i.e., the system is allowed to stay at the same state as long as the value of the event input signal does not change, and *2.)* Output triggering: $\forall q \in Q. h(q, q, 0) = 0$, i.e., the system can only output a proper event while remaining in the same state when it receives a proper input event; this is a causality condition. Henceforth, we only consider well–behaved DCESs and call them DCESs. To assign a semantics to DCESs, we introduce the notion of a run. For a finite variable function x mapping non–negative reals to values in a set, let $x(t^-)$ denote the limit from the left of x at t, i.e. $x(t^-) = \lim_{\epsilon \to 0} x(t - \epsilon)$. Then, given a DCES $D = (U, V, Q, Y, Z, f, g, h, q_0)$, we call a *run* of D over $(s_u, s_v, s_y, s_z) \in \text{beh}(U, V, Y, Z)$ any right–continuous and finite–variable function $r : \mathbb{R}_{\geq 0} \to Q$ which satisfies the following conditions for each $t \in \mathbb{R}_{\geq 0}$: *1.)* $r(t) \in f(r(t^-), s_u(t^-), s_v(t))$, *2.)* $s_y(t) = g(r(t), s_u(t))$, *3.)* $s_z(t) = h(r(t^-), r(t), s_v(t))$, and *4.)* $r(0) = q_0$.

We associate to each D the relation S_D consisting of all behaviors $(s_u, s_v, s_y, s_z) \in \text{beh}(U, V, Y, Z)$ such that there exists a run of D over (s_u, s_v, s_y, s_z).

Timed Condition/Event Systems. A timer θ is a CES whose behavior is completely determined by a *threshold vector* $T_\theta = (T_\theta^1, \cdots, T_\theta^m)$ of increasing constants called *threshold time bounds*. Depending on an input event signal s_v in $E(\text{Res})$ representing the sequence of reset points, it generates a condition signal s_y and an event signal s_z. Whereas the condition signal s_y states to which of the intervals $[T_\theta^i, T_\theta^{i+1})$, $[0, T_\theta^1)$, or $[T_\theta^m, \infty)$ the actual value of the timer belongs,

s_z shows the *alarm event* "$\theta = T_\theta^i$" exactly at the time points where the actual value of the timer equals the threshold time T_θ^i. The value of the timer grows linearly with the global time and a *reset event* results in resetting its value to 0.

The *timer function* $\tau_\theta(s_v) : [0, \infty) \to [0, \infty)$ associated to s_v is defined by $\tau_\theta(s_v)(t) = t - t'$. where t' is the greatest $t'' \leq t$ such that $s_v(t'') \neq 0$. Thus. $\tau_\theta(s_v)(t)$ is the value of the timer at time point t. Then, a *timer* θ is defined as a pair consisting of a CES (Res$_\theta$. Int$_\theta$. Ala$_\theta$, S_θ) and a vector $T_\theta = (T_\theta^1, \cdots. T_\theta^m)$ of constants $T_\theta^i \in I\!\!N_{>0}$ such that $T_\theta^1 < \cdots < T_\theta^m$ and

- Res$_\theta = \{$"$\theta := 0$"$\}$ is the input event alphabet,
- Int$_\theta = \{$"$\theta < T_\theta^1$". "$\theta \geq T_\theta^m$", "$T_\theta^1 \leq \theta < T_\theta^2$", $\cdots$, "$T_\theta^{m-1} \leq \theta < T_\theta^m$"$\}$ is the output condition alphabet.
- Ala$_\theta = \{$"$\theta = T_\theta^i$" $\mid i = 1, \cdots, m\}$ is the output event alphabet, and
- for every $(s_v. s_y, s_z) \in$ beh(Res$_\theta$. Int$_\theta$. Ala$_\theta$). $(s_y, s_z) \in S_\theta(s_v)$ iff for every $t \in I\!\!R_{\geq 0}$ and each threshold time T_θ^i the following conditions are satisfied:
$$s_y(t) = \begin{cases} \text{"}\theta < T_\theta^1\text{"} & \text{if } \tau_\theta(s_v)(t) < T_\theta^1 \\ \text{"}\theta \geq T_\theta^m\text{"} & \text{if } \tau_\theta(s_v)(t) \geq T_\theta^m \\ \text{"}T_\theta^i \leq \theta < T_\theta^{i+1}\text{"} & \text{if } i \in \{1, \cdots, m-1\} \wedge \tau_\theta(s_v)(t) \in [T_\theta^i, T_\theta^{i+1}) \end{cases},$$
$$s_z(t) = \begin{cases} \text{"}\theta = T_\theta^i\text{"} & \text{if } i \in \{1, \cdots, m\} \wedge \tau_\theta(s_v)(t) = T_\theta^i \\ 0 & \text{otherwise} \end{cases}.$$

Since a timer is completely determined by its threshold vector we identify a timer with its threshold vector. Moreover, when we consider a set $\{\theta_i \mid 1 \leq i \leq n\}$ of timers, we denote by $T_{\theta_i} = (T_{\theta_i}^1, \cdots. T_{\theta_i}^{m_i})$ the threshold vector of timer θ_i, by Res$_i$, Int$_i$, Ala$_i$ its signal alphabets. and by τ_i its timer function.

Definition 1.
A *timed condition/event system* (TCES) is a tuple $\mathcal{T} = (U, V, Y, Z, D, \Theta)$. where $\Theta = \{\theta_i \mid 1 \leq i \leq n\}$ is a set of timers and $D = (U \times \Pi_{i=1}^n \text{Int}_i, V^0 \times \Pi_{i=1}^n \text{Ala}_i^0, X, Y, Z^0 \times \Pi_{i=1}^n \text{Res}_i^0. f. g, h, x_0)$ is a DCES.

In order to assign a semantics to TCESs, we associate to each TCES $\mathcal{T}$ a CES $S_\mathcal{T}$ as follows. For any $(s_u. s_v) \in C(U) \times E(V^0)$ and any $(s_y, s_z) \in C(Y) \times E(Z^0)$, $(s_y, s_z) \in S_\mathcal{T}(s_u, s_v)$ iff there exist $(s_{y_1}, \cdots, s_{y_n}) \in \Pi_{i=1}^n C(\text{Int}_i)$, $(s_{z_1}, \cdots, s_{z_n}) \in \Pi_{i=1}^n E(\text{Ala}_i)$, and $(s_{v_1}, \cdots, s_{v_n}) \in \Pi_{i=1}^n E(\text{Res}_i)$. such that the following conditions are satisfied:

- $(s_{y_i}, s_{z_i}) \in S_{\theta_i}(s_{v_i})$, for each $i = 1, \cdots, n$. That is, the signals s_{y_i} and s_{z_i} correspond to the output signals of timer θ_i assuming the input signal s_{v_i}.
- $(s_y, (s_z, s_{v_1}, \cdots, s_{v_n})) \in S_D((s_u, s_{y_1}, \cdots, s_{y_n}). (s_v, s_{z_1}, \cdots, s_{z_n}))$.

Lemma 2. *For any timed condition/event system $\mathcal{T}$, $S_\mathcal{T}$ is a c/e system.*

To conclude this section. we remark that in contrast to the original definition of c/e timers [4], here multiple thresholds are allowed for one clock and the timer function is defined slightly differently. However, both models are equivalent. though the model introduced here allows more natural and concise specifications.

3 Timed Automata

We consider *safety timed automata* [5], that is, timed automata whose acceptance condition is trivial and every run is accepting.

Let $\mathcal{V}$ be a set of real-valued variables called *clocks*. A *constraint* ξ over $\mathcal{V}$ is a boolean combination of formulas of the form $x \# c$. where $x \in \mathcal{V}$, $\# \in \{<, \leq\}$, and $c \in I\!\!N$ is a non-negative integer. Let $\mathcal{C}(\mathcal{V})$ denote the set of constraints over $\mathcal{V}$. A *valuation* over $\mathcal{V}$ is a function $\nu : \mathcal{V} \to I\!\!R_{\geq 0}$. The satisfaction relation $\nu \models \xi$ between valuations and constraints is defined as usual. Given a valuation ν and a set $R \subseteq \mathcal{V}$, we denote by $\nu[R \mapsto 0]$ the valuation ν' which associates with each clock in R the value 0 and coincides with ν on all the other clocks. For a $t \in \mathbb{R}_{\geq 0}$ we denote by $\nu + t$ the valuation ν' such that $\nu'(x) = \nu(x) + t$, for all $x \in \mathcal{V}$.

A *safety timed automaton* (TA) over an alphabet Σ is a tuple $A = (\mathcal{Q}, q_0, \mathcal{X}, \mathcal{E})$, where $\mathcal{Q}$, a finite set of locations; q_0, the initial location; $\mathcal{X}$, a finite set of clocks; and $\mathcal{E} \subseteq \mathcal{Q} \times \mathcal{C}(\mathcal{X}) \times \Sigma \times 2^{\mathcal{X}} \times \mathcal{Q}$, a set of edges.

We call a sequence $\overrightarrow{T} = (t_i)_{i \in \omega}$ of time points $t_i \in \mathbb{R}_{\geq 0}$ a *time sequence*, if the following conditions are satisfied: 1.) $t_0 > 0$, 2.) $\forall i \in \omega.\ t_i < t_{i+1}$, and 3.) $\overrightarrow{T}$ is divergent. The set of all time sequences is denoted by TS. To define runs we introduce, for each edge $e \in \mathcal{E}$, the relation $\xrightarrow{e}$ between configurations, where a *configuration* is a triple (q, ν, t) consisting of a control location q, a valuation ν and a time point t. Thus, consider two arbitrary configurations (q, ν, t) and (q', ν', t') and $e = (q, \xi, a, R, q') \in \mathcal{E}$. Then, $(q, \nu, t) \xrightarrow{e} (q', \nu', t')$ iff there exists $\delta > 0$ such that $\nu + \delta \models \xi$, $\nu' = \nu + \delta[R \mapsto 0]$, and $t' = t + \delta$. A *run* of $A = (\mathcal{Q}, q_0, \mathcal{X}, \mathcal{E})$ is an infinite sequence $(q_0, \nu_0, t_0) \xrightarrow{e_0} (q_1, \nu_1, t_1) \xrightarrow{e_1} \cdots$ satisfying the following conditions: 1.) $\forall x \in \mathcal{X}.\ \nu_0(x) = 0$, 2.) $(t_i)_{i \in \omega}$ is a time sequence, and 3.) $(q_i, \nu_i, t_i) \xrightarrow{e_i} (q_{i+1}, \nu_{i+1}, t_{i+1})$. for each $i \in \omega$. We call the sequence $(e_i)_{i \in \omega}$ the *transition sequence associated to the run* r. Any element of $\Sigma^\omega \times$ TS is called a *timed word*. The set of all timed words over the alphabet Σ is denoted by $\mathrm{TW}(\Sigma)$. Given a run r with the associated transition sequence $(q_i, \xi_i, a_i, R_i, q_{i+1})_{i \in \omega}$, we say that r is a run over the timed word $w = (a_i, t_i)_{i \in \omega}$ and define the *language* $L(A)$ of A as the set of all timed words w such that there exists a run r of A over w.

4 Expressiveness Results

We compare the expressiveness of timed condition/event systems and timed automata. Since the behavior of a TCES is defined as a relation associating to input signals a set of possible output signals while the language of a TA is a set of timed words, we have first to relate both. Thus, consider $\mathcal{T}_R \subseteq \mathrm{beh}(U, V, Y, Z)$ describing the behavior of a TCES $\mathcal{T}$. For each $\beta = (s_u, s_v, s_y, s_z) \in \mathcal{T}_R$, let $\mathrm{Inte}(\beta)$ denote the maximal ordered sequence $(t_i)_{i \in \omega \cup \{\omega\}}$ of real numbers such that one of the following conditions is satisfied for each t_i:

- t_i is a discontinuity point of s_u or s_y, or
- $s_v(t_i)$ or $s_z(t_i)$ is a proper event, that is, it is different from 0.

Henceforth, let us only consider behaviors β such that $\mathrm{Inte}(\beta)$ is an infinite sequence. Now, we define a function *tra* which assigns to each timed word $w = (a_i, t_i)_{i \in \omega} \in \mathrm{TW}(\Sigma)$ a set of behaviors such that for each behavior $\beta \in tra(w)$ the following conditions are satisfied: 1.) $(t_i)_{i \in \omega} \supseteq \mathrm{Inte}(\beta)$, and 2.) $\forall i \in \omega.\ (\beta(t_i)) = a_i$. The mapping *tra* can be extended pointwise to sets $L(A)$ of timed words of TAs. Then, we say that a TCES $\mathcal{T}$ with the behavior $\mathcal{T}_R$ is equivalent to a TA A, if $tra(L(A)) = \mathcal{T}_R$. The main result of this section is that for each TCES one can effectively construct an equivalent TA and vice versa. Clearly, to prove this result we suppose that TCESs and TAs are given in an effective manner.

4.1 Transforming Timed C/E Systems into Timed Automata

We provide an effective function that transforms each TCES $\mathcal{T}$ into an equivalent TA $\mathit{Aut}(\mathcal{T})$ with the same structure as the discrete system underlying $\mathcal{T}$ and which contains exactly one clock for each timer of $\mathcal{T}$. The main step in the transformation is to determine the guards of the transitions. This is realized by the function *cons* mapping condition and event symbols in $\mathrm{Int}_\theta \cup \mathrm{Ala}_\theta^0$ to clock constraints. Consider a timer θ with threshold vector $(T_\theta^1, \cdots, T_\theta^m)$ and a clock x. Then, we define for each $a \in \mathrm{Int}_\theta \cup \mathrm{Ala}_\theta^0$ the constraint $cons(a, x)$ as follows:

$$cons(a, x) = \begin{cases} x \leq T_\theta^1 & \text{if } a = ``\theta < T_\theta^1" \\ x > T_\theta^m & \text{if } a = ``\theta \geq T_\theta^m" \\ x = T_\theta^j & \text{if } a = ``\theta = T_\theta^j" \\ T_\theta^j < x \leq T_\theta^{j+1} & \text{if } a = ``T_\theta^j \leq \theta < T_\theta^{j+1}" \\ \bigwedge_{j=1}^m x \neq T_\theta^j & \text{if } a = 0 \end{cases}$$

To understand this translation notice that, since s_u is a piecewise function, we have $s_u(t^-) = u$ iff there exists an interval I_t left from t and $s_u(t') = u$, for each $t' \in I_t$. Thus, given a TCES $\mathcal{T} = (U, V, Y, Z, D, \Theta)$, where $D = (U, V, \mathcal{Q}, Y, Z, f, g, h, q_0)$ is the DCES and $\Theta = \{\theta_i \mid 1 \leq i \leq n\}$ is the set of timers. we define $Aut(\mathcal{T}) = (\mathcal{Q}, q_0, \mathcal{X}, \mathcal{E})$, where

- $\mathcal{Q}$ is the set of states of the DCES D, q_0 is its initial state, $\mathcal{X} = \{x_1, \cdots, x_n\}$ is the set of clocks, one clock for each timer,
- $(q, \xi, a, R, q') \in \mathcal{E}$ if and only if there exist $\mathbf{u} = (u, u_1, \cdots, u_n) \in U \times \Pi_{i=1}^n \mathrm{Int}_i$ and $\mathbf{v} = (v, v_1, \cdots, v_n) \in V^0 \times \Pi_{i=1}^n \mathrm{Ala}_i^0$, such that the following requirements are satisfied:

  1.) $q' \in f(q, \mathbf{u}, \mathbf{v})$, 2.) $\xi = \bigwedge\limits_{i=1}^n cons(u_i, x_i) \wedge cons(v_i, x_i)$,

  3.) $a = (u, v, g(q, \mathbf{u}), h(q, q', \mathbf{v})_1)$, and 4.) $R = \{x_i \in \mathcal{X} \mid h(q, q'. \mathbf{v})_{i+1} \neq 0\}$.

Proposition 1 *There is an effective function associating to each timed c/e system $\mathcal{T}$ an equivalent timed safety automaton with the same number of control states and a number of clocks equal to the number of timers of $\mathcal{T}$.*

4.2 Transforming Timed Automata into Timed C/E Systems

Next, we prove that TCESs are **not** less expressive than TAs. To do so we provide a transformation that associates to each TA an equivalent TCES. To simplify the presentation we assume that the constraints occurring in the considered automaton are of a special form. Given a TA $\mathcal{A}$ with the set $\mathcal{X}$ of clocks, let $comp(x)$ denote the list of constants in increasing order to which x is compared in some constraint in $\mathcal{A}$. Then, we call a constraint *simple*, if it is a conjunction $\bigwedge_{x \in \mathcal{X}} \xi(x)$ such that each $\xi(x)$ is of one of the following forms: $x < c$, $x = c$, $x > c$, or $c' < x < c''$, and such that $c, c', c'' > 0$ and c' and c'' are consecutive in the sequence $comp(x)$. Henceforth, we only consider TAs such that each constraint ξ in the automaton is *simple* and such that there is always at most one edge between two control locations. It is not difficult to see that each TA can be transformed into an equivalent one satisfying this requirement. We call each TA satisfying this requirement *simple*. The TCES $\mathcal{T}_A$ associated to A contains for each clock x of A with $comp(x) = (c_1, \cdots, c_m)$ a timer $\theta(x)$ with $(c_1, \cdots, c_m)$ as threshold vector. Then. we associate for each simple constraint and clock x a condition $cond(\xi(x))$ as follows:

$$cond(\xi(x)) = \begin{cases} \text{``}\theta(x) < c_1\text{''} & \text{if } \xi(x) \equiv x < c_1 \text{ or } \xi(x) \equiv x = c_1 \\ \text{``}\theta(x) \geq c_m\text{''} & \text{if } \xi(x) \equiv x > c_m \\ \text{``}c_j \leq \theta(x) < c_{j+1}\text{''} & \text{if } \xi(x) \equiv c_j < x < c_{j+1} \\ \text{``}c_{i-1} \leq \theta(x) < c_i\text{''} & \text{if } \xi(x) \equiv x = c_i \text{ and } i \in \{2, \cdots, m\} \end{cases}$$

Given a simple TA $A = (\mathcal{Q}, q_0, \{x_1, \cdots, x_n\}. \mathcal{E})$ over the alphabet Σ we construct the corresponding TCES $\mathcal{T} = (U, V, Y, Z, D, \Theta)$, such that

- U and Y are singletons and Z is empty. $V = \Sigma$,
- $\Theta = \{\theta_i \mid 1 \leq i \leq n\}$ is the set of timers such that the threshold vector $(c_i^1, \cdots, c_i^{m_i})$ of θ_i is $comp(x_i)$.
- $D = (\Pi_{i=1}^n \mathrm{Int}_i, V^0 \times \Pi_{i=1}^n \mathrm{Ala}_i^0, \mathcal{Q}, \Pi_{i=1}^n \mathrm{Res}_i^0, f, h, q_0)$ is a DCES, where
  - For every $q, q' \in \mathcal{Q}$, $\mathbf{u} \in \Pi_{i=1}^n \mathrm{Int}_i$, and $(a, \mathbf{v}) \in V^0 \times \Pi_{i=1}^n \mathrm{Ala}_i^0$, $q' \in f(q, \mathbf{u}, \mathbf{v})$ iff there exists an edge $(q, \xi, a, R, q') \in \mathcal{E}$ such that $\mathbf{u} = (cond(\xi(x_1)), \cdots, cond(\xi(x_n)))$ and $\mathbf{v} = (v_1, \cdots, v_n)$ with $v_i = \text{``}\theta_i = c_i^j\text{''}$, if $j \in \{1, \cdots, m_i\} \wedge \xi(x_i) \equiv x_i = c_i^j$; and $v_i = 0$, otherwise.
  - $h(q, q'. (a. \mathbf{v})) = (z_1, \cdots, z_n)$ iff there exists $(q. \xi, a. R, q') \in \mathcal{E}$ and $z_i = \text{``}\theta_i := 0\text{''}$, if $x_i \in R$; and 0. otherwise.

Proposition 2 *There exists an effective function that transforms each timed automaton into an equivalent timed c/e system.*

5 Discussion and Further Work

We have presented a comparison between two different models for timed systems. To present this comparison we had to provide a precise semantics for TCESs, which was not presented in original works where this model has been introduced. From a practical point of view, we show a way how analysis tools developed for TAs can be used to analyze TCESs. Besides implementing our transformations we plan to study the modular aspects of TCESs and the typical operations used for combining CESs, namely, the cascade and feedback connections. There is also a need to develop a logical specification language for TCESs similar to those developed for TAs and to study their model checking problem.

References

1. R. Alur, C. Courcoubetis, and D.L. Dill. Model Checking for Real–Time Systems. In *LICS '90*, pages 414–425, 1990.
2. R. Alur and D. Dill. A Theory of Timed Automata. *TCS*, 126:183–235, 1994.
3. A. Bouajjani and Y. Lakhnech. Temporal Logic + Timed Automata: Expressiveness and Decidability. In *LNCS 962*, pages 531–546. Springer–Verlag, 1995.
4. S. Engell, S. Kowalewski, B. Krogh, and J. Preußig. Condition/Event Systems: A Powerful Paradigm for Timed and Untimed Discrete Models of Technical Systems. In *EUROSIM '95*, 1995.
5. T.A. Henzinger, X. Nicollin, J. Sifakis, and S. Yovine. Symbolic Model Checking for Real–Time Systems. *Information and Computation*, 111(2):193–244, 1994.
6. R.S. Sreenivas and B.H. Krogh. On Condition/Event Systems with Discrete State Realizations. In *Discrete Event Dynamic Systems 1*, pages 209–236. Boston, 1991.
7. T. Wilke. *Automaten und Logiken zur Beschreibung zeitabhängiger Systeme*. PhD thesis, Institut für Informatik und Praktische Informatik, Universität Kiel, 1994.

Design Tools for Hybrid Control Systems

M. S. Hajji, J. M. Bass, A. R. Browne and P. J. Fleming

Dept. of Automatic Control and Systems Engineering, University of Sheffield, Mappin Street, Sheffield, S1 3JD, UK.
Tel: +44-114-282-5236, Fax: +44-114-273-1729,
e-mail m.hajji@sheffield.ac.uk

Abstract. Hybrid systems exhibit a variety of discrete and continuous behaviour. Extensions to the Development Framework, an environment for real-time control system design, are described here. These extensions allow the management of discrete events such as mode switching. The Framework provides a highly automated path from a control engineering specification to a distributed system implementation. Simulink, an extension to Matlab, is used to specify control laws, and a statecharts tool, Statemate, is used to specify and model discrete-state components. The design phase of the Development Framework supports the integration of the two sets of specification. The translation of statecharts to dataflow notations is described using an example system which was implemented using the Development Framework. General translation rules are also presented.

1 Introduction

Hybrid systems are responsive control systems which incorporate discrete and continuous components. This paper describes an approach for designing such systems. A transfer function block diagram tool, Simulink [Mathworks92], and a statechart tool, Statemate [i-Logix95] are used jointly for handling hybrid systems. Statemate is used for discrete event specification and modelling, while Simulink is used for real-time control law specification.

The work is an extension to the Development Framework for the design of real-time control software. The Development Framework [Bass94, Browne96] supports three phases in the development of control software: specification, software design and implementation. It automatically converts control engineering specification into parallel code. The software design phase of the Development Framework supports the integration of the discrete and continuous components of hybrid systems.

Section 2 of this paper gives an overview of the Development Framework. Section 3 discusses related work. The design tools are described in section 4, with concluding remarks presented in section 5.

2 Development Framework Overview

The Development Framework, enables a highly automatic translation from a specification to a distributed system implementation, see Fig. 1. The specification is in a popular graphical control engineering notation, typically representing a system with stringent dependability requirements and hard real-time constraints. Purpose-built (proprietary) translation tools convert the specification into a system design represented using a software engineering notation. The design model is used to support documentation, analysis and refinement of the system under development.

The functionality of the system at the design phase is equivalent to the original specification. However, the Framework design phase enables the user to improve the design in order to meet some non-functional requirements, such as timeliness and reliability constraints. In addition, mapping and clustering tools are provided to improve the performance of code execution on parallel platform

The Development Framework provides a formal process model which uses a time Petri net representation of the process that is intended to enable temporal simulation of the system under development [Browne97]

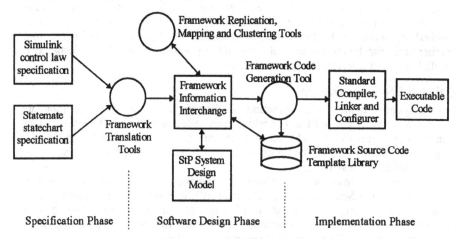

Fig. 1. Development Framework overview

Further, purpose-built translation tools convert the design model into source code that can be compiled into executable code for a network of processors. Goals of the Framework include development time reduction, improved maintainability, and support for design and/or code re-use.

3 Related Work

The Development Framework addresses a similar problem area to the ControlH design environment [Englehart95]. Both approaches involve specification in a graphical application-oriented notation and the use of an intermediate software engineering notation. The ControlH environment uses purpose-built tools throughout, while the Framework integrates commercially available tools like Simulink, Statemate and StP. Benefits of the Framework include standardised notations, an open architecture to support the addition of alternative views on the design model and an underlying formality which permits selected methods of analysis.

Taylor [Taylor96] describes work to implement a subset of hybrid systems modelling language (HSML) in Matlab. At the lowest level HSML components are "pure" continuous or discrete. These elements are assembled into composite components, and then systems. Every component has a body which describes its dynamic behaviour and an interface. The standard Matlab model schema was modified by introducing a new input variable m (mode) and two outputs variables, one to signal a state event and another to permit state reset.

Extended coloured Petri net (ECPN) is proposed in [Yang95] for hybrid system modelling. Dynamic colours, dynamic transitions and dynamic places are introduced to allow ECPN to model both aspects of hybrid systems, continuous variables and discrete events. Alla [Alla95] also proposes "continuous Petri nets" for similar purpose. Currently, Petri nets lack the necessary tools for data manipulation, are cumbersome to analyse and difficult to understand.

Bencze [Bencze95] separates the overall hybrid control system design into three subtasks: 1) design of real-time control loops; 2) synthesis of control logic, or a "meta-controller," for the real-time loop; and 3) design of the translators that allow these two elements to communicate and interact in a stable fashion. This partitioning is utilised within the Development Framework. The third subtask is, however, handled automatically by the Framework without the need for user intervention. Furthermore, the functional control loop and the control logic can be designed in parallel and then integrated in a second phase.

4 Hybrid System Design

Hybrid systems exhibit a combination of discrete and continuous components. Statemate is one example of a tool which uses statecharts [Harel90]. Unlike state-transition diagrams (STD) and finite-state machines (FSM), statecharts support both hierarchical decomposition and concurrency which lead to a higher efficiency when representing discrete state logic.

The Framework automatically converts the specification into a design represented using a CASE tool Software through Picture (StP) [IDE94]. The design phase is the suitable stage where the integration of different sets of specification can be performed. General rules were developed to translate from Statemate representation to StP representation. These rules are summarised in Table 1.

Statemate	Data/Control Flow Diagrams
Activity chart	Control and Data context.
n states S1,...Sn in a level	n processes s1,... sn in a level, Control Specification Bar and STD for states s1,...sn.
m concurrent states	m processes in one level.
Sub-states of state Sj	Sub-processes of process sj, Control Specification Bar and STD for these states.
Condition/Event	Condition/Event in STD and Check Process.
Transition S1→S2	Transition in STD s1→s2 and Action ac_s2.

Table 1 A summary of the Statemate to data/control flow diagrams translation rules

The aim is not to develop a generic translation tool from statecharts to dataflow diagrams (DFD) as much as to help the control engineer to design hybrid control systems. Consequently, the emphasise is more on the integration of the two sets of specification. The statechart implementation in Statemate has a rich set of features, many required only in exceptional circumstances. It is our view that it would be undesirable to encourage widespread use of some of these unusual features. The basic statechart features provide considerable benefits for hybrid system designers.

An inverted pendulum control system is used in this paper as a simple example to illustrate how the Development Framework handles hybrid systems. This example was fully implemented for lab demonstration. The pendulum (Fig. 2) can be in one of four different modes, off, self start, coarse control and fine control with a specific control law applied in each of these modes (more explanation in [Hajji96]).

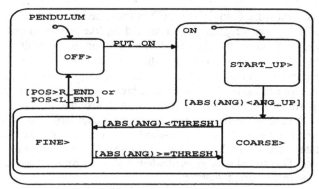

Fig. 2. The inverted pendulum states in Statemate

The hierarchy in Statemate charts is converted into a hierarchy in data/control flow diagrams in StP. For the top level states, ON and OFF in this example, two main processes *on* and *off* are created in StP. Sub-processes *start_up, coarse* and *fine* of the process *on* are created for the substates of the state ON in addition to the corresponding control specification bar and control flows, following Hatley/Pirbhai method [Hatley87]. The control flow *c_on* received from the parent level bears the control signal from the higher level to the specification bar in this level. Control specification bars "decompose" into state transition diagrams (STDs) and tables. The purpose of the control specification CSPEC is to show the sequential or combinatorial machine associated with one or more control flows. An STD is created for the states START_UP, COARSE and FINE with the correspondent transitions. In the parent diagram, an STD is created for the states ON and OFF.

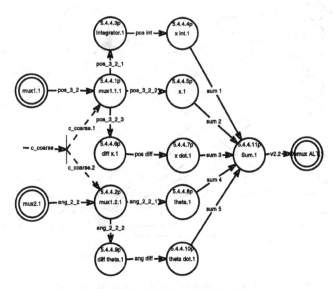

Fig. 3. StP decomposition of the process *coarse*. It concerns the *coarse control law* which runs when the system is in the state COARSE

The translation from Simulink to DFD is explained in more detail in [Bass94, Browne96]. Each process representing a basic state, like *coarse*, is decomposed into dataflow diagrams as a translation of the appropriate controller designed in Simulink. Fig. 3 presents the decomposition of the process *coarse*. The process *integrator.1* (top left of diagram), for example, corresponds to an integrator block in the Simulink representation of coarse-control law. In addition to the normal control loop decomposition, the translation tools create a control bar and control flows from this bar to each process on diagram which receives a data input. An STD containing two states, *process-enabled* and *process-disabled* is created too. The switch between these two states depends on the signal *c_coarse* received by the control bar from the parent process. This signal is originated when switching between the states *start-up*, *coarse* and *fine* in the parent diagram.

The automatic code generator takes the final StP system model and creates all the parallel source code files required to produce an executable version of the inverted pendulum controller.

A gas turbine engine controller which has a significant number of operating modes is the subject of current investigation. The controller provides the required communications for engine control, thrust management, status indication, health monitoring and maintenance data. Figure 4 is the StP representation of a part of the controller. In this diagram, when reverse thrust is selected (i.e. REVFG = true) the controller switches from *reversion* (*reversion* returns the engine to a safe operating state) to *reverse-thrust*. The control bar "learns" about the occurrence of events from the check processes. The output of this particular part of the controller initialises the fuel flow fine trim.

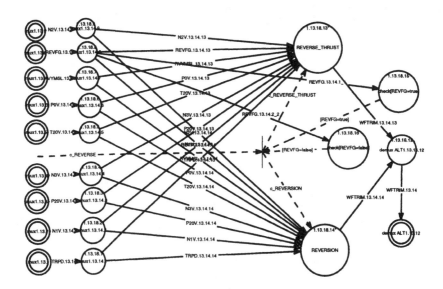

Fig. 4. StP representation of a part of an engine controller.

5 Conclusions

Hybrid control systems are a mixture of real-time control loop and discrete-state logic. An extension to the Development Framework allows the design of such systems. The discrete events and continuous components of the system can be integrated when they are translated to the software engineering domain which supports Yordon methodology with Hatley/Pirbhai real-time extension.

Generalising the approach presented in this paper can be a subject for further work. The discrete event components are not necessarily the top level components of hybrid systems. Complex reactive systems could have different operating modes in one or more subsystems.

Although the tool supports a restricted subset of statechart features, our experience suggests that this subset needs just few other features to be sufficient to represent a significant number of hybrid systems. These extra features are currently under investigation.

Acknowledgements

The authors are grateful for Peter Schroder who implemented the inverted pendulum control laws. They also acknowledge the support of UK EPSRC (under contract number GR/K64310).

References

[Alla95] Alla, H., "Petri networks: a tool adapted to modelling hybrid systems", Revue General de l'Electricite, No. 1, Jan. 1995, pp. 10-18.

[Bass94] Bass, J.M., Browne, A.R., Hajji, M.S., Marriot, D.G., Croll, P.R., and Fleming, P.J., "Automating the Development of Distributed Control Software", IEEE Parallel & Distributed Technology, 2(4), 1994, pp.9-19.

[Bencze95] Bencze, W.J. and Franklin, G.F., "A Separation for Hybrid Control System Design", *IEEE Control Systems*, 15(2), 80-85.

[Browne96] Browne, A.R., Bass, J.M., Hajji, M.S., Croll, P.R., and Fleming, P.J., "A Prototype Development Framework for Real-Time Control Software", Transputer Communications, 3(1), Jan. 1996, pp. 69-77.

[Browne97] Browne, A.R., Bass, J.M., and Fleming, P.J., "A Building-Block Approach to the Temporal Modelling of Control Software", IFAC AARTC '97, Portugal, Apr. 97. To appear.

[Englehart95] Englehart, M., and Jackson, M., "ControlH: An Algorithm Specification Language and Code Generator", IEEE Control Systems, 15(2), April 1995, pp 54-64

[Hajji96] Hajji, M.S., Browne, A.R., Bass, J.M., Schroder, P., Croll, P.R., and Fleming, P.J., "A Prototype Development Framework for Hybrid Control System Design", IFAC 13th World Cong., June 96, Vol O, pp 459-64.

[Harel90] Harel, D. et al., "Statemate: A Working Environment for the Development of Complex Reactive Systems", IEEE Trans. on Software Engineering. vol. 16, no. 4, April 1990, pp 403-414.

[Hatley87] Hatley, D.J. and Pirbhai, I.A., "Strategies for Real-Time System Specification", Dorset House, 1987.

[i-Logix95] i-Logix, (1995). *User Reference Manual Vols. 1 &2, Version 6.0*, i-Logix, Inc.

[IDE94] "Software Through Pictures Structural Environment , Release 5", Interactive Development Environments, San Francisco, U.S. 1994

[Mathworks92] "Simulink User's Guide, The Mathworks Inc., Natick, MA., U.S., 1992.

[Taylor96] Taylor, J.H. and Kebede, D., "Modeling and Simulation of Hybrid Systems in Matlab", IFAC 13th World Congress, June 96, Vol J, pp 275-280

[Yang95] Yang, Y.Y., Linkens, D.A. and Banks, S.P., "Modelling of hybrid systems based on extended coloured Petri nets", Lec. Note in Comp. Sci. 999, Hybrid Systems II, Antsaklis et. al. (eds.), Springer-Verlag, Germany, 1995, PP. 509-28

On-Line, Reflexive Constraint Satisfaction for Hybrid Systems: First Steps

(Invited Presentation)

Michael S. Branicky

Dept. of Electrical Engineering and Applied Physics, Case Western Reserve
University, 10900 Euclid Ave., Glennan 515B, Cleveland, OH 44106-7221 USA.
E-mail: branicky@eeap.cwru.edu.

Abstract. We can achieve guaranteed constraint satisfaction of a hybrid dynamical system (which takes into account the underlying continuous dynamics) in a simple, hierarchical control algorithm. Two layers of functionality, (1) piece-wise viable servo controllers and (2) a "reflex controller," are required for guaranteed constraint satisfaction. The resulting control structure allows higher-levels of "intelligence" or functionality to be added which don't have to worry about guaranteeing constraints. The structure acts as an on-line filter which approves any actions that will maintain constraints but denies any requests that would result in behavior outside that specified.

In this note, we lay down the notation and theory of this hierarchy in a broad, abstract setting. Minimal properties to assure constraint satisfaction are given. The incorporation of such a model with higher planning levels and its associated convergence properties are discussed. The philosophy has been successfully applied to a robot's maintaining collision avoidance while under higher-level control, viz. path planners and teleoperation. The system operates on-line in real-time; in executing collision-free motions, the robot uses its full mechanical bandwidth.

1 Introduction

Intelligent, autonomous systems require knowledge about themselves and their environment and the ability to integrate this knowledge to make decisions about how to act. Ultimately, these decisions may be as complex as those we make on a daily basis. But such abilities were long evolved and built up from robust underlying layers of functionality.

If we are ever to achieve such complex behaviors in systems, then we must build the robust intervening layers first: those which connect higher-level planning and lower-level servo control. One possible level of intelligence that might sit between servo control and higher layers of intelligence is called *reflex control*. Reflex control (or reflexes) obtains its name from the analogy to animal reflexes—and its functionality is almost identical. Reflexes are always there to protect us, though most of the time their influence is minimal.

Consider the act of picking up a pot on the stove. You plan to pick the pot off the burner and place it on a plate. On a lower level, your brain commands your muscles to perform the task. If all is right, you perform the task as planned. However, if the pot is too hot, your reflexes will override the "higher level planning" and command your hand to move quickly away from the danger of getting burned. A low level of intelligence (your reflexes) refuses to carry out the "higher level" plans, saving you some pain. However, they only act when it is necessary to do so. Your hand does not jerk away from a cold pot. They are transparent if no danger is present. They are minimally influential.

Two important points have been brought forward: the reflexes should be transparent unless needed; they should keep the system from violating constraints. The ramifications of these two statements are far-reaching. The reflexes must be very fast to be transparent to most control schemes. They must be able to make decisions about safe or unsafe moves. They must be aware of the system and its environment.

In this paper, we begin to deal with the problem of assuring operation of a hybrid dynamical system within specifications, taking into account the underlying continuous dynamics. The result is a hierarchical control structure which guarantees specification satisfaction with a "reflex controller," under mild conditions imposed on the plant's lower-level "servo" controllers. Essentially, the reflex controller acts like a filter of higher-level commands, passing through requests that maintain constraints and denying any that would result in behavior outside of specifications. Thus, the analogy with animal reflexes: a control layer which is always present but only initiates action when it is necessary to do so for self-preservation. Although animal reflexes are inherently parallel, we use a sequential filter implementation to mimic parallel operation because it guarantees safety even if the reflex controller fails (or takes an inordinate amount of time to make a decision).

For brevity, we refer to behavior outside of specifications as a *specification violation* or, simply, a *violation*. We call sets of violation points *obstacles*. Thus, the reflex controller only acts (denies requests) when it is necessary to do so to prevent violation (imminent because of known constraints or possible because an area has yet to be approved as satisfying specifications).

Setting. Herein, we consider as "systems" *hybrid dynamical systems* as follows [5]:

$$
\begin{aligned}
\dot{x}(t) &= f(x(t), q(t), u(t)), && \text{if}(x, q) \in S \backslash A \\
[x(t^+), q(t^+)] &= G(x(t), q(t)), && \text{if}(x, q) \in A
\end{aligned}
\tag{1}
$$

where $x(t) \in X \subset \mathbf{R}^n$, $q \in Q \simeq \{1, 2, \ldots, N\}$, and $u(t) \in U \subset \mathbf{R}^m$. The *hybrid state* of this system is $s = (x, q)$, taking values in $S = X \times Q$, with $A \subset S$ closed. We may also associate the *output equation* $y = h(s, u)$ taking values in $Y \subset \mathbf{R}^p \times \mathbf{Z}^r$. Thus, starting at $[x_0, i]$, the continuous state trajectory $x(\cdot)$ evolves according to $\dot{x} = f(x, i, u)$. If $(x(\cdot), i)$ hits A at time t_1, then the state becomes $G(x(t_1), i) = [x(t_1^+), j]$, from which the process continues.

Constraint space (C-space) is the multi-dimensional space of generalized coordinates which represent the variables of interest of a system.[1] For example, a variable of interest may be an output that must be maintained within certain (possibly time-varying) bounds. We call a point in this C-space a *configuration*.

For the moment, let us consider the C-space of the plant and its associated controllers at time t to be the set, $\mathcal{C}(t)$. Then, to use C-space for constraint satisfaction, one associates a mapping from C-space to the set $\{0, 1\}$ as follows

$$O : \mathcal{C}(t) \mapsto \{0, 1\}$$

$$O(c(t)) = \begin{cases} 0, \text{ if the system in configuration } c(t) \text{ satisfies constraints,} \\ 1, \text{ if the system in configuration } c(t) \text{ violates constraints.} \end{cases} \quad (2)$$

The Problem. With this map, constraint satisfaction is reduced to point navigation in the set $(O(t))^{-1}(0)$. In particular, if we let $c(t)$ denote the configuration of our system at time t and *free-space* at time t, $F(t)$, denote $O^{-1}(0)$ at time t, then we are interested in

Problem 1 (Constraint-Satisfaction or Obstacle-Avoidance). Control the system so that there are no violations: maintain

$$c(t) \in F(t), \qquad \text{for all } t \geq t_0.$$

The rest of this paper deals with Problem 1. Below, we will refer to any set $S \subset O^{-1}(0)$ as being *violation-free* or *obstacle-free*.

Example 1 (Anti-windup). Consider the system

$$\dot{e} = \xi,$$

where ξ is some error signal we are accumulating through integration. In many situations, one prevents "integrator windup" by limiting the value of e between some bounds L and H: $F(t) \equiv F = \{e \mid L \leq e \leq H\}$.

Example 2 (Ball in a Box). Consider the system

$$\ddot{x}(t) = u; \qquad v(t^+) = -v(t), \text{ if } x \in \{-1, 1\}$$

defined on $X = [-1, 1]$, where $v \equiv \dot{x}$, and $U = [-a_{\text{sat}}, a_{\text{sat}}]$. To specify no (non-zero velocity) collisions, set $O(t) \equiv O = \{-1, 1\}$.

For ease of presentation, we will only deal here with the case of static constraints as in the examples above, i.e., $F(t) \equiv F$ for all t, even though extensions to the time-varying case are possible.[2]

With this "constraint," it may appear that Problem 1 is a null problem—at least in Examples 1 and 2 (just set $\xi = u = 0$). Hence, completely analogous to safety and liveness in computer science [17], we refine to

---

[1] Those familiar with collision avoidance in robotics will notice the connections with configuration space [11, 12]. See below.

[2] For instance, to assure constraint specification, it is only necessary that the C-space "obstacles" arising from new or moving constraints remains outside the current valid bounding setpoint set of the reflex controller (terms defined below).

Problem 2 (Reflexive Constraint Satisfaction or Reflexive Obstacle Avoidance).
Control the system so that we have as much freedom to move in $F(t)$ as possible: any point in $F(t)$ that can be visited without irrecoverably causing violation should be attainable.

Hence, in Problems 1 and 2, we have somewhat formalized the dual nature of reflex control mentioned above: keep the system safe, but deny action only when it is absolutely necessary to do so.

Example 3 (Reflexive Anti-windup). Allow

$$\xi = \begin{cases} 0, & \text{if } \xi_{\text{request}} > 0(\text{resp.} < 0) \text{ and } e = H(\text{resp. } L), \\ \xi_{\text{request}}, & \text{otherwise.} \end{cases}$$

Example 4 (Reflexive Boxed Ball). Let $b(t) = v^2(t)/(2a_{\text{sat}})$ and $d(t) = \text{sgn}(v(t))$.
Allow

$$u = \begin{cases} -d(t)a_{\text{sat}}, & \text{if } |x(t) + d(t)b(t)| = 1, \\ u_{\text{request}}, & \text{otherwise.} \end{cases}$$

The above examples are illustrative, but quite simple. It is not hard to think of cases of a ball moving in multiple dimensions, among multi-dimensional obstacles, in which the global goal of the system can be more closely achieved by switching among different approved subsets of the C-space (cf. Figure 1). This will occur in general: different local domains of attraction will be associated with different controllers; only by stringing them together will more global goals be accomplished with safety. In general, switching may also arise due to computational considerations such as inspection time of C-space regions for obstacles or conservatism in estimating dynamic bounds.

The reader interested in more complicated examples may wait for our description of robotic collision avoidance (simplified from [3, 16]) or consider an example arising in flight control [4].

Organization. In the next section, we present a motivating example: robotic collision avoidance. In Section 3, we detail generic conditions and "algorithms" for on-line, reflexive constraint satisfaction in a hierarchical framework. In Section 4 we analyze the case of robotic collision avoidance in our general framework, providing an example of the theory we have found useful in practice. Conclusions are in Section 5.

2 Motivation: Robotic Collision Avoidance

Real-time obstacle avoidance for high-speed robots has been achieved via the *reflex control* concept [3, 13, 14, 15, 16]. The reflex layer stands between servo control and planning in the "intelligence hierarchy" of the robot. It will transparently execute the commands given by the higher level unless it is dangerous to do so. It will override commands that would result in the manipulator's colliding with itself or objects in its environment.

The reflex control layer can be introduced in either a serial or parallel manner. In a parallel implementation, the reflexes would monitor the state of the robot (positions and velocities) and the state of the world (represented in configuration space) and "step in" whenever it is necessary to specify commands that will ensure continued safe operation. That is, the reflexes are always present but intervene only when it is necessary to prevent imminent danger. It is this type of operation that prompts the analogy with biological reflexes.

The reflex control can be implemented in series with the same result: it can receive requests from higher layers and either deny or approve them based on whether a collision is imminent or not. This is the implementation adopted.

Conceptually, reflex control is consistent with layered, hierarchical control systems [2, 6]. In a hierarchical control structure, the reflex control layer is inserted between the layers of motor control and higher layers, such as path planning. They are merely a limited form of intelligence which prevents the robot from exceeding joint limits and from striking itself or objects in its environment. They are low in the control hierarchy, but that does not limit their effects. Just as motor saturation might prevent a linear control law from producing the torques it requests, so do the reflexes intervene to deny any requests from higher levels which would result in collision. And just as the effect of saturation is transparent to torque requests below the saturation level, so too should the reflexes be transparent when requested paths are not dangerous.

2.1 Analysis Preliminaries

Configuration Space. As noted before, configuration space (*C-space*) is a multi-dimensional space of generalized coordinates which represent the *degrees of freedom* (*d.o.f.*) of a system. The configuration space used in robotics is the joint space of the robot; the generalized coordinates used are the robot's joint parameters [11, 12].

For the moment, let us consider the C-space of the robot to be the set, $\mathcal{C}$, of all poses or configurations of our robot. Then, to use C-space for obstacle avoidance, one associates a mapping from C-space to the set $\{0, 1\}$ as in Equation (2), with the constraint being "the robot intersects no obstacles." Thus, obstacle avoidance is reduced to point navigation in the set $O^{-1}(0)$.

For example, consider a robotic manipulator having n links with q_i ($i = 1, \ldots, n$) representing the generalized coordinate of the ith joint. Since each joint parameter represents a degree of freedom, the configuration space of the manipulator is an n-dimensional space with orthogonal axes measuring displacements of the q_i's. Each point $q = (q_1, \ldots, q_n)$ in the configuration space corresponds to a unique pose or "configuration" of the manipulator. If each $q_i \in [\theta_{i_{\min}}, \theta_{i_{\max}}]$, then the configuration space, $\mathcal{C}$, of the manipulator is given by the following Cartesian product [7]:

$$\mathcal{C} = [\theta_{1_{\min}}, \theta_{1_{\max}}] \times [\theta_{2_{\min}}, \theta_{2_{\max}}] \times \cdots \times [\theta_{n_{\min}}, \theta_{n_{\max}}] \subset \mathbf{R}^n.$$

If the robot in question has no joint limits (or has rotational limits further than

2π apart), then one may want to associate points modulo full rotation and use S^1 instead of a closed interval in that coordinate.

Prism Notation. In a slight abuse of common notation (including that used above), we will use $[a, b]$ to denote the set of real numbers "between" a and b. If $a, b \in \mathbf{R}^1$, then

$$[a, b] = \{x | a \leq x \leq b, \text{ if } a \leq b; b \leq x \leq a, \text{otherwise}\}.$$

If $a, b \in S^1$, then $[a, b] = \{x | \text{ one of } a \leq x \leq b; b \leq x \leq a\}$, where the desired interval has been decided upon by some higher level, e.g., preference of direction or a global planner. Finally, if $a, b \in \mathcal{C}$, then

$$[a, b] = \times_{i=1}^n [a_i, b_i],$$

that is, the hyperprism generated by a and b. For convenience, we often use "prism" instead of "hyperprism."

Dynamics and Control. Let us assume dynamic decoupling of our robot, either through feedback or design. In this case, the dynamics of each subsystem may be described as

$$\dot{x} = v, \qquad \dot{v} = u, \qquad u \in [-a_{\text{sat}}, a_{\text{sat}}]. \tag{3}$$

To be specific, we assume a proportional plus derivative (PD) controller as follows:

$$u = K_p(x_{\text{setpt}} - x) - K_v(v_{\text{setpt}} - v), \tag{4}$$

where x_{setpt} is the desired position and K_p and K_v are chosen for critical- or over-damping. Hence, starting from any $(x_0, 0)$ *with* $v_{\text{setpt}} \equiv 0$, the robot will travel to x_{setpt} without overshoot, i.e., without passing to the "side" of x_{setpt} opposite from x_0. Starting from $(x(t), v(t))$, $v(t) \neq 0$, we denote the "closest" point x_{setpt} which results in no overshoot as the *braking point*, $b(t)$; we denote by $a(t)$ the "closest" point x_{setpt} which results in maximum acceleration (actuator saturation).

2.2 Implementing Reflex Control

Point-to-Point Mode. In point-to-point operation, the reflexes accept position requests from higher layers and output position setpoints to a servo controller. Assume, for well-posedness, that no obstacles are in the prism $[x(t_0), b(t_0)]$. Assume also PD control as in Equation (4) *and with* $v_{\text{setpt}} \equiv 0$, and with K_d and K_v chosen for critical damping, i.e., $K_p = \omega_n^2$, $K_v = 2\omega_n$. Taking into account velocity and actuator saturation, the maximum acceleration setpoint, $a(t)$, and the braking point, $b(t)$, may be explicitly computed. See [15] for details.

The bottom line is, given a request position, x_{request}, we must search the prism $[x(t), x_{\text{request}}]$ for obstacles if we are to approve x_{request} as a setpoint. But searching may take time. It makes no sense, however, to search farther than $a(t)$. Hence, we search may search $[x(t), a(t)]$, set $x_{\text{setpt}}(t_i) = a(t)$, and then search $[a(t_i), a(t)]$ on the next update. If we ever find an obstacle present, then the setpoint we approve is the free one "closest" to x_{request}. Since this point was outside of $b(t)$ by construction, the robot will be able to stop without overshoot—hence without hitting the obstacle—at this point.

Direct Acceleration Control Emulation. Reflex control using direct acceleration control emulation assumes the existence of some acceleration-based controller at a higher level. It assumes a servo controller that accepts position and velocity commands from the reflex module, and issues an acceleration command in terms of an anticipatory setpoint to a simple PD controller as in Equation (4).

The reflex controller invents a virtual setpoint which has the result of converting acceleration requests into acceleration commands to the linearized, decoupled plant. Specifically, acceleration commands will track acceleration requests if the reflex controller chooses

$$x_{\mathrm{setpt}} = x + \frac{K_v}{K_p}v + \frac{u_{\mathrm{request}}}{K_p}; \qquad v_{\mathrm{setpt}} = 0.$$

This transformation of acceleration commands into equivalent instantaneous position commands permits a geometric specification of the reflex algorithm. That is, an acceleration request is granted by approving the corresponding setpoint, and this setpoint is approved if it would not result in a collision under PD control (as described above). See [16] for details.

Reflex Control in Higher Dimensions. Our description of reflex control to this point has been limited to one dimension. In only one dimension, the search for potential collisions in discretized configuration-space consists of examining all C-Space elements in line between the current position and the goal position.

Reflex control in one dimension is generalizable to higher dimensions. If a decoupled plant is assumed, then each second order system may be treated like the one-dimensional system described above. The difference with higher dimensions is that the regions which must be inspected for obstacle avoidance are no longer simple distances, but areas, volumes, or hypervolumes whose vertices are given by $x_i(t)$ to $b_i(t)$ (or $a_i(t)$) in each dimension $i = 1, \ldots, n$.

Another difference with higher dimensions is the effect of *geometric coupling* among the subsystems [14, 15]. This effect is shown in Figure 1. In order to avoid choosing the overly conservative setpoint, some choice must be made as to which axis gets precedence. This choice may be made based on distance to the goal point or some other optimizing criterion. In previous implementations, however, a simple list of joint precedences was used to eliminate such conflicts.

The same figure, assuming a goal point to the upper right of the obstacle, demonstrates the need to switch among different "local" setpoints in order to accomplish global tasks.

3   Generic Conditions for Reflexive Obstacle Avoidance

In this section we detail generic properties or conditions used in the construction and analysis of our hierarchical controller.

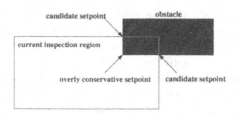

Fig. 1. Geometric Coupling in Obstacle Avoidance

We wish to build a generic hierarchical structure which provides guaranteed constraint satisfaction. For the moment, we separate the control structure of our system into three abstract levels:

1. Lower-level "servo" controller
2. Reflex controller
3. Higher-level processes

The reflex controller can be further broken down into (1) an active reflex controller and (2) a C-space inspector or approver.

These levels interact by passing objects called *bounding sets* (or, simply, *sets*). In particular, a higher-level process may, from time to time, pass a set, X, to the reflex controller.[3] We call these passings *requests*. Likewise, the reflex controller passes sets to the servo controller. We call these passings *setpoints*. Internal to the reflex controller, the C-space inspector passes the active reflex sets which it has certified as violation-free called *approvals*. Below, $S(t)$ and $R(t)$ represent the setpoint and request sets at time t, respectively.

3.1 Preliminaries

We first need a few definitions.

Definition 3 (Control Policy). A *control policy* is simply an achievable prescription for controlling the system from its current state: $u = F(s)$.

This is a general concept. The means to the end is completely arbitrary. The important thing is that the policy be physically achievable.

Example 5 (Braking Policies). In controlling a robot, we are not allowed to "stop on a dime." Typically, one has in mind only a single control policy for braking, such as "full braking in all joints," but it may be advantageous to consider several different braking policies. Another would be to use Equation (4); note we obtain a different policy for each value of K_d, K_v, and x_{request}.

---

[3] In a computer implementation a finite description of the set is passed. For instance, the prism coordinates x and y instead of the actual set $[x, y]$.

Theoretically, one could have an infinite number of control policies and we allow that in our abstract formulation: we talk about the generic control policy $\alpha \in P$, where P is the set of all implementable[4] control policies. Ultimately, though, we have to invoke particular control policies in response to given situations. Further, these invocations must be done in a timely manner if we are to guarantee obstacle avoidance while allowing behavior near the performance limitations of the system. Thus, there will typically be a tradeoff between our ability to deal with the possibilities of different control policies and the conservatism of the system's allowed dynamical behavior.

Control policies are particularly useful if the state space trajectories arising from their invocation are known, computable, or able to be bounded. In any case, however, we define the *sweep set* as follows:

Definition 4 (Sweep Set). The *sweep set corresponding to control policy α from s* is the set

$$B_\alpha(s) = \{y(s(\tau)), \tau \geq t, \text{control policy } \alpha \text{ is invoked at time } t \text{ from state } s(t)\}$$

that is, the curve in C-space resulting from initiating control policy α at s.

Just as we are interested in sets that are violation-free, we are also interested in sets which satisfy a particular dynamic property important in collision avoidance:

Definition 5 (Valid Set). A set X containing $B_\alpha(s)$, $\alpha \in P$, is said to be *valid for control policy α at s* or simply *α-valid at s*. We say that set X is *valid at s* if there exists an $\alpha \in P$ such that X contains $B_\alpha(s)$.

Note: We drop the phrase "at s" when s is clear from context.

Example 6. Consider the planar system

$$(\dot{x}, \dot{y}) = (1 - \alpha, \alpha),$$

with $\alpha(t) \in \{0, 1\}$. Then, lines parallel to the x-axis are 0-valid; lines parallel to the y-axis are 1-valid. The nonnegative x-axis is 0-valid at $(0, 0)$.

Thus, a set is valid if there exists an implementable control policy such that, if it were invoked at time t, the system remains in that set. To get a feeling for the concept, we make several easily proven remarks:

Remark. 1. If A is valid and $B \supset A$, then B is valid. Thus positive invariance, while closely related to validity, is a different concept.

2. If A is α-valid at s and B is α-valid at s, then $A \cap B$ is α-valid at s. Note that $A \cap B$ is non-empty since it must contain $y(s)$.

3. If A is valid at s and B is valid at s, $A \cap B$ need not be valid at s.

Proof. (of part 3) A could be α_1-valid, B α_2-valid, with $\alpha_1 \neq \alpha_2$. Just consider any point for the system in Example 6.

---

[4] Implementablity is an important concept. It refers not only to existence of a policy that could be physically carried out, but also to the fact that we will be able to actually compute and invoke it.

3.2 Servo Controllers

Now, we are ready to describe the functions of our three-level hierarchical control. Abstractly, there are two ways to view the servo controller:

- As receiving from the reflex controller a policy (or "code" for it) which it must enforce.
- As receiving from the reflex controller a set (or some description or "code" for it) which it must keep invariant.

As noted before, herein we think of the servo controller as receiving "commands" in the form of bounding sets called *setpoints* from the reflex controller. Thus, we assume that the servo controller, given a valid setpoint $S(t_i)$, acts in such a manner as to contain the system in that set. More specifically, it has the property:

Condition 1 (Quasi-No-Overshoot, Viable) *1. Given a valid bounding set at time t_i, $S(t_i)$, the system and at least one of its possible sweep volumes will stay within the bounding set until a new bounding set is approved, i.e.,*

for all $t_{i+1} \geq t \geq t_i$, there exists $\alpha \in P$ such that $B_\alpha(s(t)) \subset S(t_i)$,

where t_{i+1} is the time a new bounding set is approved.

2. Given an α-valid bounding set at time t_i, $S(t_i)$, the system and at its α sweep volume will stay within the bounding set until a new bounding set is approved, i.e.,

$$B_\alpha(s(t)) \subset S(t_i), \qquad for\ all\ t_{i+1} \geq t \geq t_i$$

where t_{i+1} is the time a new bounding set is approved.
In such as case, we say that $S(t_i)$ is viable (α-viable).[5]

Note: In the present work, the time t_{i+1} is not known in advance and could be infinite.

Example 7. Suppose we invoke the linear PD law of Equation (4) with K_d and K_v chosen to be critically- or over-damped. Suppose further that from (x_0, v_0) we can stop at $x_d = x_b$ without overshoot. Then, one can show that we may stop without overshoot using any point $x_d = x_b + \mathrm{sgn}(x_b - x_0)\beta^2$.

The idea of quasi-no-overshoot is simply that there exist implementable control inputs such that the bounding set is positively invariant with respect to the property of being valid: once valid, (can be kept) always valid. Since the servo need only invoke the valid policy to insure this, this condition is feasible. We allow the servo more freedom than this, though; we allow it to "use" as much of the setpoint set as it wants, as long as it can guarantee that it does not exit it. The situation is thus akin to the two steps of solving a linear program: the reflex controller proves the constraints are feasible—and perhaps gives a feasible point. The servo may, however, pick any other *feasible* point as a solution.

---

[5] Cf. viability theory [1].

3.3 Reflex Controller

Keeping with our abstraction, we take the view that the reflex controller receives requests for bounding sets from higher-level processes. It then decides whether to approve these requests based on the system's state and the presence of constraints. The reflex controller may then:

1. Deny the request outright because it is not valid, leaving the servo controller with its last bounding set.
2. Approve the requested set.
3. Approve a subset of the requested set.

To achieve constraint satisfaction, then, one merely needs to give to the servo controller valid bounding sets which are violation-free: This leads immediately to guaranteed constraint satisfaction if

Condition 2 (Obstacle Avoidance, α-Obstacle Avoidance) *The reflex controller only approves valid (α-valid) bounding sets which are violation-free.*

Finally, we combine these two properties into what we term the Obstacle-Avoidance-State:

Condition 3 (Obstacle-Avoidance-State) *The system is operating in a state consistent with constraint satisfaction if $S(t_i)$ is viable (α-viable) and violation-free for all $t_{i+1} \geq t \geq t_i$. Here, t_{i+1} is the next time that the setpoint set is updated.*

For well-posedness of the obstacle-avoidance problem, we require that the system is initially in the Obstacle-Avoidance-State (at $i = 0$). Given this, the following is an update procedure which guarantees obstacle avoidance for all time based on keeping track of only one control policy at a time:

Algorithm 1 (α-Update) *Assume we are in the α-Obstacle-Avoidance-State.*

0. If a new request is received, go to Step 1. Otherwise go to Step 0.
1. At time T ($t_{i+1} \geq T \geq t_i$), the reflex controller receives a request set $R(T)$.
   (a) If $R(T)$ is not α-valid (at $s(T)$) we do not update the setpoint and continue to operate in the Obstacle-Avoidance-State. Go to Step 0.
   (b) If it is found to be α-valid, we send $S(T)$ to the servo controller where

$$S(T) = S(t_i) \cap R(T)$$

   Note that this set is α-valid because it is the intersection of two α-valid sets. It is violation-free because $S(t_i)$ is violation-free.[6] We set $t_{i+1} = T$ and increase i in the Obstacle-Avoidance-State. Go to Step 2.
2. Now, we check the set $R(T)$ (actually, we need only search the set $R(T) - S(T)$) for obstacles. Let us assume that the inspection task is completed at time $t_{i+1} \geq T$.

---

[6] Note that we have also assumed that the check for α-validity is instantaneous. We can relax this condition.

(a) If the request set is violation-free, then we approve it: Set

$$S(t_{i+1}) = R(T),$$

increase i, go back to the α-Obstacle-Avoidance-State, and go to Step 0.
(b) If it has obstacles, then we approve instead some violation-free set, $S(t_{i+1})$, such that

$$S(T) \subset S(t_{i+1}) \subset R(T)$$

increase i, and go back to the Obstacle-Avoidance-State. Such a set is guaranteed to exist because $S(T) \subset R(T)$ and $S(T)$ is violation-free. It is α-valid because it is a superset of the α-valid set $S(T)$. Go to Step 0.

Note that we may switch the control policy in the above algorithm from α to α' whenever $S(t_i)$ is α'-valid. The following algorithm allows updates without reference to a specific control policy:

Algorithm 2 (Update) *Assume we are in the Obstacle-Avoidance-State.*

0. If a new request is received, go to Step 1. Otherwise go to Step 0.
1. At time T ($t_{i+1} \geq T \geq t_i$), the reflex controller receives a request set $R(T)$.
   (a) If $S(T) = R(T) \cap S(t_i)$ is not valid we do not update the setpoint and continue to operate in the Obstacle-Avoidance-State. Go to Step 0.
   (b) If it is found to be valid, we send $S(T)$ to the servo controller. It is violation-free because $S(t_i)$ is violation-free. We set $t_{i+1} = T$ and increase i in the Obstacle-Avoidance-State. Go to Step 2.
2. [Same as Step 2 for α-Update, but with "valid" replacing "α-valid."]

Note: The two-step approval process in the algorithms above is used because of the computation time required to check if a region is obstacle-free.

4 Analysis of Robotic Collision Avoidance

The control commands consist of prisms replacing abstract sets for sweep (hereafter, braking), request, and setpoint above. We assume decoupled joints as in Equation (3), a no-overshoot servo controller in each joint as in Equation (4), and a point-to-point reflex implementation.

We would like the robot to go (if possible) to a requested goal point g without colliding with any obstacles. In order to maintain constraint satisfaction (viz., obstacle avoidance in this example), the active reflex generates setpoints. In some cases, the reflex controller may approve the goal outright. In most cases, however, the setpoints it generates will be intermediate (as described below) and we call such points *subgoals*.

4.1 Subgoals and Convergence of Subgoals

Let's discuss in detail one particular algorithm for picking subgoals. Consider the hyperprism, $[q(t), g]$ obtained by the current configuration $q(t)$ and the current goal g. Similarly, consider the braking volume, $[q(t), b(t)]$, and maximum acceleration prism, $[q(t), a(t)]$, where $b(t)$ and $a(t)$ were defined in Section 2.4. Finally, consider the hyperprism $[q(t), o(t, p)]$, where $o(t)$ is the "closest" obstacle point in the "direction" of g obtained by *extending from the prism* $[q(t), p]$. To make things rigorous, we add joint limits as obstacles.

Note: *we only consider the case where we are "heading in the direction of the goal" in all joints*, in terms of the velocities $\dot{q}(t)$. If this were not the case, we could expand the request prism to include the braking volume. We also assume that $[q(t_0), b(t_0)] \subset [q(t_0), g] \cap [q(t_0), o(t_0, b(t_0))] \cap [q(t_0), a(t_0)]$ for well-posedness of the obstacle-avoidance task (that is, we begin in the Obstacle-Avoidance-State).

We update subgoals as follows. Given that we are heading in the direction of the goal in all joints, it makes sense that subgoals are always chosen to be within $[q(t), g]$. With all our notation, the kth subgoal, $s(t_k) = s^k$, chosen at time $t = t_k$ is simply the point in $\mathbf{R}^n$ such that

$$[q(t), s^k] = [q(t), g] \cap [q(t), o(t, s^{k-1})] \cap [q(t), a(t)], \tag{5}$$

where s^{k-1} is the previous subgoal. A different, useful scheme is to pick

$$[q(t), s^k] = [q(t), g] \cap [q(t), o(t, b(t))] \cap [q(t), a(t)], \tag{6}$$

but the convergence analysis requires more assumptions. It is clear that the update law of Equation (5) is such

$$[q(t), b(t)] \subset [q(t), s(t)], \qquad \text{for all } t \geq t_0. \tag{7}$$

4.2 The Discretized Analysis

Let the coordinates of the robot at time t be given by the n-tuple $q(t) = (q_1(t), \ldots, q_n(t))$. Similarly, let the current goal of the robot be given by the n-tuple $g = (g_1, \ldots, g_n)$. Finally, let the sequence of subgoals of the robot be given by the n-tuple $s^k = (s_1^k, \ldots, s_n^k)$. The distance from subgoal to goal may be computed with any suitable norm on $\mathbf{R}^n$, e.g., $\|s^k - g\|_1$, $\|s^k - g\|_2$, $\|s^k - g\|_\infty$.

Our above algorithm is such that $s_i^{k+1} \in [s_i^k, g]$. Thus, our algorithm for picking subgoals is such that $|s_i^{k+1} - g_i| \leq |s_i^k - g_i|$; it also has the property that $s_i^{k+1} - g_i$ and $s_i^k - g_i$ are both either non-negative or non-positive. Note that this immediately implies $\|s^{k+1} - g\| \leq \|s^k - g\|$, for the example norms above. Indeed, it holds for any l_p- or weighted l_p-norm.[7] Finally, let us consider the sequence of subgoals $\{s^j\}$, and its component sequences, $\{s_i^j\}$, $i = 1, \ldots, n$. Each component sequence is monotonically converging to g_i, a finite number. Elementary analysis [18, Thm 3.14, p. 55] gives $\{s_i^j\}$ converges to some s_i, $i = 1, \ldots, n$. This componentwise convergence implies that $\{s^j\}$ converges to $s = (s_1, \ldots, s_n)$, the convergence holding for any $\mathbf{R}^n$ norm [19, Fact 3.1(56), p. 58].

4.3 The Continuous Analysis

This is immediate. The update rules and PD control law ensure that we are viable for each subgoal. The PD control law further ensures convergence towards each subgoal.

5 Conclusions and Future Work

We have presented a means to achieve guaranteed constraint satisfaction of a hybrid dynamical system (which takes into account the underlying continuous dynamics) in a simple, hierarchical control algorithm. Two layers of functionality, (1) piece-wise viable servo controllers and (2) a "reflex controller," are required for guaranteed constraint satisfaction.

The philosophy has been successfully applied to a robot's maintaining collision avoidance while under higher-level control, viz. a variety of control algorithms, path planners, and teleoperation. [16, §7]. In [9], a means to string goals together in order to accomplish global planning (with local obstacle avoidance guaranteed by the reflex and servo layers) is presented. Basically, graph search on "dynamically visible" points is used.

In future work, we will examine different ways to combine constituent dynamical systems (control policies) together in order to achieve higher-level goals. Finally, the reflex concept presented herein should be just as effective in an on-line environment using bounding polytopes (given by linear inequalities) or ellipses. In these cases, the speed and efficacy of linear programming and linear matrix inequalities (LMIs) should prove advantageous.

---

[7] Note that it does not hold for all $\mathbf{R}^n$ norms: consider the counterexample $x = [1,1]^T$, $y = [1,-2]^T$ and the basis vectors $b_1 = [1,-1]^T$, $b_2 = [-1,2]^T$. In this basis, $x = 3b_1 + 2b_2$, $y = -b_2$, so that

$$\|x\|_{\mathcal{B},\max} = 3 > 1 = \|y\|_{\mathcal{B},\max},$$

where $\| \cdot \|_{\mathcal{B},\max}$ is the $\| \cdot \|_\infty$ norm applied to the coefficients in the new basis. See [8, pp. 77–78] for the validity of $\| \cdot \|_{\mathcal{B},\max}$ as a norm. It does not even hold when x_i and y_i are constrained to both be non-negative (or non-positive) for each i: consider the counterexample $x = [1,1]^T$, $y = [1,2]^T$ and the basis vectors $b_1 = [-1,3]^T$, $b_2 = [1,2]^T$. In this basis, $x = b_1 + 2b_2$, $y = b_2$, so that

$$\|x\|_{\mathcal{B},\max} = 2 > 1 = \|y\|_{\mathcal{B},\max}.$$

References

1. Jean-Pierre Aubin. *Viability Theory*. Birkhauser, Boston, 1991.

2. Anthony J. Barbera, M. L. Fitzgerald, and J. S. Albus. Concepts for a real-time sensory-interactive control system architecture. In *Proc. Fourteenth Southeastern Symp. on System Theory*, pp. 121–126, April 1982.

3. Michael S. Branicky. Efficient configuration space transforms for real-time robotic reflexes. Master's thesis, Case Western Reserve University, Dept. of Electrical Engineering and Applied Physics, January 1990.

4. Michael S. Branicky. Analyzing continuous switching systems: theory and examples. In *Proc. American Control Conf.*, pp. 3110–3114, Baltimore, June 1994.

5. Michael S. Branicky. *Studies in Hybrid Systems: Modeling, Analysis, and Control*. PhD thesis, Massachusetts Institute of Technology, Dept. of Electrical Engineering and Computer Science, June 1995.

6. Rodney A. Brooks. A robust layered control system for a mobile robot. *IEEE Journal of Robotics and Automation*, RA-2(1):14–23, March 1986.

7. Yao-Chon Chen and Mathukumalli Vidyasagar. Optimal trajectory planning for planar n-link revolute manipulators in the presence of obstacles. In *Proc. IEEE Intl. Conf. on Robotics and Automation*, pp. 202–208, 1988.

8. Morris W. Hirsch and Stephen Smale. *Differential Equations, Dynamical Systems, and Linear Algebra*. Academic Press, Inc., San Diego, CA, 1974.

9. Vinay K. Krishnaswamy and Wyatt S. Newman. On-line motion planning using critical point graphs in two-dimensional configuration space. In *Proc. IEEE Intl. Conf. on Robotics and Automation*, pp. 2334–2340, Nice, France, May, 1992.

10. Jean-Claude Latombe. *Robot Motion Planning*. Kluwer Academic Publishers, Boston, 1991.

11. Tomás Lozano-Pérez. Automatic planning of manipulator transfer movements. *IEEE Trans. Systems, Man, and Cybernetics*, SMC-11(10):681–698, October 1981.

12. Tomás Lozano-Pérez. Spatial planning: A configuration space approach. *IEEE Trans. Computers*, C-32(2):108–120, February 1983.

13. Wyatt S. Newman. High speed robot control and obstacle avoidance using dynamic potential functions. In *Proc. IEEE Intl. Conf. on Robotics and Automation*, pp. 14–24, March 1987.

14. Wyatt S. Newman. *High-Speed Robot Control in Complex Environments*. PhD thesis, Massachusetts Institute of Technology, Dept. of Mechanical Engineering, October 1987.

15. Wyatt S. Newman. Automatic obstacle avoidance at high speeds via reflex control. In *Proc. IEEE Intl. Conf. on Robotics and Automation*, pp. 1104–1109, May 1989.

16. Wyatt S. Newman and Michael S. Branicky. Experiments in reflex control for industrial manipulators. In *Proc. IEEE Intl. Conf. on Robotics and Automation*, pp. 266–271, May 1990.

17. Amir Pnueli. The temporal logic of programs. In *Proc. 18th IEEE Foundations of Computer Science*, Providence, RI, pp. 46–57, October 31–November 2, 1977.

18. Walter Rudin. *Principles of Mathematical Analysis*. McGraw-Hill, New York, third edition, 1976.

19. M. Vidyasagar. *Nonlinear Systems Analysis*. Prentice-Hall, Englewood Cliffs, 1978.

Hybrid Control Issues in Air Traffic Management Systems (Invited Presentation)*

Shankar Sastry

Department of Electrical Engineering and Computer Sciences
University of California
Berkeley, U.S.A.
sastry@eecs.berkeley.edu

Abstract. The modeling and control of multi-agent, scarce resource systems requires both discrete protocols which coordinate the interplay among agents, and continuous control laws which regulate each individual agent. In a new research project at Berkeley, we are studying such systems in the context of Air Traffic Management Systems (ATMS). In ATMS, a large number of aircraft are required to converge from many directions to a few well defined airspace routes which lead to the airport runways.

The need for a new ATMS arises from the overcrowding of large urban airports and the need to more efficiently land and take off larger numbers of aircraft, without building new runways. In the current air traffic system, aircraft are routed and sequenced manually by a centralized, ground-based controller and the resulting system is fault intolerant and plagued by delays. Technological advances that make a more advanced air traffic management system a reality include the availability of relatively inexpensive and fast real time computers both on board the aircraft and in the control tower. We propose an architecture for an automated, decentralized ATMS, in which much of the control functionality exists in the flight vehicle management system on board each aircraft rather than at the ground-based air traffic control. The aircraft are therefore autonomous agents competing for runway space and time.

The complexity of this system suggests the need for a hierarchical control structure in which the high level discrete protocols control the relative positioning of the aircraft and the spacing between them, and the low level continuous control laws guide the aircraft along the given route. We discuss in this talk control strategies for both the discrete and continuous layers of the hierarchy. A game theoretic approach is used to generate a set of discrete protocols for the coordination among aircraft, as well as for control of the flight vehicle management systems of individual aircraft. The interface controllers between controllers built using different models is built using notions of "consistent abstractions" of hybrid control systems. Proofs of the safety and some measures of liveness and fairness of the control schemes will be discussed.

---

* Joint work with Claire Tomlin, George Pappas, John Lygeros, and Datta Godbole.

Multiobjective Hybrid Controller Synthesis [*]

John Lygeros, Claire Tomlin and Shankar Sastry

Intelligent Machines and Robotics Laboratory
University of California, Berkeley
Berkeley, CA 94720
lygeros, clairet, sastry@eecs.berkeley.edu

Abstract. The problem of systematically synthesizing hybrid controllers that satisfy multiple requirements is considered. We present a technique, based on the principles of optimal control, for determining the class of least restrictive controllers that satisfy the most important requirement (which we refer to as safety). The system performance with respect to the lower priority requirement (which we refer to as efficiency) can then be optimized within this class. We motivate our approach by three examples, one purely discrete (the problem of reachability in finite automata) one hybrid (the steam boiler problem) and one primarily continuous (a flight vehicle management system).

1 Introduction

In this paper we concentrate on the problem of controlling hybrid systems, that is "steering" them using continuous and discrete inputs in an attempt to ensure that the system behavior satisfies certain requirements. For most real systems multiple requirements are imposed on the design. For example, for purely discrete systems the requirements usually considered are those of safety (encoded by requirements over the finite runs of the system) and liveness or fairness (encoded by requirements over the infinite runs), while for conventional control problems the requirements considered are usually safety (encoded by stability or constraints on the system trajectories) and efficiency (the requirement for small inputs or bounds on the speed of convergence for example). In such a multi-objective setting some of the requirements are usually assumed to be more important than others, either explicitly or implicitly. This priority is important from the point of view of controller synthesis, as one would like to ensure that high priority specifications are not violated in favor of low priority ones.

We present a methodology for designing hybrid controllers for hybrid systems in such a multi-objective setting. For simplicity we restrict our attention to two performance criteria and will use *safety* to refer to the high priority criterion and *efficiency* to refer to the low priority one. Using optimal control tools we attempt to classify the controllers that can be used to guarantee safety. Efficiency

---

[*] Research supported by the Army Research Office under grant DAAH 04-95-1-0588, the PATH program, Institute of Transportation Studies, University of California, Berkeley, under MOU-238, and by NASA under grant NAG 2-1039.

can then be optimized within this class of *least restrictive safe controls*. The resulting controller will typically be hybrid (even if the plant dynamics are purely continuous) as it involves switching between the safe and efficient controllers.

Our analysis is based on the hybrid system model introduced in [1], which is outlined in Section 2. The theoretical framework (presented in Section 3) is motivated by three examples. The first is purely discrete and involves the control of finite automata. The second is the well known steam boiler benchmark problem [2]. This is a hybrid problem in that a continuous process (the level of water in the boiler) is to be controlled using discrete controls (pumps being switched on and off). Finally, the third example is continuous and is motivated by the design of a flight vehicle management system.

2 Hybrid System Modeling

The basic entity of our models will be the *hybrid dynamical system* or *hybrid automaton* (the terms will be used interchangeably). Hybrid automata are convenient abstractions of systems with phased operation and they appear extensively in the literature [3, 4, 5, 6]. The model we consider will be similar to models used primarily in computer science; we take an input/output approach, along the lines of the reactive module paradigm [7]. For an overview of hybrid models from the dynamical systems point of view see [8].

We consider a finite collection of variables of two distinct kinds, discrete and continuous. A variable is called *discrete* if it takes values in a countable set and it is called *continuous* otherwise. We will assume no special algebraic structure for the values of the discrete variables. The only operations we will allow are assigning a value to a variable and checking whether the value of a variable and a member of the value set (or the values of two variables that take values in the same set) are equal. We assume that continuous variables take values in subsets of $\mathbb{R}^n$ for some value of n. The variables in our model will be split into three classes: *inputs*, *outputs* and *states*. We will denote the input space (set where the input variables take values) by $U = U_D \times U_C$, the output space by $Y = Y_D \times Y_C$ and the state space by $X = X_D \times X_C$. The subscripts D and C indicate whether the variable is discrete or continuous. To avoid unnecessary subscripts we denote an element of U by u, an element of Y by y and an element of X by (q, x). To simplify the notation we will omit X_D and q when there is only one discrete state and X_C and x when there are no continuous states.

Our model evolves in continuous time, so we will assume a set of times of interest of the form $T = [t_i, t_f] \subset \mathbb{R}$. The variables will evolve either continuously as a function of time or in instantaneous jumps. Therefore the evolution of the system will be over sets of the form:

$$\mathcal{T} = \{[\tau_0', \tau_1][\tau_1', \tau_2], \ldots [\tau_{n-1}', \tau_n]\} \tag{1}$$

with $\tau_i \in T$ for all i, $\tau_0' = t_i, \tau_n = t_f$ and $\tau_i = \tau_i' \le \tau_{i+1}$ for all $i = 1, 2, \ldots, n-1$. The implication is that τ_i are the times at which discrete jumps of the state or input occur. We will use τ to denote an element of $\mathcal{T}$.

Definition 1 *A* **hybrid dynamical system**, *H, is a collection* (X, U, Y, I, f, E, h), *with* $X = X_D \times X_C$, $U = U_D \times U_C$, $Y = Y_D \times Y_C$, $I \subset X$, $f : X \times U \to TX_C$, $E \subset X \times U \times X$ *and* $h : X \times U \longrightarrow Y$. X_C, U_C, Y_C *are respectively open subsets of* $\mathbb{R}^n, \mathbb{R}^m, \mathbb{R}^p$, *for some finite values of* n, m, p *and* X_D, U_D, Y_D *are countable sets.*

Here TX_C represents the tangent space of the space X_C. We assume that f is time invariant[2] and satisfies the standard existence-uniqueness assumptions.

Definition 2 *A* **run** *of the hybrid dynamical system* H *over an interval* $T = [t_i, t_f]$ *consists of a collection* (τ, q, x, y, u) *with* $\tau \in T$, $q : \tau \to X_D$, $x : \tau \to X_C$, $y : \tau \to Y$ *and* $u : \tau \to U$ *which satisfies the following properties:*

1. **Initial Condition**: $(q(\tau_0'), x(\tau_0')) \in I$.
2. **Discrete Evolution**: $(q(\tau_i), x(\tau_i), u(\tau_i), q(\tau_i'), x(\tau_i')) \in E$, *for all* i.
3. **Continuous Evolution**: *for all* i *with* $\tau_i' < \tau_{i+1}$ *and for all* $t \in [\tau_i', \tau_{i+1}]$:

$$\dot{x}(t) = f(q(t), x(t), u(t)), \quad q(t) = q(\tau_i'), \quad (q(t), x(t), u(t), q(t), x(t)) \in E$$

4. **Output Evolution**: *for all* $t \in \tau$, $y(t) = h(q(t), x(t), u(t))$.

It can be shown [1] that the definitions introduced here are rich enough to model continuous dynamical systems, finite state systems, rectangular automata, autonomous jumps, controlled jumps, etc. Note that the set E summarizes the information contained in the invariants, the transition guards and the transition reset relations that appear in the hybrid models of [3, 4, 5].

A number of operations can be defined on hybrid dynamical systems [1]. Here we restrict our attention to just one, called interconnection, which allows us to form new hybrid systems out of collections of existing ones. Let $\{H_i\}_{i=1}^N$ be a collection of hybrid automata, $H_i = \{X_i, U_i, Y_i, I_i, f_i, E_i, h_i\}$. We can write the inputs and outputs in vector form as $u_i = [u_{i,1} \quad \cdots \quad u_{i,m_i}]^T \in U_i$ and $y_i = [y_{i,1} \quad \cdots \quad y_{i,p_i}]^T \in Y_i$. Let:

$$\hat{U} = \{(1,1), (1,2), \ldots, (1, m_1), (2,1), \ldots, (2, m_2), \ldots, (N,1), \ldots, (N, m_N)\}$$
$$\hat{Y} = \{(1,1), (1,2), \ldots, (1, p_1), (2,1), \ldots, (2, p_2), \ldots, (N,1), \ldots, (N, p_N)\}$$

Definition 3 *An* **interconnection**, $\mathcal{I}$, *of a collection of hybrid automata is a partial map* $\mathcal{I} : \hat{U} \longrightarrow \hat{Y}$.

An interconnection of hybrid automata can be thought of as a pairing $(u_{i,j}, y_{k,l})$ of inputs and outputs. An interconnection is only a partial map (some inputs may be left free), need not be surjective (some outputs may be left free) and need not be injective (an output may be paired with more than one input). Let $Pre(\mathcal{I})$ be the subset of $\hat{U}$ for which the partial map $\mathcal{I}$ is defined and let Π_α denote the projection of a vector valued quantity to the element with index α.

---

[2] With some additional notation the same definitions can be given in terms of the flow of the vector field. The advantage would be that the definitions would directly extend to time varying vector fields, discrete time systems, etc.

Definition 4 *Given a collection of hybrid automata $\{H_i\}_1^N$ and an interconnection $\mathcal{I}$, the symbolic operation* **substitution**, *denoted by $\rightsquigarrow$, assigns to each input, $u_{i,j}$, a map on $X_1 \times \ldots \times X_N \times U_1 \times \ldots \times U_N$, according to:*

$$u_{i,j} \rightsquigarrow \begin{cases} u_{i,j} & \text{if } (i,j) \notin Pre(\mathcal{I}) \\ h_{\mathcal{I}(i,j)} : X_{\Pi_1(\mathcal{I}(i,j))} \times U_{\Pi_1(\mathcal{I}(i,j))} \to Y_{\Pi_1(\mathcal{I}(i,j))} & \text{if } (i,j) \in Pre(\mathcal{I}) \end{cases}$$

If for all $(i,j) \in Pre(\mathcal{I})$, $Y_{\mathcal{I}(i,j)} \subset U_{i,j}$, operation $\rightsquigarrow$ can be repeatedly applied to the right hand side by appropriate map compositions. The construction terminates for each $u_{i,j}$ if the right hand side either contains $u_{i,j}$ itself or contains only $u_{k,l} \notin Pre(\mathcal{I})$. The resulting map will be denoted by $(u_{i,j} \rightsquigarrow^*)$.

Because there are a finite number of inputs, the construction of $(u_{i,j} \rightsquigarrow^*)$ terminates in a finite number of steps. To ensure that an interconnection is well defined as an operation between hybrid automata we impose the following technical conditions:

Definition 5 *An interconnection, $\mathcal{I}$, of a collection of hybrid dynamical systems, $\{H_i\}_{i=1}^N$, is* **well posed** *if for all $(i,j) \in Pre(\mathcal{I})$, $Y_{\mathcal{I}(i,j)} \subset U_{i,j}$ and there are no algebraic loops, i.e. for all $(i,j) \in Pre(\mathcal{I})$ the map $(u_{i,j} \rightsquigarrow^*)$ does not involve $u_{i,j}$.*

Fact 1 *Every well posed interconnection, $\mathcal{I}$, of a collection of hybrid dynamical systems, $\{H_i\}_{i=1}^N$, defines a new hybrid dynamical system.*

3 Multiobjective Controller Design

We assume that the plant is modeled by a hybrid automaton of the form described in Section 2. We further divide the inputs into two classes, control inputs denoted by u and disturbances denoted by d. The input space is accordingly split into two subspaces, $(u, d) \in U \times D$. The interpretation is that the designer can exercise control over the inputs u but not over the disturbances. Let PC denote the space of piecewise continuous and PC^1 the space of piecewise differentiable functions of the reals and define the set of acceptable inputs by $\mathcal{U} = \{u \in PC | u(t) \in U \ \forall t\}$ and the set of acceptable disturbances by $\mathcal{D} = \{d \in PC | d(t) \in D \ \forall t\}$.

The controller design should be such that the desired performance is achieved despite the actions of the disturbances. We are interested in a situation where more than one requirement is imposed on the system performance. Here for simplicity, we restrict our attention to the case of two requirements, which we refer to as *safety* and *efficiency*. We assume that these requirements can be encoded by a pair of cost functions, J_1 and J_2 respectively, on the runs of the hybrid automaton with $J_i : PC \times PC^1 \times \mathcal{U} \times \mathcal{D} \to \mathbb{R}$. The cost functions map a run of the automaton $(q(), x(), u(), d())$ to a real number. Here we restrict our attention to the case where each pair of inputs (u, d) generates a unique state trajectory for a given initial condition (q^0, x^0). We informally refer to hybrid automata that possess this property as *deterministic hybrid automata*. In this case the cost function can be thought of as a map:

$$J_i : I \times \mathcal{U} \times \mathcal{D} \longrightarrow \mathbb{R} \tag{2}$$

To distinguish acceptable from unacceptable runs we can impose thresholds, C_1 and C_2, on the final costs. A run is acceptable if $J_i(q(), x(), u(), d()) \leq C_i$ for $i = 1, 2$. We also assume that the performance criteria come with an implicit ranking, safety being more important than efficiency.

In order to guarantee that the performance specifications are met despite the action of the disturbances we cast the design problem as a zero sum dynamic game. The two players in the game are the control u and the disturbance d and they compete over the cost functions J_1 and J_2. We seek to determine the best possible control action and the worst possible disturbance. If the performance specifications are met for this pair, then they can also be met for any other choice of the disturbance.

As higher priority is given to safety, the game for J_1 is solved first. If we assume that the game accepts a saddle solution, i.e. there exist input and disturbance trajectories, u_1^* and d_1^* such that:

$$J_1^*(q^0, x^0) = \max_{d \in \mathcal{D}} \min_{u \in \mathcal{U}} J_1(q^0, x^0, u, d) = \min_{u \in \mathcal{U}} \max_{d \in \mathcal{D}} J_1(q^0, x^0, u, d) = J_1(q^0, x^0, u_1^*, d_1^*)$$

then the set $V_1 = \{(q, x) \in X | J_1^*(q, x) \leq C_1\}$ contains all states for which there exists a control such that the objective on J_1 is satisfied for any allowable disturbance. If u_1^* is used as a control law it will guarantee that J_1 is minimized for the worst possible disturbance and, moreover, if the initial state is in V_1 it will also guarantee that the safety requirement is satisfied.

u_1^* however does not take into account the requirements on J_2. To include efficiency in the design let $\mathcal{U}_1(q^0, x^0) = \{u \in \mathcal{U} | J_1(q^0, x^0, u, d_1^*) \leq C_1\}$. Clearly:

$$\mathcal{U}_1(q^0, x^0) \begin{cases} = \emptyset & \text{for } (q^0, x^0) \notin V_1 \\ \neq \emptyset & \text{for } (q^0, x^0) \in V_1, \text{ as } u_1^* \in \mathcal{U}_1(q^0, x^0) \end{cases}$$

$\mathcal{U}_1$ can be thought of as a feedback map $\mathcal{U}_1 : X \to 2^U$, that maps to each state the subset of admissible controls which guarantees that the requirement on J_1 is satisfied; in other words, the least restrictive class of safe controls. Within this class we would now like to select the control that minimizes the cost function J_2. We again pose the problem as a zero sum dynamic game. Assume that a saddle solution (u_2^*, d_2^*) exists and let $J_2^*(q^0, x^0)$ be the corresponding cost. Then the set $V_2 = \{(q, x) \in X | J_2^*(q, x) \leq C_2\}$ contains the initial conditions for which there exists a control such that for any allowable disturbance the requirements on both J_1 and J_2 are satisfied. As the minimax problem can only be posed when $\mathcal{U}_1(q^0, x^0) \neq \emptyset$ we assume that $V_2 \subset V_1$. The control law u_2^* and the set V_2 are such that for all $(q^0, x^0) \in I \cap V_2$, for all $d \in \mathcal{D}$ and for $i = 1, 2$, $J_i(x^0, u_2^*, d) \leq C_i$.

Note that as $V_2 \subset V_1$ there may still be states for which the requirement for safety can be satisfied whereas that for efficiency can not. The controller can be extended to these states using the simple switching scheme:

$$u^*(q,x) = \begin{cases} u_2^*(q,x) & \text{if } (q,x) \in V_2 \\ u_1^*(q,x) & \text{if } (q,x) \in X \setminus V_2 \end{cases} \tag{3}$$

This makes the operation of the controller hybrid, even in the case where the plant is purely continuous. Such an extension may be particularly useful when one is trying to design a fault tolerant controller. The occurrence of a fault significantly alters the dynamics and may lead to severe shrinking of the set V_2. In cases like these one would like to resort to a controller that guarantees safety, even if the requirements for efficiency are violated.

Two special cases of the above algorithm deserve explicit mention. The first is the case in which there is no disturbance. The algorithm then calls for the solution to a pair of optimal control problems (rather than games). The optimal solution for J_1 will produce a set of states and classify the least restrictive set of controllers for which the safety requirement can be satisfied. The optimal control problem for J_2 will then attempt to determine the best possible control in terms of efficiency within this class. Application of this special case will be demonstrated in Section 6 on the flight vehicle management system example. The second special case is one where there is no control. This is for example the case when a controller has already been designed and we are asked to verify its operation or determine the sets of initial conditions for which the specifications are satisfied. The verification problem also reduces to a pair of optimal control problems. For further discussion of this special case the reader is referred to [9].

4 Reachability in Finite Automata

Consider a standard, deterministic finite automaton $G = (Q, \Sigma, \delta, Q_0)$ where Q is a finite set of states, Σ a finite set of events, $\delta : Q \times \Sigma \to Q$ a transition relation and $Q_0 \subset Q$ a set of initial states. Let $L(G)$ denote the string of events (language) generated/accepted by G. Following [10] we assume that the set of events is partitioned into two disjoint subsets, $\Sigma = \Sigma_u \cup \Sigma_c$, where the events in Σ_c are controllable (in the sense that they can be disabled) while the events in Σ_u are uncontrollable. In this setting problems of safety are usually cast as questions of reachability: can the designer ensure that the automaton state will stay away from a "bad" set of states $Q_B \subset Q$. Efficiency typically corresponds to questions of fairness or liveness. The distinction made is that safety questions can be decided by reasoning over strings of finite length in $L(G)$ while questions of fairness require reasoning over infinite strings. For our example we will only consider how reachability questions can be addressed using the techniques of Section 3.

We first cast the finite automaton G into the modeling formalism of Section 2. In the set up [10], uncontrollable events are given "priority" over controllable ones, in the sense that they can always take place independent of the action of the controller. To capture this effect (and motivated by a discussion in [5]) we assume that the evolution of the system takes place in rounds where a controllable event is followed by an uncontrollable one. To ensure that the resulting automaton will not block and that the priority of d over u is preserved we add two new states, q_G and q_B, and a new event, ϵ, and (re)define $X = Q \cup \{q_G, q_B\}$, $Q_B = Q_B \cup \{q_B\}$, $I = I \cup \{q_G\}$, $U = \Sigma_c \cup \{\epsilon\}$, $D = \Sigma_u \cup \{\epsilon\}$ and $Y = Q$. We then form a complete transition relation by defining:

$$E = \{(q_1, (d, u), q_2) \in X \times (D \times U) \times X |$$

$$q_1 \in Q \Rightarrow \begin{cases} q_2 = \delta(q_1, d) & \text{if } d \neq \epsilon, u = \epsilon, \delta(q_1, d)! \\ q_2 = q_G & \text{if } d \neq \epsilon, u = \epsilon, \delta(q_1, d) \not! \\ q_2 = \delta(q_1, u) & \text{if } d = \epsilon, u \neq \epsilon, \delta(q_1, u)! \\ q_2 = q_B & \text{if } d = \epsilon, u \neq \epsilon, \delta(q_1, u) \not! \\ q_2 = q_G & \text{if } d \neq \epsilon, u \neq \epsilon, \delta(q_1, d) \not! \\ q_2 = q_B & \text{if } d \neq \epsilon, u \neq \epsilon, \delta(q_1, d)! \\ q_2 = q_1 & \text{if } d = \epsilon, u = \epsilon \end{cases}$$

$$q_1 = q_B \Rightarrow q_2 = q_B$$
$$q_1 = q_G \Rightarrow q_2 = q_G\}$$

Here $\delta(q, e)!$ is used to denote that the map δ is defined for the pair $(q, e) \in Q \times (\Sigma_c \cup \Sigma_u)$ and $\delta(q, e) \not!$ that it is not.

To cast the problem in the setting of Section 3 consider a discrete metric, m, on Q, defined by $m(q_1, q_2) = 0$ if $q_1 = q_2$ and $m(q_1, q_2) = 1$ if $q_1 \neq q_2$. It is easy to check that m satisfies the axioms of a metric. The metric can be extended to subsets of Q in the usual way (i.e. $m(Q_1, Q_2) = 1$ if $Q_1 \cap Q_2 \neq \emptyset$ and $m(Q_1, Q_2) = 0$ otherwise). Let $d = \{d_1, d_2, \ldots\} \in D^*$ denote a sequence in D and $u = \{u_1, u_2, \ldots\} \in U^*$ denote a sequence in U and define their interleaving as $(d, u) = \{(d_1, u_1), (d_2, u_2), \ldots\} \in (D \times U)^*$. As G is assumed to be deterministic, the transition structure defines a unique state trajectory $x = \{q_0, q_1, \ldots\} \in X^*$ for every $q_0 \in I$ and every $(d, u) \in (D \times U)^*$. The defining relation is $(q_i, (d_{i+1}, u_{i+1}), q_{i+1}) \in E$. The metric can be used to assign a cost to this run by:

$$J(q_0, (d, u)) = -\min_{q \in x} m(q, Q_B)$$

The reachability problem can now be thought of as a game between u and d over the cost function J. Consider "feedback" maps $\hat{D} : X \to 2^D$ and $\hat{U} : X \to 2^U$. The following algorithm produces the least restrictive class of safe controls:

Step 0: Set $i = 1$. Define $Q'_B = Q_B$, $\hat{D}(q) = \{\epsilon\}$, $\hat{U}(q) = U$ for all $q \in Q'_B$.

Step i: Define $\text{NewQ}_B = \{q \in Q \setminus Q'_B | \exists d_i \in D, q' \in Q'_B \ni (q, (d_i, \epsilon), q') \in E\}$. If $\text{NewQ}_B \neq \emptyset$ increment i and define for all $q \in \text{NewQ}_B$ $\hat{D}(q) = \{d_i \in D| \exists q' \in Q'_B \ni (q, (d_i, \epsilon), q') \in E\}$ and $\hat{U}(q) = U$. Redefine $Q'_B = Q'_B \cup \text{NewQ}_B$ and return to step i. If $\text{NewQ}_B = \emptyset$, then for all $q \in X \setminus Q'_B$ define $\hat{D}(q) = D$ and $\hat{U}(q) = \{u_i \in U| (q, (\epsilon, u_i), q') \in E \Rightarrow q' \notin Q'_B\}$.

Lemma 1 *The algorithm terminates in at most $|X|$ steps. The system is safe if and only if $I \subset V_1 = Q \setminus Q'_B$.*

Corollary 1 $\hat{U}$ *defines the least restrictive class of controls that can guarantee that the system stays safe whenever it starts safe.*

Clearly, the least restrictive class of safe controls is already in feedback form. The above construction can also be used for reachability verification in finite automata, by letting $\Sigma_u = \Sigma$, $\Sigma_c = \emptyset$. $\hat{D}$ provides an *error trace* starting at any state $q_0 \in I \cap V_1$. In this special case the ϵ construction is not necessary.

5 The Steam Boiler

Our analysis of the steam boiler problem is based on the description of [2], which is simpler than the original specification of [11] in that the effect of faults on the system is not considered. The steam boiler consists of a tank containing water and a heating element that causes the water to boil and escape as steam. The water is replenished by two pumps which at time t pump water into the boiler at rates $\dot{p}_1(t)$ and $\dot{p}_2(t)$ respectively. At every time pump i can either be on ($\dot{p}_i(t) = P_i$) or off ($\dot{p}_i(t) = 0$). There is a delay T_{p_i} between the time pump i is ordered to switch on and the time $\dot{p}_i$ becomes P_i. There is no delay when the pumps are switched off. The requirement is that the water level remains between two values M_1 and M_2. We will use three hybrid automata to describe the system, one for the boiler and one for each of the pumps. The specification of [2] also includes a valve that, together with the pumps, can be used to bring the water level to a desirable initial condition before the heating element is turned on and the boiling starts. As the valve is only used to set the initial condition, its operation will be ignored in our safety calculations.

5.1 System Model

The boiler is modeled by a hybrid automaton, $H_B = \{X_B, U_B, Y_B, I_B, f_B, E_B, h_B\}$, with a single discrete state and two continuous states, the water level, w, and the rate at which steam escapes, r. We assume that both states are available for measurement. The system evolution is influenced by two discrete inputs, $\dot{p}_1$ and $\dot{p}_2$, and one continuous input, the derivative of the steam rate, d. The physical properties of the boiler impose the bounds $r(t) \in [0, W]$ and $d(t) \in [-U_2, U_1]$ for all t. Following [2] the dynamics are given by:

$$f_B(x_B, u_B) = \begin{bmatrix} \dot{p}_1 + \dot{p}_2 - r \\ d \end{bmatrix}; \quad E_B = \bigcup_{\substack{x_B \in X_B \\ u_B \in U_B}} (x_B, u_B, x_B); \quad h_B(x_B, u_B) = x_B$$

Note that the set E does not allow any discrete jumps of the state. It is assumed that W, U_1, U_2, P_1 and P_2 are positive constants.

Each pump can also be modeled by a hybrid automaton, $H_{p_i} = \{X_{p_i}, U_{p_i}, Y_{p_i}, I_{p_i}, f_{p_i}, E_{p_i}, h_{p_i}\}$, with two discrete states $q_i = 0$ and $q_i = P_i$ that reflect if the pump is on or off and one continuous state, T_i, that reflects the time that has elapsed since the pump was ordered to switch on. The evolution of the state is affected by a discrete input that takes the value 0 if the pump is ordered to switch off and 1 if the pump is ordered to switch on. The dynamics are given by $f_{p_i}(x_{p_i}, u_i) = u_i$, $h_{p_i}(x_{p_i}, u_{p_i}) = x_{p_i}$ and

$$E_{p_i} = \left(\bigcup_{T_i \leq T_{p_i}} ((0, T_i), 0, (0, 0)) \right) \cup \left(\bigcup_{T_i \leq T_{p_i}} ((0, T_i), 1, (0, T_i)) \right) \cup$$
$$((0, T_{p_i}), 1, (P_i, T_{p_i})) \cup$$
$$\left(\bigcup_{T_i \geq T_{p_i}} ((P_i, T_i), 1, (P_i, T_i)) \right) \cup \left(\bigcup_{T_i \geq T_{p_i}} ((P_i, T_i), 0, (0, 0)) \right)$$

The combined system automaton can be obtained as the interconnection $\mathcal{I}(\dot{p}_i) = q_i$ for $i = 1, 2$ of H_B, H_{p_1} and H_{p_2}. The resulting automaton will have four discrete and four continuous states. If the outputs q_1 and q_2 of the pumps are suppressed, the definition of the interconnection automaton gives:

$$x = ((q_1, q_2), [w \ \ r \ \ T_1 \ \ T_2]^T) \in X = \{0, P_1\} \times \{0, P_2\} \times \mathbb{R} \times [0, W] \times \mathbb{R}_+^2$$
$$u = ((u_1, u_2), d) \in U = \{0, 1\}^2 \times [-U_2, U_1]$$
$$y \in Y = \mathbb{R} \times [0, W] \times \mathbb{R}_+^2$$
$$x^0 \in I = I_B \times I_{p_1} \times I_{p_2}$$

with dynamics:

$$f(x, u) = \begin{bmatrix} q_1 + q_2 - r \\ d \\ u_1 \\ u_2 \end{bmatrix}; \quad E = E_B \times E_{p_1} \times E_{p_2}; \quad h(x, u) = (w, r, T_1, T_2)$$

By a slight abuse of notation the value of the discrete states and inputs is used in the vector field definition. Without loss of generality assume that all runs of the automaton begin at $t = 0$. Our goal is to design a feedback controller for u_1 and u_2 that keeps the water level in the interval $w(t) \in [M_1, M_2]$ for all $t \geq 0$. This requirement can be encoded by two cost functions $J_1(x^0, u_1, u_2, d) = -\inf_{t \geq 0} w(t)$ and $J_1'(x^0, u_1, u_2, d) = \sup_{t \geq 0} w(t)$. For a given run (x^0, u) the requirements are satisfied if and only if $J_1(x^0, u) \leq -M_1$ and $J_1'(x^0, u) \leq M_2$.

5.2 Saddle Solutions, Safe States and Safe Controls

For any initial condition $x^0 = ((q_1^0, q_2^0), [w^0 \ \ r^0 \ \ T_1^0 \ \ T_2^0]^T)$, consider the following candidate saddle solutions:

$$u_i^*(t) = 1 \text{ for all } t; \quad d^*(t) = \begin{cases} U_1 & \text{if } t \leq \frac{W - r^0}{U_1} \\ 0 & \text{if } t > \frac{W - r^0}{U_1} \end{cases} \tag{4}$$

$$u_i'^*(t) = 0 \text{ for all } t; \quad d'^*(t) = \begin{cases} -U_2 & \text{if } t \leq \frac{r^0}{U_2} \\ 0 & \text{if } t > \frac{r^0}{U_2} \end{cases} \tag{5}$$

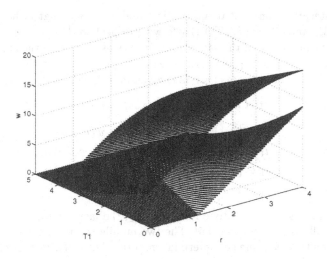

Fig. 1. Lower limit on w to avoid draining

Lemma 2 (u_1^*, u_2^*, d^*) *and* $(u_1'^*, u_2'^*, d'^*)$ *are saddle solutions for the game between* (u_1, u_2) *and* d *over cost functions* J_1 *and* J_1'.

It should be noted that the saddle solution is not unique in the case of J_1', as far as the u_i are concerned. In particular, any u_i such that $\dot{w}(t) \leq 0$ for all t will produce the same maximum water level (equal to the initial water level).

The saddle solution calculation for J_1 allows us to determine the water levels, w^0, that are safe with respect to M_1 as a function of r^0, T_1^0 and T_2^0. In particular, the boundary between safe and unsafe states can be thought of as a function $\hat{w} : [0, W] \times \mathbb{R}_+^2 \rightarrow \mathbb{R}$, which maps (r^0, T_1^0, T_2^0) to the minimum water level required for safety. The level sets of $\hat{w}$ for $T_2^0 = 0$ (pump 2 is initially off and for $T_2^0 \geq T_{p_2}$ (pump 2 is initially fully on) are shown in Figure 1. The safety boundary for any other value of T_2 will be a similar surface lying between the two surfaces of the figure. Safety $(w(t) \geq M_1)$ can be maintained as long as the current value of the water level is on or above the corresponding surface. As expected the higher the value of T_2 the more states are safe (the surface moves down). The parameters used in the figure were $M_1 = 0$, $U_1 = 0.5$, $W = 4$, $P_1 = P_2 = 2.5$ and $T_{p_1} = T_{p_2} = 5$.

The states that are safe with respect to M_2 can be similarly determined. On the boundary between safe and unsafe states, $w^0 = M_2$ (the only situation where J' becomes safety critical) and:

$$r^0 = \hat{r}(T_1^0, T_2^0) = \begin{cases} 0 & \text{if } T_1^0 < T_{p_1} \text{ and } T_2^0 < T_{p_2} \\ P_1 & \text{if } T_1^0 \geq T_{p_1} \text{ and } T_2^0 < T_{p_2} \\ P_2 & \text{if } T_1^0 < T_{p_1} \text{ and } T_2^0 \geq T_{p_2} \\ P_1 + P_2 & \text{if } T_1^0 \geq T_{p_1} \text{ and } T_2^0 \geq T_{p_2} \end{cases}$$

The interpretation is that any initial condition such that either $w^0 < M_2$ or $w^0 = M_2$ and $r^0 \geq \hat{r}(T_1^0, T_2^0)$ is safe with respect to J'.

The calculation of the safe set also allows us to classify the controls that can keep the system safe provided it starts safe (w^0 and r^0 in the ranges discussed above). The class of safe controls is given in a state feedback form.

Lemma 3 *A control law for (u_1, u_2) is safe with respect to M_1 if and only if:*

$$u_1 \in \{0,1\} \quad u_2 \in \{0,1\} \quad \text{if } w > \hat{w}(r,0,0)$$
$$u_1 = 1 \quad u_2 \in \{0,1\} \quad \text{if } \hat{w}(r,0,0) \geq w > \hat{w}(r,T_1,0)$$
$$u_1 \in \{0,1\} \quad u_2 = 1 \quad \text{if } \hat{w}(r,0,0) \geq w > \hat{w}(r,0,T_2)$$
$$u_1 = 1 \quad u_2 = 1 \quad \text{if } w \leq \hat{w}(r,T_1,T_2)$$

Note that, as $\hat{w}$ is monotone in T_1 and T_2, the last condition is satisfied if and only if all other conditions fail. The two middle conditions may in fact overlap, therefore there is some nondeterminism in the choice of safe controls: some states may be safe with either one or the other pump on, but not neither. Similarly:

Lemma 4 *A control law for (u_1, u_2) is safe with respect to M_2 if and only if:*

$$u_1 \in \{0,1\} \quad u_2 \in \{0,1\} \quad \text{if } w < M_2 \text{ or } r > \hat{r}(T_1, T_2)$$
$$u_1 = 0 \quad u_2 \in \{0,1\} \quad \text{if } w \geq M_2 \text{ and } \hat{r}(T_1, T_2) \geq r > \hat{r}(0, T_2)$$
$$u_1 \in \{0,1\} \quad u_2 = 0 \quad \text{if } w \geq M_2 \text{ and } \hat{r}(T_1, T_2) \geq r > \hat{r}(T_1, 0)$$
$$u_1 = 0 \quad u_2 = 0 \quad \text{if } w \geq M_2 \text{ and } r \leq \min\{\hat{r}(T_1, 0), \hat{r}(0, T_2)\}$$

6 Flight Vehicle Management System

The flight vehicle management system (FVMS) example is based on the dynamic aircraft equations and the design specification of [12]. The equations model the speed and the flight path angle dynamics of a commercial aircraft in still air. The inputs to the equations are the thrust T, accessed through the engine throttle, and the pitch angle θ, accessed through the elevators, and the outputs are the speed V and the flight path angle γ. There are three primary modes of operation. In **Mode 1**, T is between its specified operating limits ($T_{min} < T < T_{max}$), both T and θ can be used as inputs and both V and γ can be controlled as outputs. In **Mode 2**, T saturates ($T = T_{min} \vee T_{max}$) and is no longer available as an input. The only input is θ, and the only controlled output is V. Finally, in **Mode 3**, T also saturates, the input is again θ but the controlled output is γ. Within Modes 2 and 3 there are two submodes depending on whether $T = T_{min}$ (idle thrust) or $T = T_{max}$ (maximum thrust).

Safety dictates that V and γ must remain within specified limits. For ease of presentation we simplify this *safety envelope*, S, of [12] to $S = \{(V, \gamma) | (V_{min} \leq V \leq V_{max}) \wedge (\gamma_{min} \leq \gamma \leq \gamma_{max})\}$, where $V_{min}, V_{max}, \gamma_{min}, \gamma_{max}$ are constants. We would like to design a control scheme which will cause the aircraft to reach a target operating point $(V, \gamma)_{target}$ in S from any initial operating point in S. The resulting trajectory $(V(t), \gamma(t))$ must not exit the envelope at any time and must satisfy acceleration constraints imposed for passenger comfort.

6.1 System Model

The flight path angle dynamics of the aircraft can be summarized using two state variables, $x = [V \ \gamma]^T \in \mathbb{R} \times S^1$, where V (m/s) is the airspeed and γ (rad) is the flight path angle. The dynamics of the system are given by:

$$\dot{V} = -\frac{a_D V^2}{m} - g \sin \gamma + (\frac{1}{m})T \tag{6}$$

$$\dot{\gamma} = \frac{a_L V(1 - c\gamma)}{m} - \frac{g \cos \gamma}{V} + (\frac{a_L V c}{m})\theta \tag{7}$$

where T (N) is the thrust, m (kg) is the mass of the aircraft, g (m/s^2) is gravitational acceleration, a_L and a_D are the lift and drag coefficients, c is a small positive constant and θ is the aircraft pitch angle. For these equations to be meaningful we need to assume that $X \subset (0, \infty) \times [-\pi/2, \pi/2]$. Clearly this will be the case for realistic aircraft. Physical considerations also impose constraints on the inputs: $U = [T_{min}, T_{max}] \times [\theta_{min}, \theta_{max}]$.

To guarantee safety we need to ensure that $x(t) \in S$ for all t. Let ∂S denote the boundary of S. The requirement that the state stays within S can be encoded by a cost function:

$$J_1(x^0, u) = -\min_{t \geq 0}(x(t) - \partial S) \tag{8}$$

by defining:

$$x(t) - \partial S = \begin{cases} \min_{y \in \partial S} \|x(t) - y\| & \text{if } x \in S \\ -\min_{y \in \partial S} \|x(t) - y\| & \text{if } x \notin S \end{cases}$$

Here $\| \cdot \|$ denotes the Euclidean metric on $\mathbb{R}^2$. To ensure that the state stays within S we impose the threshold $J_1(x^0, u) \leq 0$.

Cost functions involving the linear and angular accelerations can be used to encode the requirement for passenger comfort:

$$J_2(x^0, u) = \max_{t \geq 0}(\dot{V}(t)) \text{ and } J_2'(x^0, u) = \max_{t \geq 0}(V(t)\dot{\gamma}(t)) \tag{9}$$

The requirement that the linear and angular acceleration remain within the limits determined for comfortable travel are encoded by thresholds $J_2(x^0, u) \leq 0.1g$ and $J_2'(x^0, u) \leq 0.1g$.

6.2 Safe States and Least Restrictive Safe Controls

To find the controls that keep the state within the safety envelope we solve the optimal control problem $J_1^*(x^0) = \min_{u \in \mathcal{U}} J_1(x^0, u)$, $u^*(x^0) = \arg\min_{u \in \mathcal{U}} J_1(x^0, u)$.

Proposition 1 (Optimally Safe Controls) *The optimally safe input is:*

$$u^*(x^0) = \begin{cases} (T_{max}, \theta_{min}) \ \forall x^0 = (V, \gamma) \in S \cap \{(V, \gamma) : \frac{\gamma - \gamma_{min}}{\gamma_{max} - \gamma_{min}} > \frac{V - V_{min}}{V_{max} - V_{min}}\} \\ (T_{min}, \theta_{max}) \ \forall x^0 = (V, \gamma) \in S \cap \{(V, \gamma) : \frac{\gamma - \gamma_{min}}{\gamma_{max} - \gamma_{min}} < \frac{V - V_{min}}{V_{max} - V_{min}}\} \end{cases}$$

The optimal control calculation allows us to determine the set of safe states and the class of controls that renders this set safe. If $J_1^*(x^0) > 0$ there is no control that will keep the trajectory starting at $x^0 \in S$ within S. If, however, $J_1^*(x^0) \leq 0$ there exists at least one (and maybe multiple) such safe controls. Our goal is to determine $V_1 = \{x^0 \in S | J_1^*(x^0) \leq 0\}$ and $\mathcal{U}_1(x^0) = \{u \in \mathcal{U} | J_1(x^0, u) \leq 0\}$.

We start by analyzing the system equations (6, 7) along ∂S. Consider an arbitrary point $x^0 \in \partial S$. We can distinguish three cases. If $f(x^0, u)$ points "inside" S for all $u \in U$ then all controls are safe for the given point x^0, i.e. $\mathcal{U}_1(x^0) = U$. If $f(x^0, u)$ points "outside" S for some u, let $\hat{U} \subset U$ be the controls for which this happens. These inputs are unsafe for the point x^0, i.e. $\mathcal{U}_1(x^0) = U \setminus \hat{U}$. Finally, if $f(x^0, u)$ points outside S for all $u \in U$ then all controls are unsafe for the given point x^0, i.e. $\mathcal{U}_1(x^0) = \emptyset$. A special case of the second situation is one where $f(x^0, u)$ is tangent to ∂S for some u and points outside for all others. In this case, the set of controls that make $f(x^0, u)$ tangent to ∂S will be exactly u^*. This allows us to extend the safe set construction to the interior of S. The system equations are integrated backwards using u^* from that point to determine the boundary of the safe set in the interior of the envelope.

Consider, for example, the left hand edge of ∂S (Figure 2). The complete set of controls moves from being safe to unsafe as γ varies from γ_{min} to γ_{max}. We can determine which values of (T, θ) in U are unsafe along ∂S by determining where along this boundary the vector field is tangent to ∂S. We calculate this by setting $\dot{V} = 0$, $T = \hat{T}$ in equation (6). Solving for $\hat{T}$ leads to $\hat{T}(\gamma) = a_D V_{min}^2 + mg \sin \gamma$. $\hat{T}(\gamma)$ does not depend on θ, so the safe set of inputs are all (T, θ) for which $T(\gamma) \geq \hat{T}(\gamma)$. When γ is such that $\hat{T}(\gamma) = T_{min}$, the cone of vector fields points completely "inside" S; when γ is such that $\hat{T}(\gamma) = T_{max}$, the cone of vector fields points completely "outside" S, and T_{max} is the unique thrust input which keeps the system trajectory inside S. We define γ_1 and γ_2 to be such that $\hat{T}(\gamma_1) = T_{max}$ and $\hat{T}(\gamma_2) = T_{min}$ and calculate the boundary of the safe set of states on the interior of the envelope by integrating the system equations backward in time from (V_{min}, γ_1) using the constant control (T_{max}, θ_{min}). We denote this part of the safe set boundary in the interior of S as ∂V_1^1, and the point of intersection of ∂V_1^1 with the upper edge of ∂S as (V_1, γ_{max}).

A similar calculation along the upper edge of ∂S using equation (7) yields the values of θ for which the vector field becomes tangent to ∂S: $\hat{\theta}(V) = \frac{m}{a_L V_c} \left(\frac{g \cos \gamma_{max}}{V} - \frac{a_L V (1 - c \gamma_{max})}{m} \right)$. The set of safe inputs in this case is all (T, θ) for which $\theta(V) \leq \hat{\theta}(V)$. When V is such that $\hat{\theta}(V) = \theta_{min}$, θ_{min} is the unique pitch angle input which keeps the system trajectory inside S. The calculations may be repeated for the right hand side and lower boundaries of S.

We are now in a position to describe explicitly the safe set of states V_1 and the safe controls $\mathcal{U}_1(x^0)$. Define the boundary of V_1 as

$$\partial V_1 = \{(V, \gamma)|\ (V = V_{min}) \wedge (\gamma_{min} \leq \gamma \leq \gamma_1) \vee (V, \gamma) \in \partial V_1^1 \vee$$
$$(\gamma = \gamma_{max}) \wedge (V_1 \leq V \leq V_{max}) \vee (V = V_{max}) \wedge (\gamma_4 \leq \gamma \leq \gamma_{max}) \vee$$
$$(V, \gamma) \in \partial V_1^2 \vee (\gamma = \gamma_{min}) \wedge (V_{min} \leq V \leq V_2)\}$$

V_1 is defined as the set enclosed by ∂V_1. This is depicted in Figure 2. $\mathcal{U}_1(x^0)$ is

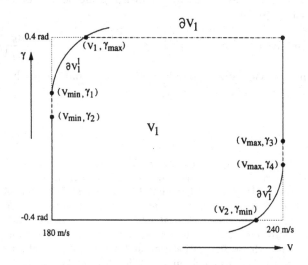

Fig. 2. The safe set of states, V_1, and its boundary ∂V_1

defined by the feedback map $G : S \rightarrow 2^U$:

$$G(V, \gamma) = \{\ \emptyset, (V, \gamma) \in S \backslash V_1$$
$$\theta \leq \hat{\theta}(V) \wedge T_{min} \leq T \leq T_{max}, (V, \gamma) \in (\gamma = \gamma_{max}) \wedge (V_1 \leq V \leq V_{max})$$
$$\theta_{min} \leq \theta \leq \theta_{max} \wedge T \geq \hat{T}(\gamma), (V, \gamma) \in (V = V_{min}) \wedge (\gamma_2 \leq \gamma \leq \gamma_1)$$
$$\theta_{min} \leq \theta \leq \theta_{max} \wedge T \leq \hat{T}'(\gamma), (V, \gamma) \in (V = V_{max}) \wedge (\gamma_4 \leq \gamma \leq \gamma_3)$$
$$\theta = \theta_{min} \wedge T = T_{max}, (V, \gamma) \in \partial V_1^1$$
$$\theta = \theta_{max} \wedge T = T_{min}, (V, \gamma) \in \partial V_1^2$$
$$\theta_{min} \leq \theta \leq \theta_{max} \wedge T_{min} \leq T \leq T_{max}, \text{else}\}$$

This map defines the *least restrictive* control scheme which satisfies the safety requirement and it determines the mode switching logic. On ∂V_1^1 and ∂V_1^2, the system must be in **Mode 2** or **Mode 3**. Anywhere else in V_1, any of the three modes is valid as long as the input constraints of G are satisfied. In the region $S \backslash V_1$ no control inputs are safe.

6.3 Additional Constraints for Passenger Comfort

Within the class of safe controls, a control scheme which addresses the passenger comfort (efficiency) requirement can be constructed. To do this, we solve the optimal control problem for J_2 and J'_2, for $x^0 \in V_1$. From this calculation, we determine the set of "comfortable" states and controls:

$$V_2 = \{x^0 \in V_1 : J_2^*(x^0) \leq 0.1g \wedge J_2'^*(x^0) \leq 0.1g\} \tag{10}$$

$$\mathcal{U}_2(x^0) = \{u \in \mathcal{U}_1 : J_2(x^0, u) \leq 0.1g \wedge J_2'(x^0, u) \leq 0.1g\} \tag{11}$$

These sets may be easily calculated by substituting the bounds on the accelerations into equations (6, 7):

$$T \leq 0.1mg + a_D V^2 + mg\sin\gamma \text{ and } \theta \leq \frac{0.1mg}{a_L V^2 c} - \frac{1 - c\gamma}{c} + \frac{mg\cos\gamma}{a_L V^2 c} \tag{12}$$

These constraints provide upper bounds on the thrust and the pitch angle which may be applied at any point (V, γ) in V_2.

References

1. J. Lygeros, *Hierarchical Hybrid Control of Large Scale Systems*. PhD thesis, Department of Electrical Engineering, University of California, Berkeley, 1996.
2. T. A. Henzinger and H. Wong-Toi, "Using HyTech to synthesize control parameters for a steam boiler," in *Formal Methods for Industrial Applications: Specifying and Programming the Steam Boiler Control*, LNCS, Springer Verlag, 1996.
3. R. Alur, C. Courcoubetis, T. A. Henzinger, and P. H. Ho, "Hybrid automaton: An algorithmic approach to the specification and verification of hybrid systems," in *Hybrid System*, no. 736 in LNCS, pp. 209–229, Springer Verlag, 1993.
4. A. Deshpande, *Control of Hybrid Systems*. PhD thesis, Department of Electrical Engineering, University of California, Berkeley, 1994.
5. A. Puri, *Theory of Hybrid Systems and Discrete Event Systems*. PhD thesis, Department of Electrical Engineering, University of California, Berkeley, 1995.
6. N. Lynch, R. Segala, F. Vaandrager, and H. Weinberg, "Hybrid I/O automata," in *Hybrid Systems III*, no. 1066 in LNCS, pp. 496–510, Springer Verlag, 1996.
7. R. Alur and T. A. Henzinger, *Computer-Aided Verification*. 1996. to appear.
8. M. S. Branicky, *Control of Hybrid Systems*. PhD thesis, Massacussets Institute of Technology, 1994.
9. J. Lygeros, D. N. Godbole, and S. Sastry, "Optimal control approach to multi-agent, hierarchical system verification," in *IFAC World Congress*, 1996.
10. P. J. G. Ramadge and W. M. Wonham, "The control of discrete event dynamical systems," *Proceedings of the IEEE*, vol. Vol.77, no. 1, pp. 81–98, 1989.
11. J.-R. Abrial, "The steam-boiler control specification problem," in *Formal Methods for Industrial Applications: Specifying and Programming the Steam Boiler Control*, LNCS, Springer Verlag, 1996.
12. C. S. Hynes and L. Sherry, "Synthesis from design requirements of a hybrid system for transport aircraft longitudinal control." (preprint), NASA Ames Research Center, 1996.

Modelling a Time-Dependent Protocol Using the Circal Process Algebra

Antonio Cerone,
Alex J. Cowie,
George J. Milne
and
Philip A. Moseley*

Advanced Computing Research Centre
School of Computer and Information Science
University of South Australia
Adelaide SA 5095
Australia
{cerone,cowie,milne,moseley}@cis.unisa.edu.au

Abstract. A time-dependent protocol is specified within Circal, a mature process algebra which has been used extensively for the description and verification of concurrent systems. Although Circal is a discret formalism, it permits the analysis of interesting aspects of timed systems. We utilise the Circal System, a mechanisation of Circal, to automatically verify the correctness of an audio control protocol.

1 Introduction

In previously reported work [2, 13] the Circal process algebra has been used successfully to describe and verify the behaviour of digital hardware, an application domain in which the modelling of time is a key concern. This paper extends this experience to the description and analysis of an audio control protocol developed by Philips in which timing considerations are central to its operation. This protocol encodes an input bitstream message into timed voltage transitions on a single wire bus and subsequently reconstructs an output message from the wire signals, and is thus hybrid in nature, involving both discrete events and a time continuum.

A version of the protocol with only one sender and one receiver was previously specified and analysed by Bosscher et al. [4] using linear hybrid systems, a hybrid formalism which employs discrete events and continuous variables. The properties stated and proved by hand in [4] have been automatically verified by Griffioen [8] using the Larch Prover, a proof-checker for first order predicate logic based on rewriting, and by Daws and Yovine [7], who translated the specification given in [4] into multirate timed automata and then transformed it in

---

* now at Mortorola Australia Software Centre, 2 Second Avenue, Technology Park, Adelaide SA 5095, Australia, email: A12414@email.mot.com

timed automata [1] in order to apply the symbolic techniques implemented in the Kronos tool. In [9] the protocol has been automatically verified by Ho and Wong-Toi using HyTech, a symbolic model checker for linear hybrid systems, which allows also the automatic synthesis of the bound on the maximum clock drift, and suggests a design modification for a more robust protocol. Larsen et al. [11] used Uppaal, a symbolic model checker for networks of timed automata, to debug an early draft version of the protocol description given in [9]. A description of the protocol with two senders and bus collision handling has been analysed in [3] using an improved version of Uppaal.

In the present work we consider a description of the protocol with only one sender and one receiver, as in [4, 7, 8, 9, 11]. We handle such description within the framework of the Circal process algebra, in which time is modelled discretely without any special extension to the formalism. In this way we take up the challenge set by Bosscher et al. [4] "to redo the verification ... within a process algebraic setting". Our purpose is to analyse the power of the Circal process algebra in the verification of a model that is hybrid in nature. It is interesting to understand which aspects of hybrid systems can be captured by such a simple process algebra framework and how accurately they can be captured.

We want to show that it is possible to develop particular application-oriented modelling styles which also make Circal useful within contexts where process algera approaches are not normally considered applicable. In particular we use the *constraint-based modelling style* [14]: the behaviour of a process may be constrained simply by composing it with another process which represents the constraints. This modelling style permits a natural representation of time and an algebraic characterization of logical properties which then result in a methodological approach to the automatic verification of real-time systems. This specific verification technique differs from the normal process algebra approach of showing behavioural equivalence between specification and implementation of a system. Rather, we show within the process algebra framework that the implementation protocol satisfies certain properties without recourse to model-checking within a modal or temporal logic formalism.

We claim neither that our approach can always be used in the verification of hybrid systems nor that it captures all the features of the audio control protocol. However, we believe that the approach allows the analysis of interesting aspects of timed systems, as has previously been shown with its extensive application to digital hardware, and of *some* hybrid systems.

2 Description of the Protocol

The protocol forms part of the physical layer of an interface bus that interconnects the components of stereo systems. Messages which consist of finite sequences of 0 and 1 bits are encoded by a sender into timed transitions of the voltage between two levels on the single wire bus connecting the components. Receivers in components attached to the bus interpret the voltage transitions and reconstruct the bitstream messages. The senders and receivers are run on different microprocessors. Since the microprocessors run code in addition to the protocol software, sender and receiver clocks may not be synchronised in frequency.

We follow [4] and restrict our example to only two processors communicating over a bus, each with its own independent clock. In this way problems of bus collisions due to different senders sending at the same time are ignored. We also suppose that the message delay on the bus is negligible. Figure 1 represents this

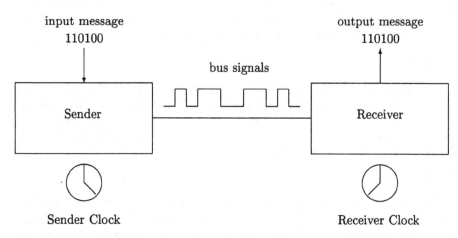

Fig. 1. The Audio Control Protocol

view of the protocol. The sender processor of the audio control protocol accepts messages from a user which it then tries to send to the receiver processor over the bus. A message is encoded by the sender according to a Manchester encoding scheme which is an algorithm for constructing a bus signal equivalent of the message. The time flow is represented as a sequence of slots of the same length, $4Q$ (where $Q = 222\mu s$ in the Philips protocol), and each bit of the sequence is sent by the sender as a voltage level transition in the middle of a slot: a bit 0 is a downgoing transition and a bit 1 is an upgoing transition. In this way, when an identical bit is sent twice in a row an additional transition is added at the boundary between the corresponding slots (see Figure 2). The precise timing required by this encoding is measured relative to the sender clock.

Since downgoing transitions take a significant time to change from high to low level, they do not appear to the receiver as edges. So the receiver can observe only upgoing edges and has to reconstruct the bitstream message from the lengths of the time intervals between successive upgoing edges. Between messages the sender holds the voltage at the low level and the encoding of each message starts with an upgoing edge and ends with a downgoing edge. There is a potential ambiguity in interpreting the start of a message. The receiver cannot distinguish

Fig. 2. Manchester encoding

between an initial 0 which would be encoded as a slot with an upgoing edge at the start and a downgoing edge in the middle and an initial 1 which would be encoded with a upgoing edge in the middle of the slot. This problem is resolved by imposing the following constraint on the input:

C0 Every message starts with the bit 1

A further ambiguity arises from the loss of downgoing edges during transmission. The receiver cannot distinguish 10 from 1 at the end of a message. In this case the only difference is in the timing of the final downgoing edge, which cannot be observed by the receiver. This problem is overcome by imposing the following constraint on the input

C1 Every message either has an odd length or ends with 00

Because of the drift between the sender and receiver clocks, the distance between successive upgoing edges measured by the receiver is in general different from that measured by the sender. The protocol has been designed to achieve reliable communication even in presence of this significant *timing uncertainty*. The receiver interprets the time distance between successive upgoing edges in different ways according to its length. However, the correct interpretation will still be made provided that the drift is held within limits that depend on the protocol specification. In our analysis of the audio control protocol, we suppose that the rate of the sender and receiver clocks are steady and therefore we consider a fixed rate of the drift.

3 The Circal Process Algebra

This section is intended to familiarize the reader with the Circal process algebra [2, 14, 13], which we shall use to formally describe the audio control protocol. To describe and analyze a system, the user works with the Circal System [14] which automates the Circal process algebra. The syntax of Circal processes is summarized by the following BNF expressions, where P is a process, D a process definition, Act is the set of possible actions (its sort), $m \subseteq$ Act, $a, b \in$ Act and I a process variable:

$$P ::= /\backslash \mid mP \mid P + P \mid P \& P \mid I \mid P * P \mid P - m \mid P [a/b]$$
$$D ::= I < - P$$

The sort corresponds to a set of labelled ports through which the process interacts with its environment. The role of each Circal operator can be described as follows:

Termination. /\: used to denote a process which has terminated or which has deadlocked.

Guarding. Single or simultaneous guards are permitted. For example:

$$P < - (a\ b)\ P'$$

represents the process P which will perform a and b simultaneously, and then evolve into P'. The fact that processes may be guarded by sets of simultaneously occurring actions is a key feature of Circal which greatly enriches the modelling potential of the algebra in contrast to process algebras such as CCS [15] and CSP [10] which only permit a single action to occur at one computation instant. An example of a single guard is:

$$P < - (a)\ P'$$

in which process P performs the single action a and evolves to P'.

Definition. A name can be given to a Circal process with the definition operator. Recursive process definitions are permitted; for example:

$$P < - (a\ b)\ P$$

is interpreted as a process that continuously repeats the simultaneous actions $(a\ b)$.

Choice. The following process can perform a or b or c:

$$P < - a\ P + b\ P + c\ P'$$

The choice is decided by the environment in which the process is executed.

Non-determinism. The term $P\&Q$ represents the process which can perform either the actions of the subterm P or the actions of the subterm Q. The computation path is decided autonomously by the process itself without any influence from its environment.

Concurrent Composition. Given processes P and Q, the term $P*Q$ represents the process which can perform the actions of the subterms P and Q together. Any synchronization which can be made between two terms, due to some atomic action being common to the sorts of both subterms, must be made, otherwise the actions of the subterms may occur asynchronously.

Abstraction. The term $P - a$ represents the process that is identical to P except that the port a is invisible to the environment.

Relabelling. The term $P\ [a/b]$ is the new version of P in which the action b replaces a wherever a occurs in P.

We have given the syntax of Circal and described informally the operational semantics, which associates a behaviour with a Circal process. Processes that are syntactically different may have the same behaviour, as long as no well-defined experiment conducted from "outside" can distinguish between them.

Many different theories exist for providing the rules for distinguishing processes, so defining many different notions of equivalence between processes [12]. The Circal System implements a testing equivalence defined in [16] giving to the expression

$$P == Q$$

the result `true`, if P and Q are equivalent and `false`, otherwise.

4 Modelling with the Circal Process Algebra

As with any process algebra, the Circal process algebra is a low level language containing the primitive constructs to model concurrency. For this reason it is a very flexible language and it is possible to develop modelling styles that allow the description of high level aspects of concurrent systems, as well as application-oriented modelling styles.

One modelling style that is very useful in several application domains, and in particular in the description of the audio control protocol, is the *constraint-based* modelling style [14]. This modelling style allows also a natural representation of time and an algebraic characterization of logical properties.

4.1 Constraint-based Modelling

The constraint-based modelling style is supported by the following distinctive features of the Circal process algebra:

- guarding of processes by sets of simultaneous actions;
- sharing of events over arbitrary numbers of processes;
- the particular nature of the composition operator which provides synchronised processes without removal of the synchronising events in the resultant behaviour.

The behaviour of a process may be constrained simply by composing it with another Circal process which represents the constraint. This approach can be used repeatidly to synthesise complex behaviours from a set of constraints.

4.2 Model of Time

In all process algebras there is an implicit model of time. The guarding operator generates a term mP which represents the behaviour in which event m occurs followed by the behaviour denoted by the process P. This implies that event occurrences are related by a temporal order relation. The temporal order is in general a partial order, due to the presence of choice operators which induces a branching behaviour. However, while this permits a natural modelling of the temporal ordering of actions, an explicit representation of time is not always possible.

In order to describe explicit time, the above notion of time is augmented by assuming the existence of time processes which perform timing signals at regular intervals. Time in process algebras is effectively discrete and so the *regular interval* property of such signals cannot be expressed directly in the process algebra formalisms, but rather is an interpretation made when using the process algebra to model time in this particular way.

The critical point is now to relate the signals describing the time to the actions performed by other processes. In Circal this can be done in a natural way, without any special extension to the formalism. The simple inclusion of a signal in a guard is used to describe the concept of *all* the actions in the guard occurring synchronously with the signal. Moreover, because of the particular nature of the composition operator, when used to compose several processes, every time a component performs synchronizations on a signal any other component that can perform that signal must also do it, and the signal occurrences are still visible after the synchronization without removal.

In this way, if a process has the timing signal in its sort, then any action performed by the process occurs either synchronously with a signal occurrence or between two successive signal occurrences and is therefore timed. On the other hand, it is possible to define processes which do not contain the timing signal in their sort. Such processes perform actions whose occurrence is independent of time; this is the case of a process describing constraints that are satisfied at all times.

In Circal it is also possible to use different timing signals for different processes and to apply the constraint-based modelling style by defining an additional process which describes the particular relation between these different signals. Such a process can then be composed with the components that describe the physical system, imposing in this way timing constraints on the whole system. We follow this approach in the analysis of the audio control protocol and define a process Drift to describe the relation between the timing signals of the sender and the timing signals of the receiver.

4.3 Characterization of Properties in Circal

A correctness concept that can be readily characterized in Circal is the behavioural equivalence between processes. The Circal expression P == Q has value true if P and Q have the same behaviour and value false otherwise. In verification, however, equivalence is often too strong a property. For certain systems, verifying their correctness consists of determining that certain properties hold, where these properties do not constitute a complete specification. This cannot be done in terms of a mere equivalence, but involves the notion of behavioural inclusion. When the behaviour of a process P is included in the behaviour of a process Q and P is deterministic, the composition of P and Q can be seen as a constraint P which restricts Q to behave as P. Therefore

$$P * Q == P$$

characterizes the behavioural inclusion of P in Q.

5 Modelling the Protocol in Circal

5.1 Bitstream Messages

A sequence of messages is readily modelled in Circal by a nested series of guarded processes where the guards are events consisting of a single actions that can describe a bit 0, or a bit 1 or the "end of the message". Different actions are used for input and output messages. We represents 0, 1 and "end of the message" by inbit0, inbit1 and inend, respectively, in the input, and by outbit0, outbit1 and outend, respectively, in the output. For example the sequence of 2 messages, given by the message 110 followed by the message 1 is represented in the input as

$$\text{inbit1 inbit1 inbit0 inend inbit1 inend } /\backslash$$

and in the output as

$$\text{outbit1 outbit1 outbit0 outend outbit1 outend } /\backslash$$

5.2 Protocol Components

In this section we give a short description of the protocol in Circal. The details and the complete Circal code are given in [6]. The main protocol components are represented by the two processes Sender and Receiver. The upgoing and downgoing edges on the bus are modelled in Circal by the events up and down, respectively.

The process Sender consists of the composition of three processes. An interface process transforms every input action inbit$y$ into an action in$z y$, where z is the value 0 or 1 of the previous input bit. Analogously every input action inend is is transformed into an action in$z$E, where z is the value 0 or 1 of the previous input bit. Then a second process both defines the sender clock by means of a sender timing signal ts belonging to each event and generates the up and down events. The time between two successive timing signals is nominally Q ($222\mu s$).

Apart from the first bit of a message, where the input in0y occurs always synchronously with an up, the input in$z y$ is read at the beginning of a slot and, if $z = y$ an up or down is generated in the same event, according to the value of $z = y$, 0 or 1, respectively. After two timing signals (after a time $2Q$), i.e. in the middle of the slot, an up or a down, according to the value of y, 1 or 0, respectively, is generated to encode the input bit. When the input in$z$E is read (at the beginning of a slot), if $z = 1$ a down is generated to reset the voltage at the low level and, in all cases, a time gap is generated by a third process to separate the two messages.

The receiver can observe only upgoing edges, i.e. occurrences of the action up. The reconstruction of the transmitted message is done interpreting the time x between the current occurrence of an up and the previous one as in Figure 3. The event error is output when the two up are so close that some transmission fault must have occurred; the events out0 and out1 define output bits 0 and 1,

previous output bit	time interval	up event	current output event
	$0 \leq x < 3Q$	yes	error
0	$3Q \leq x < 5Q$	yes	out0
	$5Q \leq x < 7Q$	yes	out01
	$x = 7Q$	no	outEND
	$0 \leq x < 3Q$	yes	error
1	$3Q \leq x < 5Q$	yes	out1
	$5Q \leq x < 7Q$	yes	out0
	$7Q \leq x < 9Q$	yes	out01
	$x = 9Q$	no	outEND

Fig. 3. Interpretation of the up actions in the receiver

respectively; the event out01 defines the output of the bits 0 and 1 in sequence at the same time; the event outEND defines the end of the message and is generated when the time reaches the end of the last interval without the occurrence of an up.

The process Receiver consists of the composition of five processes. A first process waits for an event up, then it starts measuring the time by timing signals tr, till the next up occurs, and for each tr generates an action characterizing the current time interval. This process is reset to the initial state (to wait for an up), when the end of a message is detected from the size of the interval, and to the state generating the first tr, when an up is observed. A second process interprets the incoming up events according to the previous output bit and the current time interval as shown in Figure 3. The behaviour of the composition of the two processes, after abstracting away the actions characterizing the time intervals and other auxiliary actions, is given in part in Figure 4. For example the last two terms of state S3 correspond to $x = 3Q$, the first term of state S4 corresponds to $3Q < x < 4Q$, the last two terms of state S4 correspond to $x = 4Q$ and the first term of state S5 corresponds to $4Q < x < 5Q$. So, all these terms together characterize the time interval $3Q \leq x < 5Q$, when the previous bit was 1, and therefore the occurrence of an up in each term is always simultaneous with the event out1, according to the interpretation given in Figure 3. A third process characterizes by means of events odd and **even** the length of the message, so that a fourth process can resolve the decoding ambiguity adding an additional 0 at the end when the detected message has an even length or ends with 0. This additional 0 is defined by an action outADD0 occurring in the same action set as outEND. The fifth process is an interface to generate the output events outbit0, outbit1 and outend from the internal events out0, out1, out01, outEND and outADD0.

5.3 Modelling the Clock Drift between the Sender and Receiver

In the previous section, the behaviour of the sender and receiver components has been outlined. However, it still remains to specify the relationship between the sender and receiver clocks. This is captured in the Drift process which is not

```
S0 <- (up out1) S1
S1 <- (((up error outEND) S0 +
        (up error tr outEND) S0) +
        tr S2)
S2 <- (((up error outEND) S0 +
        (up error tr outEND) S0) +
        tr S3)
S3 <- (((up error outEND) S0 +
        (up tr out1) S1 +
        tr S4
S4 <- (((up out1) S1 +
        (up tr out1) S1 +
        tr S5
S5 <- (((up out1) S1 +
        (up tr out0) S6) +
        tr S7)
        .
        .
        .
```

Fig. 4. Part of the receiver behaviour

part of the protocol as such but is an external constraint imposed on the protocol behaviour and in particular on the temporal ordering of the ts and tr events. In this section we will outline the principle used to construct the behaviour of this process. Full details can be found in [6].

Figure 5 shows the relative ordering of sender and receiver clock timing sig-

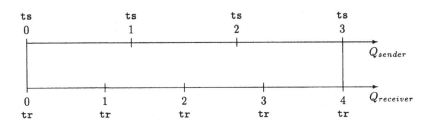

Fig. 5. Drift corresponding to NS=3 and NR=4

nals that corresponds to the the case in which the relative sender and receiver clock rates are NS=3 and NR=4 respectively. This particular ordering is expressed in Circal as a process

$$D \text{ <- (ts tr) tr ts tr ts tr } D$$

which can be generated automatically using the algorithm in Figure 6.

```
set  rt = 0
set  st = 0
loop
     if  rt < st
         emit  tr
         rt = rt + NR
     elseif  st < rt
         emit  ts
         st = st + NS
     else
         emit  (tr ts)
         rt = rt + NR
         st = st + NS
endloop
```

Fig. 6. Algorithm to generate the relative ordering of clock signals

The full behaviour of process Drift is more complex than the behaviour of process D. Each up action forces the synchronisation of ts and tr. In addition, action outEND causes a transition to an idling state which can perform only ts signals during the gap between successive messages. Since up events occur synchronously with the sender timing signals, the relationship between the up events and receiver timing signals can result in a wrong interpretation of an upgoing edge being made by the receiver. A graphical interface used for the simulation and verification of the protocol [5] permits the definition of the relative sender and receiver clock rates and the automatic synthesis of the appropriate Drift process for the rates specified [6].

5.4 Simulation of the Protocol

In Circal it is possible to perform a simulation using a given specification. This is done by composing the process that defines the specification with a process that describes the *test pattern*. In this way the test pattern constrains the specification and only the output corresponding to the test pattern appears in the behaviour of the composition. In our case the specification is given by the process

```
Protocol <- Sender * Receiver * Drift
```

and the test patterns are given by processes describing sequences of input messages. This technique is useful for developing and exploring the behaviour of systems such as the protocol.

The graphical interface to the Circal model of the protocol has been constructed using the simulation technique [5]. The graphical interface accepts

– a single message given as a sequence of characters 0 and 1

– a pair of nonnegative integers defining the relative rates of the sender and the receiver,

generates a process Message corresponding to the input message and a process Drift corresponding to the two rates, runs Circal on the composition of these processes with the Sender and the Receiver and parses the Circal output generating a graphical representation of the voltage level transitions on the bus and of their interpretation.

6 Verification of the Protocol

The simulation technique described in the previous section allows the behaviour of the protocol to be examined with a particular sequence of messages. To establish the correctness of the protocol operation would require all possible messages to be tested. Although this is infeasible, the following verification technique is adopted. We define the notion of correctness as the property **P** where

P: Each sequence of messages that satisfies constraints **C0** and **C1** is accepted as an input and is output unchanged by the protocol

Property **P** can be seen as the conjuction of two properties, $P0 \land P1$, where

P0: Each sequence of messages that satisfies constraints **C0** and **C1** is accepted as an input by the protocol

P1: Each sequence of messages that satisfies constraints **C0** and **C1** and is accepted as an input by the protocol is output unchanged by the protocol

Note that the notion of "being accepted as an input" is moved from the consequence of property **P** to the premise of property **P1** since it holds due to property **P0**.

Property **P0** is satisfied if and only if the behaviour of Constraints is included in the behaviour of the Protocol process which has the output events abstracted away. This is expressed in the Circal System by

```
(Constraints * (Protocol - outbit0 outbit1 outend)) ==
                    Constraints.
```

The fact that the input is output unchanged can be described by a process Buffer which accepts three kinds of inputs (inbit0, inbit1 and inend) and can accept an input and generate an output simultaneously. Therefore, property **P1** is satisfied if and only if the behaviour of the constrained protocol, that is the composition of Constraints and Protocol, is included in the behaviour of Buffer. This is expressed in the Circal System by

```
((Constraints * Protocol) * Buffer) == (Constraints * Protocol)
```

In practice the Circal process Protocol in our description is parameterised by the relative clock rates of the sender and receiver (expressed in the value of

$\frac{NS}{NR}$) and by the length of the gap between successive messages (specified as the number of sender time units that elapse between the final upgoing edge in one message and the first upgoing edge in the next message). The process **Buffer** is parameterised on the capacity of the buffer. Verification of the protocol in the Circal System is carried out by testing whether the two equivalences given above as characterisations of properties **P0** and **P1** hold for particular values of the three parameters. It was found that the equivalences are true only if the buffer size was taken as ≥ 2 and the protocol parameters were chosen such that

$$\begin{cases} \frac{8}{9} < \frac{NS}{NR} \leq \frac{8}{7} \\ \text{gap} > 9 \cdot \frac{NS}{NR} \end{cases}$$

These conditions are represented graphically by the shaded area in Figure 7.

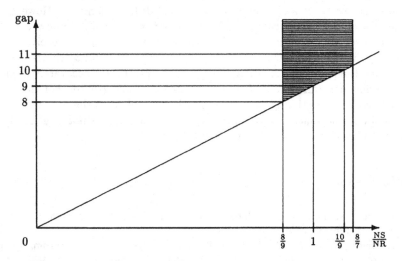

Fig. 7. Correctness Region

7 Discussion

In this paper we have shown how to use the inherently discrete Circal process algebra to model a simplified version of the audio control protocol developed by Philips and which is hybrid in nature, with both discrete and continuous behaviour.

Three distinctive features of the Circal process algebra contribute to its modelling power in such applications, namely, simultaneous event guards, sharing of events over arbitrary numbers of processes and the nature of the composition operator which provides synchronisation of processes without the removal of the synchronising events in the resultant behaviour. These permit a natural representation of timing properties and support a constraint-based modelling style, used in the representation of clock drift, in the simulation of the protocol operation and in the specification of necessary properties for protocol correctness.

The automated verification of the protocol operation was performed by showing that necessary properties were satisfied by the protocol description. A significant feature is that this proof was performed within the process algebra framework without recourse to temporal logic or model-checking techniques.

In our analysis of the audio control protocol we have neglected the problems of bus collision and delay on the bus, as in [4, 7, 8, 9, 11], and have considered a fixed rate of drift. Use of the Circal System, the mechanisation of Circal, permitted an automated verification of the protocol with respect to a given drift value and a given length of the gap between messages. In practice, the drift is due to the fact that the sender and receiver processors run code in addition to the protocol software and, therefore, the rate of drift is changable. However, the value of the drift affects the protocol behaviour only when an up event occurs. Since up occurrences force the synchronisation of the sender and receiver clocks, it is only the average values of the rate of the drift between up occurrences that determine the protocol operation. Therefore the protocol is accurately modelled by assuming constant values of the drift between up occurrences. For this reason we believe that the timing properties which are central to the operation of the protocol have been accurately captured, even if the result is not as general as in the case with certain hybrid systems.

References

1. R. Alur and D. L. Dill. A Theory of Timed Automata. *Theoretical Computer Science*, 126:183–235, 1994.
2. A. Bailey, G. A. McCaskill, and G. J. Milne. An Exercise in the Automatic Verification of Asynchronous Designs. *Formal Methods in System Design*, 4(3):213–242, May 1994.
3. J. Bengtsson, W.O.D. Griffioen, K.J. Kristoffersen, K.G. Larsen, F. Larsson, P. Pettersson, and W. Yi. Verification of an Audio Control Protocol with Bus Collision Using UPPAAL. In *Proceedings of the 8th International Conference on Computer-Aided Verification*, New Brunswick, USA, 31 July–3 August, 1996, Lecture Notes in Computer Science 1102, pages 244–256, Springer, Berlin, Germany, 1996.
4. D.J.B. Bosscher, I. Polak, and F.W. Vaandrager. Verification of an Audio Control Protocol. In H. Langmaack, W.-P. de Roever, and J. Vytopil, editors, *Proceedings of the 3rd School and Symposium on Formal Techniques in Real-Time and Fault tolerant Systems*, Lübeck, Germany, September 1994, Lecture Notes in Computer Science 863, pages 170–192, Springer, Berlin, Germany, 1994. Full version as Report CS-R9445, CWI, Amsterdam, The Netherlands, July 1994.
5. A. Cerone, A. J. Cowie, and G. J. Milne. The Circal System In C. Priami, editor, *DEMOS at CONCUR96*, pages 4-5, Technical Report TR-96-29, University of Pisa, Department of Computer Science, Pisa, Italy, 1996.
   http://lite.ncstrl.org:3803/Dienst/UI/2.0/Describe/ncstrl.unipi_it%2fTR-96-29

6. A. Cerone, A. J. Cowie, G. J. Milne, and P. A. Moseley. Description and Verification of a Time-Sensitive Protocol. Technical Report CIS-96-009, University of South Australia, School of Computer and Information Science, Adelaide, Australia, October 1996.
http://www.cis.unisa.edu.au/cgi-bin/techreport?CIS-96-009

7. C. Daws, and S. Yovine. Two Examples of Verification of Multirate Timed Automata with Kronos. In *Proceedings of the 7th 1995 IEEE Real Time Systems Symposium*, Pisa, Italy, December 1995, IEEE Computer Society Press.

8. W.O.D. Griffioen. Proof-checking an Audio Control Protocol with LP. Technical Report CS-R9570, CWI, Department of Software Technology, Amsterdam, The Netherlands, October 1995.

9. P.-H. Ho, and H. Wong-Toi. Automated Analysis of an Audio Control Protocol. In *Proceedings of the 7th International Conference on Computer-Aided Verification*, 1995, Lecture Notes in Computer Science 939, pages 381–394, Springer, Berlin, Germany, 1995.

10. C. A. R. Hoare. *Communication Sequential Processes*. International Series in Computer Science. Prentice Hall, 1985.

11. K.G. Larsen, P. Pettersson, and W. Yi. Diagnostic Model-checking for Real-Time Systems. In *Proceedings of the 4th DIMACS Workshop on Verification and Control of Hybrid Systems*, New Brunswick, USA, 22–24 October, 1995.

12. Wenbo Mao. *Verification of Concurrent Finite State Systems*. PhD thesis, University of Strathclyde, Department of Computer Science, Glasgow, UK, 1992.

13. G. J. Milne. The Formal Description and Verification of Hardware Timing. *IEEE Transactions on Computers*, 40(7):811–826, July 1991.

14. G. J. Milne. *Formal Specification and Verification of Digital Systems*. McGraw-Hill, 1994.

15. R. Milner. *Communication and Concurrency*. International Series in Computer Science. Prentice Hall, 1989.

16. F. Moller. The Semantics of Circal. Technical Report HDV-3-89, University of Strathclyde, Department of Computer Science, Glasgow, UK, 1989.

Using HYTECH to Verify an Automotive Control System

Thomas Stauner and Olaf Müller*
Department of Computer Science
Munich University of Technology
80290 Munich
{stauner,mueller}@informatik.tu-muenchen.de

Max Fuchs
BMW AG
FIZ EG-K-3
80788 Munich
maximilian.fuchs@bmw.de

Abstract. This paper shows how HYTECH, a symbolic model checker for linear hybrid systems, can be used to verify a part of an abstracted automotive control system. The system controls the height of an automobile by a pneumatic suspension system and has been proposed by BMW AG as a case study taken from a current industrial development. For a system which controls one wheel we verify safety properties, such as that the height of the car maintains within desired bounds or that the height is not changed in curves, by reachability analysis. Furthermore, a property related to stability in the sense of control theory is verified. We believe that the case study can serve as a real-life benchmark problem for the formal analysis of embedded reactive systems.

1 Introduction

Our research partner BMW AG proposed a case study taken from a current industrial development describing a system that controls the height of a chassis by pneumatic suspension. The goal of the paper is to evaluate a representative of the existing tools for the analysis of embedded reactive systems w.r.t. to the degree of automation, scalability and practicability by applying it to this case study. We choose the model of hybrid automata [ACH+95] because a state based model is more familiar to a design engineer than a logical approach. Among the existing hybrid model checkers we select HYTECH [HHWT95], as it provides the most general input language, *linear hybrid automata*. Nevertheless, the case study incorporates continuous activities which cannot be described directly using linear hybrid automata. Therefore we have to employ approximation techniques [HWT96a, HH95].

The Electronic Height Control System. The main aims of the electronic height control (EHC) are to increase driving comfort, to keep the headlight load-independent and to adjust the chassis level to off-road and on-road conditions by providing a high and a low chassis level, selectable by the driver.

This is achieved by a pneumatic suspension at each of the four wheels. The chassis level can be increased by pumping air into the suspension of the wheels

---

* Research supported by "KorSys", BMBF.

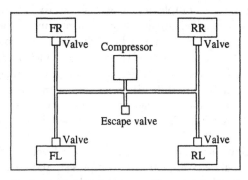

FR: front right wheel   RR: rear right wheel   sp: set point
FL: front left wheel   RL: rear left wheel   iti: inner tolerance interval
                                                 oti: outer tolerance interval

Fig. 1.   Fig. 2.

and decreased by blowing air off. As there is only one compressor and one escape valve for all four wheels, to e. g. increase the level at the front right wheel, the compressor has to be turned on, the escape valve must be closed and the valve connected to the suspension of the front right wheel must be opened (Fig. 1).

Sensors measure the deviation of the actual chassis level from a defined zero level at each of the wheels. Disturbances of high frequency, caused by road holes or block pavement for example, are filtered out by low-pass filters before the values are passed on to the controller. The controller reads them with a given sampling rate and decides how valves and compressor must be operated during the next sampling interval. Escape valve or compressor respectively are turned on if the filtered chassis height is outside an outer tolerance interval for one of the wheels (Fig. 2). For example, if the height exceeds the outer tolerance interval for one wheel, the valve connected to this wheel's suspension and the escape valve are opened. The valves are closed again if the measured chassis levels at all four wheels are below the upper limit of an inner tolerance interval, in addition the filters are reset to the set point, i. e. the desired chassis level. To increase the chassis level the compressor and the valves connected to the suspensions are operated in a similar way.

As compressor and escape valve must not be used simultaneously, priority is given to the compressor in those cases where both would have to be used.

The actual values of the tolerance intervals are determined by the mode the car is in. The modes we model in this case study are "car is stopped, engine off" and "car is driving, engine on". In section 4.3 we will regard a further mode which provides that control is suspended when the car is driving through a bend. The original study of BMW contains some more modes which mainly differ in the actual values for the inner and outer tolerance intervals.

Results. We analyze a restricted model with only one wheel and one chassis level. This seems to be a natural abstraction and allows for the verification of several interesting properties without exceeding the efficiency limits of HYTECH.

There are two factors which blow up the complexity of the case study. Firstly approximation techniques have to be applied to the filters with nonlinear[1] behavior. We argue that the complexity increase caused by such approximations are a major disadvantage w.r.t. to the practicability of the existing model checkers that are all restricted to linear hybrid systems. Secondly the environment of the EHC, namely the road conditions, are difficult to model, as they are of an inherently complex and unpredictable nature. Therefore we restrict ourselves to a road model that allows only such disturbances the compressor is able to compensate immediately.

Within this model we prove that our system keeps the chassis level within certain bounds and identify these bounds. We verify that compressor and escape valve are never used at the same time. Furthermore we analyze the property that the EHC does not try to change the chassis level when the car is driving through a bend. Inspired by the notion of stability in control theory we also examine the system response to a step-like disturbance in order to guarantee that the EHC cannot continue to use compressor and escape valve endlessly. For a particular disturbance we prove that usage of valves and compressor terminates within certain time bounds and identify these bounds.

A major problem of the existing hybrid model checkers seems to be their insufficient scalability. Our real-world example reached the efficiency limits of HyTech, where arithmetic overflows seem to be more serious in practice than nontermination. On the other hand, we appreciated the automation of the analysis and the ease of use of HyTech.

Related Work. Several case studies have been performed using HyTech. But only the steam boiler [HWT96b] and the audio control protocol [HWT95] case studies aimed at applying automated analysis techniques to real-life applications. Our case study seems to be distinctly more complex than previous ones because of the nonlinear filter and the complex environment. As far as we know, this paper is the first to describe a case study with HyTech not performed by the developers of the tool themselves. Other hybrid model checkers, such as KRONOS or UPPAAL, have been developed and applied to several examples, as well as non-algorithmic approaches. For a survey, see [AHS96].

2 Hybrid Automata

2.1 Hybrid Automata

A hybrid automaton A is a 10-tuple $(X, V, inv, init, flow, E, upd, jump, L, sync)$ with the following meaning (for a detailed introduction see [ACH+95]):

X is a finite ordered set $\{x_1, ..., x_n\}$ of variables over $\mathbb{R}$. A valuation s is a point in $\mathbb{R}^n$, the value of variable x_i in valuation s is the i-th component of s, s_i.

---

[1] *Nonlinear* in the literature on hybrid automata means that the system cannot be described by a linear hybrid automaton without approximation. This is different from the notion of (non)linear differential equations.

When Φ is a predicate over the variables in X, we write $[\![\Phi]\!]$ to denote the set of valuations s for which $\Phi[X := s]$ is true.

V is a finite set of locations. A state of automaton A is a tuple (v, s) consisting of a location $v \in V$ and a valuation s. A set of states is called a region.

inv is a mapping from the set of locations V to predicates over the variables in X. $inv(v)$ is the invariant of location v. When control is in v, only valuations $s \in [\![inv(v)]\!]$ are allowed. In our graphical representation of hybrid automata invariants $True$ are omitted.

$init$ is a mapping from the locations in V to predicates over X. $init(v)$ is called the initial condition of v. (v, s) is an initial state if $init(v)$ and $inv(v)$ are true for s. Initial conditions are expressed as incoming arrows marked with the condition in our automata diagrams. If the initial condition is $False$, the arrow is omitted.

$flow$ is a mapping from the locations in V to predicates over $X \cup \dot{X}$, where $\dot{X} = \{\dot{x}_1, ..., \dot{x}_n\}$ and $\dot{x}_i$ denotes the first time derivative of x_i, dx_i/dt. When control is in location v the variables evolve according to differentiable functions which satisfy the flow condition $flow(v)$.

Formally the δ time-step relation $\overset{\delta}{\to}$ is defined as $(v, s) \overset{\delta}{\to} (v, s')$ iff there is a differentiable function $\rho : [0, \delta] \to \mathbb{R}^n$ with

- $\rho(0) = s$ and $\rho(\delta) = s'$
- the invariant of v is satisfied: $\rho(t) \in [\![inv(v)]\!]$ for $t \in [0, \delta]$
- the flow condition is satisfied: $flow(v)[X, \dot{X} := \rho(t), \dot{\rho}(t)]$ is true for $t \in [0, \delta]$, where $\dot{\rho}(t) = (\dot{\rho}_1(t), ..., \dot{\rho}_n(t))$.

The filter automaton of Fig. 4 for example has flow condition $\dot{f} = \frac{1}{T}(h - f)$.

$E \subseteq V \times V$ is a finite multiset of edges, called transitions. A transition $(v, v') \in E$ has source location v and target location v'. In our automata diagrams transitions are represented by arrows between the corresponding locations.

upd is a mapping from E to the power set of X. $upd(e)$ is called the update set of e. Variables in $upd(e)$ can change their value when transition e is taken.

$jump$ is a mapping from E to jump conditions. The jump condition $jump(e)$ of transition e is a predicate over $X \cup Y'$, where $Y = upd(e) = \{y_1, ..., y_k\}$ and $Y' = \{y_1', ..., y_k'\}$. y_i' stands for the value of variable y_i after the transition is taken, while y_i refers to the value before the transition.

The transition-step relation $\overset{e}{\to}$ is defined as $(v, s) \overset{e}{\to} (v', s')$ iff

- $e = (v, v')$
- the invariants are satisfied: $s \in [\![inv(v)]\!]$ and $s' \in [\![inv(v')]\!]$
- the variables in $upd(e)$ change according to the jump condition: $jump(e)[X, Y' = s, s'[Y]]$ is true, where $s'[Y]$ is the restriction of s' to the variables in Y.
- the variables not in $upd(e)$ remain constant: $s_i = s_i'$ for $x_i \in X \setminus Y$.

Transition steps are discrete, no time passes while they are taken.

In our automata diagrams we write guarded assignments $\Phi \to y_i := c$ for update set $\{y_i\}$ and jump condition $\Phi \land y_i' = c$. Guards $True$ are omitted.

L is a finite set of synchronization labels. Parallel automata which have a common synchronization label must take transitions with this label simultaneously.

sync is a mapping from E to $L \cup \{\tau_A\}$. *sync(e)* is the synchronization label of transition e. The label τ_A may only be used in automaton A and marks transitions of A which are not synchronized with transitions in parallel automata. In automata diagrams each transition e is labeled with *sync(e)* and *jump(e)*, the label τ_A is omitted.

All of the above mappings are total.

Parallel Composition. Parallel composition of hybrid automata can be conveniently used for specifying larger systems. A hybrid automaton is given for each part of the system, communication between the components may occur via shared variables and synchronization labels. The parallel composition of hybrid automata is obtained by a product construction.

2.2 Hybrid Automata and the Tool HYTECH

HYTECH can automatically analyze a subclass of hybrid automata, namely linear hybrid automata. A hybrid automaton A is linear if all its invariants and initial conditions are convex linear predicates over X, all flow conditions are convex linear predicates over $\dot{X}$ and all jump conditions are convex linear predicates over $X \cup X'$. A predicate over variables in a set Y is linear if it is an (in)equality between linear terms over Y. A linear term over Y is a linear combination over Y with rational coefficients. A predicate is a convex linear predicate if it is a finite conjunction of linear predicates.

HYTECH can compute the set of states of a parallel composition of linear hybrid automata which is reachable from a set of initial states by repeatedly applying time- and transition-steps. It can also perform backward reachability analysis in which the region is computed from which a given final region can be reached by iterating time- and transition-steps. Monitor automata may be used to prove complex properties of hybrid systems. Sequences of time- and transition-steps leading from one region to another can be generated, which is very useful for finding out why given predicates are violated. Furthermore HYTECH allows the analysis of models containing parameters and can thereby be used to synthesize constraints on these parameters which are necessary for correctness of the system [HHWT95].

Approximation techniques can be used to analyze nonlinear hybrid systems [HH95]. Such techniques ensure that properties which are verified for the approximating system also hold for the real system. In section 3.2 we will apply an approximation technique to the filter of the EHC.

3 System Description

3.1 The Environment

The measured chassis level is influenced by disturbances from the outside world, e. g. the road, and by escape valve and compressor.

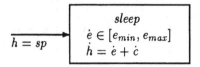

Fig. 3. Hybrid automaton for environment.

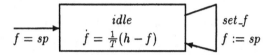

Fig. 4. Hybrid automaton for the filter.

As we cannot make any predictions on when a disturbance will occur and how strong it will be, we say that disturbances of limited strength may occur anytime. There is no sense in allowing disturbances of unlimited strength, since the real system, i. e. the stability of the car, is limited.

The exact effect escape valve and compressor have on the chassis level depends on many factors which we do not want to model explicitly. We only state that usage of the compressor increases the chassis level while usage of the escape valve decreases it. In our system the compressor can lift the chassis with a rate between $cp_{min} = 1\frac{mm}{s}$ and $cp_{max} = 2\frac{mm}{s}$, the escape valve can lower it with a rate between $ev_{min} = -2\frac{mm}{s}$ and $ev_{max} = -1\frac{mm}{s}$.

The rate of change of the chassis level at one wheel $\dot{h}$ now is the sum of the changes due to disturbances, denoted by $\dot{e}$, and the changes due to compressor and escape valve, denoted by $\dot{c}$ (Fig. 3). sp denotes the set point, i. e. the desired chassis level. In order to be able to prove upper and lower bounds for the chassis level in the presence of disturbances it is necessary to demand that disturbances cannot be stronger than compressor and escape valve respectively, i. e. $e_{min} \geq ev_{max}$ and $e_{max} \leq cp_{min}$. If the disturbances were stronger, the controller could not ensure bounds for the chassis level even if it operated correctly. In our system we set $e_{min} = ev_{max}$ and $e_{max} = cp_{min}$.

The influence of the environment is of course seriously limited this way. In reality disturbances stronger than compressor and escape valve are certainly possible. To get a more realistic environment it would therefore be necessary to model it in a way which allows strong disturbances, but provides that their influence on the system is not stronger than that of the controller over a longer period of time. We believe that the limits of the expressiveness of hybrid automata are reached with statements of this kind.

3.2 A Filter as Linear Hybrid Automaton

A filter is described by the hybrid automaton of Fig. 4. h is the chassis level as it comes from the sensor of a wheel, f is the filtered level. T is the time constant of the filter, i. e. the time it takes for f to reach $0.63\,h$ starting from $f = 0$, where

Location	Invariant	Flow condition
A	$h - f \in (-\infty, -10]$	$\dot{f} \in (-\infty, -5]$
B	$h - f \in [-10, -6]$	$\dot{f} \in [-5, -3]$
C	$h - f \in [-6, 0]$	$\dot{f} \in [-3, 0]$
D	$h - f \in [0, 6]$	$\dot{f} \in [0, 3]$
E	$h - f \in [6, 20]$	$\dot{f} \in [6, 10]$
F	$h - f \in [20, \infty)$	$\dot{f} \in [10, \infty)$

Table 1. Partitioning of location *idle*.

$h \neq f$ is constant [PH88]. We use a time constant of $2s$. The synchronization label *set_f* is used by the control logic to reset the filter to the set point *sp*.

Linear Phase-Portrait Approximation for a Filter. Unfortunately the filter of Fig. 4 is not a linear hybrid automaton, because its flow condition is not a convex linear predicate over $\{\dot{f}, \dot{h}\}$. To get a linear hybrid automaton we apply linear phase-portrait approximation, defined in [HWT96a], to the filter.

To do so we have to choose a partitioning of $[\![inv(idle)]\!] = \mathbb{R}^2$. The approximating automaton has one location for each partition then. The flow condition of such a new location is obtained by computing the upper and lower limits of the original flow condition in the respective partition. Every new location has a *set_f* transition to every other new location which corresponds to *idle* and to itself. Finally transitions have to be added which allow control to pass freely between the locations of the approximating filter which correspond to *idle*. The partitioning we use for location *idle* is shown in Table 1. $h - f$ has unit millimeters, $\dot{f}$ has unit millimeters per second.

In Table 1 locations C and D with $\dot{f} = sp$ are initial states. There is a *set_f* transition with jump condition $f' = sp$ from every location in $\{A, ..., F\}$ to every location in $\{A, ..., F\}$. Furthermore there are τ_{filter} transitions with jump condition *True* from every location of the approximating filter to every other location of it to allow control to pass freely between them.

The choice of this partitioning is guided by the desire to keep the state space small and to make locations A and F unreachable which is achieved by locations B and E. The invariants of C and D are chosen in a way to avoid arithmetic overflows during HYTECH's analysis of the EHC. The flow conditions of the locations are determined by the invariants and the original filter's flow condition: e.g. for $h - f \in [0, 6]$, $\dot{f} = \frac{1}{T}(h - f)$ is in $[0, 3]$.

A and F should be unreachable for two reasons. First it is not desirable that the distance between h and f is unbounded. Secondly in composition with the other parts of our model of the EHC the flow conditions of A and F lead to flow conditions which are not permitted by HYTECH for efficiency reasons. We therefore use $+/-100$ as upper and lower bounds for $\dot{f}$ in A and F instead of $+/-\infty$ in Table 1. Unless the contrary is mentioned, reachability analysis proves that A and F are unreachable from the initial states in all models presented in this paper. Therefore these bounds do not affect system behavior.

Linear Phase-Portrait Approximation and Parallel Composition. Although phase-portrait approximation is not defined for parallel compositions of automata in [HWT96a], our approximation of the filter is correct. To see this we have to examine what can go wrong when the approximation technique is only applied to one automaton of a parallel composition, the filter in our case.

An automaton running in parallel to the approximation of the filter might want to take a transition which would falsify the invariant of the current location of the approximating filter automaton. It is therefore not allowed to take this transition. However in the unpartitioned filter the same transition would be possible, because $inv(idle) = True$ can never be violated. So the partitioned system would not be a correct approximation, because the reachable region of the original system would contain states which would not have corresponding states in the reachable region of the approximating system. This problem can be solved by forcing all transitions which change the value of variables appearing in the invariants of filter locations, i. e. all transitions which change f or h, to occur simultaneously with new transitions in the filter which allow control to pass between the locations $A, ..., F$. To make the transitions occur simultaneously a synchronization label must be used. We will encounter this situation in section 4.4 and apply the presented method.

Time-steps cannot affect the correctness of the approximating system. Due to the definition of time-steps which demands all variables to follow differentiable trajectories, $h - f$ must also follow a differentiable trajectory. When $h - f$ moves from one interval of the partitioning of $idle$ into another in a time-step, it can therefore only do so by passing a value which is in both intervals. As soon as this value is reached, the approximating filter automaton can take a τ_{filter} transition which lets control pass freely between the locations of the corresponding intervals. Time can progress in the newly reached location then.

Linear phase-portrait approximation in the context of parallel composition is examined in detail in [Sta97].

3.3 The Controller

To focus on the main points, we restrict ourselves to model only one wheel. The principal architecture of the system is the same with four wheels. The only real simplification is that situations in which escape valve and compressor would have to be used at the same time, because the chassis level is too high at one wheel and too low at another, can no longer occur. If more than one wheel was regarded, in theory a situation is possible in which h diverges to $+\infty$ at one wheel due to the compressor being permanently busy with stabilizing the chassis level at another wheel. Our model of the environment would have to be modified to exclude this unrealistic case of a permanent disturbance in one direction.

Furthermore we only model the two basic modes *stopped* (with engine off) and *driving* (with engine on). In section 5.3 we add a further mode in which control is suspended when the car is in a bend. In addition, we abstract from the possibility to select different chassis levels and use only one level.

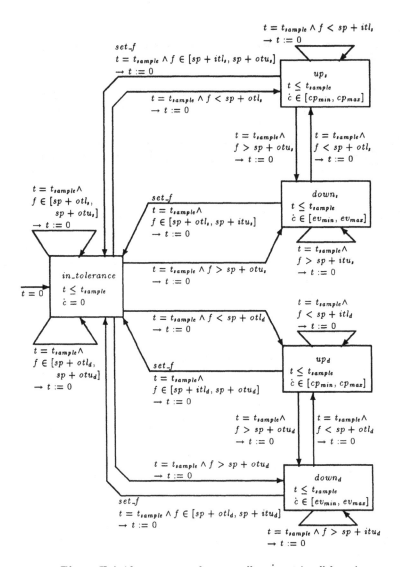

Fig. 5. Hybrid automaton for controller. $\dot{t} = 1$ in all locations.

In Fig. 5 we give the hybrid automaton for the simplified controller. It consists of two symmetric parts corresponding to the two modes, *stopped* and *driving*, denoted by indices s and d. In locations up_s and up_d the compressor is on and the valve connected to the suspension of the wheel is open, the chassis level is therefore increased by $\dot{c} \in [cp_{min}, cp_{max}]$. In locations $down_s$ and $down_d$ the escape valve and the valve to the wheel's suspension are open, the chassis level is decreased by $\dot{c} \in [ev_{min}, ev_{max}]$. Locations up_s and $down_s$ only differ from up_d and $down_d$ in the constants which are used in their incoming and outgoing transitions. In *in_tolerance* the compressor is off and both valves are closed, the controller does not influence the chassis level, $\dot{c} = 0$. t is the controller's clock. As its flow condition is $\dot{t} = 1$ in all locations, we omitted it in Fig 5. t_{sample} determines the controller's sampling rate $\frac{1}{t_{sample}}$. $[sp + otl, sp + otu]$ is the outer tolerance interval, $[sp + itl, sp + itu]$ is the inner tolerance interval. These constants have the following values provided by BMW AG: $sp = 0mm$, $otl_s = -40mm$, $otu_s = 20mm$, $otl_d = -10mm$, $otu_d = 10mm$, $itl_s = -6mm$, $itu_s = 16mm$, $itl_d = -6mm$, $itu_d = 6mm$.

Starting from the initial state, the controller has to wait in *in_tolerance* until t_{sample} expired. Then it reads the filtered chassis height f and goes to location *up* or *down* if f is outside the outer tolerance interval, otherwise it reenters *in_tolerance*. Whenever *in_tolerance* is left the controller makes a non-deterministic choice between one transition with constants from mode *stopped* and one with constants from mode *driving*. This non-determinism models that the mode may change anytime. In the real system the controller reads a sensor to get the present mode.

The locations up_s and up_d are left after t_{sample} time units if f is no longer smaller than the lower limit of the inner tolerance interval $sp + itl$, otherwise they are reentered. If f is greater than the outer upper tolerance limit $sp + otu$ now, $down_s$ or $down_d$, respectively, has to be entered. Otherwise the filter is reset to the set point by taking the *set_f* transition from up_s or up_d to *in_tolerance*. Locations $down_s$ and $down_d$ work in a similar way. In all the *up* and *down* locations a mode change is ignored until *in_tolerance* is entered.

4 Verification

In the following we will show that our system keeps the chassis level within certain bounds and identify these bounds (Section 4.1). Apart from that we prove that compressor and escape valve are never used at the same time (Section 4.2). Furthermore it is safety critical that the EHC does not try to change the chassis level when the car is driving through a bend. To treat this property we have to model bends and modify the controller appropriately (Section 4.3). Finally we are interested in stability of the EHC. Stability in a very general sense means that the system responds in some controlled manner to applied inputs [PH88]. In our case we want to know whether the controller reaches *in_tolerance* again some time after a disturbance or whether it can continue to use compressor and escape valve infinitely by e.g. operating them alternately . To examine this we regard the system response to a step-like disturbance in Section 4.4.

4.1 Bounds for the Chassis Level

Using the model of section 3, HYTECH computes that the chassis level h always is in $[-47mm, 27mm]$ for a sampling rate of once every second, $t_{sample} = 1s$. The computation takes 62 minutes[2] of CPU time. This means that the outer

---

[2] All performance figures were obtained on a Sun Sparcstation 20 with 128 MB RAM and 350 MB swap space.

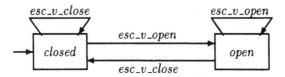

Fig. 6. Hybrid automaton for escape valve.

tolerance limits of mode *stopped*, $-40mm$ and $20mm$, are never exceeded by more that $7mm$. We believe that this result could be significantly improved by using a finer partitioning for the filter and a smaller t_{sample}.

Unfortunately the values of the constants in a model can lead to arithmetic overflows during the analysis with HyTech. When repeatedly computing successor regions in the model checking procedure increasing values can occur in the equations which determine the range of the variables of a system. These values can cause arithmetic overflow in the libraries HyTech uses for computing successors and can thereby abort analysis of a system. In our case the presented partitioning of the filter and the specific choice of the sampling rate turned out to be crucial for the avoidance of overflows. Therefore we are using this slow sampling rate in all our models although reading the sensor values more frequently would be more realistic.

We also tried to use HyTech to compute the slowest sampling rate which guarantees that the chassis level always stays within given bounds, but we did not succeed due to overflows. As soon as the filter is omitted in our model, this question can easily be solved using HyTech's feature to perform parametric analysis, but it is rather trivial then.

4.2 Escape Valve and Compressor

To prove that escape valve and compressor are never used at the same time it is first of all necessary to include them in our model of the EHC. Fig. 6 shows the hybrid automaton for the escape valve. It is opened by a *esc_v_open* transition and closed by a *esc_v_close* transition. The automata for the compressor and the valve connected to the suspension of the wheel are similar. The compressor is operated via the synchronization labels *comp_on* and *comp_off*, the valve to the suspension via *v_open* and *v_close*.

Whenever control changes from one of the locations of the controller of Fig. 5 to another one, the controller must open or close the valves and turn on or off the compressor correspondingly. For example, when control passes from up_d to *in_tolerance*, the compressor must be switched off (label *comp_off*) and the valve connected to the wheel's suspension must be closed (label *v_close*). As a transition cannot have two synchronization labels, we insert transient locations, i. e. locations in which no time passes, in the controller. Thereby one transition can be split into two, with a transient location in between them.

HyTech can analyze the resulting model in 58 minutes. It proves that all states with escape valve open and compressor on at the same time are unreachable from the initial region. In roughly the same CPU time it is possible to

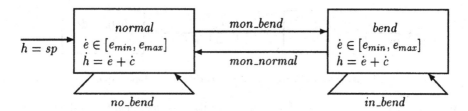

Fig. 7. Hybrid automaton for environment with curves.

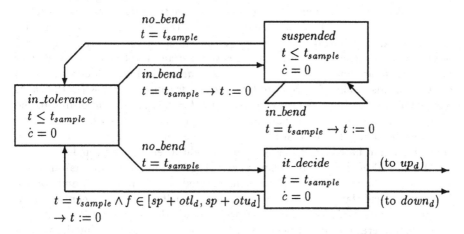

Fig. 8. Priority of the *_bend transitions. $\dot{t} = 1$ in all locations.

treat the property of the previous section together with verifying correct usage of escape valve and compressor in one analysis of the presented model.

4.3 Suspending Control in Curves

In this section we extend the environment by modeling curves and modify the controller of Fig. 5 to prevent it from changing the chassis level when the car is in a bend. Valves and compressor are omitted again.

The new model for the environment is presented in Fig. 7. The transitions with synchronization labels *in_bend* and *no_bend* allow the controller to determine whether the car is in a bend. The transitions between locations *normal* and *bend* have guard *True* and model that a bend can occur anytime. The synchronization labels *mon_bend* and *mon_normal* are matched by a monitor automaton which we use to determine whether the system works correctly. The locations *normal* and *bend* are derived from location *sleep* of Fig. 3.

As far as the controller is concerned, we need a new location, *suspended*, in which the controller does not change the chassis level and which is entered whenever the car is in a curve and left when the curve is over. It is important that *suspended* is entered even if the chassis level is outside the outer tolerance interval, since the controller may not use compressor or escape valve in a curve.

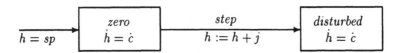

Fig. 9. Hybrid automaton for disturbance of step shape.

We must therefore find a way of modeling priority of transitions. Fig. 8 shows how this is achieved for *in_tolerance*. A new transient location, *it_decide*, is inserted. It can only be entered by a *no_bend* transition. The outgoing transitions of *in_tolerance* in the automaton of Fig. 5 now emerge from *it_decide*. The decision whether the chassis level has to be changed during the next time period is thereby postponed after ensuring that the car is not in a curve. Locations up_d and $down_d$ also get an *in_bend* transition to *suspended* and locations up_d_*decide* and $down_d$_*decide* are added in the same way as *it_decide* in Fig. 8. Mode *stopped* is omitted in this model of the controller since bends cannot occur when the car is not driving. Transitions can only be taken for $t = t_{sample}$, i. e. according to the sampling rate. The clock t is not reset to 0 when *suspended* is left, *in_tolreance* therefore becomes transient in this case and the controller is forced to make a control decision based on the chassis height right after *suspended* is left.

Our model of the system is now augmented by a monitor automaton which measures the time between a bend being entered (label *mon_bend*) and control being suspended (label *in_bend*) or the bend being left again (label *mon_normal*), whatever occurs first. As we also want to know whether the controller continues its normal operation after a bend, we furthermore measure the time between a *mon_normal* transition and the corresponding *no_bend* or *mon_bend* transition. If the elapsed time in one of these cases is greater or equal t_{sample}, the monitor enters an *error* location.

HYTECH can analyze the resulting model in 73 seconds and verifies that *error* with a measured time greater than t_{sample} is unreachable from the system's initial states. This means that *suspended* is always entered at most t_{sample} time units after a bend which lasts longer than t_{sample} started. It is left again after at most t_{sample} time units after the bend ended, if no other bend started within t_{sample} time units after the first. As we have used backward reachability analysis starting from location *error* in this verification, locations A and F of the filter may be visited during the analysis. Their flow conditions which are not justified by the nonlinear filter automaton can influence the analysis by giving incorrect values for f. We argue that the actual values of f are not crucial in this verification task since they do not affect entry and exit of *suspended*.

4.4 Step Response of the EHC

In this section we examine the response of the EHC to a disturbance of step shape. Disturbances of this kind are typical test functions in control systems engineering, they can be used to examine stability of control systems. We use this specific environment scenario to analyze whether the activation of compressor and escape valve terminates after a step-like disturbance.

Fig. 9 shows the hybrid automaton for the environment we use to model such a disturbance. This automaton replaces the environment model we used in the previous sections. Starting with a chassis level $h = sp$, h may make a jump by j units anytime. No other disturbances are possible besides this single jump, i. e. h can only be manipulated by the controller in locations *zero* and *disturbed*, denoted by flow condition $\dot{h} = \dot{c}$ in these locations. As the *step* transition changes the value of h, the approximation technique we used for the filter in section 3.2 demands that it is taken simultaneously with a filter transition which allows control to pass to another filter location. This is achieved by adding new transitions with jump condition *True* and synchronization label *step* form every filter location in $\{A, ..., F\}$ to every location in $\{A, ..., F\}$.

To do a complete analysis of the system response to a step-like disturbance, we would like to regard steps of arbitrary height j, or at least all step heights which are allowed by the mechanics of the suspension system. Unfortunately the disturbances we can analyze with HyTech are restricted, as the values we use in our system can cause an arithmetic overflow thereby aborting analysis. Therefore we only analyze the controllers response in mode *driving* to steps with $j \in (16mm, 18mm]$ as an example. Values of j in this interval are interesting, as we expect the controller to use the escape valve after disturbances of this size.

Apart from that we modify the controller of Fig. 5 by not giving intervals for the performance of compressor and escape valve in up_d and $down_d$. Instead we use the conservative assumption that they operate at their minimum rates, $\dot{c} = cp_{min}$ in up_d and $\dot{c} = ev_{max}$ in $down_d$. This restriction was necessary to avoid arithmetic overflows using HyTech.

By adding a monitor automaton to our model we use HyTech's parametric analysis facility to compute that the controller leaves location *in_tolerance* at most $4.3s$ after the disturbance and reenters it after at most $22.3s$. The chassis level then lies in $[-1mm, 6mm]$ and *in_tolerance* is not left again. The analysis takes 370 seconds of CPU time.

5 Conclusion

We have shown that HyTech can successfully analyze several interesting properties of an abstract version of the electronic height control system. Nevertheless, we reached the efficiency limits of the tool. We believe that the nonlinear behavior of the filter and the necessarily applied approximation technique is the main reason for the obtained complexity. The major efficiency problem of HyTech are overflows caused by the arithmetic libraries which are very sensitive to the specific values of the constants of the model. As the user does not get any feedback on which constants have caused the overflow, trial and error must be used to identify them. In our models the use of fractions turned out to be disadvantageous as far as overflows are concerned. Therefore we only use integer constants in the current models. The problem of overflows could perhaps be weakened by developing routines which are optimized for the use with HyTech instead of using standard libraries. Running out of memory is less problematic, because the reasons for this can be found easier than those for overflows.

Modeling the influences of the road, i.e. the environment of the system, was a major challenge of the case study. On the one hand the disturbances of the environment can be stronger than the controller for a short time, on the other hand they should be weaker than the controller when the average influence over a longer time period is regarded. We believe that the limits of expressiveness of hybrid automata are reached with complex environment models of this kind. Furthermore the high degree of non-determinism in such a model makes automated analysis extremely difficult if not impossible.

We investigated the interference of parallel composition and linear phase-portrait approximation, identified criteria for an approximation of a single component of a composition and applied them to the filter in our case study.

This case study may be a first step towards more industrial interest in the modeling and analysis of hybrid systems. We are interested in how other methods and tools can handle this case study and believe that it can serve as a real-life benchmark for the analysis of embedded systems.

Acknowledgement. We thank Howard Wong-Toi for his numerous hints on the use of HYTECH.

References

[ACH+95] R. Alur, C. Courcoubetis, N. Halbwachs, T. A. Henzinger, P.-H. Ho, X. Nicollin, A. Olivero, J. Sifakis, and S. Yovine. The algorithmic analysis of hybrid systems. *Theoretical Computer Science*, 138:3–34, 1995.

[AHS96] R. Alur, T.A. Henzinger, and E.D. Sontag. *Hybrid Systems III.* Lecture Notes in Computer Science 1066. Springer-Verlag, 1996.

[HH95] T.A. Henzinger and P.-H. Ho. Algorithmic analysis of nonlinear hybrid systems. In P. Wolper, editor, *CAV 95: Computer-aided Verification*, Lecture Notes in Computer Science 939, pages 225–238. Springer-Verlag, 1995.

[HHWT95] T.A. Henzinger, P.-H. Ho, and H. Wong-Toi. A user guide to HYTECH. In E. Brinksma et al., editors, *Tools and Algorithms for the Construction and Analysis of Systems*, LNCS 1019, pages 41–71. Springer-Verlag, 1995.

[HWT95] P.-H. Ho and H. Wong-Toi. Automated analysis of an audio control protocol. In P. Wolper, editor, *CAV 95: Computer-aided Verification*, Lecture Notes in Computer Science 939, pages 381–394. Springer-Verlag, 1995.

[HWT96a] T.A. Henzinger and H. Wong-Toi. Linear phase-portrait approximations for nonlinear hybrid systems. In R. Alur, T.A. Henzinger, and E.D. Sontag, editors, *Hybrid Systems III*, LNCS 1066. Springer-Verlag, 1996.

[HWT96b] Thomas A. Henzinger and H. Wong-Toi. Using HYTECH to synthesize control parameters for a steam boiler. In J.-R. Abrial et al., editors, *Formal Methods for Industrial Applications: Specifying and Programming the Steam Boiler Control*, LNCS 1165. Springer-Verlag, 1996.

[PH88] Charles L. Phillips and Royce D. Harbor. *Feedback Control Systems.* Prentice-Hall, 1988.

[Sta97] Thomas Stauner. *Specification and Verification of an Electronic Height Control System using Hybrid Automata.* Master thesis, Munich University of Technology, 1997.

Safety Verification for Automated Platoon Maneuvers:
A Case Study

Ekaterina Dolginova and Nancy Lynch

MIT Laboratory for Computer Science
Cambridge, MA 02139, USA
{katya, lynch}@theory.lcs.mit.edu

Abstract. A system consisting of two platoons of vehicles on a single track, plus controllers that operate the vehicles, plus communication channels, is modeled formally, using the hybrid input/output automaton model of Lynch, Segala, Vaandrager and Weinberg [7]. A key safety requirement of such a system is formulated, namely, that the two platoons never collide at a relative velocity greater than a given bound v_{allow}. Conditions on the controller of the second platoon are given, designed to ensure the safety requirement regardless of the behavior of the first platoon. The fact that these conditions suffice to ensure safety is proved. It is also proved that these conditions are "optimal", in that any controller that does not satisfy them can cause the safety requirement to be violated. The model includes handling of communication delays and uncertainty. The proofs use composition, invariants, levels of abstraction, together with methods of mathematical analysis.

This case study is derived from the California PATH intelligent highway project, in particular, from the treatment of the platoon join maneuver in [3].

1 Introduction

Increasing highway congestion has spurred recent interest in the design of intelligent highway systems, in which cars operate under partial or total computer control. An important new effort in this area is the California PATH project (see, for example, [9]), which has developed a design for automating the operation of cars in several lanes of selected California highways. In this design, cars become organized into *platoons* consisting of a leader car and several following cars; the followers do not operate independently, but follow the control instructions of the leader.

An important maneuver for the proposed PATH system is the *platoon join* maneuver, in which two or more adjacent platoons combine to form a single platoon. The design of such a maneuver is described and analyzed in [3]. This maneuver involves both discrete and continuous behavior: discrete behavior appears in the form of synchronization and agreement among the controllers about the join process, plus communication among the various system components, whereas continuous behavior appears in the motion of the cars. The combination forms a hybrid system of considerable complexity.

A key issue for the platoon join maneuver is its safety, represented by the requirement that cars never collide at too great a relative speed. In [3], a proof of such a safety property is outlined, for the specific platoon join maneuver given in that paper. The key to the proof turns out to be that the given maneuver always ensures that either (a) the platoons are sufficiently far apart that the second platoon can slow down sufficiently before hitting the first platoon, or (b) the relative speeds of the two platoons are already close enough.

Although the outline [3] gives the key ideas, from our point of view, it is incomplete as a safety verification. It does not include a complete model of all system components – in particular, the discrete components are not modeled. It does not seem to cover all cases that could arise: for instance, only some types of communication delay are handled, and uncertainties in the values of some parameters are not considered. The analysis contains informal "jumps" in which certain types of behavior are claimed to be the "worst possible", and then only these cases are analyzed carefully; however, it is not made clear how one can be sure that the claimed worst cases are in fact the worst. Another problem is that the analysis is presented for just the single maneuver, and is intertwined with the proofs of other properties for that maneuver (successful join, optimality of join time). However, it seems that the analysis should be decomposable, for example, proving the safety requirement in a way that allows the proof to apply to other maneuvers besides just the platoon join.

In previous work [7], Lynch, Segala, Vaandrager and Weinberg have developed a formal model, the *hybrid input/output automaton model*, for hybrid systems, together with associated proof techniques. These techniques include methods based on automaton composition, on invariant assertions, on levels of abstraction, and on mathematical analysis for reasoning about continuous behavior. They have developed methods of incorporating standard methods of analysis into automaton-based proofs. So far, these methods have been used to model and verify a variety of simple real-time systems, including several very simple maneuvers arising in automated transportation systems ([11], [10], [6]).

In this case study, we apply the hybrid I/O automaton model and its associated proof methods to the task of describing and verifying safety for the PATH platoon join maneuver. This is a more complex example than those previously considered using hybrid I/O automata. We aim for an accurate, complete model of the system, plus proofs that cover all cases and accommodate all realistic variations, including delays and uncertainties. Our safety proofs should apply as generally as possible, for instance, to other maneuvers besides platoon join. Our model should also be usable for proving other properties, such as successful join and optimality. The system and its proofs should admit decomposition into separate parts, as far as possible, and should be easy to extend.

In the work we have completed so far, we have made certain simplifications. Namely, we consider the case of two platoons only (as in [3]), and we consider uncertainties in only some of the parameter values. Moreover, we pretend that the controllers control the cars' acceleration rather than their jerk (derivative of the acceleration). We intend to remove these restrictions in later work, and are designing our models and proofs to make such extensions easy.

For this simplified setting, we have succeeded in modeling the complete system, which consists of two platoons of cars on a single track, plus controllers that operate the cars, plus communication channels. We have formulated the safety requirement, namely, that the two platoons never collide at a relative velocity greater than a given bound v_{allow}. We have given conditions on the controller of the second platoon, designed to ensure the safety requirement regardless of the behavior of the first platoon, and we have proved that these conditions suffice to ensure safety. Our proofs cover all cases, and are sufficiently general to apply to other maneuvers besides platoon join. The proofs use discrete systems techniques, such as composition, invariants, and levels of abstraction. Additionally, the methods of mathematical analysis developed for proving invariance of state-space sets in [2] are used for reasoning about the continuous parts of the system.

In addition to proving safety, we also give results showing that the given conditions on the controllers are "optimal", in the sense that any controller that does not satisfy them can cause the safety requirement to be violated. The optimality results are proved using the same techniques (in particular, invariants and composition) that are used for the safety proof. Again, the optimality results apply to other maneuvers besides platoon join.

An alternative approach to proving safety for the platoon join maneuver, based on game theory, is presented in [5], [4]. There has been a large amount of prior work on modelling and verification of hybrid systems, as represented, for example, in the six previous workshops on hybrid systems. Nearly all of this work differs from ours in using either control theory methods, or else algorithmic techniques (e.g., decision procedures based on finite-state analysis). Other formal models for hybrid systems appear in [8], [1]; these differ from ours primarily in placing less emphasis on issues of external behavior, composition and abstraction.

We consider the research contributions of this paper to be: (a) The model and proof of safety for the platoon join (and other maneuvers). (b) The optimality result and its proof. (c) A demonstration of the power of hybrid I/O automata and its associated proof methods for reasoning about interesting hybrid systems. (d) A demonstration of the use of abstraction levels as a means of handling complexity.

2   HIOA Model

The Hybrid I/O Automata model presented in [7] is capable of describing both continuous and discrete behavior. The model allows communication among components using both shared variables and shared actions. Several HIOA techniques make them particularly useful in modeling and reasoning about hybrid systems. These include composition, which allows to form complex automata from simple building blocks; implementation relations, which make it easy to use levels of abstraction when modeling complex systems; invariant assertions, which describe the non changing properties of the system.

A *state* of a HIOA is defined to be a valuation of a set of variables. A *trajectory* w is a function that maps a left-closed interval I of the reals, with left endpoint equal to 0, to states; a trajectory represents the continuous evolution of the state over an interval of time. An HIOA A consists of:

- Three disjoint sets of *input, output and internal* variables. Input and output variables together are called *external* variables.
- Three disjoint sets of *input, output and internal* actions.
- A nonempty set of *start states*.
- A set of *discrete transition*, i.e. (state, action, state) triples.
- A set of trajectories over the variables of A.

We now define executions of HIOAs. A *hybrid execution fragment* of A is a finite or infinite alternating sequence of trajectories and actions, ending with a trajectory if it is finite. An execution fragment records all the discrete changes that occur in an evolution of a system, plus the continuous state changes that occur in between. Hybrid execution fragments are called *admissible* if they are infinite. A *hybrid execution* is an execution fragment in which the first state is a start state. A state of A is defined to be *reachable* if it is the last state of some finite hybrid execution of A. A *hybrid trace* of a hybrid execution records only the changes to the external variables. Hybrid traces of an HIOA A (hybrid trace that arise from all the finite and admissible hybrid executions of A) describe its visible behavior.

HIOA A *implements* HIOA B if every behavior of A is allowed by B. A is typically more deterministic than B in both the discrete and the continuous level. Formally, if A implements B, then 1) A and B are *comparable* HIOA, meaning that they have the same external actions and external variables; 2) all the hybrid traces of A are included in those of B. To prove the second part, we need to show that there exists a *simulation relation* from A to B. A simulation relation from A to B is a relation R from states of A to states of B satisfiying:

- If s_A is a start state of A, then there exists s_B, $s_A R s_B$, such that s_B is a start state of B.
- If a is an action of A, (s_A, a, s'_A) is a discrete transition of A, $s_A R s_B$, and both s_A and s_B are reachable, then B has a finite execution fragment starting with s_B, having the same trace as the given step, and ending with a state s'_B with $s'_A R s'_B$.
- If w_A is a trajectory of A from s_A to s'_A, $s_A R s_B$, and both s_A and s_B are reachable, then B has a finite execution fragment starting with s_B, having the same trace w, and ending with a state s'_B with $s'_A R s'_B$.

Another technique for reducing complexity is HIOA *composition*. HIOAs A and B can be composed if they have no output actions or output variables in common, and if no internal variable of either is a variable of the other. The composed HIOA C's input variables/actions are the union of A and B's input variables/actions minus the union of A and B's output variables/actions; all the other components (output and internal variables/actions, start states, discrete actions, trajectories) are the unions of the corresponding components of A and B. The crucial result is that the composition operator respects the implementation relation: if A_1 implements A_2 then A_1 composed with B implements A_2 composed with B. Finally, *invariant assertions* state system properties that are true in any reachable state of the system.

3   System Model

We consider two platoons of vehicles, moving along a single track. While the behavior of the leading platoon is arbitrary, the second platoon's controller must make sure that no "bad" collisions occur. "Bad" collisions are collision at a high relative speed. This is called the *Safety* requirement for the second controller. This *Safety* requirement is general for all platoon maneuvers, and is independent of the particular algorithm used. We devise the most nondeterministic safe controller, so that later we can use this controller as a correctness check: a controller implementing any platoon maneuver must implement our safe controller in order to be correct. This should be very useful in formally proving correctness of complicated algorithms.

3.1   *Controlled-Platoons*

We compose our system of a piece modeling the real world (the physical platoons) and two pieces modeling the controllers of each platoon (which are described in the next subsection). Each piece is modeled by a hybrid automaton. The real world piece is called *Controlled-Platoons*, shown in Figure 1. It consists of two platoons, named 1 and 2, where platoon 1 precedes platoon 2 on a single track. Positions on the track are labeled with nonnegative reals, starting with 0 as a designated beginning point. We pretend for simplicity here that the platoons have size 0. In the full version of the paper this restriction is relaxed. Note that the velocities of the platoons are always nonnegative – the vehicles will never go backwards, and the platoons are not allowed to bypass each other.

Only single collisions are modeled here. A special *collided* variable keeps track of the first occurrence of a collision. Before a collision, the platoons obey their respective controllers by setting the given acceleration. After a collision occurs, the platoons are uncoupled from the controllers and their velocities are set arbitrarily.

We use the constants $v_{allow} \in \mathsf{R}^{\geq 0}$ to represent the largest allowable velocity when a collision occurs, and $a_{min} \in \mathsf{R}^{\geq 0}$ to represent the absolute value of the maximum emergency deceleration. The platoons' position, velocity, and acceleration data is modeled by x_i, $\dot{x}_i$, and $\ddot{x}_i$, respectively. The dots are used as a syntactic device only. The differential relationships between these variables is a consequence of the trajectory definitions; however, this differential relationship

Actions:

> Internal: *collide*

Variables:

> Input: $\ddot{x}_i \in \mathsf{R}$, $i \in \{1,2\}$, initially arbitrary
>
> Output: $\dot{x}_i \in \mathsf{R}^{\geq 0}$, $i \in \{1,2\}$, initially arbitrary
>
> $x_i \in \mathsf{R}^{\geq 0}$, $i \in \{1,2\}$; initially $x_2 = 0$ and x_1 is arbitrary
>
> *collided*, Boolean, initially *false*
>
> *now*, initially 0

Discrete Transitions:

> *collide*
>
>> Pre: $x_1 = x_2$
>>
>> $collided = false$
>>
>> Effect: $\ddot{x}_i :=$ arbitrary value, $i \in \{1,2\}$
>>
>> $collided = true$

Trajectories:

> an *I*-trajectory w is included among the set of nontrivial trajectories exactly if
>
> *collided* is unchanged in w
>
> for all $t \in I$ the following hold:
>
>> if $collided = false$ in w then
>>
>> $w(t).\ddot{x}_i = w(0).\dot{x}_i + \int_0^t w(u).\ddot{x}_i du$, $i \in \{1,2\}$.
>>
>> $w(t).now = w(0).now + t$
>>
>> $w(t).x_2 \leq w(t).x_1$
>>
>> $w(t).x_i = w(0).x_i + \int_0^t w(u).\dot{x}_i du$
>>
>> if $w(t).x_1 = w(t).x_2$ and t is not the right endpoint of I then
>>
>>> $collided = true$.

Fig. 1. The *Controlled-Platoons* Hybrid I/O Automaton

is partly lost after a collision occurs. The acceleration data is received from the controllers which are defined below. This will be used in our statement of the correctness property, below — we only want to assert what happens the first time a collision occurs. The second conditions on the trajectories of *Controlled-Platoons* guarantees that the platoon only executes the controller's decisions until the first collision occurs.

3.2 Controllers

Controller$_1$ is described in Figure 2. It is an arbitrary hybrid automaton with the given interface, restricted only by physical limitations. Note that the controller does not have any actions. The last restriction on continuous trajectories, for example, guarantees that the controller does not make the platoon to have negative velocity. The internal velocity and position variables ($\dot{x}_{int1}$ and x_{int1}) are used to keep track of the platoon's own data. This data is obtained by integrating their acceleration settings. Since there are no delays or uncertainties, these variables should correspond exactly to the actual position and velocity of the platoon.

The *Controller*$_2$ hybrid automaton is the same as *Controller*$_1$, except that

Variables:
> Input: $\ddot{x}_2 \in \mathsf{R}^{\geq 0}$
>
> $x_2 \in \mathsf{R}^{\geq 0}$
>
> Output: $\ddot{x}_1$
>
> Internal: $\dot{x}_{int1} \in \mathsf{R}^{\geq 0}$, initially $\dot{x}_{int1} = \dot{x}_1$
>
> $x_{int1} \in \mathsf{R}^{\geq 0}$, initially $x_{int1} = x_1$

Trajectories:
> an I-trajectory w is included among the set of nontrivial trajectories exactly if
>> $\ddot{x}_1$ is an integrable function
>>
>> for all $t \in I$, at $w(t)$
>>> $\dot{x}_{int1} = w(0).\dot{x}_{int1} + \int_0^t w(u).\ddot{x}_1 \, du$
>>>
>>> $x_{int1} = w(0).x_{int1} + \int_0^t w(u).\dot{x}_{int1} \, du$
>>>
>>> $\ddot{x}_1 \geq -a_{min}$

Fig. 2. *Controller$_1$* Hybrid I/O Automaton

it inputs x_1 and $\dot{x}_1$, and outputs $\ddot{x}_2$.

Compose *Controlled-Platoons*, *Controller$_1$* and *Controller$_2$* using hybrid I/O automata composition rules to obtain an automaton that models our platoon system with each platoon having its own controller.

3.3 Safety Condition

We place a safety condition on states of *Controlled-Platoons*. The safety condition guarantees that if the platoons ever collide, then the first time they do so, their relative velocity is no more than v_{allow}. We formulate this condition formally as an invariant assertion:

> *Safety* : If $x_1 = x_2$ and *collided* = *false*, then $\dot{x}_2 \leq \dot{x}_1 + v_{allow}$.

We define a new automaton, *Safe-Platoons*, to serve as a correctness specification. *Safe-Platoons* is exactly the same as *Controlled-Platoons* except that all the states are restricted to satisfy the safety condition.

We are supposed to design *Controller$_2$* so that when it is composed in this way with arbitrary *Controller$_1$*, the resulting system satisfies the safety condition. Then we can say that it implements the *Safe-Platoons* automaton, using a notion of implements based on preserving hybrid traces. Here, the hybrid trace includes the output variables, which are the positions, velocities and accelerations of both platoons plus the *collided* flag. That is enough to ensure that the *Safety* condition of the spec carries over to the implementation.

4 The Ideal Case

4.1 The Model

We start with a treatment of the safety property in the ideal setting. This allows us to prove some important properties of the simpler model first, and then extend them to the more complicated models via simulation mappings. By ideal setting we mean that there are no delays and/or uncertainties in either the sensor's data or the controller's directives. In the next few sections we will make the model more realistic by relaxing these restrictions. Also, in this abstract, we make the simplifying assumption that the platoons have size 0. In the full paper we show how to relax this restriction easily.

We define and prove correctness of a specific *Controller₂*, called C_2, which implements our safety condition, in Figure 3. This controller is very nondeterministic.

Definition:

$$safe\text{-}measure = \max(x_1 - x_{int2} - \frac{(\dot{x}_{int2})^2 - (\dot{x}_1)^2 - (v_{allow})^2}{2a_{min}}, \dot{x}_1 + v_{allow} - \dot{x}_{int2})$$

Variables:
  Input: $\dot{x}_1 \in \mathsf{R}^{\geq 0}$
      $x_1 \in \mathsf{R}^{\geq 0}$
  Output: $\ddot{x}_2$, initially if *safe-measure* ≤ 0, then $\ddot{x}_2 = 0$
  Internal: $\dot{x}_{int2} \in \mathsf{R}^{\geq 0}$, initially $\dot{x}_{int2} = \dot{x}_2$
      $x_{int2} \in \mathsf{R}^{\geq 0}$, initially $x_{int2} = x_2$

Trajectories:
  an *I*-trajectory w is included among the set of nontrivial trajectories exactly if
    w is a trajectory of *Controller₂*
    if *collided = false* in $w(0)$ then $\forall t \in I$
      if *safe-measure* ≤ 0 then $\ddot{x}_2 = -a_{min}$

Fig. 3. C_2 Hybrid I/O Automaton

C_2 ensures that if the position and velocity parameters are on the boundary defined by *safe-measure*, then platoon 2 is guaranteed to be decelerating as fast as possible. This is guaranteed by the second condition on the trajectories of C_2.

4.2 Correctness of C_2

We will now prove correctness of our controller. This means that any controller that implements C_2, will be correct (safe).

We define a predicate S on states of *Platoons*, as follows:

Predicate S: If *collided = false* then *safe-measure* ≥ 0, where

$$safe\text{-}measure = \max(x_1 - x_{int2} - \frac{(\dot{x}_{int2})^2 - (\dot{x}_1)^2 - (v_{allow})^2}{2a_{min}}, \dot{x}_1 + v_{allow} - \dot{x}_{int2})$$

This says (from the definition of *safe-measure*, see Figure 3) that if the platoons have not collided yet, then either (a) the distance between the two platoons is great enough to allow platoon 2 to slow down sufficiently before hitting platoon 1, even if platoon 1 decelerates at its fastest possible rate, or (b) the relative velocities of the two platoons are already close enough.

We define a new automaton *C-Platoons*, which is exactly like *Controlled-Platoons*, with the additional restriction that in all the initial states *safe-measure* ≥ 0 (thus all the initially states satisfy Predicate S, since initially *collided* = *false*). The system *Implemented-Platoons* that we are considering is the composition of *C-Platoons*, an arbitrary *Controller$_1$*, and C_2. C_2 is designed to guarantee explicitly that if S is ever violated, or even if it in danger of being violated (because equality holds), platoon 2 is decelerating as fast as possible. We claim that this strategy is sufficient to guarantee that S is always true:

Lemma 1. *S is true in every reachable state of the* Implemented-Platoons.

As a simple consequence of Lemma 1, we obtain the safety condition:

Lemma 2. *In any reachable state of Implemented-Platoons, if $x_1 = x_2$ and collided = false, then $\dot{x}_2 \leq \dot{x}_1 + v_{allow}$.*

Now we use Lemma 2 to prove that the system is in fact safe, i.e., that it implements *Safe-Platoons*. We prove this using a simulation relation f. This simulation is trivial – the identity on all state components of *Safe-Platoons* (velocities, positions, and the *collided* flag).

Lemma 3. *f is a forward simulation from the composed system* Implemented-Platoons *to Safe-Platoons.*

Proof: By induction on the number of steps in the hybrid execution. Lemma 2 deals with trajectories; the proofs for the start states and discrete steps are relatively simple.

Theorem 4. *The Implemented-Platoons system implements Safe-Platoons, in the sense that for every hybrid execution α of Implemented-Platoons, there is a hybrid execution α' of Safe-Platoons that has the same hybrid trace – here, means same positions, velocity and collided flag values.*

Proof. *Implemented-Platoons* and *Safe-Platoons* are comparable and by Lemma 3, there is a simulation relation f from *Implemented-Platoons* to *Safe-Platoons*. Therefore, this composed system implements *Safe-Platoons*.

4.3 Optimality

We will now prove optimality of *safe-measure* using the analysis theorem about non-increasing functions. Informally, we want to prove that any *Controller$_2$* that does not implements C_2 is unsafe. The formal definition of this optimality property appears in Theorem 7. Combined with the correctness result of the previous subsection, this will allows to decide whether any given controller is safe, since it is safe *if and only if* it implements C_2.

Define *Controller₂* to be bad (and call it *Bad-Controller₂*), if there exists some *Controller₁*, such that in any admissible hybrid execution α of an automaton composed of *Controlled-Platoons*, *Controller₂* and *Controller₁*, $\exists s \in \alpha$, which does not satisfy Predicate S.

Define *Bad-Controller₁*, given *Bad-Controller₂*, so that in the system composed of *Bad-Controller₁*, *Bad-Controller₂* and *Controlled-Platoons* (the system called *Bad-Platoons*), for any admissible hybrid execution β, the following hold:

- $\exists s \in \beta$, s does not satisfy Predicate S;
- strictly after the occurrence of s, $\ddot{x}_1 = -a_{min}$.

The first lemma shows that once Predicate S is violated, it will remain violated, given some "bad" *Controller₁*. Formally,

Lemma 5. *If a given Controller₂ is bad, then in any Bad-Platoons system with this Controller₂, Predicate S is violated in all the states $\in \beta$ that occur strictly after s, in which collided = false. (Bad-Platoons, β, s are as defined above.)*

The next lemma shows that if Predicate S is violated in some state, then *safety* will also be violated eventually. Formally,

Lemma 6. *If a given Controller₂ is bad, then in any Bad-Platoons system with this Controller₂, in any admissible execution γ, $\exists s' \in \gamma$, which does not satisfy safety).*

Theorem 7. *For any Bad-Controller₂, there always exists such Controller₁ (C'₁, which is not necessarily the same as C₁), that a Bad-Platoons system composed with these controllers has in its hybrid trace a state s', in which safety is violated.*

The last theorem shows that our controller is optimal, i.e., any *Controller₂* that does not implement it, might lead to an unsafe state, given some "bad" *Controller₁*.

5 Delayed Response

Now we consider the case where there is a delay between the receipt of information by the controller for platoon 2 and its resulting action. There appear to be two distinct types of delay to consider — the inbound and the outbound delay; we model them separately. The inbound delay is due to delays in communicating sensor information to the controllers. The outbound delay comes from the fact the controller's decision are implemented by the platoons after some delay.

We use levels of abstractions to deal with the complexity of the delayed case. The use of simulation relations enables us to build correctness and optimality proofs based on the previous ideal case results. This makes all the proofs significantly easier.

5.1 The System with Inbound and Outbound Delays

We model both the inbound and the outbound delays by special delay buffers. To obtain the delayed system, we then compose our new controller with the delay buffers. First, we introduce the inbound delay buffer B_i (lag time in communicating sensor information) in Figure 4.

Variables:
    Input: $\dot{x}_1 \in R^{\geq 0}$, $x_1 \in R^{\geq 0}$
    Output: $\dot{x}_{i1} \in R^{\geq 0}$, $x_{i1} \in R^{\geq 0}$
    Internal: *saved* - maps from an interval $(0, d_i)$ to $(\dot{x}_1, x_1)$
Trajectories:
    an I-trajectory w is included among the set of nontrivial trajectories exactly if
        for all $t \in I$, $t > 0$ the following hold:

$$w(t).(\dot{x}_{i1}, x_{i1}) = \begin{cases} w(0).saved(t) & \text{if } t < d_i \\ (w(t-d_i).\dot{x}_1, w(t-d_i).x_1) & \text{otherwise} \end{cases}$$

$$\forall t' \in (0, d_i),$$

$$w(t).saved(t') = \begin{cases} w(0).saved(t'-t) & \text{if } t' > t \\ (w(t-t').\dot{x}_1, w(t-t').x_1) & \text{otherwise} \end{cases}$$

Fig. 4. B_i Hybrid I/O Automaton

B_i acts in such a way that the output variables have exactly the values of the input variables, exactly time d_i earlier, where d_i is the maximum "information delay" – the longest time that it can take for a controller to receive the velocity and position sensor data. This delay buffer actually implements the more realistic version, in which the length of the delay varies nondeterministically within known bounds. Initially, the buffer (*saved*) is "prefed" with information that could have happened in that initial time period (so that the last position and velocity values in the buffer match up the initial position and velocity values of the platoons). Setting the maximum deceleration for that "imaginary" time period lets the controller be the most flexible (and thus optimal, as will be proven later), in the initial d_i time period.

Formally, the initial value of the *saved* variable is determined as follows. For any start state s of the system, construct a trajectory w of length d_i of *Controlled-Platoons* so that $w(0).\dot{x}_1 = s.\dot{x}_1 + d_i a_{min}$, $w(0).x_1 = s.x_1 + s.\dot{x}_1 d_i + \frac{a_{min} d_i^2}{2}$, and $\forall t \in (0, d_i)$, $w(t).\ddot{x}_1 = -a_{min}$. The second platoon's state components can be arbitrary, as long as no collisions occur. Now, $\forall t \in (0, d_i)$, let $saved(t) = w(t).(\dot{x}_1, x_1)$.

Next, we add an outbound delay buffer B_o (lag time in communicating control information). An outbound delay buffer B_o, is almost like the inbound buffer B_i, but with input variable $\ddot{x}_{o2}$, and output variable $\ddot{x}_2$. The *saved* variable is the same as in B_i, except that it now "saves" $\ddot{x}_{o2}$ and the length of the interval is d_o, where $d_o \in R^{\geq 0}$ is the maximum "action delay" – the longest time that it can take for a platoon to react to the controllers directives. Again, the delay time-length is exact. Initially, $\forall t \in (0, d_o)$, $saved(t) = -a_{min}$. This makes the platoons safe in the initial d_o time interval even if the first platoon starts decelerating.

Definition:

$$safe\text{-}measure_d = \max(x_{i1} + \dot{x}_{i1}t' - \frac{a_{min}t'^2}{2} - x_{new2} - \frac{\dot{x}_{new2}^2 - (\dot{x}_{i1} - a_{min}t_1)^2 - (v_{allow})^2}{2a_{min}},$$

$$\dot{x}_{i1} - a_{min}t' - \dot{x}_{new2} + v_{allow}),$$

where $t' = \min(d_i + d_o, \frac{\dot{x}_{i1}}{a_{min}})$

Variables:

Input: $\dot{x}_{i1} \in R^{\geq 0}$

$x_{i1} \in R^{\geq 0}$

Output: $\ddot{x}_{o2}$, initially if $safe\text{-}measure \leq 0$, then $\ddot{x}_2 = 0$

Internal: internal variables of $Controller_2$ ($\dot{x}_{int2}$ and x_{int2})

a_2 - maps from an interval $(0, d_o)$ to $\ddot{x}_{o2}$,

initially, $\forall t \in (0, d_o), a_2(t) = -a_{min}$, otherwise – arbitrary

$x_{new2}, \dot{x}_{new2}$ - the position and velocity of the second platoon after time d_o passes, provided *collided* still equals *false*

Trajectories:

an I-trajectory w is included among the set of nontrivial trajectories exactly if

w is a trajectory of $Controller_2$

if *collided* = *false* in $w(0)$ then for all $t \in I$, $t > 0$:

if $safe\text{-}measure_d \leq 0$ then

$\ddot{x}_2 = -a_{min}$

$\forall t' \in (0, d_o)$,

$$w(t).a_2(t') = \begin{cases} w(0).a_2(t' - t) & \text{if } t' > t \\ w(t - t').\ddot{x}_{o2} & \text{otherwise} \end{cases}$$

$$w(t).\dot{x}_{new2} = w(t).\dot{x}_{int2} + \int_0^{d_o} w(t).a_2(u)du$$

$$w(t).x_{new2} = w(t).x_{int2} + \int_0^{d_o} (\int_0^u w(t).a_2(u')du' + w(t).\dot{x}_{int2})du$$

Fig. 5. D_2 Hybrid I/O Automaton

Finally, we modify the controller so that it handles the delays correctly. The controller D_2 (see Figure 5) implements $Controller_2$. It is similar to C_2 in that it also tries to keep the second platoon within the bounds set by $safe\text{-}measure_d$, which is $safe\text{-}measure$ redefined for the delayed case. The new controller gets its inputs from the inbound delay buffer B_i and its output variable goes into the outbound delay buffer B_o. Additional internal variables (x_{new2} and $\dot{x}_{new2}$) are added to store the "future" position and velocity data, as calculated from the acceleration settings. A buffer for storing acceleration settings that the controller has output, but that has not been executed yet (a_2) is used for this purpose. Also, $safe\text{-}measure_d$ is defined instead of $safe\text{-}measure$; the only changes from $safe\text{-}measure$ are that the 1st platoon's parameters are exchanged by their "worst-case" values after $d_i + d_o$ time units pass; and the 2nd platoon's parameters are exchanged by their projected values after d_o time units pass.

5.2 Correctness of D_2

We will now compose B_i, D_2, B_o using the hybrid I/O automata composition rules to obtain the delayed controller, which we call *Buffered-Controller*. A straightforward simulation relation shows that this composed system implements C_2. This simulation relation f is the identity on all the external state components of C_2. The use of simulation relations will allows us to prove correctness of our more complicated delayed controller relatively easily, since we have already proven correctness in the simple (ideal) case.

First we prove that if the old *safe-measure* (the one used in the ideal case) is non-positive in some state of a trajectory of *Buffered-Controller*, then the new controller D_2 (the one that has both the inbound and the outbound delays), will also output maximum deceleration, just as the old (ideal) controller would. Formally,

Lemma 8. *If collided = false in $w(0)$ of a trajectory w of Buffered-Controller, then $\forall t \in I$, such that $w(t).safe\text{-}measure \leq 0$, $\ddot{x}_2 = -a_{min}$.*

Lemma 9. *f (an identity relation on all the external components of C_2) is a forward simulation from the composed system Buffered-Controller to C_2.*

Proof. By induction on the number of steps in the hybrid execution. Start states and discrete steps are proven trivially; Lemma 8 is used to prove the simulation relation on continuous trajectories.

This lemma proves that *Buffered-Controller* implements C_2, since the two automata are comparable and there is a simulation relation from the first one to the second one. Therefore, we are now able to prove correctness of our delayed controller:

Theorem 10. *The doubly-delayed hybrid automaton composed of C-Platoons, Buffered-Controller implements Safe-Platoons.*

Proof. We have proved that the system *Buffered-Controller* implements C_2 in Lemma 9. But by Theorem 4, C_2 composed with *C-Platoons* (the *Implemented-Platoons* system) implements *Safe-Platoons*. Thus, the doubly-delayed hybrid automaton composed of *C-Platoons*, and *Buffered-Controller* also implements *Safe-Platoons* by the hybrid I/O automata composition rules!

5.3 Optimality

We will now prove optimality of D_2. Again, we will be basing our proofs on the optimality property of the controller C_2, which was proven in section 4.3. We want to prove that any controller with both inbound and outbound delays that does not implement controller D_2, is unsafe given some "bad" controller C_1. However, knowing that C_2 is optimal makes the proof much easier: we only need to show that a controller that would let *safe-measure$_d$* get negative, will eventually lead to a state in which *safe-measure* itself is negative. Then we can use optimality of C_2 to show that any such controller would not be correct.

Define *Buffered-Controller$_2$* to be bad (*Bad-Buffered-Controller$_2$*), if there exists some *Controller$_1$* (C_{d1}), such that in any admissible hybrid execution α of an automaton composed of *Controlled-Platoons*, C_{d1} and *Bad-Buffered-Controller$_2$*, $\exists s_d \in \alpha$, which does not satisfy Predicate S_d.

Lemma 11. *Any Bad-Buffered-Controller$_2$ implements Bad-Controller$_2$.*

Since we have just shown that the delayed automaton implements the non-delayed one, we can use the optimality property of the ideal case controller, to prove the optimality of the delayed controller easily:

Theorem 12. *Given any* Bad-Buffered-Controller$_2$, *we can always construct* Controller$_1$ *(C_{d1}) such that a system composed of* Controlled-Platoons *and these controllers has in any admissible hybrid execution a state s', in which safety is violated.*

Proof. Take any *Bad-Buffered-Controller$_2$*. By Lemma 11, it implements *Bad-Controller$_2$*. Then by Theorem 7, there exists such *Controller$_1$* (C_1'), that a system composed of *Controlled-Platoons*, C_1' and this *Bad-Buffered-Controller$_2$* has in its hybrid trace a state s' that violates *safety*.

Therefore, the delayed *Controller$_2$* is also optimal.

6 Uncertainty

Our model already includes both the inbound and the outbound delays in sending and receiving information between the controller and *Controlled-Platoons*. Now we will introduce an extra complexity which will make the model even more realistic: the uncertainty in information that the controller receives. This inbound uncertainty arises from inexact sensors that communicate the position and velocity data to the controllers. We will use similar methods to the ones used in the delay case. A special "uncertainty buffer" automaton will be defined, similar to the previous delay buffers. Then, the uncertainty will be implemented by adding this new automaton to the model and modifying the controller slightly. We will then prove correctness using the simulation relation to the delayed case which we have already worked out. This use of levels of abstraction makes the proofs for the complicated case, which involves both the delays and the uncertainties, relatively easy to both write and understand.

6.1 The System

We implement the delayed controller D_2 with a composition of two hybrid automata: another controller U_2, and an inbound uncertainty buffer U_i. We call

Variables:
    Input: $\dot{x}_{i1}, x_{i1}$
    Output: $\dot{x}_{iu1}, x_{iu1}$
Trajectories:
    an I-trajectory w is included among the set of nontrivial trajectories exactly if
    for all $t \in I$, $t > 0$ the following hold:
$$\dot{x}_{iu1} \in [\dot{x}_{i1} - \dot{\delta}, \dot{x}_{i1} + \dot{\delta}]$$
$$x_{iu1} \in [x_{i1} - \delta, x_{i1} + \delta]$$

Fig. 6. U_i Hybrid I/O Automaton

this composed system *Uncertain-Controller*. The uncertainty buffer U_i nondeterministically garbles the position and velocity data within the given bounds (see Figure 6). The bounds are predefined constants $\delta \in \mathsf{R}^{\geq 0}$ — the maximum absolute value of uncertainty in position sensor data, and $\dot{\delta} \in \mathsf{R}^{\geq 0}$ — the maximum absolute value of uncertainty in velocity sensor data. The controller U_2 is the same as D_2 except that it now takes its inputs from the inbound uncertainty buffer U_i and that *safe-measure$_u$* (see below) is defined to account for the uncertainties. Same as in the delay case, the only changes from *safe-measure$_d$* are that the first platoon's data is adjusted to the "worst case" behavior of the first platoon.

$$\text{safe-measure}_u = \max((x_{iu1} - \delta) + (\dot{x}_{iu1} - \dot{\delta})t'' - \frac{a_{min}t''^2}{2} - x_{new2}$$

$$- \frac{(\dot{x}_{new2})^2 - ((\dot{x}_{iu1} - \dot{\delta}) - a_{min}t'')^2 - (v_{allow})^2}{2a_{min}},$$

$$(\dot{x}_{iu1} - \dot{\delta}) - a_{min}t'' - \dot{x}_{new2} + v_{allow}),$$

where $t'' = \min(d_i + d_o, \frac{\dot{x}_{iu1} + \dot{\delta}}{a_{min}})$.

6.2 Correctness of U_2

A straightforward simulation relation shows that the *Uncertain-Controller* system implements D_2. This simulation relation f is the identity on all state components of D_2.

First we show that if the old *safe-measure$_d$* (the one used in the delayed case) is non-positive in some state of a trajectory of *Uncertain-Controller*, then the new controller (the one that has an inbound uncertainty), will also output maximum deceleration, same as the old (delayed only) controller would. Formally,

Lemma 13. *Let w be an I-trajectory of* Uncertain-Controller. *If collided = false in $w(0)$, then $\forall t \in I$, such that safe-measure$_d \leq 0$, $\ddot{x}_{o2} = -a_{min}$ and safe-measure$_u \leq 0$.*

Then we are able to prove that f is the simulation relation from the composed system *Uncertain-Controller* to D_2.

Theorem 14. f *(an identity relation on all the external state components of D_2) is a forward simulation from the Uncertain-Controller system to D_2.*

We have already proven correctness of the delayed controller D_2, and we have also shown that *Uncertain-Controller* implements D_2; thus, our new uncertain system is also correct in a sense that it implements our correctness specification, *Safe-Platoons*.

7   Conclusion

The system consisting of two platoons moving on a single track has been modeled using hybrid I/O automata, including all the components (physical platoons, controllers, delay and uncertainty buffers), and the interactions between them. Safety conditions were formulated using invariant assertions. Correctness and optimality of controllers were proved using composition, simulation mappings and invariants, and the methods of mathematical analysis. Complexity (delays and uncertainty) was introduced gradually, using the levels of abstraction, which significantly simplified the proofs.

The case study describes formally a general controller that would guarantee the safety requirement regardless of the behavior of the leading platoon. Such a controller can be later reused to prove correctness of complicated maneuvers, such as merging and splitting, where the setup is similar.

In future work, we will extend the model to handle outbound uncertainty; use jerk instead of acceleration; motion in 2D planes. Also, we will consider cases with several platoons operating independently. Additional properties of the join maneuver, such as successful join, time optimality, and passenger comfort, will be studied; other maneuvers arising in this setting will be investigated using the same models.

References

1. R. Alur, C. Courcoubetis, T.A. Henzinger, P.H. Ho, X. Nicollin, A. Olivero, J Sifakis, and S. Yovine. The algorithmic analysis of hybrid systems. *Theoretical Computer Science*, 138:3–34, 1995.
2. Michael S. Branicky, Ekaterina Dolginova, and Nancy Lynch. A toolbox for proving and maintaining hybrid specifications. Submitted for publication. To be presented at *HS'96: Hybrid Systems*, October 12-16, 1996, Cornell University, Ithacs, NY.
3. Jonathan Frankel, Luis Alvarez, Roberto Horowitz, and Perry Li. Robust platoon maneuvers for AVHS. Manuscript, Berkeley, November 10, 1994.
4. John Lygeros. *Hierarchical Hybrid Control of Large Scale Systems.* PhD thesis, University of California, Department of Electrical Engineering, Berkeley, California, 1996.
5. John Lygeros, Datta N. Godbole, and Shankar Sastry. A game theoretic approach to hybrid system design. Technical Report UCB/ERL-M95/77, Electronic Research Laboratory, University of California Berkeley, October 1995.

6. Nancy Lynch. A three-level analysis of a simple acceleration maneuver, with uncertainties. In *Proceedings of the Third AMAST Workshop on Real-Time Systems*, pages 1–22, Salt Lake City, Utah, March 1996.

7. Nancy Lynch, Roberto Segala, Frits Vaandrager, and H. B. Weinberg. Hybrid I/O automata. In R. Alur, T. Henzinger, and E. Sontag, editors, *Hybrid Systems III: Verification and Control* (DIMACS/SYCON Workshop on Verification and Control of Hybrid Systems, New Brunswick, New Jersey, October 1995), volume 1066 of *Lecture Notes in Computer Science*, pages 496–510. Springer-Verlag, 1996.

8. O. Maler, Z. Manna, and A. Pnueli. From timed to hybrid systems. In J.W. de Bakker, C. Huizing, W.P. de Roever, and G. Rozenberg, editors, *REX Workshop on Real-Time: Theory in Practice*, volume 600 of *Lecture Notes in Computer Science*, pages 447–484, Mook, The Netherlands, June 1991. Springer-Verlag.

9. Pravin Varaiya. Smart cars on smart roads: Problems of control. *IEEE Transactions on Automatic Control*, AC-38(2):195–207, 1993.

10. H. B. Weinberg and Nancy Lynch. Correctness of vehicle control systems: A case study. In *17th IEEE Real-Time Systems Symposium*, pages 62–72, Washington, D. C., December 1996. Complete version in Technical Report MIT/LCS/TR-685, Laboratory for Computer Science, Massachusetts Institute of Technology, February 1996. Masters Thesis.

11. H. B. Weinberg, Nancy Lynch, and Norman Delisle. Verification of automated vehicle protection systems. In R. Alur, T. Henzinger, and E. Sontag, editors, *Hybrid Systems III: Verification and Control* (DIMACS/SYCON Workshop on Verification and Control of Hybrid Systems, New Brunswick, New Jersey, October 1995), volume 1066 of *Lecture Notes in Computer Science*, pages 101–113. Springer-Verlag, 1996.

Verifying Hybrid Systems Modeled as Timed Automata: A Case Study*

Myla Archer and Constance Heitmeyer

Code 5546, Naval Research Laboratory, Washington, DC 20375
{archer, heitmeyer}@itd.nrl.navy.mil

Abstract. Verifying properties of hybrid systems can be highly complex. To reduce the effort required to produce a correct proof, the use of mechanical verification techniques is promising. Recently, we extended a mechanical verification system, originally developed to reason about deterministic real-time automata, to verify properties of hybrid systems. To evaluate our approach, we applied our extended proof system to a solution, based on the Lynch-Vaandrager timed automata model, of the Steam Boiler Controller problem, a hybrid systems benchmark. This paper reviews our mechanical verification system, which builds on SRI's Prototype Verification System (PVS), and describes the features we added to handle hybrid systems. It also discusses some errors we detected in applying our system to the benchmark problem. We conclude with a summary of insights we acquired in using our system to specify and verify hybrid systems.

1 Introduction

Researchers have proposed many innovative formal methods for developing real-time systems [9]. Such methods can give system developers and customers greater confidence that real-time systems satisfy their requirements, especially their critical requirements. However, applying formal methods to practical systems requires the solution of several challenging problems, e.g., how to make formal descriptions and formal proofs understandable to developers and how to design software tools in support of formal methods that are usable by developers.

We are building a mechanized system for specifying and reasoning about real-time systems that is designed to address these challenging problems. Our approach is to build formal reasoning tools that are customized for specifying and verifying systems represented in terms of a specific mathematical model. In [2], we describe how we are using the mechanical proof system PVS [18, 19] to support formal specification and verification of systems modeled as Lynch-Vaandrager timed automata [15, 14]. Reference [2] also presents the results of a case study in which we applied our method to prove invariant properties of a solution to the Generalized Railroad Crossing (GRC) problem [7].

Our system provides mechanical assistance that allows humans to specify and reason about real-time systems in a direct manner. To specify a particular timed automaton, the user fills in a template provided by our system. Then, he or she uses our system, a version of PVS augmented with a set of specialized PVS proof strategies, to verify properties of the automaton. Use of our system for specification and verification is usually quite straightforward because our system provides both a structure and a set of specialized theories useful in constructing

---

* This work is funded by the Office of Naval Research. URLs for the authors are http://www.itd.nrl.navy.mil/ITD/5540/personnel/{archer, heitmeyer}.html

timed automata models and proving properties about them. By focusing on a particular mathematical model—the timed automata model—our system allows a user to reason within a specialized mathematical framework. The user need not master the base logic and the complete user interface of the underlying proof system, PVS.

The results of our initial study proved encouraging. All of the hand proofs of invariant properties in the GRC solution were translated into corresponding PVS proofs with a very similar structure. Moreover, most of the PVS proofs could be done using our specialized strategies alone. Neither the time required to enter the specifications using our template nor the time required to check the proofs of state invariants was excessive.

While our initial study only involved deterministic automata, we recently demonstrated that the same approach could be applied to hybrid automata, where the effects of time passage and other events can be nondeterministic. To study the utility of our proof techniques for verifying hybrid systems, we applied them to a solution of the Steam Boiler Controller problem described in [12]. In verifying this application, we investigated the following issues:

- How should nondeterminism be modeled?
- Is it practical to use an automatic theorem prover to reason about nonlinear real arithmetic?
- Can the timed automata template and proof strategies developed in our earlier case study be extended to handle the more general problem of hybrid automata?
- How suitable is PVS for verifying properties of hybrid systems?

In verifying the solution in [12], we detected several errors. While most of these errors are minor and easily fixed, at least two of the errors are errors in reasoning, and their correction is nontrivial.

Like other approaches to verifying real-time systems, such as SMV [16, 5], HyTech [10], and COSPAN [11], our approach is based on a formal automata model. Moreover, like these other approaches, our methods can be used to prove properties of particular automata and, like COSPAN, to prove simulations between automata.[1] However, our approach is different from other approaches in several ways. First, the properties we prove are expressed in a standard logic with universal and existential quantification. This is in contrast to most other approaches, where the properties to be proved are expressed either in a temporal logic, such as CTL or ICTL, or in terms of automata. By using standard notations and standard logics and by providing templates for developing specifications, we largely eliminate the need for the special notations, logics, etc., required by other verification systems. Second, unlike other automata-based methods, the generation of proofs in our method is not completely automatic. Rather, our method uses a mechanical proof system to check the validity of hand proofs that use deductive reasoning.

Interaction with an automatic proof system does demand a higher level of sophistication from the user. But by supporting reasoning about automata at a

---

[1] We have designed support for verifying simulation proofs, but implementation of this support requires some improvements to PVS.

high level of abstraction, we can prove more powerful results than is possible with tools requiring more concrete descriptions of automata. Moreover, our approach avoids the state explosion problem inherent in other automata-based approaches and also provides considerable feedback when an error is detected. Such feedback is extremely useful in correcting errors.

Section 2 reviews the timed automata model, PVS, and the template and tools we developed in our earlier verification of the GRC. Section 3 introduces the Steam Boiler Controller problem, presents the main hybrid automaton specified in [12], and discusses the techniques we used to adapt our template to proving properties of hybrid automata. Section 4 presents our major results—the errors we detected in applying our system to the Boiler Controller problem and the time and effort we needed to adapt and apply our methods. Section 5 presents an example of a hand proof and the corresponding PVS proof for a state invariant whose proof involves both nondeterminism and nonlinear real arithmetic. Section 6 presents some conclusions about the usefulness of our methods for verifying hybrid automata.

2 Background

2.1 The Timed Automata Model

The formal model used in the specification of the Boiler Controller problem and its solution [12] represents both the computer system controller and its environment as Lynch-Vaandrager *timed automata*. A timed automaton is a very general automaton, i.e., a labeled transition system. It need not be finite-state: for example, the state can contain real-valued information, such as the current time, the boiler's water level, and the boiler's steam rate. This makes timed automata suitable for modeling not only computer systems but also real-world quantities, such as water levels and steam rates. The timed automata model describes a system as a set of timed automata, interacting by means of common actions. In the Boiler Controller solution presented in [12], separate timed automata represent the boiler and the controller; the common actions are sensors reporting the water level, steam rate, and number of active pumps in the boiler, and actuators controlling whether to stop the boiler and when the next sensor action is scheduled. The definition of timed automata below, which is based on the definitions in [8, 7], was used in our earlier case study [2]. It is a special case of Lynch-Vaandrager timed automata, which requires the next-state relation, *steps(A)*, to be a function. How to modify the definition to handle the nondeterminism inherent in hybrid automata was one of the issues addressed in our current study.

A *timed automaton A* consists of five components:

- *states(A)*, a (finite or infinite) set of states.
- *start(A)* $\subseteq$ *states(A)*, a nonempty (finite or infinite) set of start states.
- A mapping *now* from *states(A)* to $R^{\geq 0}$, the non-negative real numbers.
- *acts(A)*, a set of actions (or events), which include special *time-passage* actions $\nu(\Delta t)$, where Δt is a positive real number, and *non-time-passage* actions, classified as *input* and *output* actions.
- *steps(A)* : *states(A)* $\times$ *acts(A)* $\rightarrow$ *states(A)*, a partial function that defines the possible steps (i.e., transitions).

2.2 PVS

PVS (Prototype Verification System) [19] is a specification and verification environment developed by SRI International's Computer Science Laboratory. In contrast to other widely used proof systems, such as HOL [6] and the Boyer-Moore theorem prover [4], PVS supports *both* a highly expressive specification language and an interactive theorem prover in which most low-level proof steps are automated. The system provides a specification language, a parser, a type checker, and an interactive proof checker. The PVS specification language is based on a richly typed higher-order logic that permits a type checker to catch a number of semantic errors in specifications. The PVS prover provides a set of inference steps that can be used to reduce a proof goal to simpler subgoals that can be discharged automatically by the primitive proof steps of the prover. The primitive proof steps incorporate arithmetic and equality decision procedures, automatic rewriting, and BDD-based boolean simplification.

In addition to primitive proof steps, PVS supports more complex proof steps called *strategies*, which can be invoked just like any other proof step in PVS. Strategies may be defined using primitive proof steps, applicative Lisp code, and other strategies, and may be built-in or user-defined.

2.3 A Template For Specifying Timed Automata in PVS

Our template for specifying Lynch-Vaandrager timed automata provides a standard organization for an automaton. To define a timed automaton, the user supplies the following six components:
- declarations of the non-time actions,
- a type for the "basic state" (usually a record type) representing the state variables,
- any arbitrary state predicate that restricts the set of states (the default is **true**),
- the preconditions for all transitions,
- the effects of all transitions, and
- the set of start states.

In addition, the user may optionally supply
- declarations of important constants,
- an axiom listing any relations assumed among the constants, and
- any additional declarations or axioms desired.

To support mechanical reasoning about timed automata using proof steps that mimic human proof steps, we also provide an appropriate set of PVS strategies, based on a set of standard theories and certain template conventions. For example, the induction strategy, which is used to prove state invariants, is based on a standard automaton theory called **machine**. To reason about the arithmetic of time, we have developed a special theory called **time_thy** and an associated simplification strategy called **TIME_ETC_SIMP** for time values that can be either non-negative real values or ∞.

3 Specifying and Reasoning About Hybrid Automata

Like others, we use the term *hybrid automaton* to describe a state machine which has both continuous and discrete components. Since hybrid automata are used to model physical systems controlled by a discrete computer system, the laws of physics affect their behavior. Because changes in measurable properties of the physical systems depend on environmental factors, automata models of these systems typically have some nondeterministic transitions. Thus, hybrid automata differ from deterministic, discrete automata in two major ways: their behavior is nondeterministic and reasoning about their transitions often leads to complex computations involving nonlinear real arithmetic.

Strictly speaking, the automata in our case study of the GRC problem [2] are hybrid automata by the above definition. However, their description involved neither nondeterminism nor nonlinear arithmetic. For the example in the current case study, it was necessary to extend our previous methods to handle both of these features.

3.1 Reasoning About Hybrid Automata

Handling Nondeterminism. To specify and to reason about nondeterministic automata using an automated theorem prover, one needs to describe the nondeterministic transitions. An obvious approach is to represent the transitions as a relation. In an initial experiment, we encoded the transitions in a deterministic automaton from our initial study [2] as a relation and redid the proofs of state invariants up to and including the Safety Property, the major theorem. Because the selected timed automaton is deterministic, the transition relation was expressed as an equality between the result state and a case expression involving the action and the initial state. The new versions of the state invariant proofs were quite similar to the corresponding proofs in [2] , except for two important differences. First, they took about twice as long to execute. Second, frequent human intervention was required to substitute the value of the result state where it was needed. Human intervention was required even in the "trivial" branches of induction proofs that our system previously handled automatically. Clearly, this situation would only worsen in the case of true nondeterminism.

In the case of hybrid automata, not all transitions are nondeterministic, and among those that are, not all parts of the state change nondeterministically. Our ultimate solution, which took advantage of this fact, uses PVS's implementation "epsilon" of Hilbert's choice operator ϵ in expressing the transition relation as a function. The choice operator ϵ is defined as follows. Let $P : \tau \rightarrow$ **bool** be any predicate on a nonempty type τ. Then $\epsilon(P)$ has two known properties. First, $\epsilon(P)$ is an element of type τ. Second, the ϵ-*axiom* holds, namely: if there is an element of type τ that satisfies P, then P holds for $\epsilon(P)$, i.e., $P(\epsilon(P)) =$ **true**.

Two features of $\epsilon(P)$ should be noted. First, it is always defined, whether or not P can be satisfied, and second, it is deterministic. These characteristics need to be remembered when one is using ϵ to reason about nondeterministic automata. Reasoning about such automata using ϵ is sound, provided 1) one does not try to draw conclusions about existence of an action's result state *satisfying the specified constraints* from the fact that the action is defined, and 2) one does not draw conclusions about the equality of states reached by identical action sequences. If such inferences are avoided, any specification or proof errors uncovered using ϵ will be genuine errors.

A careful use of ϵ that avoids these two dangers will simulate what we consider to be a better solution: implement "ANY" and "SOME" quantifiers in PVS that can be skolemized and instantiated in fashions analogous to "FORALL" and "EXISTS" in PVS. An "ANY" quantifier would avoid at least problem 2), since constants arising from its multiple skolemization could never be proved equal. A solution to problem 1) is more difficult in PVS, since PVS does not allow

```
real_thy: THEORY
   BEGIN
      nonnegreal: TYPE = {r:real | 0 <= r};
   % posreal_mult_closed: LEMMA (FORALL (x,y:real): (x > 0 & y > 0) => x*y > 0);
      nonnegreal_mult_closed: LEMMA (FORALL (x,y:real): (x >= 0 & y >= 0) => x*y >= 0);
      greater_eq_nonnegmult_closed: LEMMA (FORALL (x,y,z:real): (x >= 0 & y >= z) => x*y >= x*z);
   END real_thy
```

Fig. 1. A (Partial) Theory of Real Numbers

partial functions; "ANY" would have to behave more or less like ϵ with respect to definedness.

Handling Nonlinear Real Arithmetic. Computation in several of our PVS proofs involved nonlinear real expressions. Existing PVS decision procedures are able to handle some simple cases of such reasoning by expanding terms into a normal form and matching identical terms. When the reasoning is more complex, PVS needs help. Though it does not always happen, the simplifications done by the PVS decision procedures can sometimes confuse the argument in the proof. There are two approaches available to solve these problems. First, we can provide PVS with a list of facts about real numbers. Second, we can name subterms in a manner that prevents PVS from overdoing simplification.

Providing support for convenient use of the second approach must await planned enhancements to PVS. However, implementing the first approach is more straightforward. We have added a line to our template that imports the theory **real_thy**, which contains useful standard definitions and lemmas about real numbers. Figure 1 shows the subset of that theory that was helpful in completing the real arithmetic reasoning in the proofs in the current case study. We note that, unlike [20], our goal is *not* to encode and use deep facts about real analysis (such as the properties of integrals), since 1) these properties are well understood, and 2) it is usually possible to isolate any application of them (and the proof that it was done correctly) from the rest of the specification correctness proof. Instead, we aim to compile a list of facts that can eventually be used automatically in an improved "real-arithmetic" PVS strategy.

3.2 The Steam Boiler Controller Problem

The Steam Boiler Controller problem, which is intended to provide a realistic benchmark for comparing different real-time formalisms, was defined by J.-R. Abrial et al. in [1] and is derived from a competition problem previously posed by Lt-Col. J.C. Bauer for the Institute for Risk Research at the University of Waterloo in Canada. Below is the condensed and informal description of the problem that appears in [12]:

> The physical plant consists of a steam boiler. Conceptually, this boiler is heated (e.g., by nuclear fuel) and the water in the boiler evaporates into steam and escapes the boiler to drive, e.g., a generator (this part is of no concern to the problem). The amount of heat and, therefore, the amount of steam changes without any considered control. Nevertheless, the safety of the boiler depends on a bounded water level (q) in the boiler and steam rate (v) at its exit. A set of four equal pumps may supply water to compensate for the steam that leaves the boiler. These four pumps can be activated or stopped by the controller system. The controller reacts to the information of two sensors, the

water level sensor and the steam rate sensor, and both may fail. Moreover, the controller can deduce from a pump monitor whether the pumps are working correctly. Sensor data are transferred to the controller system periodically. The controller reacts instantaneously with a new setting for the pumps (**pr_new**) or decides to shut down the boiler system (**stop**).

There are two basic time constants: First, the time between two consecutive sensor readings (denoted **I**) and, second, the delay time (**S**) until the reaction of the controller causes consequences in the boiler. The latter delay time usually represents a worst case accumulation of sensor reading delay, calculation time in the controller, message delivery time, reaction time of the pumps, and other minor factors.

More precisely, the water level has two safety limits, one upper (denoted M_2) and one lower limit (denoted M_1). If the water level reaches either limit, there is just time enough to shut down the system before the probability of a catastrophe gets unacceptably high. The steam rate has an upper limit (denoted **W**) and, again, if this limit is reached the boiler must be stopped immediately. In addition the human operator has the possibility to activate the shutdown anytime.

Several automata are specified in [12]: a boiler, a simple controller, the combination of these into a combined system, and a combined system with a fault-tolerant controller. Figure 2 presents, essentially verbatim, the specification of the simple combined system given in [12].[2] In our study, we began by entering the details of this specification into our timed automaton template.

Figure 3 shows a fragment of our template's version of this specification. This fragment includes the definitions of two (parameterized) predicates **steam_rate_pred** and **water_level_pred** (shown bold face for readability) that capture the constraints on the new steam rate v' and water level q' in the effect of the time passage action $\nu(\Delta t)$, together with the $\nu(\Delta t)$ part of the definition of the template's transition function *trans* that encodes the effects of the actions defined in Figure 2. Appropriate applications of the ϵ operator in a LET construct allow the effect of the action $\nu(\Delta t)$ on state s to be expressed both compactly and *carefully*.[3] All the nondeterministic state components in either the start state or in the transitions were similarly handled.

The major proof goal in [12] for the simple combined system was to establish two properties: either the steam rate has an acceptable value or the system is in the stop mode, and either the water level has an acceptable value or the system is in the stop mode. These two properties are expressed formally in [12] as:

Theorem 1: In all reachable states of the boiler system, $v < \mathbf{W}$ *or stop* = **true**.

Theorem 2: In all reachable states of the boiler system, $M_1 < q < M_2$ *or stop* = **true**.

Although neither property mentions time explicitly, the correctness of the specification with respect to these properties is established by proving a sequence of state invariants that depend on time and thus relies heavily upon timing restrictions.

---

[2] The meanings of several of the constants and variables are found in the above problem description. The Appendix defines the remaining constants and variables.

[3] The LET construct allows ϵ to be used in encoding the definition of any state in which two nondeterministic components are related (as is the case in the start state—where q and wl, which satisfy the same predicate, must be equal—as well as in the result state of $\nu(\Delta t)$) without the determinism of the ϵ operator being taken for granted.

State:
> *now*, a nonnegative real, initially 0
> *do_sensor* : boolean, initially *true*
> *do_output, stopmode, stop* : boolean, initially *false*
> *pr, pr_new, pumps, px* : [0 .. *#pumps*], initially 0
> *error* : [0 .. *pr_new*], initially 0
> *v, read* : nonnegative real, initially 0
> *set* : nonnegative real, initially S
> *sr* : [0,W], initially 0
> *q, wl* : [0,C], initially equal and $\gg M_1$ and $\ll M_2$

Transitions:

actuator(e_stop,pset)
> **Precondition:**
> *do_output = true*
> *pset = px*
> *e_stop = stopmode*
> **Effect:**
> *do_output'* = *false*
> *do_sensor'* = *true*
> *pr_new'* = *pset*
> *stop'* = *e_stop*
> *read'* = *now* + *I*

controller
> **Precondition:**
> *true*
> **Effect:**
> $0 \leq px' \leq \#pumps$

v(Δt)
> **Precondition:**
> *stop = false*
> $now + \Delta t \leq read$
> $now + \Delta t \leq set$
> **Effect:**
> $v - U_2 * \Delta t \leq v' \leq v + U_1 * \Delta t$
> $q + pr * P * \Delta t - \delta_{HIGH}(v, v', \Delta t) \leq q'$
> $q' \leq q + pr * P * \Delta t - \delta_{LOW}(v, v', \Delta t)$
> $now' = now + \Delta t$

sensor(s,w,p)
> **Precondition:**
> *now = read*
> *do_sensor = true*
> *stop = false*
> *w = q*
> *s = v*
> *p = pr*
> **Effect:**
> *pumps'* = *p*
> *do_sensor'* = *false*
> *do_output'* = *true*
> *sr'* = *s*
> *wl'* = *w*
>
> if $sr' \leq W - U_1 * I$ or
> $wl' \geq M_2 - P*(pumps'*S + (max\_pumps\_after\_set)$
> $*(I - S)) + min\_steam\_water(sr)$ or
> $wl' \leq M_1 - P*(pumps'*S + (min\_pumps\_after\_set)$
> $*(I - S)) + max\_steam\_water(sr)$
> then *stopmode'* = *true*
> else *stopmode'* = *{true, false}* arbitrary

activate
> **Precondition:**
> *now = set*
> *stop = false*
> **Effect:**
> *set'* = *read* + *S*
> $0 \leq error' \leq pr\_new$
> *pr'* = *pr_new* − *error*

Fig. 2. Initial Specification of the Boiler System

4 Results

Extensions Required to Previous Methods. Once the decision to model nondeterminism using the Hilbert ϵ was made, just two modifications to the system used in our earlier study [2] were required before the system could be applied to hybrid systems. First, some necessary lemmas about real arithmetic were identified and included in a new auxiliary theory **real_thy**. The other major modification involved ϵ: we developed a technique for using ϵ in specifications (illustrated in Figure 3) and in proofs (illustrated in Figure 6).

Errors Discovered. Using our proof system to analyze the Boiler, Controller, and combined Boiler System specifications in [12] identified approximately a

---

steam_rate_pred(v_old:nonnegreal,delta_t:(fintime?))(v_new:nonnegreal):bool =
    v_old − U_2*dur(delta_t) <= v_new & v_new <= v_old + U_1*dur(delta_t);

water_level_pred(q_old:water_level,pr:num_pumps,v_old,v_new:nonnegreal,delta_t:(fintime?))
        (q_new:water_level):bool =
    q_old + pr*P*dur(delta_t) − delta_HIGH(v_old,v_new,delta_t) <= q_new
  & q_new <= q_old + pr*P*dur(delta_t) − delta_LOW(v_old,v_new,delta_t);

trans (a:actions, s:states):states =
    CASES a OF
      nu(delta_t): LET new_v_part = epsilon(steam_rate_pred(v(s),delta_t)),
                       new_q_part = epsilon(water_level_pred(q(s),pr(s),v(s),new_v_part,delta_t))
          IN s WITH [now := now(s) + delta_t,
                       basic := basic(s) WITH [v_part := new_v_part, q_part := new_q_part]],
    <... other action cases ...>
    ENDCASES;

Fig. 3. Specifying Nondeterminism Using ϵ

dozen errors.[4] These errors were discovered during each of the three basic stages of the mechanization process: filling in the template, applying the PVS type checker, and checking the hand proofs of the invariants. The errors detected in filling in the template were usually due to vagueness in the original specification. For example, what is the meaning of the expressions "$q << M_2$" and "$q >> M_1$"? How is the water level variable wl in the Controller assigned the value of the actual water level q in the Boiler when the Controller is considered as a separate automaton?

In checking the hand proofs, several types of errors were found. The most common were missing assumptions about the constants. The second most common were simple typographical errors. All of these errors, as well as the errors due to vagueness, are easily fixed. To correct them, we determined what changes were needed to successfully complete the hand versions of the invariant proofs. While trivial, detecting and correcting these easily fixable errors does clearly lead to better specifications and better proofs.

Figure 4 shows an example of the type of feedback that occurs in the presence of an error. It shows the encoding of invariant lemma 3.3 of [12], and the goal of the sensor action proof branch generated by the induction strategy, with the precondition in hypothesis {−4} expanded. The goal is in the form of a PVS Gentzen-style *sequent*, where the conjunction of the hypotheses (numbered negatively) must be shown to imply the disjunction of the conclusions (numbered positively). Clearly, the proof on this branch would be valid if the inequality "sr_1 <= W − U_1 * I" in the condition in conclusion [2]—which is easily traceable to the definition of the sensor action in Figure 2—were reversed. A little thought shows that the first inequality in this definition is indeed backwards: one wants to prevent the steam rate sr' from becoming too *large*, not too small.

In addition to the minor errors described above, two more serious errors—errors in reasoning—were discovered. How to correct these errors was not obvious. Like many of the minor errors, errors in reasoning were found when a dead

---

[4] A complete description of these errors can be found in [3].

Inv_3_3(s: states):bool = (sr(s) + U_1*I < W) OR stopmode(s) = true;
lemma_3_3: LEMMA (FORALL (s:states): reachable(s) => Inv_3_3(s));

lemma_3_3.2 :

[−1] reachable(s_1)
[−2] (((sr_part(basic(s_1)) + U_1 * I) < W) OR stopmode_part(basic(s_1)))
[−3] enabled_general(sensor(sr_1, w_1, p_1), s_1)
{−4} now(s_1) = fintime(read(s_1)) AND do_sensor(s_1) AND w_1 = q(s_1) AND sr_1 = v(s_1)
    AND p_1 = pr(s_1) AND NOT(stop(s_1))
├─────────
[1] (U_1 * I + sr_1 < W)
[2] IF (sr_1 <= W − U_1 * I
    OR ((w_1 >= (((max(0, sr_part(basic(s_1)))*I − U_2*I/2*I) + M_2) − (I*P*num_pumps)) − P*S*p_1)
        + (P*S*num_pumps)))
    OR (w_1 <= (M_1 + sr_part(basic(s_1)))*I + U_1*I/2*I − P*S*p_1))))
    THEN TRUE ELSE epsilon(bool_pred) ENDIF

Fig. 4. Feedback Example.

end was reached on some branch of an induction proof. But, unlike the minor errors, making simple corrections to the specifications or to the statements of the lemmas would not remedy the problems. In these cases, close examination of the corresponding parts of the original hand proofs revealed errors in reasoning.

We note that the discovery of an error in the hand proof does not necessarily mean that a given assertion is false. It merely means that the assertion does not hold for the stated reasons. At least one invariant lemma in [12] (lemma 6) in fact had a *simpler* PVS proof than the hand proof supplied. When assertions hold for reasons other than the expected ones, one has to wonder whether the specification has in fact captured the required concepts.

Of the three modes of error discovery, the typechecking process produced the most varied results. Remedying the error in one case required extending the effect of an operator; in another, it required introducing a new invariant. A third "error" was one uncovered by the typechecker during a proof via the TCC (type-correctness condition) mechanism: when the ϵ-axiom was invoked to assert the corresponding predicate on the nondeterministic water-level component q' of the result state of the time passage action, a TCC was generated that required proving that q' belongs to the correct subrange of values, namely $[0, C]$. "Error" is in quotes because this problem can be avoided by letting the water level range over all real values.[5] Provided that it can be established, Theorem 2 will then show that the water level does in fact stay within the physical limits.

Time Required to Check the Boiler Controller. Checking the Boiler Controller solution required the checking of 14 lemmas and 2 theorems. The entire process of encoding three specifications and checking the proofs (with backtracking when corrections were made in the specification or statements of the lemmas) took less than 3 work weeks. Part of this time was spent designing and implementing modifications to our system to support reasoning about nonlinear real arithmetic and nondeterminism. In a few proofs, we did not complete the check

---

[5] However, the Boiler automaton as specified is then not a priori constrained in the same way as the physical Boiler.

of the real arithmetic inequalities when this became too tedious in PVS and when their validity became obvious.

When all went smoothly, checking the proof of an invariant required on average half an hour to an hour. Exceptions occurred in three induction proofs involving significant reasoning about complex inequalities. For these, our basic induction strategy typically took between 10 and 15 minutes to reach the stage where the user could continue the proof by completing the base case and action cases, as opposed to times more on the order of 30 seconds for other induction proofs of invariants.[6] The reason for this inefficiency is a question for further study.

5   A Proof Example

Most of the invariant proofs in our case study are very similar to the corresponding hand proofs. However, the PVS proofs required more detail when they involved either nondeterminism or sufficiently complicated nonlinear arithmetic. An example of a state invariant for the Boiler Controller specification in Figure 2 that involves both reasoning about a nondeterministic action and nonlinear arithmetic is Lemma 11 in [12]. Figure 5 provides the statement and hand proof of Lemma 11 from [12]. Figure 6 shows the corresponding PVS proof using our specialized strategies.

This is one case in which no special help was required to allow PVS to reason about the real arithmetic: this aspect of the proof can be done by applying equalities, expanding products, and matching the resulting terms, all of which the primitive proof steps of PVS can do automatically.

On the other hand, PVS does need help at the point where the constraints on the nondeterministic value v', the steam rate in the result state, are introduced. The step in the PVS proof which does this is the **USE_EPSILON** step. The effect of this step is to invoke the ϵ-*axiom* for the (parameterized) predicate $P =$ **steam_rate_pred(v(s_1), t_1)** and the type $\tau = nonnegreal$, thus creating the two proof branches represented by the cases "1" and "2" that follow this step. In the first branch, the required constraints on $v' = \epsilon(P)$ have been introduced into the hypotheses of the current goal. In the second branch, it is required to prove that there does indeed exist a possible value for v'—that is, that there is indeed a value of type $nonnegreal$ satisfying P. This is done by supplying the instance $v(s\_1)$. This branch is necessary because of the form of the ϵ-*axiom* (see Section 3.1).

6   Conclusions

Applying our proof system to a solution of the Steam Boiler Controller problem led to several insights about the utility of our approach to specifying and verifying hybrid systems. We discuss these insights below.

Mechanizing Specifications and Proofs. The results of our study clearly demonstrate the utility of mechanized tools for detecting *and* correcting errors in both formal specifications and formal proofs. While verification systems that use totally automatic proof methods can provide some limited feedback, the generality of interactive theorem proving provides more specific, and therefore more useful, feedback. The feedback we obtained from the PVS typechecker and

---

[6] These times are on a SPARCstation 20 running SunOS 5.4.

Lemma 11: In all reachable states of the combined steam boiler system,
$$v + U_1*(read - now) < W \; or \; stop = \textbf{true}$$

Proof. The basis is vacuously satisfied. We distinguish on the cases for the action a. For $a \in \{sensor, activate\}$ this lemma is trivially true. Otherwise we get:

A) a = actuator (v, $stop$, and now are unchanged):

We know $sr + U_1*I < W \; or \; stopmode = \textbf{true}$ (Lemma 3.3), $do\_output = \textbf{true}$ from the precondition and if do_output then $now = read$ and $sr = v$ (Lemma 4). From this we can infer $v + U_1*(now + I - now) < W \; or \; stopmode = \textbf{true}$. Moreover, we get $stop' = e\_stop = stopmode$ and $read' = now + I$ from the effect and thus, we know $v + U_1*(read' - now) < W \; or \; stop = \textbf{true}$.

B) a = time-passage ($read$ and $stop$ are unchanged):

We know from the precondition $stop = \textbf{false}$ and $v + U_1*(read - now) < W$ from the assumption. This is equivalent to $v + U_1*(read - now - \Delta t + \Delta t) < W$ and it follows $v + U_1*\Delta t + U_1*(read - now - \Delta t) < W$. Since we know from the effect $v' \leq v + U_1*\Delta t$ and $now' = now + \Delta t$, finally, this is equivalent to $v' + U_1*(read - now') < W$. □

Fig. 5. Hand Proof of Lemma 11

Inv_11(s: states): bool = (v(s) + U_1*(read(s) – dur(now(s)))) < W OR stop(s) = true);

lemma_11: LEMMA (FORALL (s: states): reachable(s) => Inv_11(s));

```
("" (AUTO_PROOF_BOILERSYS "Inv_11")
 (("1" (APPLY (THEN (EXPAND "enabled_specific") (BOILERSYS_SIMP))
       "Case nu(t_1)." "Invoke the precondition.")
      (APPLY (THEN (USE_EPSILON "nonnegreal" "steam_rate_pred" "v(s_1)" "t_1") (BOILERSYS_SIMP))
      "Invoke the restrictions on v'.")
     (("1" (TIME_ETC_SIMP))
      ("2" (APPLY (THEN (EXPAND "steam_rate_pred" 1) (BOILERSYS_SIMP))
          "Doing the existence proof for epsilon.")
         (APPLY (THEN (INST 1 "v(s_1)") (BOILERSYS_SIMP)))
         (TIME_ETC_SIMP))))
  ("2" (APPLY (THEN (APPLY_INV_LEMMA "3_3") (BOILERSYS_SIMP))
      "Case actuator(e_1,p_1).")
     (APPLY (THEN (EXPAND "enabled_specific") (BOILERSYS_SIMP))
     "Invoke the precondition.")
     (APPLY (THEN (APPLY_INV_LEMMA "4") (BOILERSYS_SIMP))))))
```

Fig. 6. PVS Proof of Lemma 11

from reaching dead ends in proofs not only localized the source of an error but often made it easy for us to to correct the error. Moreover, the time required to check the specifications and the proofs was not significant.

Using Specialized Tools. We have applied the system that we developed for our initial study [2] to two additional problems, the Boiler Controller benchmark and the proof of a timed version of Fischer's algorithm [13]. In each of the later applications, our earlier-developed system greatly simplified the specification and proof process. Each new application of the system led to some additions: the automata in Fischer's algorithm were a special class of timed automata that represent MMT automata [17]; those in the current study involved nondeterminism and nonlinear real arithmetic. However, the needed enhancements required the addition of only a very small number of new proof strategies, sometimes with specification conventions to support them. The total time required for solving each of the two later problems, which included the time needed to add the required enhancements, was less than three work weeks.

It is not clear when we will reach the point where applying our proof system requires no additions. Our experience so far is that one can build upon the previously developed system. Further, while we have not yet reached the point where our system can be applied by users who lack knowledge of PVS, we believe that use of our system reduces the time and effort that are required to check proofs, while providing meaningful feedback about errors.

On Reasoning About Nondeterministic Automata. For efficient reasoning about nondeterministic automata in PVS, our current solution is to represent the transition relation as a function by making (careful) use of the Hilbert ϵ. We believe that incorporation of "ANY" and "SOME" constructs with accompanying skolemization and instantiation would be useful additions to PVS for reasoning about nondeterminism, since this would help solve the problem of the results of actions being represented in PVS as deterministic (though nonspecific).

The desire to represent the transition relation as a function also partly arises from the intuition that the intention in specifications such as the one in this study is that transitions should be enabled when their preconditions are satisfied: that is, that they can then "fire". Thus, the transition relation should be considered to be a many-valued (but never no-valued) function. Note, this interpretation involves an implicit proof obligation for any such specification: one must demonstrate the existence of at least one result state satisfying the constraints of every legal transition. We note that PVS naturally enforces this requirement, since it demands that all functions be total.

On Using PVS. While there are no complete decision procedures for reasoning about inequalities involving nonlinear real arithmetic, our experience so far indicates that the human guidance PVS is likely to need in reasoning about hybrid automata is the application of a small number of facts about the real numbers, together with control over the grouping of terms into factors. It should be possible to design specialized PVS strategies that will provide much of the needed guidance. These strategies fall into two categories: those that simplify application of lemmas about real arithmetic, and what could be called "naming" strategies, strategies that simplify the use of names for terms. The second type can also be helpful in simplifying the job of the human user of PVS, since they could also be used to manage expressions, such as **epsilon** expressions, that do not necessarily involve a sum of terms, but are simply long. The implementation of at least some of these strategies will require the use of certain enhancements to PVS that are in progress, such as the ability to name and track assertions. Whether strategies can be designed that execute with acceptable speed remains an open question.

Acknowledgments

We wish to thank Gunter Leeb and Nancy Lynch for sharing a challenging example with us; Steve Garland, Gunter Leeb, Victor Luchangco, and Nancy Lynch for insightful discussions; and the anonymous reviewers for helpful comments.

References

1. J.-R. Abrial, E. Boerger, and H. Langmaack. Preliminary report for the Dagstuhl-Seminar 9523: Methods for Semantics and Specification. Dagstuhl, June 1995.
2. M. Archer and C. Heitmeyer. Mechanical verification of timed automata: A case study. In *Proc. 1996 IEEE Real-Time Technology and Applications Symp. (RTAS'96)*. IEEE Computer Society Press, 1996.
3. M. Archer and C. Heitmeyer. Verifying hybrid systems modeled as timed automata: A case study. Technical report, NRL, Wash., DC, 1997. In preparation.
4. R. Boyer and J. Moore. *A Computational Logic*. Academic Press, 1979.
5. S. Campos, E. Clarke, and M. Minea. Analysis of real-time systems using symbolic techniques. In *Formal Methods for Real-Time Computing*, chapter 9. John Wiley & Sons, 1996.
6. M. J. C. Gordon and T. Melham, editors. *Introduction to HOL: A Theorem Proving Environment for Higher-Order Logic*. Cambridge University Press, 1993.
7. C. Heitmeyer and N. Lynch. The Generalized Railroad Crossing: A case study in formal verification of real-time systems. In *Proc., Real-Time Systems Symp.*, San Juan, Puerto Rico, Dec. 1994.
8. C. Heitmeyer and N. Lynch. The Generalized Railroad Crossing: A case study in formal verification of real-time systems. Technical Report MIT/LCS/TM-51, Lab. for Comp. Sci., MIT, Cambridge, MA, 1994. Also TR 7619, NRL, Wash., DC 1994.
9. C. Heitmeyer and D. Mandrioli, editors. *Formal Methods for Real-Time Computing*. Number 5 in Trends in Software. John Wiley & Sons, 1996.
10. T. Henzinger and P. Ho. Hytech: The Cornell Hybrid Technology Tool. Technical report, Cornell University, 1995.
11. R. P. Kurshan. *Computer-Aided Verification of Coordinating Processes: the Automata-Theoretic Approach*. Princeton University Press, 1994.
12. G. Leeb and N. Lynch. Proving safety properties of the Steam Boiler Controller: Formal methods for industrial applications: A case study. In J.-R. Abrial, et al., eds., *Formal Methods for Industrial Applications: Specifying and Programming the Steam Boiler Control*, vol. 1165 of *Lect. Notes in Comp. Sci.* Springer-Verlag, 1996.
13. V. Luchangco. Using simulation techniques to prove timing properties. Master's thesis, Massachusetts Institute of Technology, June 1995.
14. N. Lynch and F. Vaandrager. Forward and backward simulations – Part II: Timing-based systems. To appear in *Information and Computation*.
15. N. Lynch and F. Vaandrager. Forward and backward simulations for timing-based systems. In *Proc. of REX Workshop "Real-Time: Theory in Practice"*, volume 600 of *Lecture Notes in Computer Science*, pages 397–446. Springer-Verlag, 1991.
16. K. L. McMillan. *Symbolic Model Checking*. Kluwer Academic Publishers, 1993.
17. M. Merritt, F. Modugno, and M. R. Tuttle. Time constrained automata. In J. C. M. Baeten and J. F. Goote, eds., *CONCUR'91: 2nd Intern. Conference on Concurrency Theory*, vol. 527 of *Lect. Notes in Comp. Sci.* Springer-Verlag, 1991.
18. S. Owre, J. Rushby, N. Shankar, and F. von Henke. Formal verification for fault-tolerant architectures: Prolegomena to the design of PVS. *IEEE Transactions on Software Engineering*, 21(2):107–125, Feb. 1995.
19. N. Shankar, S. Owre, and J. Rushby. The PVS proof checker: A reference manual. Technical report, Computer Science Lab., SRI Intl., Menlo Park, CA, 1993.
20. J. Vitt and J. Hooman. Assertional Specification and Verification Using PVS of the Steam Boiler Control System. In J.-R. Abrial et al., editors, *Formal Methods for Industrial Applications: Specifying and Programming the Steam Boiler Control*, volume 1165 of *Lect. Notes in Comp. Sci.* Springer-Verlag, 1996.

A Appendix. Constant and Variable Definitions

Name	Type	Unit	Description
I	real, >0	s	time between the periodical sensor reading
S	real, >0	s	delay to activate pumps after the last sensor reading
U_1	real, >0	l/s^2	maximum gradient of the increase of the steam rate
U_2	real, >0	l/s^2	maximum gradient of the decrease of the steam rate
M_1	real, >0	l	minimum amount of water before boiler becomes critical
M_2	real, >0	l	maximum amount of water before boiler becomes critical
W	real, >0	l/s	maximum steam rate before boiler becomes critical
P	real, >0	l/s	exact rate at which one active pump supplies water to the boiler
#pumps	int, >0		Number of pumps that can supply water to the boiler in parallel
C	real, >0	l	Capacity of the boiler

Table 1. Constants for the Boiler and Controller Models

Name	Initial Value	Type	Values Range	Unit	Description
now	0	real	$[0, \infty)$	s	current time
pr	0	integer	$\{0, ..., \textbf{#pumps}\}$		number of pumps actively supplying water to the boiler
q	$\gg M_1$, $\ll M_2$	real	$[0, C]$	l	actual water level in the boiler
v	0	real	$[0, \infty)$	l/s	steam rate of the steam currently leaving the boiler
pr_new	0	integer	$\{0, ..., \textbf{#pumps}\}$		number of pumps that are supposed to supply water after the activation delay
error	0	integer	$\{0, ..., pr\_new\}$		number of pumps that fail to supply water to the boiler after activation
do_sensor	true	boolean	{true, false}		enable a single sensor reading
set	S	real	$[0, \infty)$	s	next time the pumps change to the new settings
read	0	real	$[0, \infty)$	s	next time the sensors will be read
stop	false	boolean	{true, false}		flag whether emergency shut down is activated
do_output	false	boolean	{true, false}		flag that enables the output. This represents a kind of program counter.
stopmode	false	boolean	{true, false}		flag to activate the shut down
wl	q,	real	$[0, C]$	l	current water level reading
sr	0	real	$[0, W]$	l/s	current steam rate reading
now	0	real	$[0, \infty)$	s	current time
pumps	0	integer	$\{0, ..., \textbf{#pumps}\}$		number of currently active pumps suppling water to the boiler
px	0	integer	$\{0, ..., \textbf{#pumps}\}$		number of pumps that shall supply water next

Table 2. Variables for the Boiler and Controller Models

Name	Type	Unit	Value	Description
max_pumps_after_set	integer		#pumps	maximum number of pumps that can supply water to the boiler after the delay considering the pump failure model.
min_pumps_after_set	integer		0	minimum number of pumps that can supply water to the boiler after the delay considering the pump failure model.
max_steam_water(sr)	real	l	$\max(0, (sr - U_2 * I/2) * I)$	minimum amount of water that can evaporate into steam until the next sensor reading
min_steam_water(sr)	real	l	$(sr + U_1 * I/2) * I$	maximum amount of water that can evaporate into steam until the next sensor reading

Table 3. Additional Definitions for the Controller Model

Using an Object-Oriented Methodology to Bring a Hybrid System from Initial Concept to Formal Definition

David Sinclair

School of Computer Applications,
Dublin City University, Glasnevin,
Dublin 9, Ireland.
email: David.Sinclair@compapp.dcu.ie

Abstract. A real-time embedded system is characterised by the interaction between a discrete control program and an environment with continuous and discrete nature. Because a real-time embedded system interacts and controls a real world system, issues such as *liveness* and *safety* can be critically important. The failure of such a system to meet it *liveness* and *safety* constraints can result in large economic costs and the loss of life. The behaviour of such systems needs to verified. The first step to verifying the behaviour of a system is to formally specifying the system. Hybrid automata provide a formalism for specifying and modelling both the continuous and discrete nature of the entire system. In common with other formal methods, deriving a hybrid automata description of anything but a toy system is not a trivial task. Two factors that make the derivation a formal specification difficult are:

- bridging the "detail gap" between the initial descriptions of a system and its formal specification; and
- managing the scale of real world applications.

This paper presents a methodology that addresses these problems. This methodology OHMS (**O**bject-oriented **M**ethodology for **H**ybrid **S**ystems) assists the user in capturing the initial system requirements, identifying the component parts of the system, defining the interfaces between these components, and incrementally developing these into a hybrid automata specification of the system. The major aspects of this methodology are illustrated with a simple case study.

1 Introduction

A real-time embedded system consists primarily of a program that controls a real world system. The program has a discrete nature, moving from one state to another depending on the signals it receives from its environment and the values of the discrete state variables within the program. The system being controlled by the program may have a distinctly different nature. The behaviours of real world systems are not limited to discrete changes of state variables under the control of a master clock. The state variables that describe the current state of the real world system may change continuously as well as in discrete steps. It is this interaction between a discrete control program and an environment with continuous and discrete nature that defines a real-time embedded system. Because a real-time embedded system interacts and controls a real world system, issues such as *liveness* and *safety* can be critically important. The failure of such a system to meet it *liveness* and *safety* constraints can result in large economic costs and the loss of life. The behaviour of such systems needs to verified. The first step to verifying the behaviour of a system is to formally specify the system.

While techniques such as StateCharts [1], ModeCharts [2], Petri Nets [3], Timed Automata [4], process algebras [5][6], and proof theoretic reasoning can model the discrete program and its behaviour, they do not model the continuous nature of the program's environment. Verification of the complete system, using these methods, requires the environment's behaviour to be "discretised". Continuous systems modelling languages, such as Omola [7], allow the user to model the continuous nature of the system. Within these continuous time models discrete events can be modelled, but these systems are not intended to be fully fledged continuous and discrete modelling languages. Hybrid automata [8] provide a mechanism for formally specifying and modelling both the continuous and discrete nature of the entire system. In common with other formal methods, deriving a hybrid automata description of anything but a toy system is not a trivial task. Two factors that make the derivation a formal specification difficult are:

- bridging the "detail gap" between the initial descriptions of a system and its formal specification; and
- managing the scale of real world applications.

Once the system has been formally specified, this formal description has two distinct uses, verification and implementation. The formal specification can be verified, using tools such as HyTech [9] and UPPALL [10], to see if the safety properties of the complete system are violated. Since the formal specification completely defines the behaviour of the control program, it can also be used as a basis for implementation.

This paper presents a methodology that addresses these problems. This methodology OHMS (**O**bject-oriented Methodology for **H**ybrid **S**ystems) assists the user in capturing the initial system requirements, identifying the component parts of the system, defining the interfaces between these components, and incrementally developing these into a hybrid automata specification of the system. This formal specification can then be used as a base for verification and implementation.

2 Methodology

2.1 Object Orientation

The methodology presented in this paper uses an object-oriented model rather than the more traditional process-oriented model. Within the software community object-orientation is receiving a lot of positive attention. Focusing on the identification and definition of objects makes more sense in real-time embedded system than focusing on processes since it is relatively easy to identify the physical components of a system and their behaviour. Finding the processes of a system can be harder since many physical components may be involved in a process. While object-orientation provides a natural view of the system, it is not the sole reason for using object-orientation. The main advantages of an object-oriented model are a) that provides mechanisms for hierarchical structuring and component abstraction, which is essential in order to handle the scale of industrial real-time embedded system; and b) it encourages the reuse of components, which makes the large systems easier to modify and maintain.

Each object in the system consists of a set of variables and an object behaviour which may be the composition of behaviours. The object variables can be private (local to the object) or public (visible by other objects) and they hold the state of the object. The object's behaviour, which defines the object's dynamic evolution, is the composition of a set of behaviours that belong to the object. These behaviours can be composed into the object's behaviour using parallel composition, sequential composition and choice operators. Since the real-time embedded system is a collection of interacting objects, each with its own dynamic behaviour, the different object behaviours are synchronised by passing signals between objects. These signals can carry values or simply signify that a specified event has occurred within an object.

2.2 Phases

The methodology proposed in this paper addresses the first three phases of the production and verification of a real-time embedded system, as identified in fig. 1, namely the *Analysis Phase*, the *System Design Phase* and the *Detailed Design Phase*. The methodology, as yet, does not address the verification and implementation phases, though it is envisaged that the output of the *Detailed Design Phase* could be translated into a form suitable for hybrid automata verification tools.

2.2.1 Analysis Phase

The goal of this phase is to identify the objects in the system and arrange them in an object hierarchy if possible. This hierarchy may due to the structural nature of the system, or because some objects may be specialisations of existing objects. In addition to identifying the object in the system, it is necessary to identify the signals passed between the objects. These signals are used to pass data between objects and to synchronise the behaviour of interacting objects. The result of this phase is an *Analysis Model* that identifies the objects in the system, their structural hierarchy and the signals that pass between them.

2.2.2 System Design Phase

The goal of the *System Design Phase* is to identify the state variables of each object and to define the interface for each object behaviour. The type of each state variable must be specified, as well as any initial value. Each object behaviour may be the composition of several other behaviours. How these behaviours are combined is defined (parallel, sequential or choice composition), as well as signals sent and received by each behaviour. The result of this phase is a *System Design Model* that identifies the state variables of each object in the system, the behaviours that specify each object's dynamic evolution and the interfaces of these behaviours.

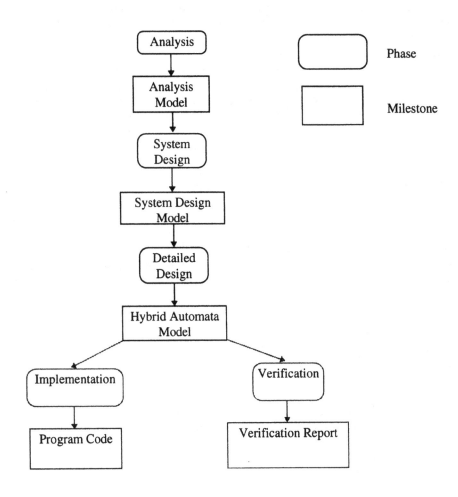

Fig. 1: Phases and Milestones in the Production and Verification of a Real-Time Embedded System

2.2.3 Detailed Design Phase

The goal of the *Detailed Design Phase* is to specify, for each object, the behaviours that are composed to define the object's dynamic evolution. Each behaviour is specified as a hybrid automaton. The result of this phase is the *Hybrid Automata Model* that completely and formally specifies the real-time embedded system using a system of hybrid automata.

2.3 Notation

2.3.1 Analysis and System Level Notation (Object Diagram)

System/Subsystem
System_identifier

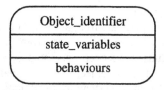

A rectangular box defines the boundaries of a system. Systems can be nested within other systems. Such subsystems are useful in abstracting a large specification. The outermost system represents the environment in which the hybrid system is situated.

Object

Object_identifier
state_variables
behaviours

A rectangular box with rounded corners represents an object. An object consists of an identifier (*Object_identifier*), a set of state variable definitions and a composition of concurrent interacting behaviour declarations. The state variables and behaviours are required at *System Level Design*, but are optional during the *Analysis Phase*. A state variable has a type, either **int**, **real**, or **bool**, and may have initial value.

Inheritance

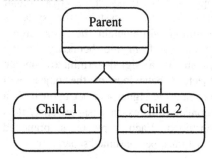

A child object inherits state variables and behaviours of it parent object. These state variables and behaviours can be overloaded by local state variables and local behaviours with the same identifier. A parent object may define a state variable, or behaviour, as *virtual*. These virtual state variables and behaviours must be overload with a non-virtual definition by objects in its inheritance tree such that each object at the lowest level of the inheritance tree has a non-virtual definition for all its state variables and behaviours. It is the lowest level objects that represent entities in the system being modelled. The higher level objects in the inheritance tree represent generalisations of these objects to allow for re-use and structured design.

Signals

A signal is an event, a continuous value or a message, that is sent between objects, systems and signal sinks. Signals can carry data as a parameter list. Unparameterised signals are typically use during the *Analysis Phase*, and then parameters are associated with signals during the *System Design Phase*. A continuous signal has a continuous value. The value of a discrete signal is the time at which it occurred. A signal may branch indicating that multiple copies of the signal are produced.

signal_identifier (parameter_list)

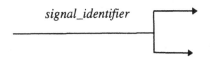

signal_identifier

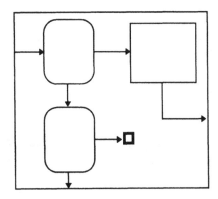

A signal can originate from an object or a system. If a signal originates from an object, it must have been generated by a behaviour of that object. If a signal originates from a system, it must have been generated by an object contained in that system. Signals originating from the boundary of a system are generated by the enclosing system or by the environment of the hybrid system.

A signal can terminate on a system, an object or a signal sink. If a signal terminates on a system, the system's definition will contain the same signal originating from the boundary of the system. If a signal terminates on an object, a behaviour of the object will consume the signal and act upon it. If a signal terminates on a signal sink, it is consumed. Each copy of a branched signal must be consumed separately.

Signal Sinks

□

A signal sink consumes signals.

Behaviour Declarations

Each object has an object behaviour that defines the object's dynamic evolution. This object behaviour may be the composition of several behaviours. These behaviours may be composed in parallel, sequentially or by guarded choice (|||, ; or   ). Each behaviour has a list of signal mappings. A signal mapping (denoted →) associates a local signal generated within the behaviour to a signal terminating on or emanating from the object. It is through these terminating and emanating signals that different objects interact. Each local signal is mapped onto one or more signals terminating on or emanating from the object or one of its predecessors (objects that are higher than it in the inheritance tree containing the object). A signal can be qualified by appending an additional identifier in square brackets. If a local signal is mapped onto a signal terminating on the object, the local signal occurs when the terminating signal occurs, plus an optional delay which may be one of the parameters of the terminating signal. If a local signal is mapped onto a signal emanating from the object, the emanating signal occurs at the local signal occurs. Each signal mapping specifies whether the signal is discrete or continuous, and whether it terminates or emanates from the object (**s_in**, **c_in**, **s_out** and **c_out** respectively).

2.3.2   Detailed Design Notation (Dynamic Behaviour Diagrams)

Behaviour
Behaviour_identifier

A rectangular box encloses the states and transitions on a behaviour of an object.

State

State_identifier
[initial_conditions;]
[evolutionary equations]

A state, represented by an oval, consists of an identifier, an optional set of initial conditions on state variables, and an optional set of differential equations defining the evolution of the state variables within this state. The state variables must be defined in the object *System Level* specification.

Transition

A transition represents a behaviour changing from one state to another state. A transition may be guarded, in which case the transition is traversed (fired) if and only if the guard evaluates to true. A transition can optionally contain a list of actions to be executed when a transition is fired. These actions are separated from the guard by $/$, and may assign a value to a variable or generate a signal. Signal generation is denoted a $\uparrow$ followed by a signal name.

Start States

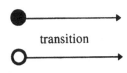

A start state is the state in which a behaviour exists when it is created. Start states are represented by circles. If the circle is a solid circle the transition emanating from is fired immediately. If the circle is a hollow circle then the transition emanating from the start state is guarded.

Termination States

A terminating state consumes signals and terminates the behaviour.

3 Example: Railroad Crossing Gate

In order to illustrate some of its features, the methodology was applied to the Railroad Crossing Gate problem. Due to size limitations it is not possible to present a large case study as an example. Even though this example is a small system it includes some of the features essential for dealing with large real world systems, such as abstraction, reuse and hierarchical structuring, and demonstrates how they can be used.

Analysis

System

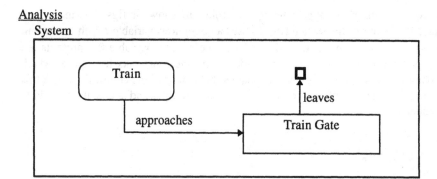

Fig. 2: Analysis view of the System

Train Gate

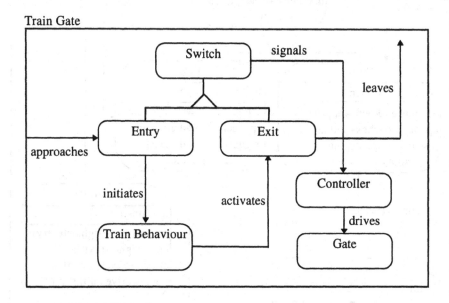

Fig. 3: Analysis view of the Train Gate

The two major entities in this example are the train and the train gate. The train gate subsystem is further expanded into an entry and exit switch, the physical gate, the gate controller and an object to model the behaviour of the train within the train gate subsystem. The entry and exit switch are specialisations of a more general switch object and inherit their behaviour from this object. The next phase is *System Design* where the state variables and object behaviours are identified. The signal mappings relate signals that originate and terminate on objects with signals generated and consumed within the composed behaviours. The *Detailed Object Behaviour Design* for the switches, the physical gate and the controller are shown in figs. 5, 6 and 7. The debounce behaviour of the switch (fig. 5) uses a continuous variable *t* with a constant 1st order derivative to model time. In fig. 5 the continuous variable θ represents the angular position of the gate and it 1st order derivative represents the rate at which gate opens and closes. Fig. 7 shows the detailed behaviour of the controller. Since this will be implemented in software it only has discrete variables and is essentially a finite state machine.

System-Level Design
Train Gate

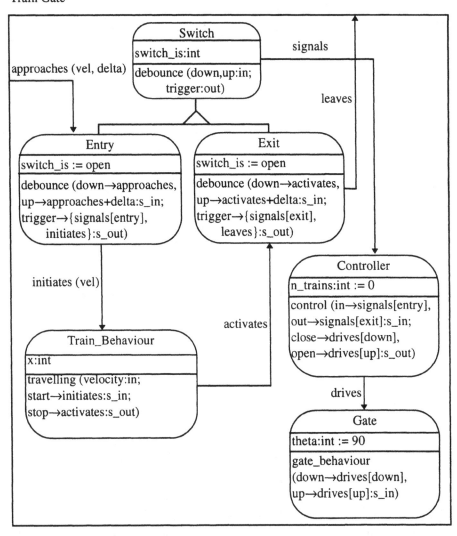

Fig. 4: System Design view of the Train Gate

Detailed Object Behaviour Design
Switch:debounce

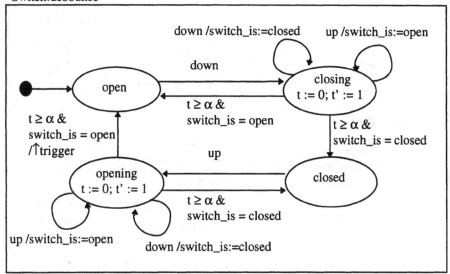

Fig. 5: Behaviour debounce of object Switch

Gate:gate_behaviour

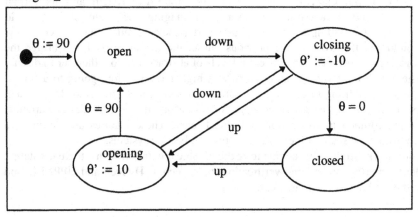

Fig. 6: Behaviour gate_behaviour of object Gate

Controller:control

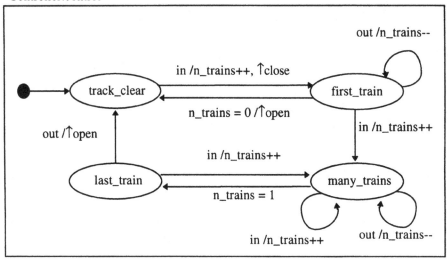

Fig. 7: Behaviour control of object controller

4 Conclusions and Further Work

The methodology presented in this paper is focused on a) bridging the "detail gap" between the initial description of a real-time embedded system and its formal specification; and b) assisting the user in managing the scale of real world applications. By providing a series of phases and associated milestones, in which one phase builds on the work of the previous phase, the methodology helps bridge the "detail gap" incrementally. The different levels of abstraction allow the user to discuss the system with the customer and others, at a higher level, before going to a formal description. In addition to providing a natural view of a real-time embedded system and its environment, the use of concepts of object-orientation provides hierarchical structuring, component abstraction and object reuse. These features are essential in order to handle the scale of industrial real-time embedded systems.

Future work will involve augmenting the graphical notation with a textual notation in order to interface to existing verification tools, such as HyTech and UPPALL, and testing the methodology on large case studies.

References

1. D. Harel. Statecharts: a visual formalism for complex systems. *Science of Computer Programming*,8(1):231-274, 1987.

2. F. Jahanian and A.K. Mok. Modechart: A specification language for real-time systems. *IEEE Trans. on Software Eng.*, 20(10):879-889, October 1994.
3. P. Merlin and D. Farber. Recoverability of communications protocols. *IEEE Trans. on Communications*, 24(9):1036-1043, September 1976.
4. N. Lynch and F. Vaandrager. Forward and backward simulations for timing-based systems. *Proceedings of REX Workshop "Real-TIme: Theory in Practice", LNCS*, volume 600, pages 397-446, Mook, The Netherlands, June 1991. Springer-Verlag.
5. I. Lee, P. Brémond-Grégoire and R. Gerber. A Process Algebraic Approach to the Specification and Analysis of Resource-Bound Real-Time Systems. *Proceedings of the IEEE*, pages 158-171, January 1994.
6. G. Reed and A. Roscoe. Metric spaces as models for real-time concurrency. In *Proceedings, Mathematical Foundations of Computer Science, LNCS*, volume 298, New York, 1987, Springer-Verlag.
7. S.E. Mattsson and M. Andersson. The Ideas Behind Omola. *Proceedings of the 1992 IEEE Symposium on Computer-Aided Control System Design*, pages 23-29, March 1992.
8. R. Alur, C. Courcoubetis, T. Henzinger, P. Ho, X. Nicollin, A. Olivero, J. Sifakis and S. Yovine. The algorithmic analysis of hybrid systems. *Theoretical Computer Science*, 138:3-34, 1995.
9. T, Henzinger and P. Ho. HyTech: The Cornell hybrid technology tool. Technical Report, Cornell University, 1995.
10. J. Bengtsson, K. Larsen, . Larsson, P. Pettersson, W. Yi. UPPAAL - A tool suite for automatic verification of real-time systems. *Proceedings of the 4th Workshop on Verification and Control of Hybrid System.*New Brunswick, New Jersey, 22-24 October, 1995.

A Digital Real-Time Simulator for Rail-Vehicle Control System Testing

Peter Terwiesch   Erich Scheiben   Anders Jenry Petersen

Thomas Keller

ABB Corporate Research
Communication & Control Group
CH-5405 Baden-Dättwil, Switzerland.
Fax: +41-56-486-7365.
E-mail: Peter.Terwiesch@chcrc.abb.ch

ABB Daimler-Benz Transportation
Propulsion Systems
Dept. BAA
CH-8050 Zürich, Switzerland

Abstract

Modern electrical locomotives are powered by controlled power converters. Integration testing of the vehicle control system is performed using real-time simulation. The present contribution presents an evaluation of a digital real-time simulator for this purpose, with particular emphasis on switching and events, which is one of the more demanding tasks in the real-time simulation of hybrid systems. A pragmatic solution to the difficult problem of simulating hybrid systems in real-time is proposed.

Keywords: digital real-time simulation, hardware-in-the-loop, control system testing, integration methods, discrete events, hybrid system, variable causality.

1 Introduction

Today's rail propulsion systems deliver a functionality in terms of energy efficiency and power factor that was unheard of before the days of computer-controlled power converters. This revolutionary increase in functionality is accompanied by an ever growing control system complexity. Due to the power converters, modern locomotives are hybrid systems. This paper focuses on the integration testing of a locomotive's dedicated computer control system using digital real-time ("hardware-in-the-loop") simulation.

Traditionally, most technical systems are tested by putting them into their designated working environment and seeing whether they work according to expectations. Given today's importance and complexity of digital control systems, this traditional *ad hoc* approach is no longer adequate, and concern must be given to system integration, testing, and verification.

This demand is driven by the following factors:
• *risk* (loss of human life or capital)
• *cost* (test in target system can be prohibitively expensive)
• *availability* (target system or designated working environment are not available)
• *coverage* (not all test states can be reached during regular operation)

One of the most powerful, but also most demanding, tests for dedicated real-time control systems (frequently also referred to as *embedded systems*) is to connect the inputs and outputs of the control system under test to a real-time simulation of the target process. This implies that all control loops are closed via the simulator, so this method is often called *hardware-in-the-loop* simulation.

A particular merit of this approach is that it even permits a gradual change-over from simulation to actual application, as it allows to start from a pure simulation and to gradually integrate real electrical and mechanical subsystems into the loop as they become available.

Depending on requirements and process dynamics, three conceptual alternatives for hardware-in-the-loop simulation exist:
* embedded system with built-in self-test/self-simulation mode
* use of another control system as a simulator
* dedicated real-time simulation system

The advantage of the first approach is that it requires little or no additional hardware and that it can be activated during maintenance stops without further external tools. However, a built-in self-test alone is hardly sufficient for finding communication bottle-necks or even design flaws, and its fault coverage may often be insufficient. The second approach is relatively popular in an industrial development environment, as it delivers the advantages of full-scale hardware-in-the-loop simulation at comparatively low cost. It saves on wiring, interfacing, and conversion, as controller and simulator are made of identical hardware and are thus usually easily connected. One limitation of this approach is that it typically fails to work for extremely fast systems, where the computing power sufficient for running a controller may be an order of magnitude too slow for executing a detailed process simulation. Another limitation lies in the programming of this simulation, since elementary control function blocks are rather ill-suited for the implementation of complex simulation systems.

For very demanding requirements a dedicated real-time system can offer the required computing power requiring a special hardware interface to the control system and the largest initial investment of the three approaches. This paper describes such a real-time system. It is used for testing the line and the propulsion converter control system of a locomotive. The interface to the simulator consists of discrete firing pulses from the control system to the power converters and of measured currents, voltages, and drive speeds from the simulated vehicle to the control system. The discrete and continuous dynamics of the main electric circuits and main mechanical modes of a locomotive are simulated in a real-time closed loop with a hybrid control system, resulting in a challenging hybrid control and simulation problem.

2 System Overview

2.1 Overall Locomotive

The "Loc 2000", the pride of Swiss Federal Railways, employs 3-phase propulsion and GTO converters to reach maximum power of 6100 kW and top speed of 230 km/h. Weighing in at 81 tons, it is the first member of a family of locomotives on which future European and overseas railway projects will be based [8]. Together with the power transformer, the converters and their controllers are really the "gut" of the locomotive. They have the task of converting the 16 2/3 Hz single-phase AC supply into a three-phase AC source of variable amplitude and frequency, which is suitable for supplying the induction motor drives. The flow of energy conducted by the line converter does not equal that conducted by the propulsion converter at every instant. The difference is either stored or supplied by the DC link circuit, in banks of metallised paper capacitors.

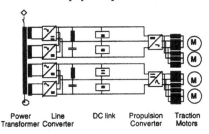

Fig. 1. Loc2000 electrical system

Fig. 1 gives a schematic overview of the Loc 2000's electrical system. Electrical energy is fed in from the catenary through the pantograph, and is then transformed and rectified to feed an intermediate-link capacitor. In Lok2000 Re460, this capacitor is split in two symmetric halves, providing the voltages $+0.5\ V_d$, 0, $-0.5\ V_d$, V_d being the DC link voltage. A computer-controlled GTO inverter bridge (propulsion converter) converts this DC energy to a three-phase AC system that feeds the driving motor. Note that this flow of energy can also be reversed for regenerative braking, which converts energy of motion back to electrical energy that is fed back into the line.

2.2 Control Structure

The control structure of Loc 2000 is sketched in Fig. 2. The control signals, which determine whether the GTO thyristors are conducting or not, are generated by the SLG for the line converter and by the ALG for the inverter. The SLG control signals are generated by a pulse width modulator. By varying the widths of the different voltage impulses, the amplitude and phase-angle of the 4-quadrant controller voltage can be adjusted to fit the needs of stability and energy. The ALG produces GTO firing pulses to provide the motors with a three-phase supply. To generate the corresponding pattern of control impulses and driving the motor under optimal conditions at all times the motor frequency range is divided into 3 control ranges: indirect self control, direct self control and full impulse method. Each control range has its own requirements with regards to hardware-in-the-loop simulation and they must therefore be examined individually.

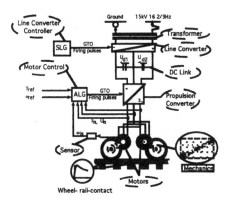

Fig. 2. Control structure

2.3 Simulator Requirements

The goal of the proposed simulator is to reduce the number of tests required on the real vehicle, which are very expensive. The real control system of a locomotive will be connected to the simulator in order to test the dynamic behaviour of the closed-loop-system. The structure of the hardware-in-the-loop simulation in shown in Fig. 3. A typical feature tested by the simulator is the amplitude of the torque ripple produced by the induction motor at different operating points. Another typical measurement is the transient motor and line currents when switching from acceleration to braking.

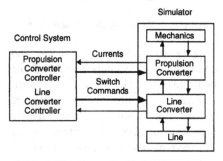

Fig. 3. Hardware-in-the-loop system structure

There are very stringent real-time requirements. The control system is running with a cycle time of 40-60μs. This together with the need to simulate the switched currents accurately leads to a required *frame time* of about 30μs. Here, frame time is the calculation time needed for one simulation step with the entire model. At the end of each frame time the simulator communicates with the connected control system, meaning that the frame time is the granularity of real-time synchronization between simulator and connected control system. Another requirement is that the system should be user-friendly and configurable even by non-simulation experts.

Several commercial vendors are addressing this market with platform solutions. One of them is dSPACE, a German company that offers a fast,

modular real-time hardware. dSPACE systems are widely used for rapid prototyping and hardware-in-the-loop simulation in many application areas, especially in the automotive industry. Based on DEC Alpha processors and TMS320C40 signal processors they offer scalable computing power. The processors are connected by fast point-to-point connections, avoiding the bottleneck of a common bus. Every C40 processor has its own peripheral bus for connecting I/O boards, allowing parallel I/O on different processors. A number of very powerful I/O boards is available.

dSPACE offers a user-friendly development environment based on Simulink as a graphical modelling environment [15]. This is one of the reasons for using Simulink in spite of its deficiencies mentioned below. Another reason is the fact that Simulink is widely used for off-line simulations, resulting in the existence of adequate model libraries for our purposes. A drawback of Simulink and similar languages is the implication of fixed causality in its block diagrams, and that it has no built-in support for event handling. In the following sections we describe our hybrid simulation problem and a pragmatic solution to circumvent the above mentioned deficiencies of fixed-causality simulation languages.

3 Hybrid Simulation Problem

When dealing with hybrid physical systems in which discrete mode transitions are due to idealised models of switches, we have to deal with at least two separate issues [12,13]:

- The causality of the model will change for every mode-transition.

- State events will have to be detected and continuous mode-equations re-arranged and re-initialised for every mode-transition.

Since we are using Simulink, a modelling environment not allowing for variable causality, and since real-time constraints prohibit to perfectly incorporate causality, the first issue is naturally going to be a problem and is discussed in section 3.3.

The second issue is discussed in section 3.1-3.2 from a general point of view. In our simulator, state events are handled by using small frame times combined with event time registration and correction.

3.1 Discrete and Continuous Subsystem

The Loc 2000 simulation model, as many others, consist of an *analog block* with a number of first-order ordinary differential equations (ODEs) for the continuous-valued system states x and a *binary block* with combinatorial and sequential equations for the discrete-valued system states X. The two blocks are coupled through switches (change of continuous input variables or equation structure by digital signal) and comparators (binary result from comparison of continuous variable with threshold) as shown in Fig. 4. While there is a wealth of methods available for simulation of either block [3,4,5], a great challenge both in non-real-time and real-time simulation is to properly account for the links between continuous and discrete simulation blocks.

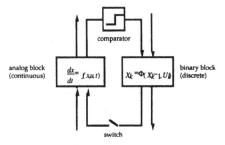

Fig. 4. Hybrid model structure

This coupling between discrete and continuous subsystems is one of the main reasons why *digital* real-time simulation is taking so long to replace *hybrid* simulation computers: hybrid computers contain comparators (analog to discrete) and switches (discrete to analog) that implement this link very naturally [1,2]. In digital simulation, on the other hand, three types of problems resulting from the link of discrete to continuous blocks can be distinguished: time events, state (or internal) events, and external events. The first two have in common that a continuous variable passes a threshold, so that a comparator generates an event. This event will typically operate on a switch, so that it may significantly change the continuous system in its structure, causality, or parameters. In the third case, the event typically already has a digital form.

Example (time event): A switch is opened a pre-programmed time after a shutdown command.

Example (state event): Ideal diode. An ideal diode works as an open switch as long as neither voltage nor current are positive. In an electric circuit, together with other elements such as capacitors or inductances, passing of this limit will structurally change the state equations, i.e., the differential equations governing the dynamics of the system.

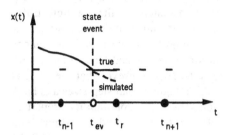

Fig. 5. Internal Event

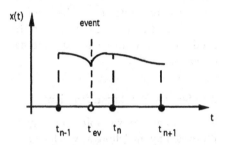

Fig. 6. External Event

Example (external event): Trigger pulses for controlled electrical converters switch an electrical circuit from one configuration to another and thus also alter the state equation structure.

State events and external events represent the major problems of real-time simulation, since the limited frame time, corresponding to a fixed external step size, normally does not permit adaptive step-sizing or iterations around the event. These two types of events thus call for a specific treatment depending on the exact requirements. Time events, on the other hand, are usually known in advance and can be incorporated by using adequate synchronization instants, so that they are normally not the main problem of real-time simulation.

There are different ways to approach the handling of state and external events in a digital hardware-in-the-loop simulation. External event problems can often be solved by one of the following approaches:
- small frame time
- interrupt-driven integration
and state event problems by
- small frame times
- event time registration and correction

The best approach to event problems is in all cases small frame times. It is therefore always advisable to examine the exact requirements of the system before implementing different approaches, as they may lead to much bigger frame times due to computational effort spent on iterations and corrections. However, when the dynamics of the system are very fast in comparison to the computational qualities, one of the other approaches must be used, see section 3.3.

3.2 Integration of Continuous Part

The choice of the right numerical integration method for the continuous part may greatly contribute to the power of a digital real-time simulator. This section summarizes some considerations for selecting appropriate algorithms, compare [9,11].

Mathematically, the continuous part of the system that needs to be simulated in real-time is usually described by a set of first-order ordinary differential equations

$$\frac{dx}{dt} = f(x, u, t)$$

Here x is the system state vector (containing elements such as position, speed, voltage of a capacitor, current through an inductance etc.), u represents external inputs (such as manual switches, or signals coming from the control hardware), and t stands for time.

In general, this type of system can be integrated using several approaches, which are distinguished by the following properties [9]:

• *explicit / implicit methods:* Explicit methods only use past values of x (up to the present time t_k), whereas implicit methods also use the new value $x(t_{k+1})$, which they find by iteration. Implicit methods are particularly suitable for stiff systems, i.e., for systems in which very slow and very fast dynamics need to be considered simultaneously.

• *single-step / multi-step methods:* Single-step methods only use one past value, $x(t_n)$, for computing the new value, $x(t_{n+1})$. Multi-step methods use more than one past value, e.g., $x(t_n)$ and $x(t_{n-1})$.. As single-step methods need smaller frame times or more evaluations of the model in the interval $[t_n , t_{n+1}]$ to achieve the same accuracy as multi-step methods (as long as the state trajectories can be approximated by polynomials), they are also computationally more expensive.

Note that real-time simulation does not permit to use predictor/corrector methods beyond the current frame, since no knowledge of u_{k+1} is available. As long as there are no abrupt changes in the state trajectories, multi-step

methods, such as the Adams-Bashforth algorithm, offer better performance and/or accuracy than single-step methods. However, as soon as switching needs to be accounted for, single-step methods such as the explicit Euler algorithm must be used, since an extrapolation of the past is no longer desirable and would lead to inaccurate results. Note however that these algorithms require small step-sizes and have comparatively poor stability properties [9]. Note that modern algorithms capable of event detection (e.g. DASSL) are not suited for real-time simulation without modifications (such as bounds on the number of iteration, which in turn would effect their accuracy) due to the need to synchronize with an external cycle time of 30µs.

3.3 Switch Simulation Problem

The problem of variable causality is illustrated by modelling a switch located between the transformer and the rectifier of a locomotive. Without the switch the connection between transformer and rectifier would be modelled as shown in Fig. 7. A switch between the two blocks (Fig. 8) would have inputs U_d and I_q and outputs U_{st} und I_d (Fig. 9).

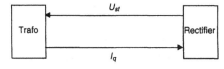

Fig. 7. Transformer and Rectifier

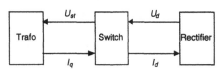

Fig. 8. Transformer and Rectifier with a Switch

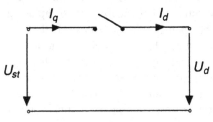

Fig. 9. Switch

Now let us consider the change of causality depending on the state of the switch. When the switch is *closed*, we have $U_{st} = U_d$ and $I_d = I_q$. This is modelled easily with the given structure:

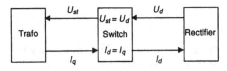

Fig. 10. Closed Switch

But when the switch is *open*, the causality changes: We have $I_d = 0$ and $I_q = 0$. Instead of U_{st}, I_q is given and the voltage U_{st} is undefined (Fig. 11). But I_q is also an output of the transformer. In order to solve this conflict in a clean way, the transformer equations have to be solved for U_{st} instead of I_q.

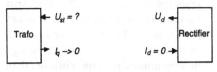

Fig. 11. Open Switch

Implicit simulation languages, such as ACSL [7] are able to describe events, but are not directly suited for real-time simulation. Explicit simulators such as Simulink do not support such equations. One can obviously program all cases

manually, but a system with n switches would then have 2^n different representations, depending on the state of the switches [14,6,10].

Alternatively, we propose a pragmatic method to estimate the undefined open switch voltage U_{st}.

4 Switch Approximation for Fixed-Causality Simulation

4.1 Open Switch Voltage Estimation

An open switch has to set the voltage U_{st} in such a way that the current I_q becomes zero. We try to achieve this with an estimator (Fig. 12).

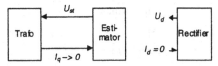

Fig. 12. Open Switch with Estimator

The estimator is based on a simplified model of the open circuit, e.g. an R-L-link for a transformer (Fig. 13). U_q, R, and L represent the model of the transformer. Assuming that R and L are approximately known, the estimator guesses the unknown voltage U_q by applying a voltage U_{st} and evaluating the resulting change in the current I_q.

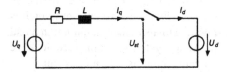

Fig. 13. Open Switch with R-L-Link

One problem is the choice of an adequate structure for the internal model of the estimator. An R-L-link is a reasonable choice when an inductive circuit like a transformer or an induction motor is connected to the switch. For the values of R and L we choose the resistance and leakage inductance of the corresponding coil, neglecting the coupling of other coils. As shown later, modelling errors result in a leakage current.

The R-L-link in Fig. 13 is described by the following discrete equation:

$$L\frac{I_q(k)-I_q(k-1)}{\Delta t} = -RI_q(k-1) + \qquad (1)$$
$$U_q(k-1)-U_{st}(k-1)$$

At time k we know:

- $I_q(k)$ New current from the transformer

- $I_q(k-1)$ Old current from the transformer

- $U_{st}(k-1)$ The voltage of the switch from the last time step.

With these values the unknown source voltage U_q during the last time step can be estimated in the following way:

$$\hat{U}_q(k-1) = L\frac{I_q(k)-I_q(k-1)}{\Delta t} + \qquad (2)$$
$$RI_q(k-1)+U_{st}(k-1)$$

At zero current this is the desired open switch voltage. But if the switch is opened with non-zero current (which is possible in a simulation, but not in a real physical system), a non-zero leakage current remains. A means must be provided to force this current to zero. This is achieved by adding a correcting term to the estimated source voltage, which draws the current to zero:

$$U_{st}(k) = \hat{U}_q(k-1) + k_{pso}I_q(k) \qquad (3)$$

The parameter k_{pso} is the gain of the estimator. It is equivalent to replacing the open circuit with an extremely large resistor, which is a well known method for simulating open switches. Using an estimator, however, the voltage over the resistor can be considerably reduced, resulting in a much smaller leakage current.

The leakage current will decrease with the time constant

$$\tau = \frac{L}{R+k_{pso}}. \qquad (4)$$

We normalize the estimator gain k_{pso} with

$$k_{pso0} = \frac{L}{\Delta t} - R, \qquad (5)$$

the gain for $\tau = \Delta t$. The estimator gain will now be described through the relative factor k_{corr}:

$$k_{corr} = \frac{k_{pso}}{k_{pso0}} \qquad (6)$$

The following facts contribute to the leakage current:

- *Time delay:* The voltage estimation is delayed by one time step.

- *Switching off a non-zero current:* Opening the switch with a non-zero current flowing results in voltage peaks, since the current through the inductivity L cannot change discontinuously. The current then fades away with a time constant given by the estimator parameters, as shown above.

- *Error in L:* A mismatch between the inductivity L in the estimator and the true inductivity in the circuit leads to oscillations or even instability. The effect can be reduced by allowing a larger time constant in the estimator (section 4.2).

Looking at Fig. 12 we see that even if the leakage current I_q is not zero, the current I_d on the other side of the open switch can be set to zero. The leakage current therefore does not have any disturbing effects on that side.

4.2 Stability and Choice of Estimator Parameters

Two kinds of parameters must be chosen for the estimator:
• Internal model parameters
• Estimator gain

In the following discussion we assume an R-L-link as an internal model of the estimator, as described above (Fig. 13).

The choice of R is not critical. It will be added to k_{pso} and can normally be neglected, namely if $R\,\Delta t / L \ll 1$.

The choice of L remains critical. Assuming that the real circuit is an R-L circuit according to Fig. 13 with L_{eff} instead of L, it can be shown that the estimator is only stable, if L_{eff} is in the range

$$\frac{k_{corr} + 2}{4} L < L_{eff} < 2L \qquad (7)$$

When the inductivity of the open circuit is known exactly, the estimator can be tuned to make the leakage current disappear after one time step ($L = L_{eff}$, $k_{corr} = 1$). However, in many cases L_{eff} is not known a priori, since it depends on the state of other switches. Then for simplicity an ad hoc decision is taken for choosing L. Often the inductance of the branch which is directly connected

to the switch is used, e.g. the leakage inductance of a transformer coil.

The stability range (7) can be extended at the lower end by choosing a small value for k_{corr}, This makes the system more robust against changes in L_{eff} at the cost of a slower decreasing leakage current. As we see, the choice of k_{corr} is a trade-off between stability and performance and must be examined experimentally. Resonable values of k_{corr} lie in the range 0..1.

Typical values in our transformer-rectifier example (Fig. 13) are $R = 0.02\Omega$, $L = 0.005H$, $\Delta t = 25e\text{-}6s$. With R $\Delta t / L = 1e\text{-}4 \ll 1$ we can neglect R in equations (5) and (6). For differenct choices of k_{corr} we have the following results (see also section 5.2):

$k_{corr} = 1$: fast decreasing leakage current ($\tau = \Delta t$), but not robust (oscillations when error in L).

$k_{corr} = 0.25$: still fast ($\tau = 4\Delta t$), no oscillations, robust until $L_{eff} \approx L/2$.

$k_{corr} = 0$: very slow ($\tau = L/R = 0.25s$), no oscillations, robust until $L_{eff} = L/2$

We see that the choice of k_{corr} is not very critical.

4.3 Extensions to the Method

By modelling a diode as an ideal switch, the same method can also be applied to diodes. We have modelled the diodes of the line converter in this way. With the proposed method the behavior of the locomotive is simulated qualitatively correct even with only the diodes working (switching pulses off). There is a small leakage current $I_q \neq 0$ in the transformer which normally can be neglected, especially when the diodes are used only for charging the DC link capacitor at startup.

The method can be applied in a similar way to other structures of the internal model of the estimator than R-L-links, since it basically involves a first-order lag approximation of a passive network that is changing due to different combinations of switches.

An alternative approach would be to zero the current in other parts of the model (e.g. transformer block) when the switch is open. This decreases the leakage current by at least an order of magnitude, but depends on changes in the model blocks (e.g. transformer) and additional, artificial, connections between them. We have decided against this approach, since it would come at the cost of giving up the current modularity, since artificial control signals would be needed to transmit the state of the switches to the blocks.

Instead of setting the current to zero it is also possible to design an internal event correction method which calculates the exact time of the event and adds a correcting term in the next simulation step. This method has the same drawbacks concerning modularity as setting the current to zero.

5 Results

5.1 Implementation

Fig. 14 shows the top layer of our Matlab/Simulink model of the locomotive. The controller on the left side is used in off-line simulations and sensitivity studies only. In on-line

simulations the controller model will be exchanged against the real hardware controller, and a real-time hardware-in-the-loop simulation is performed.

As the locomotive consists of two almost symmetrical parts, only half of the locomotive (half the line converters and inverters, one intermediate circuit and one bogie) need to be considered.

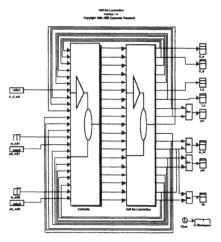

Fig. 14. Benchmark Controller/Locomotive

A benchmark test has been done together with dSPACE. The model was carefully partitioned on more processors to improve the frame time. Note that there is a natural bound on the number of processors due to communication needs, which will increase the frame and reaction time. Fig. 15 shows how the model is partitioned on 7 different C40 digital signal processors. Not all parts of the model actually require a high sampling rate. The mechanics, for instance, have very slow dynamics in comparison to the rest of the model and can therefore be sampled slower, thereby increasing the computational availability.

Benchmarking different configurations, the final hardware architecture has been chosen. With two 300Mhz, 64bit, DEC alpha processors and six C40 digital signal processors a frame time of less than 30 µs has been achieved.

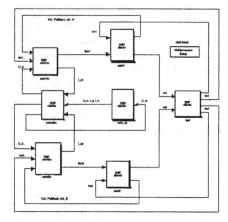

Fig. 15. Partitioning the Model on 7 DSPs

The converter switching pulses are read in with a time stamping hardware with a resolution of 25ns. The time stamp information is used in the simulation in order to increase the accuracy. More than one pulse per integration step can be handled. This makes the simulator ready for future applications with higher switching frequences, such as IGBT converters.

5.2 Numerical Results for Switches

The following simulation shows the development of the voltage U_{st} and the current I_q during switching in an R-L-link with $U_d = 0$. With the chosen value $k_{corr} = 0.25$ the leakage current decreases with a time constant $\tau = 4\Delta t$. Notice the voltage peaks at switching off due to forcing the current in the inductance to

zero. Fig. 17 shows that the system is robust to a leakage in L ($L = 0.003$ instead of 0.005).

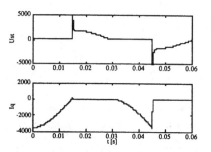

Fig. 16. Switch in R-L-Link. Switching times: 0.015 off, 0.03 on, 0.045 off. $k_{corr} = 0.25$

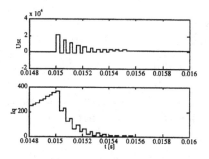

Fig. 17. Error in L with $k_{corr} = 0.25$

5.3 Numerical Results for Vehicle Simulation

The following figures show simulations of the whole locomotive. Fig. 18 shows the stator currents in alpha and beta components (space coordinates) and Fig. 19 shows the observed electrical torque generated by the asynchronous machine. The currents are of special interest, since they are measured and used by the control system to observe the electrical torque. The currents depend, as earlier mentioned, critically on the handling of switching events.

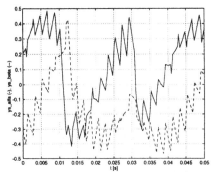

Fig. 18. Stator Currents (Normalized)

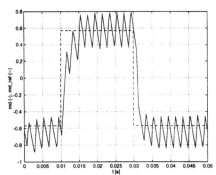

Fig. 19. Generated Electrical Torque (Normalized)

6 Conclusions and Perspective

We have presented selected results from a digital real-time simulation study of traction vehicles. Particular emphasis has been placed on the simulation of converter switching phenomena in the otherwise continuous system. Since this hybrid nature of the problem does not permit an exact solution in real time, we have presented an approximation method based on an estimator, which has satisfied our requirements. Distributing the locomotive model over 8 processors yields a sampling time of 30 μs for the continuous part, while discrete switching is resolved even more

accurately. Although a difficult task, we hope that simulation platform suppliers will make hybrid real-time simulation even more user-friendly in future versions. One promising approach can be to do the modeling on a higher level, without intially fixing the causalities, as is supported by object-oriented modeling tools such as Dymola [16]. An intelligent translator can then resolve the causalities analytically for a small number of discrete combinations, or, knowing the network and possible siwtch combinations, resolve them approximatively by finding an appropriate first-order lag approximation as demonstrated in our example.

We would like to take the opportunity to thank a number of people who have supported us throughout this work: R. Graf, P. Wenk, P. Kulli, B. Wixinger, and P. Häse from ABB Daimler-Benz Transportation (Suisse) AG; S. Menth from ABB Corp. Research, and U. Kiffmeier from dSPACE (Germany).

Constructive comments from anonymous conference reviewers are gratefully appreciated.

References

[1] H. Bühler: A driving simulator for investigating the static and dynamic characteristics of speed regulating systems for traction vehicles, Bulletin Oerlikon No.364/365, pp15-22, 1965.

[2] H.P. Wenk: SIMSTAR-Ein modernes Simulationswerkzeug. ABB Verkehrssysteme AG, *Roll-on*, no.3, 1992.

[3] G.A. Korn, J.V. Wait: Digital continuous-system simulation, Prentice-Hall, 1978.

[4] F.E. Cellier: Continuous System Modeling. Springer-Verlag, 1991.

[5] A. Naim: Systems modeling and computer simulation. M. Dekker, 1988.

[6] O. Ruhle: Echtzeitsimulation schneller transienter Vorgänge mit Hilfe von Parallelrechnern, VDI-Fortschrittsber., Reihe 20, Nr. 127, 1994.

[7] Technical Committee on Continuous Simulation Languages, Simulation Councils, Inc. (SCi): The SCi Continuous System Simulation Language", *Simulation*, no.9, pp.281-303, 1967.

[8] Locomotive 2000 product description, ABB Transportation Systems Ltd, Schweizer Eisenbahn-Revue 10/1991.

[9] G.H. Golub, J.M. Ortega: Scientific computing and differential equations: an introduction to numerical methods. Academic Press, 1992.

[10] O. Rathjen: Digitale Echtzeitsimulation: Simulation einer Hochspannungs-Gleichstrom-Übertragung. Vieweg, 1993.

[11] T. Hopkins, C. Phillips: Numerical methods in practice: using the NAG Library. Addison-Wesley, 1988.

[12] G.M. Asher, V. Eslamdoost: A novel causality changing method for the bond graph modelling of variable topology switching circuits. In Proc. IMACS Symposium, Lille 1991, pp371-376.

[13] J.E. Strömberg, J. Top, U. Södermann: Variable causality in bond graphs caused by discrete effects. In Proc. of the First Int. Conf. on Bond Graph Modeling (ICBGM '93), SCS, San Diego, 1993.

[14] W. Borutzky, J.F. Broenink, K.C. Wijbrans: Graphical description of physical system models containing discontinuities. In Proc. ESM'93, Lyon, 1993, pp203-207.

[15] H. Hanselmann: DSP in Control: The Total Development Environment. Int. Conf. on Signal Processing Applications and Technology, Boston, MA, 1995.

[16] H. Elmqvist, F.E. Cellier, M. Otter: Object-oriented modeling of hybrid systems. ESS'93, European Simulation Symposium, Delft, NL, Oct 25-28, 1993.

Hybrid Flow Nets for Hybrid Processes Modelling and Control

Jean-Marie Flaus and Guy Ollagnon

Laboratoire d'Automatique de Grenoble
ENSIEG, BP 46
F-38402 St Martin d'Hères
FRANCE
Email: flaus@lag.grenet.fr

Abstract. This paper presents a new approach for modelling and control of industrial transformation processes for which the dynamics behavior has an hybrid nature, that is to say, when the continuous dynamics interacts closely the event-driven evolution of the system. This methodology called hybrid flow nets is based on Petri Nets interacting with what we called Continuous Flow Nets and that can be seen as an continuous version of Petri nets with an evolution rule defined by a continuous non linear differential equation. In this work, we address the problem of controlling the flow through a continuous transition and propose an hybrid controller for keeping this flow within some specified bounds. The proposed methodology has been used to design the control system of a real industrial process used for the filtration in the production of beer.

1 Introduction

Industrial manufacturing processes can generally be classified as continuous, discrete or batch, depending on whether material are processed as a continuous flow or in discrete batch and quantities. In a continuous process, products are made by passing continuous material flow through pieces of equipment operating at steady state. However, in the start-up of shut down phases, the control of such processes requires sequential actions combined with classical regulation and exhibits an hybrid behavior.. In a discrete process, production is achieved by manufacturing parts on successive workstations, in a sequential way. Batch processes are, like discrete ones, discontinuous in nature. Material is processed sequentially following a recipe. However, each phase of the recipe is a continuous transformation, such as mixing, cooking or chemical reaction. So batch processes, that are more and more used nowadays, are of hybrid nature. They combine features of the continuous as well as the discrete world, that have rather different frameworks for system modelling, analysis and control design. Up to now the continuous control problem and the sequential control of a process were considered in a separate way. However, in order to prove that the system satisfies some specifications, a global approach must be used.

Hybrid systems have received a lot of attention recently [10] [12] [2] [14] and new modelling frameworks have been proposed, that can be classified in three classes [15]: (i) models that can be seen as extension of continuous models, as for example, differential-algebraic models with parameters jumps, or hybrid bond graphs, (ii) models based on a discrete framework approach, such as hybrid Petri nets [7] or hybrid automata [1], (iii) mixed models made up of a discrete part, a continuous part and an interface [6].

Models of the first class are not very well suited for batch processes modelling where the discrete aspect is important and not due only to parameters or state jumps. The drawback of the second class is that they are restricted to a class of linear systems while the last class includes models that are complex and difficult to analyze.

So the approach we have chosen is to propose an extension of models of the second class in order to be able to represent systems with a complex continuous part. Indeed, we have found that hybrid Petri nets are very attractive for processes modelling: discrete Petri nets have an inherent quality in representing logic in an intuitive and visual way, and the continuous part can be used to represent the evolution of continuous variables in the case of a linear system.

In this work, we use a new modelling methodology, called Hybrid Flow Nets [8], able to handle continuous flows with non linear characteristics and relying on Petri nets for describing sequential behavior. Our goal is to propose a modelling framework simple enough in order to work with, but not too simple in order to represent real life problems. This modelling approach is well suited for transformation processes modelling. In this paper, we focus on the problem of the control of a system modeled by an HFN. We first define an Input/Output interpretation of the model and propose a structure for an hybrid controller. Then, we design a control law in order to keep the flow through transition within some bounds. The modelling methodology and the control approach is illustrated on an industrial process of filtration for beer production for which the control law has been implemented on a Modicon PLC.

This paper is organized as follows: in the first part, we introduce hybrid flow nets for modelling and simulation, then we discuss the control problem and in the last part, we present an industrial hybrid control problem that we have treated with the hybrid flow nets modelling methodology..

2 Hybrid Flow Nets for Modelling and Simulation

In this part, we are going to present the modelling tool we propose. This one is based on Petri nets, used to describe the event-driven part, and on what we have called Continuous Flow Nets (CFN) used to represent the continuous part and that can be seen as a non linear continuous extension of Petri Nets. This two parts are interacting together according to some rules that we give in the sequel to lead to what we have called Hybrid Flow Nets (HFN). This modelling approach can be seen as an improvement of Hybrid Petri Nets [4] in order to model more complex continuous dynamics and to provide an equivalent of the enabling rule of a continuous transition defined as a continuous function of the

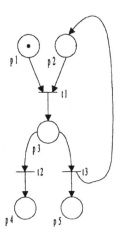

Fig. 1. Example of Petri net

input places, so that the evolution of the system can be described by a differential equation instead of a test of the input places values.

2.1 Discrete Petri Nets

Petri nets are a powerful tool for modelling of discrete events systems. A detailed presentation of Petri nets can be found in [7]. Here, we only recall the definition and the evolution rules

Definition 1. A **Petri net** is an 5-uplet $Z = (P_d, T_d, I_d, O_d, m_0)$ where:

P_d is a set of places, graphically represented by circles;

T_d is a set of transitions, graphically represented by bars, with $P_d \cup T_d \neq \emptyset$ and $P_d \cap T_d = \emptyset$,

$I_d : P_d \times T_d \rightarrow \{0, 1\}$ is the input function that specifies the arcs directed from places to transitions;

$O_d : P_d \times T_d \rightarrow \{0, 1\}$ is the output function that specifies the arcs directed from transition to places;

$m : P_d \rightarrow N$ is a marking whose i-th component is the numbers of tokens, graphically represented by dots, in the i-th place and m_0 is the initial marking.

The behavior of a Petri net is determined by exercising the enabling and the firing rules:

- A transition is enabled if the marking of its input places is strictly positive, that is to say, iff $m(p) > 0$ when $I(p, t) = 1, \forall p \in P$.
- An enabled transition t may be fired leading from marking m yielding to the new marking m' obtained by adding one token to each output places of t and substracting one token to each input places of t, which can be written $m'(p_i) = m(p_i) + O(p_i, t) - I(p_i, t)$ pour $i = 1, ... |P_d|$.

The Petri net is autonomous.

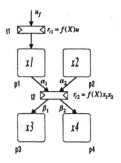

Fig. 2. Basic Continuous Flow Net

2.2 Continuous Flow Nets

We have developed the notion of continuous flow net as the counterpart of Petri Nets for continuous systems. Such an extension has already been made and, for example, the notion of continuous Petri nets has been proposed by [7]. However, the drawbacks of these approaches is that the enabling rule of transition needs some tests and cannot be expressed as a continuous function of the input places. Here, we propose a new approach which allows one to write the evolution of the CFN as a set of non linear differential equations.

We are now going to define formally the notion of continuous flow net. This is a graph structure similar to the one of Petri nets, where vertices are places and transitions. In a continuous flow net, places will be used to represent extensive variables such as volume, quantity of components or energy whose value is the marking of the place. Each place is connected to transitions (figure 2). The flow through this transition depends, on the one hand, on the value of the place, and on the other hand, on a function of the marking of the net which must be positive and bounded. More formally, we will define a continuous flow net as follows:

Definition 2. A **continuous flow net** (CFN) is defined by a n-uplet
$C = (P_c, T_c, I_c, O_c, \Phi, X_0)$ where:

P_c is a set of places represented graphically by a rounded square. To each place is associated a real number, positive, which is called *marking* or *value* of the place, and denoted x. The set of values makes a vector X where $X(i)$ is the marking of the i-th place. X_0 is the initial marking.

T_c is a set of n_t transitions (or gates), represented graphically by a specific rectangle (figure 2);

$I_c : P \times T \rightarrow \{0,1\}$ is the *input function* that specifies the arcs directed from places to transitions;

$O_c : P \times T \rightarrow \{0, 1\}$ is the *output function* that specifies the arcs directed from transitions to places;

$\Phi : T \rightarrow F_R$ associates to each transition a function $f_t(X)$ defined from $\Re^m$ to $\Re$, that is bounded $0 \leq f_t(X) \leq F_{\max}$. The flow r_t through a transition t is proportional to the value of the input places of this transition and to the value of the input flows connected to the input of this transition:

$$r_t = f_t(X) \prod_{I_c(i,t)=1} X(i) \text{ with } 0 \leq f_t(X) \leq F_{\max}$$

The output flow of a place p_i through a transition t_j to which the place is connected with a weight α_{ij} is equal to $-\alpha_{ij} r_{t_j}$. The input flow of a place p_i from a transition t_j to which the place is connected with a weight β_{ji} is equal to $+\beta_{ji} r_{t_j}$. The evolution of the value of a place is equal to the sum of the input flows and output flow with a negative sign, and is then given by:

$$\frac{dX_i}{dt} = - \sum_{O_c(i,j)=1} \alpha_{ij} r_{t_j} + \sum_{I_c(i,j)=1} \beta_{ji} r_{t_j}$$

This definition allows the description of an important number of physical systems and leads to two important properties:

1. the flow is null through a transition if at least one input place is empty (has a marking equal to zero).
2. the flow from and to a given transition for any places connected to this transition is related by a proportional coefficient..

These two properties make flow nets evolution very similar to the ones of Petri nets: (1) is equivalent to the enabling rule and (2) is equivalent to the firing rule.

Remark. If a transition has no input places, then the flow rate through this transition is defined as being equal to a nonlinear function of the state.

2.3 Hybrid flow nets

In order to model the continuous and discrete aspects of a system, we introduce what we call *hybrid flow net*. This modelling tool is made of a continuous flow net interacting with a Petri net according to a control interaction, that is to say the Petri net controls the CFN and vice versa (figure 5) . As we are going to see it, the overall philosophy of Petri nets is preserved again: the validation of transition implies that all the input places are not empty and the evolution rule is similar.

The interface of the discrete part to the continuous part is made through the control of a continuous transition by a discrete place. The flow rate of the transition is then equal to:

$$r_h(t_j) = m(p_i).r_t(t_i)$$

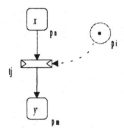

Fig. 3. Influence of the discrete to the continuous part

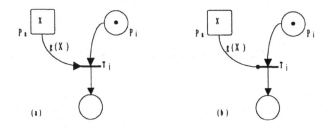

Fig. 4. Interface of the continuous part to the discrete part

where $m(p_i)$ is the marking of the discrete place used for the control and $r_t(t_i)$ is the continuous flow rate defined above.

The influence of the continuous part on the discrete one is made via some conditions on the continuous variable that are used to enabled a discrete transition. For example (figure 4.a), the transition t_j is enabled if $m(p_i) > 0$ and $X(p_n) > g(X)$. Then, the firing of the transition t_j leads to a new marking obtained in the same way as for a classical Petri net.

More formally, we will define a hybrid flow net as follows:

Definition 3. An **hybrid flow net** (HFN) is defined by a n-uplet
$H = (C, Z, \Psi_{c/d}, \Psi'_{c/d}, \Psi_{d/c})$ where
C is a continuous flow net as defined above;
Z is a Petri net;
$\Psi_{d/c} : P_d \times T_c \to \{0, 1\}$ specifies the continuous transition T_j controlled by a discrete place P_i.

$\Psi_{c/d} : P_c \times T_d \to \{0, 1\}$ specifies the discrete transitions T_j controlled by a place P_i . If $\Psi_{c/d}(i, j) \neq 0$, the transition is enabled iff $X(i) \geq \Psi(i, j)$, and if it is enabled by the rest of the net. If $\Psi_{c/d}(i, j) = 0$, there is no arc between the place and the transition T_j.

$\Psi'_{c/d} : P_c \times T_d \to \{0, 1\}$ specifies the discrete transitions T_j inhibited by a place P_i . If $\Psi'_{c/d}(i, j) \neq 0$, the transition is enabled iff $X(i) \leq \Psi(i, j)$, and if it

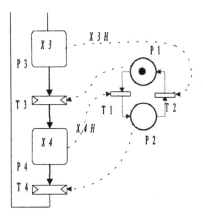

Fig. 5. Example of Hybrid Flow Net

is enabled by the rest of the net. If $\Psi'_{c/d}(i,j) = 0$, there is no arc between the place and the transition T_j.

The hybrid state of an hybrid flow net is the vector $W = \begin{bmatrix} X & M \end{bmatrix}^T$. The hybrid state space is denoted E .

As we can notice, the structure of the state equation of an hybrid flow net has a special form: it can be seen as a continuous model switched by a discrete model. The interaction between the continuous and discrete parts is not made via symbolic value as for example the hybrid model proposed by [3]. The relatively simple structure of the hybrid flow net is interesting for investigating properties, such as stability of the net. Indeed, results proposed by [5] can be applied and some properties can be obtained.

Hybrid flow nets are also of practical interest for modelling batch process in view of control design [8][9]. Continuous flow nets are useful for modeling flows of extensive variables in the process such as matter, energy or various products. Continuous transition are used to describe valves and chemical or biochemical reactions. For the discrete part of the process, recipes are described by Petri nets.

3 Control of Systems modelled by Hybrid Flow Nets

In this part, we are concerned with the problem of controlling an hybrid flow net in order to satisfy some specifications. We will define them through a subset of the hybrid state space $S \subset E$. The regulation problem that we have as an objective is to design a control law which must keep the HFN state in a set of constant specifications.

In order to control an hybrid flow net, we must first define how to act on the hybrid flow net and then give an input-output interpretation of the closed loop system in order to clearly distinguished between the system to be controlled and the controller.

3.1 Controlled hybrid flow net

Controlled Petri were initially introduced by Krogh [11]. The basic idea is to define some control places, which are connected by an oriented arc to some transitions that we want to control. The marking of the control places may be 0 or 1. A controlled transition is enabled if its control place has a marking equal to 1 and if its input places have a marking strictly positive. We will rely on this definition for controlling the discrete transition of hybrid flow nets.

In order to control the continuous part of the hybrid flow net, we are going to define what we call controlled continuous flow net in a way similar to the definition of controlled Petri Nets. The idea is to modify the flow through the non controlled transition according to the value of the external control variable. We will consider two cases:

(a) the flow modulation by a variable q (figure 6.a). In this case, the flow rate through a transition becomes:

$$r_{t_j} = q.f(X) \prod_{I_c(i,j)=1} X(i) \text{ with } 0 \leq f(X) \leq F_{\max}$$

(b) the inhibition of the flow by a variable q such that $q > 0$(figure 6.b). In this case, the flow rate through a transition becomes

$$r_{t_j} = \frac{1}{q}.f(X) \prod_{I_c(i,j)=1} X(i) \text{ with } 0 \leq f(X) \leq F_{\max}$$

This relation means that as the value of q increases, then the flow rate of the transition decreases. This kind of relation is encountered for example when we model chemical or physical phenomena where dilution occurs.

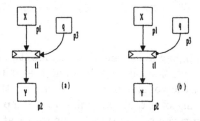

Fig. 6. Réseaux de flux commandés

The graphical representation of a controlled transition is shown on figure 6. The arcs from control variables to transitions are directed to the small sides of the transition. Modulation arcs are ended by an arrow while inhibition arcs are ended by a point. It must be noticed that these variables control the flow but are not affected by it. Only the variables with arcs connected to the large side of the transition are flowing out.

3.2 Structure of the hybrid controller

Now that the input of hybrid flow nets has been defined, we still have to choose the structure of the controller. We could imagine to impose to the controller to have to the structure of an hybrid flow net. In fact, this does not let enough freedom for the design of the control law and the structure given on the figure 7 is better suited: the controller is a static law for the continuous part interacting with a discrete event controller. It uses as an input the hybrid state which is assumed to be available.

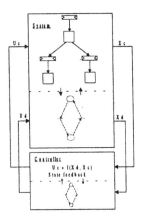

Fig. 7. Structure of the hybrid controller

3.3 Control law

The problem of controlling an hybrid flow net includes two general problems that can be separated: the control of places values and the control of flow values. For example, the control of the volume of a tank belongs to the first category while the control of the heating of a tank according to a ramp will go in the second class. Then, the objective can be to control some values at a given setpoint, or it can be to keep a value between some bounds. In this work, we have considered the control of the flow through a transition that we want to keep as high as possible but below an upper bound. From a practical point of view, this means that we try to get the faster flow rate from a tank to another but without exceeding a prescribed value because of physical limitations. Given the hybrid flow net structure, the control law is easy to derive. Let us consider a continuous transition controlled by an external variable. The flow rate through this transition is given by:

$$r_j = u_c \left[f_j(X) \prod_{I_c(i,j) \neq 0} x(i) \prod_{\Psi_{d/c}(i,j) \neq 0} m(i) \right]$$

It is easy to show that if we choose the following value for u_c then the flow rate r_j is always between 0 and $R_{j\,\text{max}}$:

$$u_c = R_{j\,\text{max}} \left[\varepsilon + f_j(X) \right] \left[\varepsilon + \prod_{I_c(i,j) \neq 0} x(i) \right] \left[\varepsilon + \prod_{\Psi_{d/c}(i,j) \neq 0} m(i) \right]$$

where ε is any small positive number.

4 Industrial Application

4.1 Statement of the problem

In this part, we will apply the methodology about hybrid systems described above to a beer production process. The problem is to control a filtration process used to produce wort. More precisely, we consider the steps when the sugar rate is increased. For this, water is first mixed with grounded cereal particles ; then the produced maische is filtered ; at the end of this filtration, the solution is sent to the heating tank. The distinctive feature of this filtration is that the filter is made of the deposit of particles on a grate. As a consequence, it is necessary to build this so called "cake" filter before filtering, during a recirculation step. The sugar is extracted during three similar washings. Between these washings, some clear water is introduced, and mixed with the particles. At any time, the filter may fill in if the pressure drop, proportional to the flow rate, is too high; when possible, it is made porous again using a crammer ; otherwise, it is thrown away. The flow rate is controlled with a pump and a modulating valve.

The objective of the controller is to execute the sequence (filling,building cake,filtration) three times in taking into account the possible fill in of the filter and in controlling the flow rate from tank 1 to tank 2 at the highest possible value that does not cause the filter to fill in.

4.2 Modelling and Simulation

The following continuous variables define the continuous state of the system:

$$X^T = \left[m_c \ L \ m_h \ M_c \ M_h \ V_f \right]$$

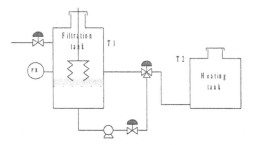

Fig. 8. Filtration process

where m_c and m_h are the mass of particles in the filtering tank $T1$ and $T2$, L is the width of the cake, M_c and M_h are the mass of flow in the tank $T1$ and $T2$, and V_f is the volume of flow which has passed through the cake. The opening of the modulating valve is denoted by u_c and we introduce the binary variable δ_1 which is equal to 1 when the fluid is recirculating and 0 otherwise. The flow rate of the outlet of the tank 1 is modelled as follows:

$$F_s = F_{s\,max}\frac{M_c}{M_c + \beta M_{c\,max}}u_c$$

where $F_{s\,max}, M_{c\,max}$ are some parameters and $\beta = 10^{-6}$. The evolution of the mass of particles in solution in the tank 1 obeys the following law:

$$\frac{dm_c}{dV_f} = -\frac{\eta m_c \rho}{M_c} \text{ (recirculation phase)}$$

$$\frac{dm_c}{dV_f} = -\frac{m_c \rho}{M_c} \text{ (transfer phase)}$$

where η is the percentage of particles retained by the filter and ρ is the density of water. The evolution of the particles mass in the tank 2 is (during the transfer phase):

$$\frac{dm_c}{dV_f} = -\frac{(1-\eta)m_c \rho}{M_c}$$

Then, from mass balances, we get the following relations:

$$\frac{dM_c}{dt} = -\rho F_s(1-\delta_1) = -\rho F_{s\,max}\frac{M_c}{M_c + \beta M_{c\,max}}u_c(1-\delta_1)$$

$$\frac{dM_h}{dt} = +\rho F_s\delta_1 = \rho F_{s\,max}\frac{M_c}{M_c + \beta M_{c\,max}}u_c\delta_1$$

$$\frac{dV_f}{dt} = \rho F_s = -\rho F_{s\,max}\frac{M_c}{M_c + \beta M_{c\,max}}u_c$$

More details are about modelling are available in [13]. From these relations, we can build the hybrid flow net modelling the process as shown on figure 9.

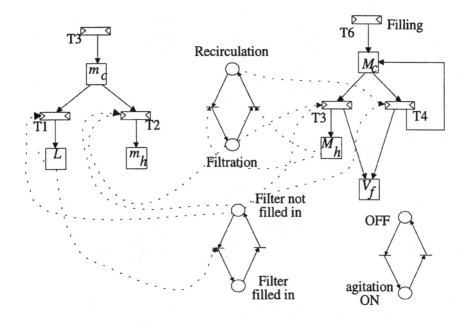

Fig. 9. Hybrid flow net of the process

4.3 Control design and implementation

The inputs of the hybrid system described in the last paragraph are of two kinds: continuous inputs, which are the information given to control the modulating valve u_c and the discrete inputs, to control the crammer, to switch between a filtration step and a recirculation step, and to fill the tank. Its outputs are its continuous state variables, the flow passing through the cake, the drop of pressure, and the discrete state of the cake (i.e. filled, or not). The control law synthesis is made according to the method proposed above. The continuous part of the controller computes the opening of the modulating valve as follows:

$$u_c = U_{\max}(L)\frac{1}{\varepsilon + f_4(X)}\frac{1}{\varepsilon + M_c}\frac{1}{\varepsilon + \prod_{\Psi d/c(i,4)=1} m(i)}$$

where $U_{\max}$ is a parameter depending on L so that no filling in occurs and ε is a small positive real number. The discrete part of the controller is described by the Petri net given in figure 10.

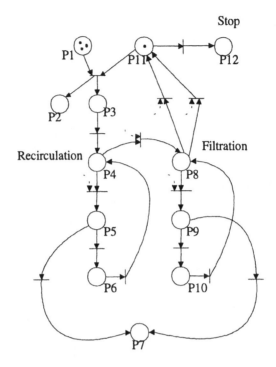

Fig. 10. Discrete event part of the controller

Place	Description
1	Number of cycles left
2	Number of cycle running
3	Filling
4	Recirculation
5	Management of filling in
6	cramming
7	Irreversible filling in
8	Filtration
9	Management of filling in
10	Cramming
11	Steps Management
12	End

The control strategy described was first tested on simulation. We have used an HFN simulation toolbox developed in the MATLAB environment..

Satisfactory results have been obtained. The figure 11 shows a typical run. As

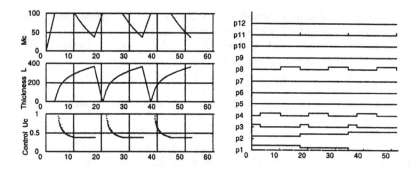

Fig. 11. Closed loop behaviour of the filtration process

we can see, the filter does not fill in (the marking of the places P_5 and P_9 stay at zero) while the flow rate through the filter was kept at the highest value possible below the upper limit.

In a second stage, the control algorithm has been implemented on a Modicon Programmable Logic Controller. The Petri net and the continuous law have been programmed using a language based on ladder diagram. More details are available in [13].

5 Conclusion

This paper presents an application of the hybrid flow net approach for modelling and control of a real life system. The problem of the control of an hybrid flow net has been discussed . A control law for the flow control has been proposed and has been applied on an industrial process of filtration. Simulation have been performed using a matlab toolbox. Then the controller has been implemented on a Modicon PLC. As a conclusion, we think that hybrid flow nets are useful and well suited for modelling a transformation process. Some important properties of the system such as extensive variables conservation, flow decomposition and positivity of variables are taking into account, which can not be made with a more general modelling approach. Then, the structure of the model can be used to design an hybrid control law, as we have seen it in this example. Future work on hybrid flow net includes an extension of the class of the systems that can be modelled and the design of controllers for more complex specifications.

References

1. R. Alur, C. Courcoubetisa, N. Halwachs, T. HenZiger, P. Ho, X. Nicollin, A. Olivero, J. Sifakis, and S. Yovine. The algorithmic analysis of hybrid systems. *Theoritical computer science*, 138:3–34, 1995.

2. P. Antsaklis, W. Kohn, A. Nerode, and S. Sastry, editors. *Hybrid Systems II*. Lecture Notes in Computer Science, Springer Verlag, 1995.

3. P. Antsaklis, J. Stiver, and M. Lemmon. Hybrid systems modeling and autonomous control systems. *Lecture Notes in Computer Science*, 736:366–392, 93.

4. J. L. Bail, H. Alla, and R. David. Hybrid petri nets. In *European Control Conference, Grenoble*, pages 1472–1477, July 2-5 91.

5. M. Branicky. *Studies in Hybrid Systems: Modelling, Analysis and Control*. PhD thesis, MIT, 95.

6. M. S. Branicky, V. Borkar, and S. Mitter. A unified framework for hybrid control. In *11th International Conference on Analysis and Optimization of Systems*. Lecture Notes in Control and Information Sciences, Springer Verlag, June 94.

7. R. David and H. Alla. *Du Grafcet au réseaux de Petri*. Hermès, 1992.

8. J. Flaus. Hybrid flow networks for hybrid process modelling. Technical report, Internal Report, LAG, 96.

9. J. Flaus. Modélisation hybride de procédés batchs. In *Journées D'étude*. SEE, February 96.

10. R. Grossman, A. Nerode, A. Ravn, and H. Rischel, editors. *Hybrid systems*. Lecture Notes in Computer Science, Springer Verlag, 1993.

11. B. H. Krogh. Controlled petri nets and maximally permissive feedback logic. In *Proceedings of the 25th Annual Alberton Conference*, University of Illinois, Urbana, 87.

12. S. Machietto, editor. *Workshop Analysis and Design of Event driven operations in process systems*. Imperial College, London, 1995.

13. G. Ollagnon and J.-M. Flaus. Modélisation et commande d'un système de filtration hybride. Technical report, LAG, 96.

14. A. Pnueli and J. Sifakis. Hybrid systems: special issue. *Theoritical Computer science*, 138, 95.

15. Y. Quenec'hdu, H. Guéguen, and J. Buisson. Les systèmes dynamiques hybrides. In *Automatisation Des Processus Mixtes: Les Systèmes Dynamiques Hybrides*. ADPM, nov 94.

Representation of Robust and Non-robust Solutions of Nonlinear Discrete-Continuous Systems*

Boris M. Miller [†]

Abstract

The discrete-continuous systems are considered as the systems with impulsive inputs, which cause the discontinuous behaviour of the system paths. The robustness for nonlinear discrete-continuous systems means the stability of the system path with respect to the approximation of pure impulsive inputs by ordinary ones. Necessary and sufficient conditions of this type of robustness give the opportunity to extract rather narrow class of robust systems. Although robust systems look like very attarctive from theoretical point of view and have many of usefull features, the great part of dynamical system which are of a practical interest are non-robust, and some nontraditional theoretical tools are necessary for their treatment. In this paper the problem of description and path representation in the form of differential equation with a measure is considered. Some specific features of non-robust systems are discussed from the point of controllability. It was shown that non-robust systems can provide the additional controllability opportunities by using the impulsive inputs.

1 Introduction and statement of the problem

Consider the evolution of discrete-continuous dynamical system, whose behaviour is described on some interval $[0, T]$ by the variable $X(t) \in R^n$, which satisfies the differential equation

$$\dot{X}(t) = F(X(t), t), \tag{1}$$

with given initial condition $X(0) = x_0 \in R^n$ and following intermediate conditions

$$X(\tau_i) = X(\tau_i-) + \Psi(X(\tau_i-), \tau_i, w_i), \tag{2}$$

which are given for some sequence of instants $\{\tau_i, i = 1, ..., N\}, N \leq \infty$, satisfying inequalities

$$0 \leq \tau_1 < \tau_2 < ... < \tau_i < ... < \tau_N \leq T. \tag{3}$$

*This work was supported in part by National Science Foundation of USA grant CMS 94-1447s and International Association for the Promotion of Cooperation with Scientists from the Independent States of the Former Soviet Union (INTAS) grants 94-697 and 93-2622

[†]Institute for Information Transmission Problems, 19 Bolshoy Karetny per., Moscow, GSP-4, 101447, Russia, FAX (095) 2090579, E-mail:bmiller@ippi.ac.msk.su.

which are given for some sequence of instants $\{\tau_i, i = 1, ..., N\}, N \leq \infty$, satisfying inequalities

$$0 \leq \tau_1 < \tau_2 < ... < \tau_i < ... < \tau_N \leq T. \tag{3}$$

In equation (2) $X(\tau_i-) = \lim_{t\uparrow\tau_i} X(t)$, and $w_i \in W \subset R^m$ is the sequence of the impulsive action intensities.

So, the state of system changes continuously in halfintervals $[0, \tau_1), ... [\tau_{i-1}, \tau_i),$ ... $[\tau_N, T]$, and undergoes a sudden change at every instant τ_i, whose value, due to equation (2), depends on the state preceeding the jump, time of jump, and intensity of the impulsive action.

Definition 1.1. *The sequence of pairs* $\mathcal{I} = \{(\tau_i, w_i), i = 1, ..., N\}$, *where* τ_i *satisfy (3), and* $N \leq \infty$ *will be called the* impulsive input.

Definition 1.2. *The right continuous function* $X(t)$ *which satisfies the integral equation*

$$X(t) = x_0 + \int\limits_0^t F(X(s), s)ds + \sum_{\tau_i \leq t} \Psi(X(\tau_i-), \tau_i, w_i) \tag{4}$$

will be called solution of the system with impulsive input.

It is easily seen that equation (4) joints equation (1) with conditions (2). Note that equation (4) can be used in its differential form, i.e.

$$\dot{X}(t) = F(X(t), t) + \sum_{\tau_i \leq t} \Psi(X(\tau_i-), \tau_i, w_i)\delta(t - \tau_i), \tag{5}$$

with initial condition $X(0) = x_0$, where $\delta(t - \tau)$ is the standard delta-function [7], satisfying the relation

$$\int\limits_{-\infty}^{+\infty} f(t)\delta(t - \tau)dt = f(t)$$

for any continuous function $f(t)$.

Various examples of systems, described by equation of a type (4) or (5), arise in flight dynamics, where impulsive controls are the thrust impulses of the orbit corrections [11], [18], in chemical and radiotheraphy of cancer deseases, where values of doses of therapeutic actions can also be treated as impulse actions [6], in macroeconomical analysis [1], and so on. The main feature of the systems, in which the control actions can be considered as the controls of impulsive type, is as follows: *the changes caused by the control action happen considerably more quickly than proper dynamic processes.* For this class of systems it becomes possible to consider the fast change as the jump, and to simplify the system description. However, to be sure that after this simplification the "new system" will have the same (or almost the same) properties as the original one, the procedure of the approximation of impulsive inputs by bounded ones has to be considered more carefully. First, it is necessary that description (5) has to be correct in following sense: *if we consider system (5) as a result of passage to a limit for some system with bounded controls, which approximate the impulsive one, the result must not depend on the way of approximation.* Therefore, if we consider the basic properties of systems with impulse control, such as correctness and stability of solution with respect to the variations of impulse control, we have to recognize the fact: *if the description of system with impulse control is the result of the correct limiting procedure, then function $\Psi(X, t, w)$ must satisfy some specific conditions, and these conditions have to be considered as the essential properties of this kind of systems.*

For futher consideration we will need some assumptions about the functions in the r.h.s. of our system with impulsive inputs.

Assumption 1.3. Let $F(X, t)$, $\Psi(X, t, w)$, and W satisfy following conditions:

1. they are continuous with respect to all variables,

2. for any $x, y \in R^n$ and $t \in [0, T]$ there exists some positive constant L_1 such that
$$\|F(x, t) - F(y, t)\| \le L_1 \|x - y\|,$$

3. for any $x, y \in R^n$, $w \in W$, and $t \in [0, T]$ there exists some positive constant L_2 such that
$$\|\Psi(x, t.w) - \Psi(y, t, w)\| \le L_2 \|x - y\| \|w\|,$$

4. the set $W \subset R^m$ contains the point $0 \in R^m$ and
$$\Psi(x, t, 0) = 0$$
for any $x \in R^n$ and $t \in [0, T]$.

For any impulse input $\mathcal{I} = \{(\tau_i, w_i), i = 1, ..., N\}$, define a function
$$U(t) = \sum_{\tau_i \le t} w_i,$$

which has a sense of the input resource spent till the time t. If the impulse input satisfies condition
$$\sum_{\tau_i \le T} \|w_i\| < \infty,$$

or the number of impulses $N < \infty$, then $U(t)$ will be a function of bounded variation and right continous.

Definition 1.4. *We will say that the sequence of impulsive inputs $\mathcal{I}^n = \{(\tau_i^n, w_i^n), i = 1, ..., N^n\}$ converges to the impulsive input $\mathcal{I} = \{(\tau_i, w_i), i = 1, ..., N\}$,* *if the associated sequence of functions $U^n(t)$ has uniformly bounded variations, i.e.,*

$$\sup_n \sum_{\tau_i \leq T} \|w_i\| < \infty,$$

and $U^n(t)$ converges to $U(t)$ at all points of continuity.

This definition corresponds to the convergence in weak - $*$ topology of the space of functions of bounded variation [7].

Let $X(t)$ be the solution of (4) with some initial condition X_0 and impulse input $\mathcal{I}$, and $X^n(t)$ be the sequence of solutions of (4) with the same initial condition, corresponding to some sequence of impulse inputs $\mathcal{I}^n$.

Definition 1.5. *System with impulse input will be called* robust or stable *with respect to variation of the impulse input or simply robust if, for any initial condition and impulse input $\mathcal{I}$, the convergence of sequence $\mathcal{I}^n$ to $\mathcal{I}$ implies the convergence of $X^n(t)$ to $X(t)$ at all points of continuity.*

Our aim is to provide necessary and sufficient conditions of robustness and to discuss properties of robust and non-robust systems.

2 Necessary and sufficient conditions of robustness

Theorem 2.1. [13] If the system with impulse input is robust, then for any $x \in R^n$, $t \in [0, T]$ and $w_1, w_2 \in W$

$$\Psi(x, t, w_1 + w_2) = \Psi(x, t, w_1) + \Psi(x + \Psi(x, t, w_1), t, w_2). \tag{6}$$

Another form of robustness condition follows from the approach based on the approximation of impulsive action by continuous one. Consider the system, described by ordinary differential equation

$$\dot{X}(t) = F(X(t), t) + B(X(t), t)w(t), \tag{7}$$

where $F(x, t)$ is the same as in (4), and $B(x, t)$ is some continuous $(n \times m)$ matrix-valued function, satisfying Lipshitz condition

$$\|B(x, t) - B(y, t)\| \leq L\|x - y\|,$$

with some $L > 0$ for any $x, y \in R^n$, and $w(t)$ is a control, taking values from the space R^m. We will consider this system as the *continuous approximation* of the system with impulse control. It means that we will consider the sequence of ordinary bounded inputs, which approximates the inputs of δ - function type. Our aim is to define the response of $X(t)$ to this kind of controls.

We are going to approximate the impulsive action $w(t) = w_0\delta(t - \tau)$ by some sequence of ordinary inputs $\{w^n(\cdot)\}$, for example,

$$w^n(t) = w_0\phi^n(t, \tau),$$

where $\{\phi^n(t, \tau)\}$ satisfyies conditions:

$$\phi^n(t, \tau) \geq 0, \quad \phi^n(t, \tau) \to 0 \quad \text{almost everywhere on} \quad [0, T] \setminus \{\tau\},$$

$$\lim_n \int_0^{\tau - \varepsilon} \phi^n(t, \tau) dt = 0, \quad \lim_n \int_{\tau + \varepsilon}^T \phi^n(t, \tau) dt = 0, \tag{8}$$

$$\lim_n \int_{\tau - \varepsilon}^{\tau + \varepsilon} \phi^n(t, \tau) dt = 1,$$

for any $\varepsilon > 0$.

Let $\{X^n(t)\}$ be the solution of (7) with some initial condition X_0, corresponding to control $w^n(t)$. To describe the behaviour of the sequence $\{X^n(t)\}$ we need the following

Definition 2.2. *Let* $\Phi(x, t, w, s)$ *denotes the solution of the equation*

$$\frac{dY(s)}{ds} = B(Y(s), t)w$$

with initial condition $Y(0) = x \in R^n$, *and with fixed parameters* $t \in [0, T]$, $w \in R^m$.

By virtue of continuity and Lipshitz condition, function $\Phi(x, t, w, s)$ will be continuous with respect to all variables.

Define a function $\bar{X}(t)$, which is solution of equation (4) with initial condition x_0 and impulse input $\{(\tau, w_0)\}$, and with the impulsive input response, described by function

$$\Psi(x, t, w_0) = \Phi(x, t, w_0, 1) - x,$$

Next theorem shows that $\bar{X}(t)$ is the limit of sequence $\{X^n(t)\}$.

Theorem 2.3. The sequence $\{X^n(t)\}$ converges to function $\bar{X}(t)$ everywhere on $[0, T]$, except maybe the point τ.

The proof of this theorem is omitted due to the lack of space. As a corollary of Theorem 2.3 one can prove the approximation result for general impulsive control $\mathcal{I} = \{(\tau_i, w_i), i = 1, ..., N < \infty\}$.

Let the sequence

$$w^n(t) = \sum_{i=1}^N w_i \phi^n(t, \tau_i),$$

where functions $\phi^n(t, \tau_i)$ satisfy (8), approximates the impulsive input

$$w(t) = \sum_{i=1}^N w_i \delta(t - \tau_i).$$

Let $X^n(t)$ denotes the solution of equation (7) with initial condition X_0, corresponding the input $w^n(t)$, and $X(t)$ is the solution of equation (4) or

$$\dot{X}(t) = F(X(t), t) + \sum_{\tau_i \leq t} \Psi(X(\tau_i-), \tau_i, w_i) \delta(t - \tau_i),$$

with initial condition x_0, where $\Psi(x,t,w) = \Phi(x,t,w,1) - x$ by Def. 2.2.

Corollary 2.4 The sequence $\{X^n(t)\}$ converges to $X(t)$ everywhere on $[0,T]$, except maybe the set of points $\{\tau_i, i = 1, ..., N\}$.

The robustness of system (4) means that the limit $X(t)$ does not depend on the way of approximation of impulse input.

Theorem 2.5. If the system with impulse input is robust, then system (7), which is continuous approximation of system (4) satisfies Fröbeneus conditions [2], i.e. for any vectors $h, g \in R^m$, and for any $x \in R^n$, $t \in [0,T]$

$$\langle (B'_x(x,t) \circ B(x,t))h, g \rangle = \langle (B'_x(x,t) \circ B(x,t))g, h \rangle.$$

Remark 2.6 Symbol $\langle \cdot, \cdot \rangle$ means a scalar product in R^m.

Remark 2.7 Symbol $(B'_x(x,t) \circ B(x,t))$ means a composition of linear maps $B(x,t) : R^m \to R^n$, and $B'_x(x,t) : M^{n \times m} \to M^{m \times n}$, where $M^{m \times n}$ is the space of $m \times n$ matrices.

This condition can be expressed also in component-wise form as

$$\sum_{k=1}^{n} \frac{\partial b_{ij}(x,t)}{\partial x_k} b_{kl}(x,t) = \sum_{k=1}^{n} \frac{\partial b_{il}(x,t)}{\partial x_k} b_{kj}(x,t),$$

$$j, l = 1, ..., m, \quad \text{for all} \quad x \in R^n, t \in [0,T]$$

where $b_{ij}(x,t)$, $i = 1, ..., n$, $j = 1, ..., m$ are the components of matrix-valued function $B(x,t)$.

Remark 2.8 If function $B(x,t)$ satisfies Fröbenious condition then $\Psi(x,t,w)$ defined as $\Psi(x,t,w) = \Phi(x,t,w,1) - x$ with Φ defined according to Def. 2.2 satisfies condition (6).

Theorems 2.1 and 2.5 provide a collection of natural and intrinsic characterization of necessary conditions of robustness. Moreover one can prove that under some additional assumptions they are sufficient for stability.

Theorem 2.9. [Sufficient condition of stability.] [13], [19], [21].

Let function $\Psi(x, t, w)$ satisfies condition

$$\Psi(x,t,w_1 + w_2) = \Psi(x,t,w_1) + \Psi(x + \Psi(x,t,w_1), t, w_2),$$

for any $x \in R^n$, $t \in [0,T]$ and $w_1, w_2 \in R^m$, and derivatives $\Psi'_x(x,t,w)$, $\Psi'_t(x,t,w)$, and $\Psi'_w(x,t,w)$ exist for any $x \in R^n$, $t \in [0,T]$, and $w \in W$, and satisfy locally Lipschitz condition with respect to (x,w), i.e., for any bounded set $C \in R^n \times R^m$ there exists a constant $L(C) > 0$, such that for any $(x,w), (y,v) \in C$

$$\|\Psi'_x(x,t,w) - \Psi'_x(y,t,v)\| \le L(C)(\|x - y\| + \|w - v\|),$$

$$\|\Psi'_t(x,t,w) - \Psi'_t(y,t,v)\| \le L(C)(\|x - y\| + \|w - v\|),$$

$$\|\Psi'_w(x,t,w) - \Psi'_w(y,t,v)\| \le L(C)(\|x - y\| + \|w - v\|).$$

Then for any impulse input $\mathcal{I} = \{(\tau_i, w_i), i = 1, ..., N\}$ and any sequence of impulse inputs $\mathcal{I}^n = \{(\tau_i^n, w_i^n), i = 1, ..., N^n\}$, which converges to $\mathcal{I}$ according to Def. 1.4, the appropriate sequence of solutions of equation (4)

$$X^n(t) = x_0 + \int_0^t F(X^n(s), s)ds + \sum_{\tau_i^n \leq t} \Psi(X^n(\tau_i^n-), \tau_i^n, w_i^n)$$

converges to the solution of equation

$$X(t) = x_0 + \int_0^t F(X(s), s)ds + \sum_{\tau_i \leq t} \Psi(X(\tau_i-), \tau_i, w_i)$$

at all points of continuity.

3 Properties of robust systems

3.1 Independence of output upon the way of input approximation

Another sufficient stability condition relates to the case of approximation of impulse input by ordinary ones.

Theorem 3.1. [13], [19], [21].

Let function $B(x, t)$ in equation (7) satisfies Fröbeneus condition for any vectors $h, g \in R^m$, and moreover, has a derivative with respect to t, which is locally Lipschitzian with respect to x.

Then for any impulse input $\mathcal{I} = \{(\tau_i, w_i), i = 1, ..., N\}$ and any sequence of ordinary inputs $w^n(\cdot)$, which approximates $\mathcal{I}$, the appropriate sequence of solutions of (7)

$$\dot{X}^n(t) = F(X^n(t), t) + B(X^n(t), t)w^n(t)$$

with initial condition $X^n(0) = X_0$ converges at all points of continuity to the solution of equation

$$X(t) = X_0 + \int_0^t F(X(s), s)ds + \sum_{\tau_i \leq t} \Psi(X(\tau_i-), \tau_i, w_i),$$

where $\Psi(x, t, w) = \Phi(x, t, w, 1) - x$ and function $\Phi(x, t, w, 1)$ satisfies Definition 2.2.

3.2 Generalized solutions and their representation by differential equations with a measure

In previous section the problem of approximation of any impulsive control action and its robustness has been considered. This problem is tightly connected with the problem of correct description of the optimal solutions in optimization problems with iimpulse controls. Indeed, when we are considering some optimization problem both in the situation of proof the existence theorem, or in the situation of using some computation procedure, we obtain some sequence of controls (impulsive or ordinary), which minimizes the value of appropriate performance criterion. This sequence may be convergent or not, so as the appropriate sequence of the system solution. However, if the set of admissible solution is compact, we can extract from the sequence of solutions some subsequence, which will converge to some function, and the later can be treated as the optimal solution [20]. In the theory of optimal control such solution called as *weak solution* [12], or *generalized curve* [20]. During the investigation of the optimization problems for systems with impulse control we need the analogue of this concept, and our previous results give us the opportunity to introduce mathematically correct description of *generalized solution*.

Definition 3.2. *Function $X(t)$, which is continuous from the right and has bounded variation on the interval $[0, T]$ will be called the* generalized solution *of system (4) if there exists a sequence of impulse controls $\mathcal{I}^n = \{(\tau_i^n, w_i^n), i = 1, ..., N^n\}$, satisfying conditions*

$$w_i^n \in W, \qquad \sup_n \sum_0^{N^n} \|w_i^n\| < \infty,$$

such that appropriate sequence $\{X^n(t)\}$ of solutions of system (4) with initial condition $X^n(0) = X(0-)$ converges to $X(t)$ at all points of continuity.

In robust case any generalized solution can be represented with aid of nonlinear differential equation with a measure [14].

Theorem 3.3. Let assumptions of Theorem 2.7 hold. Then for any generalized solution $X(t)$ of system (4) there exists a vector-valued measure $\mathcal{U}(dt)$, whose values belong to $con(W) = \{w \in R^n | \exists \{\lambda \geq 0, e \in W\} : w = \lambda e\}$, i.e., for any Borel subset $A \in [0, T]$

$$\mathcal{U}(A) \in con(W),$$

such that $X(t)$ satisfies the integral equation

$$X(t) = X(0-) + \int_0^t F(X(s), s)ds + \int_0^t B(X(s), s)dU^c(s)+ \tag{9}$$

$$\sum_{\tau \leq t} \Psi(X(\tau-), \tau, \Delta U(\tau)),$$

where $U(t)$ is the distribution function of measure $\mathcal{U}(dt)$, $U^c(t)$ and $\Delta U(\tau)$ are continuous component and jumps of function $U(t)$ in its Lebesgue decomposition as a function of bounded variation, i.e.,

$$U(t) = U^c(t) + \sum_{\tau < t} \Delta U(\tau).$$

Equations of a type (9) give the universal description of the discrete-continuos system behaviour with ordinary and impulsive inputs. If measure $\mathcal{U}(dt)$ has only a finite numbers of atoms, that is, the points τ, where $\mathcal{U}(\{\tau\}) \neq 0$, then it can be related with some impulse control $\mathcal{I} = \{(\tau, \mathcal{U}(\{\tau\}))$. When this measure be absolutely continuous with respect to Lebesgue measure [7], it can be interpreted as the ordinary controls $w(t) = \dot{U}(t)$, where $U(t)$ is the distribution function of measure $\mathcal{U}(dt)$. In terms of optimal control theory this motion can be treated as *impulsive sliding mode* [5]. In general case we will say that measure $\mathcal{U}(dt)$ defines the *generalized control*.

3.3 Stability with respect to disturbances

One of the most usefull properties of robust systems is the stability with respect to disturbances, which are bounded on average. Consider now the impulsive input $\mathcal{I} = \{(\tau_i, w_i), i = 1, ..., N\}$ as a disturbance acting on the system. We will say that this disturbance is *bounded on average* if function

$$U(t) = \sum_{\tau_i \leq t} w_i$$

is bounded on the interval $[0, T]$, i.e., exists positive constant M such that

$$\sup_{[0,T]} \|U(t)\| = M < \infty.$$

Denote by $X^U(t)$ and $X^0(t)$ solutions of systems (4) which corresponds to disturbed and undisturbed motions respectively. Next theorem give the estimate for difference $X^U(t) - X^0(t)$.

Theorem 3.5. Let assumptions of Theorem 2.7 hold, and the impulsive disturbances are bounded on the average. Then for any solution $X^U(t)$ of system (4) the following inequality takes place

$$\sup_{[0,T]} \|X^U(t) - X^0(t)\| \leq \mathcal{K}(X(0-), T, M) \sup_{[0,T]} \|U(t)\|.$$

Remark 3.6. Notice that this estimate is uniform over the class $\mathcal{B}(M)$ of disturbances, which are bounded on average with some fixed constant M, although this class contains the disturbances with total energy, which is as large as desired, so as

$$\sup_{\mathcal{B}(M)} \sum_{\tau_i \leq T} \|w_i\| = \infty.$$

It means that robust systems do not accumulate the disturbances.

Remark 3.7. For continuous inputs this result is known as vibrocorrectness property [9].

4 Impulsive controls in non-robust systems

When system (4) is non-robust it is impossible to guarantee that limit solution of discrete-continuous system does not depend on the way of approximation of the impulse input by contrast with Theorem 3.1. However the non-robustness provides an additional opportunities in the control by using impulsive-type controls. General representation of discontinuous solutions to nonlinear systems with impulse control was obtained in [15], [17], where the optimization problem was considered. Here we will demonstrate this additional opportunities by following example.

Consider continuous system with state $X(t) = \{x_1(t), x_2(t), x_3(t)\}$ and input $W(t) = \{w_1(t), w_2(t)\}$

$$\dot{x}_1(t) = w_1(t),$$

$$\dot{x}_2(t) = w_2(t), \tag{10}$$

$$\dot{x}_3(t) = w_1(t) - x_3 w_2(t).$$

The right-hand-side of (10) does not meet Frobeneus condition, hence when $W(t) = \{w_1(t), w_2(t)\}$ approximates some impulsive input, the output of system (10) approximates the output of some non-robust discrete-continuous system.

As follows from general theory of the jumps representation [15], if some sequence $W^n(t) = \{w_1^n(t), w_2^n(t)\}$ approximates impulsive input $\mathcal{I} = \{(W_1, W_2, \tau)\}$ and the corresponding sequence $X^n(t)$ converges to some function $X(t)$ everywhere, excepting maybe the point τ, then the jump of $X(t)$ satisfies following conditions:

$$\Delta x_1(\tau) = W_1,$$

$$\Delta x_2(\tau) = W_2, \tag{11}$$

$$\Delta x_3(\tau) \in [e^{-W_2}(x_3(\tau-) + W_1), e^{-W_2} x_3(\tau-) + W_1].$$

(We suppose here for simplicity that $W_1 > 0$ and $W_2 > 0$.) The last relation in (11) shows that jump of third component cannot be uniquely defined and depends on the way of the impulsive input approximation. Moreover, the choice of appropriate type of approximation gives the opportunity to realize any of the jump function, which satisfies inclusion (11). It means that non-robust system has more of the control opportunities than robust one.

5 Conclusion

Our aim was to show the application of the robustness concept to a new class of discrete-continuous dynamical system. So this paper is a new interpretation of the old results, which are known to specialists in the area of optimal control, but possibly are not well known for scientists working in the area of hybrid systems. Author hopes that his paper removes this annoying gap and gives the opportunity to look at the robustness problems for nonlinear discrete-continuous dynamical systems in a new fashion.

References

[1] A. Bensoussan, S. Sethy, R. Vickson and N. Derzko, "Stochastic production planning with production constraints," *SIAM J. Control Optim.*, **22**, 920–935 (1984).

[2] H. Cartan, *Calcul Différentiel. Formses Différentielles*, Hermann, Paris (1967).

[3] F. N. Grigor'ev, N. A. Kuznetsov, and A. P. Serebrovskii, *The Control of Observations in Automatic Systems* [in Russian], Nauka, Moscow (1986).

[4] V. K. Gorbunov and G. U. Nurakhunova, "Processes with controlled discontinuities of phase trajectories and simulation of production with moving basic funds," *Izv. Akad. Nauk SSSR, Tekh. Kibern.*, No. 6, 55–61 (1975).

[5] V. I. Gurman, "Optimal processes with unbounded derivatives," *Avtomat. Telemekh.*, No. 12, 14–21 (1972).

[6] V. K. Ivanov, B. M. Miller, P. I. Kitsul, and A. M. Petrovskii, "A mathematical model of control of the healing of an organism damaged by a malignant growth," in: *Biological Aspects of Control Theory* [in Russian], Inst. Control Sciences, Moscow, No. 8 (1976), pp. 15–22.

[7] A. N. Kolmogorov and S. V. Fomin, *Elements of Theory of Function and Functional Analysis* [in Russian], Nauka, Moscow, 1976.

[8] M. A. Krasnosel'skii and A. V. Pokrovskii, "Vibrostable differential equations with continuous right-hand side," *Tr. Mosk. Mat. Obshch.*, **27**, 93–112 (1972).

[9] M. A. Krasnosel'skii and A. V. Pokrovskii, *Systems with Hysteresis* [in Russian], Nauka, Moscow (1983).

[10] H. J. Kushner, "On the optimal timing of observation for linear control systems with unknown initial states," *IEEE Trans. Automat. Control*, **AC-9**, 144–145 (1964).

[11] D. Lawden, *Optimal Trajectories for Space navigation* Butterworth, London, 1963.

[12] E. B. Lee and L. Markus, *Foundations of Optimal Control Theory* John Wiley and Sons, Inc., New York, London, Sydney, 1967.

[13] B. M. Miller, "Stability of solutions of ordinary differential equations with measure," *Uspekhi Mat. Nauk*, No. 2, 198 (1978).

[14] B. M. Miller, "The nonlinear sampled-data control problem for systems described by ordinary measure differential equations I, II," *Automat. Remote Control*, **39**, No. 1, 57-67; No. 3, 338-344 (1978).

[15] B. M. Miller, "Optimization of dynamic systems with a generalized control," *Automat. Remote Control*, **50**, No. 6, 733-742 (1989).

[16] B. M. Miller, "Generalized optimization in problems of observation control," *Automat. Remote Control*, **52**, No. 10, 83-92 (1991).

[17] B. M. Miller., "The generalized solutions of nonlinear optimization problems with impulse controls," *SIAM J. Control Optim.*, **34**, No. 4, 1420-1440 (1996)

[18] L. Neustadt, "A general theory of minimum-fuel space trajectories," *J. SIAM. Ser. A, Control*, **3**, No. 2, 317-356 (1965).

[19] Yu. V. Orlov, *Theory of Optimal Systems with Generalized Controls*. [in Russian], Nauka, Moscow (1988).

[20] L. C. Young, *Lectures on Variational Calculus and the Theory of Optimal Control*, W. B. Saunders Company, Philadelphia, Londod, Toronto, 1969.

[21] S. T. Zavalishchin and A. N. Sesekin, *Impulsive Processes. Models and Applications* [in Russian], Nauka, Moscow (1991).

Controller Design of Hybrid Systems

Stefan Pettersson and Bengt Lennartson

Control Engineering Lab, Chalmers University of Technology
S-412 96 Gothenburg, Sweden
e-mail: sp, bl@control.chalmers.se

Abstract. In this paper we present two strategies to design a hybrid controller for a system described by several nonlinear vector fields. Besides the overall goal to find a controller that stabilizes the closed-loop hybrid system, the selection will also be made in such a way that an exponentially stable closed-loop system is obtained. The design strategies are based on stated stability and exponential stability theorems for hybrid systems. The first approach results in regions where it is possible to change vector fields guaranteeing (exponential) stability of the closed-loop hybrid system. The second design strategy utilizes the fact that a system is (exponentially) stable if it is always possible to choose a vector field that points in a direction such that the trajectory approaches the equilibrium point. These conditions can be verified by solving a linear matrix inequality (LMI) problem.

1 Introduction

This paper presents strategies that can be used for design of hybrid controllers for an open-loop hybrid plant described by nonlinear vector fields. The goal is to stabilize or exponentially stabilize the closed-loop hybrid system.

Up to this point, there are few design strategies presented in the literature. In [10] a method is proposed for a system consisting of two linear vector fields. The goal of this strategy is to select the vector fields in such a way that a stable system is obtained. It is assumed that the two linear vector fields are unstable, otherwise one stable vector field can be selected. To guarantee that this method can be used, it is assumed that there exists a stable convex combination of the two linear vector fields.

In [5] a method is proposed where there is a Lyapunov function coupled to each vector field. Also in this case, the goal is to obtain a stable closed-loop system. The strategy used is called the *min-switch strategy* since the vector field corresponding to the smallest Lyapunov function is selected. The designed controller does not contain any memory and sliding modes may occur in the closed-loop system.

In this paper we propose two design methods. If certain conditions are satisfied we can ensure that the closed-loop system becomes stable or exponentially stable. The strategies are based on stated stability and exponential stability results for hybrid systems.

The first design method is close to the *min-switch strategy* [5]. Regions are obtained where it is possible to change vector field so that (exponential) stability of the closed-loop hybrid system can be guaranteed. Sliding modes are avoided in this approach. This is an attractive property since the controller otherwise has to change vectorfields infinitely often.

The second design method utilizes the fact that a system is (exponentially) stable if it is always possible to select a vector field that points in a direction such that the trajectory approaches the equilibrium point. One version of the method is called the *min-projection strategy*, another the *min-skew-projection strategy* since we minimize the time derivative of the squared or skew squared distance from the equilibrium point. Conditions when these methods guarantee stability or exponential stability can be verified by solving a linear matrix inequality (LMI) problem.

The outline of this paper is as follows: The next section introduces the hybrid model. In Sect. 3, theorems for stability and exponential stability of hybrid systems are presented. The two presented design strategies are explained in Sect. 4.

2 Hybrid Model

The evolution of an open-loop switched system can be described by:

$$\dot{x}(t) = f(x(t), m(t)) \tag{1}$$

where $x \in \Re^n$ and $m \in M = \{m_1, \ldots, m_N\}$. Here, x is the continuous state and m is the discrete state. The vector field $f(x, m_i)$ is continuous in $x \ \forall m_i \in M$. When m changes value this usually results in abrupt changes in the vector field f.

Contrary to [7, 6], where m is changed when $x(t)$ hits given boundaries, the goal in this paper is to find the function $m(t)$ to achieve stability of the system (1). Furthermore, it may also be desirable to design $m(t)$ in such a way that an exponentially stable system is obtained. The controller function $m(t)$ depends on the continuous state x and the discrete state m according to

$$m(t) = \phi(x(t), m(t^-)) \tag{2}$$

Thus, $m(t)$ is in general a hybrid controller. If there is no dependence on m, i.e. $m(t) = \phi(x(t))$, then the controller is of variable structure type. The notation t^- indicates that $m(t)$ is continuous from the right.

Figure 1 shows the closed-loop hybrid system where the goal is to design a hybrid controller such that the closed-loop hybrid system becomes (exponentially) stable.

The function ϕ will herein be expressed by sets S_{ij} which are defined as:

$$S_{ij} = \{x \in \Re^n \mid m_j = \phi(x, m_i)\} \tag{3}$$

We will typically give these sets as hyper surfaces $s_{ij}(x) = 0$, e.g. hyper planes $s_{ij}(x) = v^T x + \nu = 0$, where v is the constant normal vector and ν is a constant vector.

Each vector field $f(x, m_i)$, $m_i \in M$, in (1) is coupled to a region $\Omega_i \subseteq \Re^n$. Region Ω_i indicates where in the state space the evolution $\dot{x} = f(x, m_i)$ is allowed or possible. There is no loss of generality in assuming that each $\Omega_i \neq \emptyset$, $i = 1, \ldots, N$, since otherwise if $\Omega_i = \emptyset$, then the discrete state m_i is excluded from the set M.

There can be several reasons for restricting $f(x, m_i)$ to a specific region Ω_i

- The vector field $f(x, m_i)$ is the result of designing a local controller for a specific region Ω_i, and different controllers are specified for different regions.

Switched system

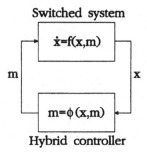

Hybrid controller

Fig. 1. Closed-loop hybrid system consisting of a switched system and a hybrid controller. The problem is to find the function $\phi(x, m)$ such that the closed-loop hybrid system becomes (exponentially) stable.

- The performance of (1) using $f(x, m_i)$ is only acceptable in region Ω_i.

The following assumptions are introduced:

A1 Coverage: $\cup_{i=1}^{N} \Omega_i = \Re^n$
A2 Overlap: $\exists q, r, q \neq r$ such that $\Omega_q \cap \Omega_r \neq \emptyset$.

The first condition guarantees that at least one vector field is allowed for every continuous state. If this was not the case, the control problem is bad stated. The second assumption is introduced to guarantee that a choice between different vector fields is possible, at least in some region. Otherwise, it may happen that there is no control problem left. Note that if all vector fields are allowed in the whole $\Re^n$, then the assumptions A1 and A2 are satisfied. Figure 2 illustrates three vector fields that are allowed or possible in different regions.

3 Stability of Closed-Loop Hybrid Systems

To be able to design the hybrid controller, we will use a minor modification of a result that can be found in [7, 6], which considers stability of hybrid systems. We will begin by assuming that the sets S_{ij} are given. In the next section it is shown how to obtain these sets utilizing the stability result.

Given the sets S_{ij}, the continuous trajectory $x(t)$ evolves from a given initial state $(x_0, m_0) \in I_0$ according to (1), (2) and exists for all $t \geq t_0$. The origin is the only equilibrium point of the hybrid system. Furthermore, it is assumed that the hybrid state space $H = \Re^n \times M$ is partitioned into ℓ disjoint hybrid regions $\tilde{\Omega}_q$, $q = 1, \ldots, \ell$. In each region $\tilde{\Omega}_q$, $q = 1, \ldots, \ell$, a scalar function V_q is used as a measure of the hybrid system's energy. The set $\tilde{\Omega}_q^x \subseteq \Re^n$ denotes the continuous part in $\tilde{\Omega}_q$. The set $\tilde{\Omega}_{qr}$ is comprized of all continuous states for which the trajectory $x(t)$, with initial states $(x_0, m_0) \in I_0$, pass from $\tilde{\Omega}_q$ to $\tilde{\Omega}_r$ at some time t, i.e. $x(t^-) \in \tilde{\Omega}_q$ and $x(t) \in \tilde{\Omega}_r$.

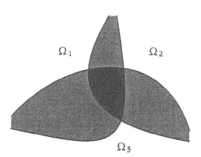

Fig. 2. Overlapping vector fields. In the white regions where there is only one allowed vector field, the corresponding vector field must be selected. In regions (grey and black) where there are several allowed vector fields, we have to construct a controller that selects one of these.

If sliding motions occur for some states in the continuous state space we define an equivalent continuous dynamic for such continuous states. Then we obtain a hybrid system with an equivalent dynamic where no sliding modes occur. This means that only a finite number of switches of the functions V_q, $q = 1, \ldots, \ell$, occur in finite time.

Theorem 1. *If there exist scalar functions* $V_q : \tilde{\Omega}_q^x \to \Re$, *each* $V_q(x)$ *differentiable in* x $\forall x \in \tilde{\Omega}_q^x$, $q = 1, \ldots, \ell$, *and class K functions* $\alpha_q : \Re^+ \to \Re^+$ *and* $\beta_q : \Re^+ \to \Re^+$, $q = 1, \ldots, \ell$, *such that*

- $\forall x \in \tilde{\Omega}_q^x$, $\alpha_q(||x||) \leq V_q(x) \leq \beta_q(||x||)$, $q = 1, \ldots, \ell$
- $\forall (x, m) \in \tilde{\Omega}_q$, $\dot{V}_q(x) \leq 0$, $q = 1, \ldots, \ell$
- $\forall x \in \tilde{\Omega}_{qr}$, $V_r(x) \leq V_q(x)$, $q = 1, \ldots, \ell$, $r = 1, \ldots, \ell$

then the equilibrium point 0 is (uniformly) stable in the sense of Lyapunov.

A class K function is a continuous function $\kappa : \Re^+ \to \Re^+$ fulfilling: $\kappa(0) = 0$, $\kappa(z) > 0$ $\forall z > 0$ and $\kappa(z_1) \leq \kappa(z_2)$, $z_1 < z_2$.

Proof. Define $\alpha(||x||) = \min_q \alpha_q(||x||)$ and $\beta(||x||) = \max_q \beta_q(||x||)$. The functions $\alpha(||x||)$ and $\beta(||x||)$ are also class K functions. We have that $\forall x \in \tilde{\Omega}_q^x$, $\alpha(||x||) \leq \alpha_q(||x||) \leq V_q(x) \leq \beta_q(||x||) \leq \beta(||x||)$, $q = 1, \ldots, \ell$. Replacing the first item with this condition results in a theorem that was proved in [6]. $\square$

The reason for writing the second condition as $\forall (x, m) \in \tilde{\Omega}_q \ldots$, is that $\dot{V}_q(x)$ depends on the discrete state according to $\dot{V}_q(x) = \frac{\partial V_q(x)}{\partial x} f(x, m)$. Note that Theorem 1 is the usual Lyapunov stability result if only one (positive definite) function $V(x)$ is used [8].

If the class K functions take the special form $k||x||^s$ where $k > 0$ and $s > 0$, then the following theorem can be stated:

Theorem 2. *If there exist scalar functions* $V_q : \tilde{\Omega}_q^x \to \Re$, *each* $V_q(x)$ *differentiable in* x $\forall x \in \tilde{\Omega}_q^x$, $q = 1, \ldots, \ell$, *and constants* $\alpha_q, \beta_q, \gamma_q > 0$, $q = 1, \ldots, \ell$, *such that:*

- $\forall x \in \tilde{\Omega}_q^x$, $\alpha_q ||x||^s \leq V_q(x) \leq \beta_q ||x||^s$, $q = 1, \ldots, \ell$
- $\forall (x, m) \in \tilde{\Omega}_q$, $\dot{V}_q(x) \leq -\gamma_q ||x||^s$, $q = 1, \ldots, \ell$
- $\forall x \in \tilde{\Omega}_{qr}$, $V_r(x) \leq V_q(x)$, $q = 1, \ldots, \ell$, $r = 1, \ldots, \ell$

then the equilibrium point 0 is exponentially stable in the sense of Lyapunov. Furthermore, an upper bound for the convergence rate of $x(t)$ in region $\tilde{\Omega}_q$ is given by

$$||x(t)|| \leq \left(\frac{\beta_q}{\alpha_q}\right)^{\frac{1}{s}} e^{-\frac{\gamma_q}{s\beta_q}(t-t_0)} ||x(t_0)|| \quad t \in [t_0 \, t_1) \qquad (4)$$

where t_1 may be infinite.

Proof. Stability is given by Theorem 1. Before we prove exponential stability, it will first be shown that (4) holds. Therefore, assume that $x(t)$ is in region $\tilde{\Omega}_q$ for $t \in [t_0 \, t_1)$, where t_1 may be infinite. Using $V_q(x) \leq \beta_q ||x||^s$,

$$\dot{V}_q(x) \leq -\gamma_q ||x||^s \leq -\frac{\gamma_q}{\beta_q} V_q(x)$$

Multiplying both sides with the positive function $e^{\frac{\gamma_q}{\beta_q}t}$, this is equivalent to:

$$\frac{d}{dt}\left(e^{\frac{\gamma_q}{\beta_q}t} V_q(x)\right) \leq 0$$

The function $e^{\frac{\gamma_q}{\beta_q}t} V_q(x)$ is obviously decreasing. Consequently for $t \in [t_0 \, t_1)$

$$e^{\frac{\gamma_q}{\beta_q}t} V_q(x(t)) \leq e^{\frac{\gamma_q}{\beta_q}t_0} V_q(x(t_0))$$

which is the same as:

$$V_q(x) \leq V_q(x(t_0)) e^{-\frac{\gamma_q}{\beta_q}(t-t_0)}$$

Using this and $\alpha_q ||x||^s \leq V_q(x) \leq \beta_q ||x||^s$ implies that

$$||x||^s \leq \frac{1}{\alpha_q} V_q(x) \leq \frac{1}{\alpha_q} V_q(x(t_0)) e^{-\frac{\gamma_q}{\beta_q}(t-t_0)} \leq \frac{\beta_q}{\alpha_q} e^{-\frac{\gamma_q}{\beta_q}(t-t_0)} ||x(t_0)||^s$$

Hence, raising both sides with $1/s$ shows that the upper convergence rate of $x(t)$ in $\tilde{\Omega}_q$ is given by (4). To prove exponential stability [8] it must be shown that there exist two positive numbers, $\delta_1 > 0$ and $\delta_2 > 0$, such that

$$||x(t)|| \leq \delta_1 e^{-\delta_2(t-t_0)} ||x(t_0)|| \quad \forall t \geq t_0$$

We will show that this is the case with $\delta_1 > 0$ and $\delta_2 > 0$ defined according to

$$\delta_1 = \max_{\substack{q \in \{1, \ldots, \ell\} \\ r \in \{1, \ldots, \ell\}}} \left(\frac{\beta_q}{\alpha_r}\right)^{\frac{1}{s}}, \quad \delta_2 = \frac{1}{s} \min_{q \in \{1, \ldots, \ell\}} \frac{\gamma_q}{\beta_q}$$

If $x(t)$ never leaves $\tilde{\Omega}_q$ then the statement is proved since

$$||x(t)|| \leq \left(\frac{\beta_q}{\alpha_q}\right)^{\frac{1}{s}} e^{-\frac{\gamma_q}{s\beta_q}(t-t_0)} ||x(t_0)|| \leq \delta_1 e^{-\delta_2(t-t_0)} ||x(t_0)|| \quad \forall t \geq t_0$$

If the trajectory $x(t)$ hits $x \in \tilde{\Omega}_{qr}$ at time t_1, then from the third condition, $V_r(x(t_1)) \le V_q(x(t_1))$. Assume that $x(t)$ is in region $\tilde{\Omega}_r$ for $t \in [t_1\ t_2)$, where t_2 may be infinite. In the same way as above for $t \in [t_1\ t_2)$

$$V_r(x) \le V_r(x(t_1))e^{-\frac{\gamma r}{\beta_r}(t-t_1)}$$

Using $V_r(x(t_1)) \le V_q(x(t_1))$ and $V_q(x) \le V_q(x(t_0))e^{-\frac{\gamma r}{\beta_r}(t-t_0)}$ it follows that for $t \in [t_1\ t_2)$

$$V_r(x) \le V_q(x(t_1))e^{-\frac{\gamma r}{\beta_r}(t-t_1)} \le V_q(x(t_0))e^{-\frac{\gamma q}{\beta_q}(t_1-t_0)}e^{-\frac{\gamma r}{\beta_r}(t-t_1)}$$
$$\le V_q(x(t_0))e^{-s\delta_2(t_1-t_0)}e^{-s\delta_2(t-t_1)} = V_q(x(t_0))e^{-s\delta_2(t-t_0)}$$

Using $\alpha_r||x||^s \le V_r(x)$ and $V_q(x(t_0)) \le \beta_q||x(t_0)||^s$ implies that

$$||x(t)|| \le \left(\frac{\beta_q}{\alpha_r}\right)^{\frac{1}{s}}e^{-\delta_2(t-t_0)}||x(t_0)|| \le \delta_1 e^{-\delta_2(t-t_0)}||x(t_0)|| \quad t \in [t_1\ t_2)$$

Thus, in the whole time interval $t \in [t_0\ t_2)$

$$||x(t)|| \le \delta_1 e^{-\delta_2(t-t_0)}||x(t_0)|| \quad t \in [t_0\ t_2)$$

This proves that the equilibrium point 0 is exponential stable since the procedure can be repeated if there are more switchings of the functions $V_q, q = 1,\dots \ell$. □

The introduction of the specific α_q, β_q and γ_q for each region $\tilde{\Omega}_q$ is vital to obtain a non-conservative upper bound of the convergence rate. The theorem shows that the convergence rate is improved in a region $\tilde{\Omega}_q$ when $\gamma_q > 0$ is large and $\beta_q > 0$ is small. It can also be concluded that even if the function $V_q(x)$ is scaled by $\lambda > 0$, resulting in $\lambda V_q(x)$, it will not change the upper bound of the exponential convergence rate.

This theorem is an obvious generalization of a result concerning exponential stability for non-autonomous nonlinear systems [4] where we here use several functions V_q, $q = 1,\dots,\ell$, as measure of the system's energy. A similar but slightly more restricted formulation of Theorem 2 can be found in [3] where $\alpha_q = \alpha$, $\beta_q = \beta$ and $\gamma_q = \gamma$, $q = 1,\dots,\ell$. However, in Lemma 1 in [3] it is stated that $\dot{V}(t) \le -\gamma||x||_2^2$, but note that $\dot{V}(t)$ is not in general defined for all t.

4 Controller Design

4.1 Problem Formulation

As mentioned in Section 2, the goal of this paper is to design ϕ such that a stable closed-loop system is obtained. We will use Theorem 2 to achieve this, if possible. The design problem can be formulated as

Problem 3. Find ϕ in (2) such that the conditions in Theorem 1 are satisfied.

Furthermore, it may also be desirable to obtain a closed-loop hybrid system that is exponentially stable. The design problem is then

Problem 4. Find ϕ in (2) such that the conditions in Theorem 2 are satisfied.

In the following, two possible design strategies are presented which results in an exponentially stable closed-loop hybrid system if additional conditions are satisfied. The first strategy results in regions where it is possible to change vector field guaranteeing exponential stability of the closed-loop hybrid system. The second strategy utilizes the fact that a system is (exponentially) stable if it is always possible to choose a vector field that points in a direction such that the trajectory approaches the equilibrium point in a defined way.

4.2 Beforehand Specified Functions V_i

In some cases, the vector fields $f(x, m_i)$, $m_i \in M$, are the result of designing local (exponentially) stabilizing control-laws $u(x, m_i) = h(x, m_i)$, $m_i \in M$, for the system $\dot{x}(t) = g(x, u)$ in the regions Ω_i, $i = 1, \ldots, N$. The result is a system on the form (1) where the different vector fields are coupled to different regions Ω_i, $i = 1, \ldots, N$. It is possible to design the control-laws $u(x, m_i) = h(x, m_i)$, $m_i \in M$, in the regions Ω_i, $i = 1, \ldots, N$, from beforehand specified functions V_i, $i = 1, \ldots, N$. This is a local design method that is related to the use of Lyapunov theory to find a control-law that globally stabilizes a closed-loop system. In nonlinear feedback control [8], adaptive control [1] and fuzzy control [9] this is a common design strategy.

Assuming that $V_i(x)$ corresponds to Ω_i and is differentiable in this region, the time-derivative of $V_i(x)$ is $\dot{V}_i(x) = \frac{\partial V_i(x)}{\partial x} g(x, u)$. Hence, a possible way to design $u(x, m_i)$ is to satisfy the condition $\frac{\partial V_i(x)}{\partial x} g(x, u) \leq \gamma_i ||x||^s$ in Ω_i for a given $\gamma_i > 0$. This means that for each region Ω_i, $i = 1, \ldots, N$, there is a specified function $V_i(x)$, $i = 1, \ldots, N$, that locally (exponentially) stabilizes the system. Sufficient conditions for global stability are given by the following theorem:

Theorem 5. *Assume that the conditions A1 and A2 in Section 2 are satisfied. Furthermore, assume that*

A3 for each pair $(f(\cdot, m_i), \Omega_i)$ there is a function V_i, differentiable in x $\forall x \in \Omega_i$, satisfying
  - *$\forall x \in \Omega_i$, $\alpha_i ||x||^s \leq V_i(x) \leq \beta_i ||x||^s$*
  - *$\forall x \in \Omega_i$, $\dot{V}_i(x) = \frac{\partial V_i(x)}{\partial x} f(x, m_i) \leq -\gamma_i ||x||^s$*

A4 Every change of vector field from $f(\cdot, m_i)$ to $f(\cdot, m_j)$ occurs in $x \in \bar{S}_{ij}$, where $\bar{S}_{ij}$ is defined according to

$$\bar{S}_{ij} = \{x \in \Re^n \mid x \in \Omega_i,\ x \in \Omega_j,\ V_j(x) \leq V_i(x)\} \tag{5}$$

A5 No sliding motions occur.

Then the closed-loop system is exponentially stable.

Proof. Define $\tilde{\Omega}_i = \{(x, m) \in H \mid x \in \Omega_i,\ m = m_i\}$. Obviously, $\tilde{\Omega}_i^x = \Omega_i$. Then the first and second condition of assumption A3 is the same as the first and second condition in Theorem 2. Furthermore, defining $\tilde{\Omega}_{ij} \subseteq \bar{S}_{ij}$, the third condition in Theorem 2 is satisfied whenever assumption A4 is fulfilled. Hence, the closed-loop system is exponentially stable. $\qquad\square$

If the first and second condition of Assumption A3 are replaced by $\forall x \in \Omega_i, \alpha_i(||x||) \leq V_i(x) \leq \beta_i(||x||)$, where $\alpha_i(||x||)$ and $\beta_i(||x||)$ are class K functions, and $\forall x \in \Omega_i, \dot{V}_i(x) = \frac{\partial V_i(x)}{\partial x} f(x, m_i) \leq 0$, then the closed-loop system is stable according to Theorem 1.

Theorem 5 can be used for the construction of switch sets $S_{ij} \subseteq \bar{S}_{ij}$ guaranteeing (exponential) stability of the closed-loop hybrid system. This is shown by the following example.

Example 1. Consider a system with two vector fields, both allowed in the entire $\Re^n$, according to

$$f(x, m_1) = \begin{bmatrix} -x_1 - 10x_2 - x_1^3 - x_1 x_2^2 \\ 100x_1 - x_2 - x_1^2 x_2 - x_2^3 \end{bmatrix}$$

$$f(x, m_2) = \begin{bmatrix} -x_1 + 100x_2 - 0.1e^{4sin(x_1)}x_1 \\ -10x_1 - x_2 - 2e^{4sin(x_1)}x_2 \end{bmatrix}$$

The functions $V_i(x) = x^T P_i x$, $i = 1, 2$, where

$$P_1 = \begin{bmatrix} 10 & -0.05 \\ -0.05 & 1 \end{bmatrix},$$

$$P_2 = \begin{bmatrix} 1 & 0.005 \\ 0.005 & 10 \end{bmatrix}$$

fulfill

$$0.99||x||_2^2 \leq x^T P_1 x \leq 10.1||x||_2^2$$
$$0.99||x||_2^2 \leq x^T P_2 x \leq 10.1||x||_2^2$$
$$\dot{V}_1(x) \leq -0.99||x||_2^2$$
$$\dot{V}_2(x) \leq -2.09||x||_2^2$$

The closed-loop hybrid system is exponentially stable if the change of vector fields occur in the regions

$$\bar{S}_{12} = \{x \in \Re^2 \mid -9x_1^2 + 0.11x_1x_2 + 9x_2^2 \leq 0\},$$
$$\bar{S}_{21} = \{x \in \Re^2 \mid -9x_1^2 + 0.11x_1x_2 + 9x_2^2 \geq 0\}$$

and that no sliding motions occur. For instance, $S_{12} \subseteq \bar{S}_{12}$ and $S_{21} \subseteq \bar{S}_{21}$, where

$$S_{12} = \{x \in \Re^2 \mid x_2 = -0.5x_1\},$$
$$S_{21} = \{x \in \Re^2 \mid x_2 = -2x_1\}$$

is a possible choice. A simulation, with this choice, from the initial state $(x_0, m_0) = ([0 \; 10]^T, m_2)$ is shown in Fig. 3.

Some remarks about using Theorem 1 for the construction of switch sets $S_{ij} \subseteq \bar{S}_{ij}$ are in order.

- Sliding motions are avoided by choosing the switch sets $S_{ij} \subseteq \bar{S}_{ij}$ in such a way that there is a minimal time $\epsilon > 0$ between the changes of the vectorfields.
- Note that the possibility to utilize Theorem 5 for the design of switch sets S_{ij} guaranteeing (exponential) stability of the closed-loop hybrid system strongly depends on the stated functions $V_i(x)$, $i = 1, \ldots, N$.

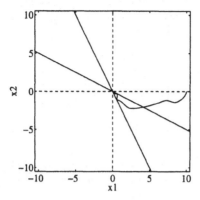

Fig. 3. Simulation of the closed-loop hybrid system. No sliding motions occur.

The construction of the switch sets $S_{ij} \subseteq \bar{S}_{ij}$ guaranteeing (exponential) stability of the closed-loop hybrid system is related to stability analysis for hybrid systems with known switch sets (3) [7, 6]. As a robustness measure acceptable uncertainties in the switch sets are obtained. These can be interpreted as *acceptable switch regions* and corresponds to the sets $\bar{S}_{ij}$ in (5).

4.3 Review of Other Control Design Methods

In [5] a stable switch strategy is proposed where the vector field corresponding to the smallest function $V_i(x)$, $i = 1, \ldots, N$ is chosen; hence the procedure is called the *min-switch strategy*. The given functions are positive definite. Every switch from $f(\cdot, m_i)$ to $f(\cdot, m_j)$ occurs when the corresponding functions are equal, i.e. $V_i(x) = V_j(x)$. Since the controller choses the vector field corresponding to the smallest function $V_i(x)$, $i = 1, \ldots, N$, it is a state-based controller implying that it does not contain any memory. Using this strategy, sliding modes may occur. Using the min-switch strategy for the systems in Example 1 results in a sliding mode at the surface $x_2 = -0.994x_1$. A simulation from the initial state $(x_0, m_0) = ([0\ 10]^T, m_2)$ is shown in Fig. 4.

If the first and second condition of Assumption A3 in Theorem 5 are replaced by $\forall x \in \Omega_i, \alpha_i(||x||) \leq V_i(x) \leq \beta_i(||x||)$, where $\alpha_i(||x||)$ and $\beta_i(||x||)$ are class K functions, and $\forall x \in \Omega_i, \dot{V}_i(x) = \frac{\partial V_i(x)}{\partial x} f(x, m_i) \leq 0$, and the smallest function V_i is chosen in the same way as in the *min-switch strategy* then Theorem 5 is the same as the *min-switch strategy* in [5] if sliding motions are allowed in Theorem 5. This is due to the fact that for every positive definite function $V_i(x)$ there exist class K functions $\alpha_i(||x||)$ and $\beta_i(||x||)$, such that $\alpha_i(||x||) \leq V_i(x) \leq \beta_i(||x||)$ [4].

In [10] a design procedure is proposed for stable switching between two unstable linear subsystems. It is assumed that there exists a stable convex combination of the two unstable subsystems. One of three possible switch strategies avoids sliding modes. The other two strategies are time average and sliding mode control. The drawback with this

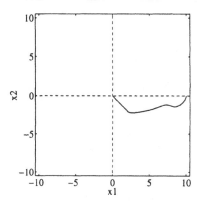

Fig. 4. Simulation of the closed-loop hybrid system using the *min-switch strategy*. A sliding mode occur at the surface $x_2 = -0.994x_1$.

design procedure is that it is restricted to switch between only two linear subsystems for which it must exist a stable convex combination. However, the advantage is that it guarantees the existence of a stable switching control law if a stable convex combination exists. Furthermore, it is then shown how to construct the switch sets S_{12} and S_{21}. In this sense the procedure in [10] is more constructive than the min-switch strategy in [5] and Theorem 5 since there is in general no guarantee that the condition A4 can be satisfied for given functions V_i, $i = 1, \ldots, N$. The problem is that it is necessary to change vector field when a region Ω_i is left. However, it is not always possible to change to another vector field such that the energy is the same or decreases. Scaling the different functions V_i by λ_i or searching for new functions V_i may sometimes help in solving this problem. However, it is not obvious how to change these functions appropriately.

4.4 Min-Projection Strategy

In this section we will present a design method which coincides with the intuition that a system is (exponentially) stable if it is always possible to choose a vector field that points in a direction such that the trajectory approaches the equilibrium point.

Min-projection strategy For a specific $x \in \Re^n$, choose vector field according to the criterion

$$f(x, m) = \arg \min_{f(x, m_i) \in F_x} x^T f(x, m_i) \tag{6}$$

where the set F_x is defined according to

$$F_x = \{f(x, m_i) \mid f(x, m_i) \text{ is allowed in } \Omega_i)\} \tag{7}$$

This design strategy says that the vector field corresponding to the smallest value $x^T f(x, m_i)$ is selected among the ones that are possible for a given state x. If all vector fields are allowed in the entire state space then they are all contained in F_x. The geometric interpretation of this strategy is that the vector field corresponding to the largest projection on the vector $-x$ is chosen, cf. Fig. 5; hence the name *min-projection strategy*.

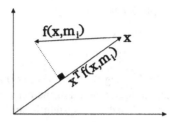

Fig. 5. Geometric interpretation of the *min-projection strategy*. The vector field $f(x, m_i)$ corresponding to the largest projection on the vector $-x$ is chosen.

The following theorem gives sufficient conditions when the *min-projection strategy* results in a stable closed-loop system.

Theorem 6. *If for all states $x \in \Re^n$*

$$\exists f(x, m_i) \in F_x \text{ such that } x^T f(x, m_i) \le 0 \tag{8}$$

then the closed-loop system is stable using the min-projection strategy. Specifically, if for all states $x \in \Re^n$ there exists a $\gamma > 0$ (independent of x) and

$$\exists f(x, m_i) \in F_x \text{ such that } x^T f(x, m_i) \le -\frac{1}{2}\gamma\|x\|^s \tag{9}$$

then the closed-loop system is exponentially stable using the min-projection strategy.

Proof. Choose the Lyapunov function $V(x) = x^T x$. Then $\dot{V}(x) = 2x^T f(x, m)$. Hence, if for all $x \in \Re^n$ it always exist a vector field $f(x, m_i) \in F_x$ such that $x^T f(x, m) \le 0$ then obviously $\min_{f(x,m_i) \in F_x} x^T f(x, m_i) \le 0$. Hence, $\dot{V}(x) = 2\min_{f(x,m_i) \in F_x} x^T f(x, m_i) \le 0$, implying that the closed-loop system is stable.

In the same way, if (9) is satisfied then $\dot{V}(x) = 2\min_{f(x,m_i) \in F_x} x^T f(x, m_i) \le -\gamma\|x\|^s$, implying that the closed-loop system is exponentially stable. $\quad\square$

Using the *min-projection strategy*, it is obvious from the proof of Theorem 6 that $\alpha = \beta = 1$. Hence, if there exists a $\gamma > 0$ such that (9) is satisfied then an upper bound of the convergence rate is $-\frac{\gamma}{s}$, cf. Theorem 2. Note from Theorem 6 that smaller values of $x^T f(x, m_i)$ implies larger $\gamma > 0$. Consequently, using the *min-projection strategy* can be interpreted as a minimization of the upper convergence rate of the closed-loop solution.

Example 2. The min-projection strategy will be illustrated on an example given in [5]. The system consists of two different vector fields which are allowed in the entire $\Re^2$

$$A(m_1) = \begin{bmatrix} -1 & 5 \\ 0 & -1 \end{bmatrix}, \quad A(m_2) = \begin{bmatrix} -1 & 0 \\ -5 & -1 \end{bmatrix}.$$

The min-projection strategy implies that

$$A(m_1) \text{ for } x_1 x_2 \leq 0$$
$$A(m_2) \text{ for } x_1 x_2 > 0$$

Theorem 6 is satisfied with $\gamma = 2$ using the Euclidean norm. Hence, the closed-loop system is exponentially stable. An upper bound of the convergence rate is $-\frac{\gamma}{s} = -1$, $i = 1, 2$. Two simulations of the closed-loop system is shown in Fig. 6. The solution obtained using the min-projection strategy converges much faster than using the min-switch strategy in [5].

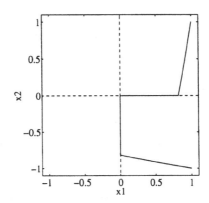

Fig. 6. Simulations of the closed-loop system from initial states $x_0 = [1 \ 1]^T$ and $x_0 = [1 \ -1]^T$ using the *min-projection strategy.*

The following example shows that if the condition in Theorem 6 is not satisfied it may happen that the closed-loop system becomes unstable applying the *min-projection strategy.*

Example 3. Consider two vector fields that are allowed in the entire $\Re^2$

$$A(m_1) = \begin{bmatrix} -1 & -10 \\ 100 & -1 \end{bmatrix}, \quad A(m_2) = \begin{bmatrix} 1 & 0 \\ 0 & 1 \end{bmatrix}.$$

Using the min-projection strategy gives for instance that $A(m_2)$ should be used on the surface $x_1 = x_2$. However, this results in an unstable solution.

Consequently, if the *min-projection strategy* is applied but Theorem 6 is not fulfilled then it is necessary to check if the obtained closed-loop system is stable. This can be done by solving a linear matrix inequality (LMI) problem corresponding to the conditions in Theorem 1, see [6].

The *min-projection strategy* results in a state-based controller where sliding modes may occur. However, if this is the case, it is possible to afterwards slightly adjust the switch surfaces such that sliding modes are avoided and then check if the obtained closed-loop hybrid system is stable in the same way as above.

Note that the difference between the *min-switch strategy* [5] and the *min-projection strategy* is that in the first case the vector field corresponding to the smallest function V_i, $i = 1, \ldots, N$, is chosen among the possible ones for a given state x, while in the latter case the vector field corresponding to the smallest function $\frac{d}{dt}||x||_2^2$ is chosen.

4.5 Using LMIs to Verify Theorem 6

If all vector fields are possible in the entire $\Re^n$ then a possible way to verify the conditions in Theorem 6 is to use the following lemma

Lemma 7. *If* $\exists \delta_i \geq 0$, $i = 1, \ldots, N$, *such that*

$$\sum_{i=1}^{N} \delta_i > 0 \quad \text{and} \quad \forall x \in \Re^n, \ \sum_{i=1}^{N} \delta_i x^T f(x, m_i) \leq 0 \tag{10}$$

then $\forall x \in \Re^n$ $\exists f(x, m_i)$ *such that* $x^T f(x, m_i) \leq 0$. *Furthermore, if for* $\gamma > 0$ $\exists \delta_i \geq 0$, $i = 1, \ldots, N$, *such that*

$$\sum_{i=1}^{N} \delta_i > 0 \quad \text{and} \quad \forall x \in \Re^n, \ \sum_{i=1}^{N} \delta_i (x^T f(x, m_i) + \frac{1}{2}\gamma||x||^s) \leq 0 \tag{11}$$

then $\forall x \in \Re^n$ $\exists f(x, m_i)$ *such that* $x^T f(x, m_i) \leq -\frac{1}{2}\gamma||x||^s$

Proof. The proof is shown by contradiction. Therefore assume that $\exists x \in \Re^n$ such that $x^T f(x, m_i) > 0$, $i = 1, \ldots, N$. Then $\sum_{i=1}^{N} \delta_i x^T f(x, m_i) > \sum_{i=1}^{N} \delta_i > 0$ since $\delta_i \geq 0, i = 1, \ldots, N$, and $\sum_{i=1}^{N} \delta_i > 0$, which is a contradiction to the statement (10). The second proof can be carried out in a similar manner. □

For linear vector fields, $f(x, m_i) = A(m_i)x$, Lemma 7 becomes

Corollary 8. *If* $\exists \delta_i \geq 0$, $i = 1, \ldots, N$, *such that*

$$\sum_{i=1}^{N} \delta_i > 0 \quad \text{and} \quad \forall x \in \Re^n, \ \sum_{i=1}^{N} \delta_i x^T A(m_i)x \leq 0 \tag{12}$$

then $\forall x \in \Re^n$ $\exists A(m_i)$ *such that* $x^T A(m_i)x \leq 0$. *Furthermore, if for* $\gamma > 0$ $\exists \delta_i \geq 0$, $i = 1, \ldots, N$, *such that*

$$\sum_{i=1}^{N} \delta_i > 0 \quad \text{and} \quad \forall x \in \Re^n, \ \sum_{i=1}^{N} \delta_i (x^T A(m_i)x + \frac{1}{2}\gamma||x||^s) \leq 0 \tag{13}$$

then $\forall x \in \Re^n$ $\exists A(m_i)$ *such that* $x^T A(m_i)x \leq -\frac{1}{2}\gamma||x||^s$

To find $\delta_i \geq 0$, $i = 1, \ldots, N$, such that (12) or (13) is satisfied can be formulated as a linear matrix inequality (LMI) problem [2].

Example 4. Verifying that the closed-loop system in Example 2 becomes exponentially stable using the *min-projection strategy* results in checking that for $\gamma > 0$ there exist $\delta_i \geq 0$, $i = 1, 2$, such that $\delta_1 + \delta_2 > 0$ and

$$\delta_1 (A(m_1) + \frac{1}{2}\gamma||x||_2^2) + \delta_2 (A(m_2) + \frac{1}{2}\gamma||x||_2^2) \leq 0 \tag{14}$$

Only if $0 < \gamma \leq 2$ there is a solution to this problem. If $\gamma = 2$ one solution is $\delta_1 = \delta_2 = 1$.

4.6   Extensions

If the conditions in Theorem 6 cannot be satisfied, it is possible to instead utilize a *min-skew-projection strategy*.

Min-skew-projection strategy For a specific $x \in \Re^n$, choose vector field according to the criterion

$$f(x, m) = \arg \min_{f(x, m_i) \in F_x} x^T P f(x, m_i) \tag{15}$$

where the set F_x is defined according to

$$F_x = \{f(x, m_i) \mid f(x, m_i) \text{ is possible in } \Omega_i)\} \tag{16}$$

and $P > 0$.

This design strategy says that the vector field corresponding to the smallest value $x^T P f(x, m_i)$ is selected among the ones that are possible for a given state x. The geometric interpretation of this strategy is that the vector field corresponding to the largest projection on the vector $-Px$ is chosen; hence the name *min-skew-projection strategy*.

By choosing the Lyapunov function $V(x) = x^T P x$ and carrying out the proof in the same way as Theorem 6, the following theorem is obtained.

Theorem 9. *If for all states $x \in \Re^n$*

$$\exists f(x, m_i) \in F_x \text{ such that } x^T P f(x, m_i) \leq 0 \tag{17}$$

then the closed-loop system is stable using the min-skew-projection strategy. Specifically, if for all states $x \in \Re^n$ there exists a $\gamma > 0$ (independent of x) and

$$\exists f(x, m_i) \in F_x \text{ such that } x^T P f(x, m_i) \leq -\frac{1}{2}\gamma||x||^s \tag{18}$$

then the closed-loop system is exponentially stable using the min-skew-projection strategy.

In the same way as in Corollary 8 we can solve an LMI problem to verify the conditions in Theorem 9. Furthermore, we can let $P > 0$ be unknown and included as a variable in the LMI formulation. Hence, we obtain a $P > 0$ that satisfies the conditions in Theorem 9, if there exists a solution.

Finally, if $P > 0$ is unknown, note that if a linear vector field that is stable always can be selected, then the condition (17) in Theorem 9 can be satisfied. This follows from the fact that it always exists a $P > 0$ satisfying $2x^T PAx \leq 0$ for a stable linear vector field Ax [8].

5 Conclusions

We have presented two strategies how to design a hybrid controller for a system consisting of several nonlinear vector fields. The goal when designing the controller is to stabilize or exponentially stabilize the closed-loop hybrid system. The first strategy results in regions where it is possible to change the vector field guaranteeing (exponential) stability of the closed-loop hybrid system.

The second strategy utilizes the fact that a system is (exponentially) stable if it is always possible to choose a vector field that points in a direction such that the trajectory approaches the equilibrium point. Sufficient conditions for (exponential) stability using this strategy are stated. These conditions can be verified by solving a linear matrix inequality (LMI) problem.

Acknowledgements Valuable discussions with Prof. Bo Egardt are very much appreciated. This work has been financially supported by the Swedish Research Council for Engineering Sciences (TFR) under the project number 92-185.

References

1. K. J. Åström and B. Wittenmark. *Adaptive Control*. Addison-Wesley, 1989.
2. S. Boyd, L. El Ghaoui, E. Feron, and V. Balakrishnan. *Linear Matrix Inequalities in System and Control Theory*. SIAM, 1994.
3. M. Johansson and A. Rantzer. Computation of piecewise quadratic Lyapunov functions for hybrid systems. Technical Report ISRN LUTFD2/TFRT–7549–SE, Department of Automatic Control, Lund Institute of Technology, June 1996.
4. H. K. Khalil. *Nonlinear Systems*. Prentice-Hall, 1996.
5. J. Malmborg, B. Bernhardsson, and K. J. Åström. A stabilizing switching scheme for multi controller systems. In *Proc. of 13th IFAC*, pages F:229–234, 1996.
6. S. Pettersson and B. Lennartson. LMI for stability and robustness of hybrid systems. Technical Report I-96/005, Control Engineering Laboratory, Chalmers University of Technology, 1996.
7. S. Pettersson and B. Lennartson. Stability and robustness for hybrid systems. In *Proc. of 35th CDC*, Kobe, 1996. To appear.
8. J.-J. E. Slotine and W. Li. *Applied Nonlinear Control*. Prentice-Hall, 1991.
9. H. O. Wang, K. Tanaka, and M. F. Griffin. An approach to fuzzy control of nonlinear systems: Stability and design issues. *IEEE Trans. Fuzzy Systems*, 4(1):14–23, 1996.
10. M. A. Wicks, P. Peleties, and R. A. DeCarlo. Construction of piecewise Lyapunov functions for stabilising switched systems. In *Proc. of the 33rd CDC*, pages 3492–97, 1994.

What Can We Learn from Synchronous Data-Flow Languages? (Invited Presentation)

Paul Caspi

Laboratoire Verimag, `Paul.Caspi@imag.fr`

Abstract. In this talk we try to explain why synchronous data-flow languages have been successful in programming critical control systems. A tentative explanation is that these languages have their background in both computer and control sciences. This supports our opinion that such a common background would be useful for other related topics, for instance formal methods and distributed programming. Furthermore, we believe that computer based control of critical systems requires engineers trained in both sciences, and finding such engineers is difficult. A common cultural background, which the hybrid community tries to establish, may help in this issue.

When I was asked to give this talk, I spent a long time hesitating: what could I say that might be useful to your Hybrid System community? And since I had no better idea, I decided simply to report our synchronous data-flow experience and then to discuss some lessons that could possibly be drawn from this experience concerning the relations between computer science and control.

A Short Story of Control Implementation

Early Technologies

Early implementations of control systems were mechanical. Then electro-mechanical, electronic, analog and discrete devices appeared, which grew up at the same time as control theory did. Thus it does not seem that any of these technological changes has induced a crisis in the minds of control system designers or, at least, if such a crisis arised, its history is so deeply buried that it seems difficult to trace it.

In particular, traditional conceptual tools of control science, such as transforms and transfer functions and subsequent graphic representations (block diagrams) perfectly matched the usual implementation techniques.

Computers

This situation was to evolve drastically when computers (especially micro-computers) began to appear: this entailed important changes in the way control people looked at their systems:

They had to sequentialize those activities that were formerly thought of as parallel.

This must have been a real concern, the more so as these people were in general not computer scientists but electrical, electronic and control engineers. At the beginning, when the first micro-computers appeared, they presumably looked at these beasts as some new kind of digital circuit, which, as other circuits, came along with some "data sheet" providing for user instructions! Thus assembly languages were for long widely used.

We must note here that the landscape is not the same for simulation: simulation tools have always been in advance with respect to implementation tools; this fact may be given several explanations linked to efficiency problems: simulation not being "real time" has benefited from the usual advances of computer science and technology: languages (SIMULA for instance was a pioneering concept in object-oriented technology), windows and mouses.

High-Level Real-Time Languages

Even when higher level languages such as FORTRAN and C were used, this sequencing "by hand" was not very satisfactory for evident reasons of cost, reliability and versatility. Thus there was an intensive research activity for real-time parallel languages and operating systems.

Yet it doesn't seem that this research has been totally successful, at least when it comes to critical control systems, i.e. the most demanding kind of application. For instance, a well-known motto in this area is: "Do not use tasking in critical systems." Why?

The phenomenon is intriguing and deserves careful examination. Such an examination is clearly behind the scope of this talk. What follows is simply a tentative explanation:

Those languages have been based mainly on some kind of "computer science parallelism" originating from time-sharing operating systems. It is in this area that such concepts as processes, synchronization primitives, monitors, rendezvous and FIFO queues appeared. Quite naturally, these concepts were then adapted to "real-time" languages and operating systems.

But it is quite clear that this parallelism is different from control parallelism: the former intends to solve such problems as sharing a printer between several user programs, while the latter addresses the problem of controlling *both* altitude *and* speed of an aircraft *at the same time*. In other words, computer science emphasizes *interleaving* while control stresses *simultaneity*.

Synchronous Languages

In this context, synchronous languages had two origins:

- Some practitioners found it possible to "compile" their parallel specifications, for instance block diagrams, into sequential "single loop" programs. This gave birth to many specific languages and tools, a typical example being the SAO graphic language of AÉROSPATIALE [5].

- Casually some computer and control scientists met and noticed that they could find, in their respective backgrounds, concepts that were quite close to each other. For instance, data-flow [8, 2], temporal logic [12, 9] and transforms, or Milner's synchronous calculus [10] and simultaneity. This gave birth to languages and tools [7, 6, 4, 3] that were somewhat more formally sound than the above tools, from both control and computer science aspects: as a consequence, they progressively replace them and become used and accepted.

The Story Continued?

Fortunately at least for researchers, synchronous languages do not solve every problem of control implementation. We can cite at least two of them which, in our opinion, require also some close interaction between control and computer science.

Formal Methods in Control

Among the many areas within the scope of computer activities, control is one where the most critical applications are found. Just think of computer controlled aircrafts, railways and nuclear plants! This has become a major question since computer implementations appeared, because computers allow control systems to become much more complex and then difficult to debug and test.

In this domain, computer scientists have proposed to use formal methods, either for designing or for formally verifying software systems and many, either experimental or more mature techniques are available [1, 11]. Yet these methods have not been primarily designed for use in control area. And, here also, control science possesses a strong and long lasting formal mathematical background. Founding computer formal methods upon this mathematical background might be promising.

Modularity and Distribution

Among the other problems not solved by synchronous languages, we can mention those arising from distributed and modular design, required for many reasons (location of sensors and actuators, fault tolerance, performance and complexity issues).

Here also, practitioners have developed methods by their own, based on typical control and signal processing techniques such as Shannon's sampling theory and more generally information theory. How these topics can be merged with general purpose distributed computing theories and practices might also be interesting.

Training

We have already stressed the importance of critical control systems. It seems sensible that these systems should be designed and implemented by engineers *trained in both control and computer sciences.*

Unfortunately, this does not seem to be the case by now. On the contrary, traditional task partitioning is currently applied: control engineers design and specify and computer engineers implement. This seems to lead us back to the old mismatches previously discussed when dealing with programming languages and this looks quite unsatisfactory.

However, having engineers trained in both sciences is also a difficult problem: How can we have high level people in some discipline, if this discipline does not truly exist? Getting this discipline alive is surely important here. It requires merging together common aspects of control and computer science and, clearly, hybrid systems are an important step in this direction.

References

1. J.-R. Abrial. *The B-Book.* Cambridge University Press, 1995.
2. E. A. Ashcroft and W. W. Wadge. *Lucid, the data-flow programming language.* Academic Press, 1985.
3. A. Benveniste and P. LeGuernic. Hybrid dynamical systems theory and the signal language. *IEEE Transactions on Automatic Control*, 35(5):535–546, 1990.
4. G. Berry and G. Gonthier. The ESTEREL synchronous programming language, design, semantics, implementation. *Science of Computer Programming*, 19(2):87–152, 1992.
5. D. Brière, D. Ribot, D. Pilaud, and J.L. Camus. Methods and specification tools for Airbus on-board systems. In *Avionics Conference and Exhibition*, London, December 1994. ERA Technology.
6. N. Halbwachs, P. Caspi, P. Raymond, and D. Pilaud. The synchronous dataflow programming language LUSTRE. *Proceedings of the IEEE*, 79(9):1305–1320, September 1991.
7. D. Harel and A. Pnueli. On the development of reactive systems. In *Logic and Models of Concurrent Systems*, volume 13 of NATO *ASI Series*, pages 477–498. Springer Verlag, 1985.
8. G. Kahn. The semantics of a simple language for parallel programming. In *IFIP 74 Congress*. North Holland, Amsterdam, 1974.
9. L. Lamport. The temporal logic of actions. *ACM Transactions on Programming Languages and Systems*, 16(3):872–923, 1994.
10. R. Milner. Calculi for synchrony and asynchrony. *Theoretical Computer Science*, 25:267–310, 1983.
11. S. Owre, J. Rushby, and N. Shankar. PVS: a prototype verification system. In *11th Conf. on Automated Deduction*, volume 607 of *Lecture Notes in Computer Science*, pages 748–752. Springer Verlag, 1992.
12. A. Pnueli. The temporal logic of programs. In *18th Symp. on the Foundations of Computer Science*. IEEE, 1977.

Verification of Real Time Chemical Processing Systems

Invited Presentation

Adam L. Turk, Scott T. Probst and Gary J. Powers
Department of Chemical Engineering
Carnegie Mellon University

Abstract

The application of Symbolic Model Verification, SMV, to the fault analysis of chemical processing systems was investigated. The objective was to measure the ability of the modeling language, employed by SMV, to capture significant logical and dynamic behaviors present in the processing systems. These behaviors originated from continuous dynamic chemical processing equipment, failure prone human operators, and control systems that are composed of relay ladder logic executed by programmable logic controllers. Also, the study measured the time and effort required to build models of the processing systems, assemble appropriate specifications for these systems, verify the system models, interpret the results, and revise the system model or original process design. We have verified systems for the transportation of multi-component solids in a conveying process for the manufacture of aluminum, leak testing of a fuel gas piping network, and batch reaction to produce fertilizer. Verification of each of these systems revealed numerous faults that lead to improved designs.

1. Introduction

Companies are increasingly interested in quality, operability, and safety of their processes since early and continued detection of faults that effect these issues are economically advantageous. These companies are using manual fault analysis methods routinely during the design and operating phase of the processing system. Current process engineering methods for fault analysis include engineering standards, check lists, prototype testing, and field testing. Engineering standards are domain accepted heuristics for design and development of processes while check lists are usually company in-house rules. Prototype testing is the evaluation of steady state simulations or pilot plant experiments that represent the proposed process. Field testing occurs before and during initial start up of a full scale process. These techniques are essential but limited by time and resources due to the large state space of typical processes. A typical small industrial process may be modeled using 50-100 binary variables. The total combinatorial space of a process model with 50 binary variables has 2^{50} (or 10^{15}) total states. Analysis of even modest size systems is difficult with current methods, thus an automated approach is greatly needed for more rigorous and complete verification.

Automated formal methods can exhaustively search a large state space, typically on the order of 10^{20} states [2], and verify that a property of the system is true. Originally, model checking was developed for the formal verification of integrated circuits and communication protocols. The question is whether these methods, in particular SMV [15] can be extended to verifying a complex chemical processing system. These processes are comprised of physical equipment, control logic, and human operators which can exhibit behavior that ranges from the continuous to the discrete. In order to capture the diversity and richness in the process behavior, the modeling language of the formal method must be flexible and concise. Recently, SMV has been used to verify several different processing systems [1. 9, 16, 17, 18, 19, 20, 21]. This work investigates the application of SMV to modeling and verifying chemical processes.

2. Symbolic Model Checking

In order for a process to be verified, a model is constructed from the process descriptions. The first task is to identify the behavior of concern or specifications. A model is subsequently constructed from the process description and the specifications. The specifications are used as guides in identifying the key behavior of the process to be included in the model. The model is then internally converted by SMV into a symbolic representation, which implicitly describes the state space.

The specifications are created from quality, operability, and safety issues that concern process engineers. The specifications are written using computational tree logic (CTL) which is a subset of time-branching temporal logic operators [3, 5, 7, 10]). CTL formulae contain propositional operators, path quantifiers and temporal operators. The following is a simple example of a CTL specification:

$$!EF(level\_tank = 8) \qquad (1)$$

This assertion states that: *"There does not exists a future state where the level of the tank will equal value 8 (overflow)"*. The specifications which represent the states of concern can be used to identify key process behavior that is included in the system model.

A logic model of the process is constructed from process information. Process flowsheets, operating procedures, relay ladder logic (RLL), human interactions, failure modes and other miscellaneous process descriptions are abstracted and reduced into a collection of significant behaviors and interactions. These behaviors can be described by various levels of complexity ranging from partial differential equations to Boolean. The continuous and dynamic nature of the process must be discretized over a finite set of states, which is then represented with binary variables in the SMV modeling language. The specifications, initial conditions, and logic model are combined together to form the overall model that is verified by SMV.

3. Modeling

Completeness and compactness of the logic model is a key issue in efficiently and correctly verifying a process. The appropriate level of complexity and behavior is abstracted from the process model [11, 12, 13, 14]. An overly complicated model may have too large of a state space for efficient verification.

The general strategy is to use the specifications and critical process behavior as a guide in modeling the process. Initial verification results will be generated as the process model is verified. These results will identify new specifications and key behavior which in turn will require the complexity level of the model to be changed. This iterative operation will insure that the model is continuously improving and that significant errors are uncovered in the process.

Building appropriate models that capture enough behavior to allow for the efficient detection of faults requires several comprehensive strategies. The strategies involve:

1. Exclusion of process parts that do not effect the specifications being tested.
2. Reducing the number of state variables required.
3. Capturing time in a manner that reduces model complexity.
4. Grouping process behavior using symmetry and modularity.
5. Using non-determinism in a selective manner.
6. Combining similar failure modes into combined failure events.

The combination of these strategies is an attempt to tame combinatorial explosion by building models and specifications that in an efficient manner are true to the underlying process physics, chemistry, control laws, and operating goals. The validity of these modeling assumptions to the actual behavior of the processing system, of course, must be tested.

3.1 Modeling Symmetry

The most common method for simplifying a model without eliminating behavior is to not include symmetrical process components [6]. This method is valid if the assumption is made that there are no significant interactions between these symmetric process components. These elements can always be added to the logic model at a later time in order to study their interactions.

3.2 Modular Modeling

The modeling of symmetry can be considered as a subset of a more general idea of modular modeling. Re-occurring process behavior or equipment can be modeled in modules and then instantiated for each occurrence. This modular approach will allow for quicker construction of the logic model. It will aid in future modeling efforts by forming a module library from which to draw on.

3.3 Modeling Control Logic

The automated functions or control logic can be implemented in a process either as software, such as relay ladder logic (RLL), hardwiring, or human operating procedures. RLL, a graphical programming language which resembles hardwired circuit diagrams, is implemented on a programmable logic controller (PLC). This graphical program defines the individual

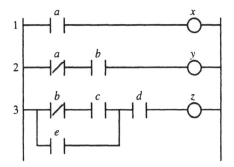

Figure 1 Sample Relay Ladder Logic

expressions of the control logic as relays which are then represented as rungs of a ladder. Figure 1 contains three example rungs, while many RLL program contain several hundred to several thousand rungs of logic.

In many PLCs, the RLL inputs are obtained and stored. The PLC sequentially evaluates each rung based upon the stored input parameters. The new output or control parameters are then updated for the process. A complete evaluation of the RLL by the PLC is referred to as a scan. The time scale of the scan commonly is on the order of milliseconds, while the time scale of the physical behavior found in the process is usually seconds, minutes, or hours. The control logic and physical behavior of the system have widely different time domains (fig 2). The transition of the states in the model is equated with physical changes in the process as a basis since time is implicit in SMV. The time domain for the control logic is included in the model by creating a variable that prevents the transition to a new state until the control logic variables have reached stable values.

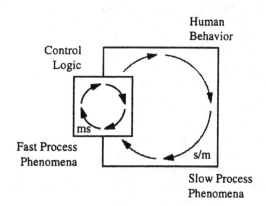

Figure 2 Dual Time Domains

$$w = \begin{cases} 1 & \text{if } (x = x') \wedge (y = y') \wedge (z = z') \\ 0 \end{cases} \qquad (2)$$

$$s' = \begin{cases} 0 & \text{if } w \wedge x \\ 1 & \text{if } w \wedge y \\ 2 & \text{if } w \wedge z \\ s \end{cases} \qquad (3)$$

where: w signals that the control logic has reached a stable state,
        x, y, and z are a control parameters,
        y is a control parameter,
        s is a state variable, and
        a primed variable refers to its next state.

Evaluation of the individual expressions in the RLL control logic can be done by sequentially evaluation [16], transitive closure [3], and design recovery [8]. Transitive closure provides a direct relationship between input and control parameters. The control logic in the model does not have to cycle until it reaches a stable state with this technique. Design recovery allows the control logic to be divided into groups that can be evaluated simultaneous while the groups are executed sequentially. The sequential technique was used by all the examples illustrated in this paper.

3.4 Modeling Continuous Behavior

The process model is transformed into a finite state machine. Continuous behavior, which can be considered as an infinite state machine, is discretized into a finite set of states. A simple example of representing the continuous behavior is shown for the tank in figure 3. The tank has a constant inlet flow, α, and a constant outlet flow, β. A linear differential equation [4] can be constructed for the tank level using these parameters.

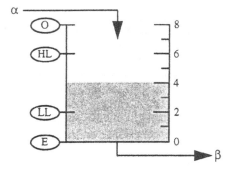

Figure 3: Liquid Level in a Tank

$$\frac{dL}{dt} = \alpha - \beta \tag{4}$$

where: L is the level in the tank,
        α is the inlet flow,
        β is the outlet flow, and
        t is time.

This differential equation is discretized into an algebraic equation

$$L = L_0 + \alpha \cdot \Delta t - \beta \cdot \Delta t \tag{5}$$

The resulting equation can further be simplified by assuming a constant time step deduced from the process and specification time scales.

$$L = L_0 + A - B \tag{6}$$

Bounds are placed on the tank level.

$$L_{min} \leq L \leq L_{max}$$
$$\qquad\qquad : L, L_0, L_{min}, L_{max} \in R \tag{7,8}$$
$$L_{min} \leq L_0 \leq L_{max}$$

The inlet and outlet flowrates are mapped into discrete integer values.

$$l = l_0 + a - b \tag{9}$$

These bounds are then normalized over an integer range based on process behavior and assertions.

$$0 \leq l \leq l_{max}$$
$$\qquad\qquad : l, l_0, l_{max} \in I \tag{10,11}$$
$$0 \leq l_0 \leq l_{max}$$

The tank level behavior described by equations 6, 7, and 8 is built into a logic model with conditional statements. The variable, l, is mapped into an integer range I. The range is selected by using the specifications and key process interactions as guides in identifying the critical states of the process and the resolution needed to observe these states. A possible specification for a tank would be that: "the tank does not over flow". The critical states needed in the model might be described by just three states; empty, full, and overflow. However, this set of states does not provide enough resolution with which to describe the levels of the tank needed by control logic. The discretization of the process into a finite state machine allows critical states to be identified and modeled with a minimum number of states.

3.5 Modeling Operating Procedures

Beside continuous behavior, dynamic processes can also contain discrete events such as operating procedures. Operating procedures are a series of tasks or functions that are normally performed in a defined order by a human operator. The tracing of the state path allows for the task ordering in the operating procedure to be monitored. A variable, which is declared as an integer range, is created to keep track of the state path.

$$step = \{0.8\} \tag{12}$$

The state variable, $step$, is defined so it can assume any discrete integer value between 0 and 8. At each step n, certain logic constraints, $pred$, need to be satisfied before $step$ can progress to the next interval, $n+1$.

$$step' = \begin{cases} 0 & \text{if } pred \vee step = 8 \\ 1 & \text{if } step = 0 \wedge pred \\ 2 & \text{if } step = 1 \wedge pred \\ \cdot \\ 8 & \text{if } step = 7 \wedge pred \\ step \end{cases} \tag{13}$$

If the order of operation or certain process behavior is not well defined by the procedure then non-determinism can be used in the model.

3.6 Modeling Non-determinism

Non-determinism allows the transitions to many states and allows the definition of many possible initial states. The ability of SMV to handle non-determinism allows multiple initial conditions and a large state space to be verified for the given specifications.

$$u' = \begin{cases} \{0,1\} & \text{if } pred \\ u \end{cases} \tag{14}$$

Non-deterministic behavior increases the complexity and in many cases the verification time of the model.

3.7 Modeling Failure Modes

It is often necessary for more complete verification to include failures of equipment, sensors, and humans in the system model. A failure mode can be included in the model by created a state variable which controls the conditions when the failure occurs or is corrected.

$$f' = \begin{cases} \{0,1\} & \text{if } pred \wedge time \\ 0 & \text{if } pred \wedge time \\ 1 & \text{if } pred \wedge time \\ f \end{cases} \tag{15}$$

where: f is a state variable,
pred are process conditions that triggers the fault, and
time is the state event during which the failure can trigger.

Instead of allowing the transition relation of the failure mode to be non-deterministic as in equation 15, the initial conditions of the failure can be left undefined. The non-determinism allows for the failure to occur arbitrarily anywhere along a state path. The failure mode, once occurred, will persist until it has been corrected at a later time period. Intermittent failures were not considered in the following examples

4. Results of Case Studies

The chemical process examples we examined are a solids transportation system for the manufacturing of aluminum [17, 20], a leak testing procedure for a combustion system [20], and a batch reactor to produce fertilizer [20]. These example processes were converted using the previously outlined approach into logic models. The construction, debugging, and verification time for each of these models ranged from 3 to 6 man months. The following sections outline a brief sample of the modeling and verification results for each chemical process example.

4.1 Solids Transport Process

The transportation system in an aluminum manufacturing process moves crushed anode and cryolite cover from storage tanks to a two compartment distribution bucket using a series of tank and transfer screws (fig. 4). The flow of material from the storage tanks and to the distribution bucket is control by knife gate valves. The level in the distribution bucket is detected with ultrasonic sensors. The transportation system is automated through the use of RLL. Human supervision is not required for normal operation. However, manual controls are available for special functions and process conditions.

The transport system model initially was simplified using symmetry by including only one crushed solids source. The RLL of the system had 112 rungs with approximately 30 rungs necessary in representing the process behavior and specifications were used. Operator or human behavior was modeled as a set of non-deterministic variables with time constraints.

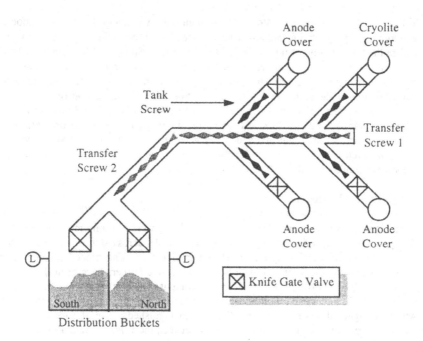

Figure 4: Distribution Bucket Fill Station

Table 1 gives the number of Boolean variables, the number of the reachable states, the number of nodes in the transition relation, and the CPU time required for the solids transport process models. The number of Boolean variables is an estimate of the model size while the size of the reachable states and transition relation provides an indication of the model complexity. The CPU time of the model includes the time for verification and counter example generation . The models, *Level 1* to *Level 4*, contain in order an increasing amount of human operator behavior. The *cryolite* model adds a cryolite source to the model from *Level 4* in order to check that two different solids can not simultaneously be feed into the transport system.

Table 1 Information on Verified Solid Transport System Models

Model Name	Boolean Variables	Reachable States	Transition Relationship (OBDD Nodes)	CPU Time (min)[a]
Level 1[b]	74	5,360	5,435	1.4
Level 1[c]	74	5,807	6,146	2.0
Level 2	75	7,049	6.144	2.9
Level 3	80	42,101	8.663	3.4
Level 4	84	1,670,390	10,207	5.7
Cryolite	93	15,351,100	19,335	45.5

[a]Computations were performed on an IBM RS6000 workstation
[b]Old Relay Ladder Logic
[c]Corrected Relay Ladder Logic

Two Level 1 models were created. One with an older version of the RLL and another with a more recently updated form. An error had been identified in the older RLL and corrected before the construction of the models. The idea was to show that SMV could be applied to the verification of a chemical process and find errors. The error in this system allowed one of the compartments in the distribution bucket to overflow. The fault involved the consolidation of the solids in a filled compartment before the second one was also filled causing the system to not halt its solids purge cycle. SMV did find this error in the old RLL and showed that it does not occur in the revised version. However, two other undetected errors were identified and corrected in the new RLL using SMV. One of these faults involved a case where both levels become full at the same time (within a scan) and the system would not halt.

4.2 Leak Test Procedure

Explosions in furnaces can be reduced if the valves in the gas train are checked for leaks. The procedure is a series of steps that check for leaks across shut off valves. The valve diagram for a combustion system is shown in figure 5. The procedure pressurizes the system by igniting the pilot and main burner. Once the burners have been ignited then the hand valves bv7 and bv15 are closed which extinguishes the burners. The blocking valves, l15, b13, ls12, and bs11 automatically close since they are linked to a safety interlock which is triggered when the burner flame is extinguished. The closure of these valves creates a series of pressurized pipe segments between the valves. Pressure in each of these segments is assumed by the test procedure to leak down stream to a lower pressure location. The main line is checked first by opening the tap valves tp3 and then tp2. Each section is checked for bubbling, which indicates no leaking downstream, and for bubbling to stop, which indicates no leaking up stream. If at any time, a leak is detected then the apparently leaking valve is replaced and the test procedure is started over.

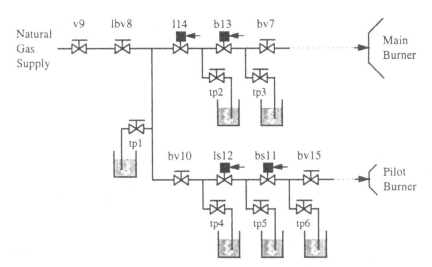

Figure 5: Piping and Valve Diagram for Combustion System

The specifications and key process behavior can be used to guide the construction of an appropriate model for the process. The main issues are the following:

- Are leaking valves detected?
- Are non-leaking valves being replaced needlessly?
- Does the procedure terminate?

Based upon these assertions and key process behavior, a logic model was constructed that contains failure modes such as valves leaking. This behavior was modeled by creating a leak variable for each valve. The initial conditions of these variables were made non-deterministic. Variables would retain their selected value until they were repaired at the correct point in the test procedure. The leak test procedure, itself, was modeled using a variable, *status*, which indicated the mode of operation in the furnace ignition procedure and variable, *step*, which described the leak testing procedure. The transition relation of the valves and pressure were not modeled with state variables but with definition variables. Definition variables are logic expressions that are evaluated at each state and can then be linked to a state variable. Pressure, in particular, was locally modeled for each pipe segment with more global definition variables that would link them together if they shared a common leaking valve.

Table 2 Information on Verified Leak Test Procedure Models

Model Name	Boolean Variables	Reachable States	Transition Relationship (OBDD Nodes)	CPU Time (sec)[a]
Test 1	24	5,944	11,221	4.81
Test 2	25	11,859	14,804	2.03
Test 3	26	22,263	18,734	3.90
Test 4	27	42,154	22,961	5.15
Test 5	28	84,313	28,215	27.88
Test 6	29	169,049	33,701	39.26
Test 7	30	340,193	39,087	51.66
Test 8	32	5,132	44,688	6.15
Test 9	32	15,984	49,601	18.3

[a]Computations were performed on a Hewlett-Packard 715/75 workstation

Performance indicators for the leak test procedure models are given in table 2. The model, *Test 1*, is the base case and models, *Test 2* to *Test 7*, contain an increasing number of failure modes for leaking tap valves. Model , *Test 8*, contains an expanded number of procedure steps, a refined *status* variable, and state variable for all the tap valves, bv10, and lbv8. Finally, the behavior in Test 9 was augmented to allow for slow valve leaks.

The initial verification indicates that the procedure did not explicitly check for both bubbling to start and stop at each of the taps. This failure could potentially lead to the procedure becoming dead locked on one of the steps or for a valve to be diagnosed as leaking and replaced unnecessarily. After this error was corrected, the verification results showed that the valves, bv7, bv15, and lbv8, because of their location in the system, could be diagnosed as leaking and replaced unnecessarily.

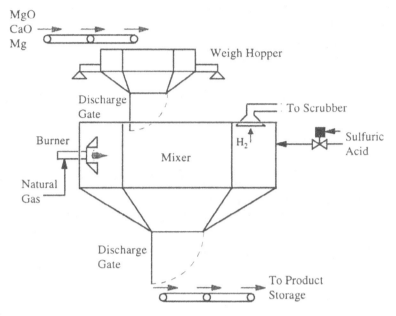

Figure 6 Batch Reaction System

4.3 Fertilizer Batch Reactor

A batch reactor produces fertilizer prills, MgO, $MgSO_4$, $CaSO_4$, $Mg(OH)_2$, and $Ca(OH)_2$, from the waste, MgO, CaO, and nascent Mg, from a magnesium plant by treating it with sulfuric acid. The general layout of the process equipment is provided in figure 6. The first step is to charge the mixer or reactor with a specific amount of waste. This material is then sprayed with sulfuric acid and mixed in order to produce the desired fertilizer products. The mixing in the reactor is done by a rotating pan, a counter-rotating inner shell, and an impeller. A by-product of the reactions is hydrogen gas which at certain concentrations can lead to an explosion. In order to avoid this situation, a natural gas flame is burning at all times during mixing. The gases in the mixer are vented to a scrubber. At the end of the mixing cycle, the fertilizer prills are discharged and conveyed to storage. In addition to the physical equipment and its behavior, the batch reactor has an operating procedure and RLL for controlling specific functions in the process such as opening and closing the mixer discharge gate.

The weigh hopper was modeled similar to a tank with the level being replaced by weight and the outlet flow being instantaneous. The mixer was modeled as a set of Boolean variables that represent the presence of material in the mixer, the position of the valve for spraying sulfuric acid, and the three mixing motors. These motors are connected to the pan, the inner shell, and the impeller of the mixer. The belt conveyers had variables for the motor and for the presence of material on the belt. The process had 176 rungs of RLL with only 28 relevant to its behavior and specifications. The model also included operator behavior and failure modes.

Results for the batch reactor models are shown in table 3. The various models have different equipment failure modes. The *Idle* model represents a process in which the human fails to complete tasks in the operating procedure.

Two faults were discovered with the model *Idle*. The specification that was violated checked the possibility of the acid being sprayed into an empty mixer. The spraying of acid into the mixer could lead to corrosion and failure of the mixer. This error occurred because the purge timer for the acid spray is started by the weigh hopper being full and by the control logic assuming that the material will automatically be discharged into the mixer.

Table 3 Information on Verified Batch Reaction System Models

Model Name	Boolean Variables	Reachable States	Transition Relationship (OBDD Nodes)	CPU Time (min)[a]
Base	74	398	65,321	0.5
Idle	75	25,378	65,357	14.6
Sensors-bc56	78	6,027	65,691	6.2
Sensors-bc57	78	14,579	69,690	61.3
Load-Flame	78	11,021	68,172	15.6
Limit-Switch-1	78	6,426	70,246	10.0
Limit-Switch-2	78	6,888	70,590	3.7
Valves-Gates	86	1,821,060	125,552	83.7
Motors-Stuck	87	791,160	133,773	33.3

[a]Computations were performed on a Hewlett-Packard 715/75 workstation

However, the discharge of the material from the weigh hopper does not occur if the operator has failed to complete a series of tasks such as shutting off the conveyor to the weigh hopper and closing the mixer discharge gate. The failure of the first task allows the hopper to become full. The second failure does not allow the material to enter the mixer since the discharge gate is open. The other discovered error which allowed material to remain in the weight hopper indefinitely was caused by the same idle operator behavior. Two other More process errors were detected and corrected with the other logic models.

5. Conclusions

Our initial conclusion is that the formal verification of modest complexity chemical processes is feasible. The faults discovered were acknowledged to be real and repaired by the owners in the actual processes. However, the synthesis of high integrity models is demanding and time consuming. The combination of insight into the underlying purpose of the process, control system, operating procedure, and strategies based on symmetry and modularity is needed to control the model size and verification times.

References

[1] R. Anderson, P. Beame, S. Burns, W. Chan, F. Modugno, D Notkin, and J. Reese, Model Checking Large Software Specifications. Proceedings of the Fourth ACM Symposium on the Foundation of Software Engineering: 156, 166, October, 1996.

[2] J. R. Burch, E. M. Clarke, K. L. McMillan, D. L. Dill, and L. J. Hawng, Symbolic Model Checking: 10^{20} states and Beyond. Information and Computation, 98(2): 142-170, June 1992.

[3] J. R. Burch, E. M. Clarke, D. E. Long, K. L. McMillan and D. L. Dill, Symbolic Model Checking for Sequential Circuit Verification. *IEEE Transactions on Computer-Aided Design of Integrated Circuits and Systems*, 13, 401- 424, 1994.

[4] C. Chiu, and B. J. Kuipers, Comparative Analysis and Qualitative Integral Representations. Presented at the Third Qualitative Physics Workshops, Stanford, CA, August 1989.

[5] E. M. Clarke, A. Emerson and A. P. Sistla, Automatic Verification of Finite-State Concurrent Systems Using Temporal Logic Specifications. *ACM Transactions on Programming Languages and Systems*, 8 (2), 244-263, 1986.

[6] E. M. Clarke, T. Filkorn, and S. Jha, Exploiting Symmetry in Temporal Logic Model Checking. *Proceedings of the Fifth Workshop on Computer-Aided Verification*, Ed. C. Courcoubetis. June/July 1993.

[7] E. M. Clarke, O. Grumberg, K. L. McMillan, and X. Zhao, Effective Generation of Counterexamples and Witnesses in Symbolic Model Checking. Technical Report No. CMU-CS-94-204, Carnegie Mellon University, PA, 1994.

[8] A. Falcione and B. H. Krogh, Design Recovery for Relay Ladder Logic. *IEEE Control Systems Magazine*, 13(2), April 1993.

[9] V. Hartonas-Garmhausen, T. Kurfess, E. M. Clarke, and D. E. Long, Automatic Verification of Industrial Designs. Proceedings of the 1995 IEEE Workshop on Industrial -Strength Formal Specification Techniques. 88-96. IEEE Comput. Soc. Press, April 1995.

[10] M. Jackson, *Software Requirements and Specifications*, ACM and Addison-Wesley, New York, 1995.

[11] B. J. Kuipers, Qualitative Simulation using Time-Scale Abstraction. *Int. J. Artificial Intelligence in Engineering*, 3(4), 185-191, 1988

[12] B. J. Kuipers, Reasoning with Qualitative Models. *Artificial Intelligence*, 59, 125-132, 1993

[13] B. J. Kuipers, and B. Shultz, Reasoning in Logic about Continuous Systems. *Principles of Knowledge Representation and Reasoning: Proceedings of the Fourth Annual International Conference*, Ed. J. Doyle, E. Sandewall, and P. Torasso, Morgan Kaufmann, San Mateo, CA, 1994

[14] D. E. Long,, *Model Checking Abstraction and Compositional Verification*, Ph.D. Thesis, Carnegie Mellon University, 1993.

[15] K. L. McMillan, *Symbolic Model Checking - An Approach to the State Explosion Problem*, Ph.D. Thesis, Carnegie Mellon University, 1992.

[16] I. Moon, *Automatic Verification of Discrete Chemical Process Control Systems*, Ph.D. Thesis, Carnegie Mellon University, 1992.

[17] S. T. Probst, G. J. Powers, D. E. Long, and I. Moon, Verification of a Logically Controlled Solids Transport System using Symbolic Model Checking. Submitted for publication in *Computers and Chemical Engineering*, 1994.

[18] S. T. Probst, and G. J. Powers, Automatic Verification of Control Logic in the Presence of Process Faults. Presented at the Annual AIChE Conference, San Francisco, CA, November 1994.

[19] S. T. Probst, A. L. Turk, and G. J. Powers, Formal Verification of a Furnace System Standard. Presented at the Annual AIChE Conference, Miami Beach, FL, November 1995.

[20] S. T. Probst, *Chemical Process Safety and Operability Analysis using Symbolic Model Checking*, Ph.D. Thesis, Carnegie Mellon University, 1996.

[21] T. Sreemani and J. Atlee, Feasibility of Model Checking Software Requirements: A Case Study. Technical Report CS96-05, Department of Computer Science, University of Waterloo, January, 1996.

Functional Specification of Real-Time and Hybrid Systems *

Olaf Müller   Peter Scholz

Institut für Informatik, Technische Universität München
D-80290 München, Germany
E-mail: {mueller,scholzp}@informatik.tu-muenchen.de

Abstract. Functional specifications have been used to specify and verify designs of a number of reactive, discrete systems. In this paper we extend this specification style to deal with real-time and hybrid systems. As mathematical foundation we employ Banach's fixed point theory in metric spaces. The goal is to show that the theory used for discrete functional specifications smoothly carries over to real-time and hybrid systems. An example of a thermostat specification illustrates the method.

1   Introduction

Hybrid systems are dynamical systems consisting of both discrete and continuous components. They are used to model the behavior of embedded real-time systems in a physical environment. Recently, a number of description and specification languages for reactive and/or real-time systems together with their proposed methodology for analysis, verification, and refinement were extended to deal with hybrid systems. For example, for model checking purposes a theory of hybrid automata has been developed [ACH+95], TLA has been extended to TLA+ [Lam93], I/O Automata have been extended to describe hybrid systems [LSVW95], and there are many other [GNRR93, AKNS95] hybrid description techniques.

In this paper we extend the formalism of *functional specification* [BDD+93, Bro93] to deal with real-time and hybrid systems. Functional specifications describe the behavior of a system as a network of functions, where every function processes infinite streams of incoming messages and yields infinite streams of outgoing messages. In the discrete setting, several approaches have been taken to give functional specifications a semantics:

- In [Bro93] domain theory is used to develop a semantic model for discrete stream processing functions together with a tailored refinement methodology.
- In [GS96] metric spaces are employed to give a semantics for functionally specified, discrete mobile data-flow networks.

We follow the second approach and extend the static parts of [GS96] to a description and specification method for hybrid systems. Our goal is to show that only

---

* This work is partially sponsored by the German Federal Ministry of Education and Research (BMBF) as part of the compound project "KorSys" and by BMW (Bayerische Motoren Werke AG).

slight modifications must be carried through, so that the whole theory smoothly carries over to the hybrid world.

The paper is organized as follows: Section 2 introduces stream processing functions and relates them to the corresponding notions in the theory of metric spaces. In Section 3 composition operators are defined that are used to build networks out of single functions. In particular, the mathematical foundation of the feedback operator is presented. Finally, Section 4 illustrates the specification method with the simple example of a thermostat.

2 Specification with Stream Processing Functions

We regard a distributed system as a network of components that exchange messages via directed channels. On every input or output channel messages are received from, or sent to, the environment. Therefore, every channel reflects an input or output communication history of the system. The system itself is described by a set of functions, where each function processes input histories and produces output histories according to its specification. To describe underspecification or nondeterminism we use sets of functions instead of single functions.

2.1 Dense Communication Histories

Communication histories of discrete systems can be modeled by sequences of messages, i.e., functions of type $I\!N \to M$, where M denotes the set of all messages [Bro93, BDD$^+$93]. For hybrid systems this model has to be extended to incorporate real time. One possibility is to add real time stamps. In the literature this is known as *sampling* semantics [MP93]. Here, instead, we develop a *super dense* semantics and therefore introduce real time or *dense* streams.

Let M be the (potentially infinite) set of all messages. A *dense stream* x over a set M is represented by a total function $x : I\!R_+ \to M$, where $I\!R_+$ denotes the set of all non-negative real numbers. Since we describe reactive systems, which continuously respond to stimuli from the environment, time never halts, and we use $I\!R_+$ as the time scale instead of time intervals. The set of all dense streams is denoted by $M^{I\!R_+}$. For every dense stream x we abbreviate the restriction $x|_{[0,t]}$ by $x{\downarrow}t$.

In order to motivate the usefulness of this definition we have adapted the example of a thermostat from [ACH$^+$95], where it is presented by means of hybrid automata.

Example 1 Dense Stream. The temperature of a room in a cool environment can be modeled by a dense stream x. We assume that without the presence of any heater, the temperature decreases according to the exponential function $x(t) = \Theta e^{-Kt}$, where t denotes the time, Θ the initial temperature, and K is a positive constant determined by the room.

A mathematical treatment of functional specifications requires dealing with feedback loops. In the discrete case, dealing with streams of type $I\!N \to M$, the semantics of such loops has been successfully described as least fixed points of functions over domains [Bro93, BDD$^+$93]. The underlying mathematical model is Scott's domain theory [SG90, Win93]. Fixed points of stream processing functions over dense streams, however, are more naturally and elegantly described by the fixed point theory of Banach. It is based upon the mathematical background of metric spaces. In order to specify loops of stream processing functions in Section 3, we therefore introduce the main concepts of metric space theory.

Definition 1 Metric Space. A *metric space* is a pair (D, d) consisting of a nonempty set D and a mapping $d : D \times D \to I\!R$, called a *metric* or a *distance*, which has the following properties:

(1) $\forall x, y \in D : \quad d(x, y) = 0 \quad \Leftrightarrow \quad x = y$
(2) $\forall x, y \in D : \quad d(x, y) = d(y, x)$
(3) $\forall x, y, z \in D : d(x, y) \leq d(x, z) + d(z, y)$.

We need a metric for dense streams, which is defined in the sequel.

Definition 2 The Baire Metric of Streams. The Baire metric space of dense streams $(M^{I\!R+}, d)$ is for all $x, y \in M^{I\!R+}$ defined as follows (see [Eng77]):

$$d(x, y) = inf\{2^{-t} \mid t \in I\!R_+ \wedge x{\downarrow}t = y{\downarrow}t\}.$$

From this definition a metric $d^{(n)}$ for n-tuples of streams $(M^{I\!R+})^n$ can be easily derived. Let $n \in I\!N$ and $x, y \in (M^{I\!R+})^n$ then $d^{(n)}(x, y)$ is defined as

$$d^{(n)}(x, y) = max\{d(x_i, y_i) \mid 1 \leq i \leq n\}.$$

A metric space (D, d) is called *complete* whenever each Cauchy sequence converges to an element of D [Eng77]. The Baire metric space on stream tuples $((M^{I\!R+})^n, d^{(n)})$, we consider in this paper, is complete [Eng77]. Complete metric spaces are a presupposition for Banach's fixed point theorem. This theorem, which will be explained later on, guarantees — under certain assumptions — the existence of a unique fixed point of loops in functional specifications.

2.2 Stream Processing Functions

Components of real time or hybrid systems can be functionally specified by stream processing functions over dense streams. First ideas in this area come from system theory [MT75]. Components are connected by directed channels to form a network. Each channel links an *input port* to an *output port*. A (m, n)-ary stream processing function with m input and n output ports is a function f with

$$f : (M_1^{I\!R+})^m \to (M_2^{I\!R+})^n$$

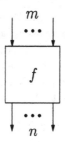

Fig. 1. Stream Processing Function

where M_1 and M_2 represent two (not necessarily different) sets of messages. The graphic notation of f is pictured in Fig. 1. If we want to express some kind of nondeterminism we describe components by a set of stream processing functions rather than by a single function.

Our operational understanding that stream processing functions model interacting components leads to a basic requirement for them. An interactive component is not capable to take back an output message that it has already emitted. This requirement can be fulfilled by a certain kind of stream processing functions, namely behaviors.

A stream processing function is said to be a *behavior* if its input until time t completely determines its output until time t. It is said to be a *delayed behavior* if its input until time t completely determines its output until time $t + \delta$ for $\delta > 0$. In other words, a delayed behavior imposes a delay of at least an arbitrarily small real value between input and output. Here, δ denotes the delay of f. It is quite realistic to assume components to be delayed because reactive systems always need a certain time to react. Instantaneous reactions, however, can be expressed by (non-delayed) behaviors.

Definition 3 (Delayed) Behavior. A (m,n)-ary stream processing function f is called a *behavior* if

$$\forall x, y \in (M^{R+})^m,\ t \in R_+ : x{\downarrow}t = y{\downarrow}t \Rightarrow f(x){\downarrow}t = f(y){\downarrow}t$$

and *a delayed behavior (with delay $\delta > 0$)* if

$$\forall x, y \in (M^{R+})^m,\ t \in R_+ : x{\downarrow}t = y{\downarrow}t \Rightarrow f(x){\downarrow}(t + \delta) = f(y){\downarrow}(t + \delta).$$

Note that the operator ${\downarrow}$ is overloaded to stream tuples in a point-wise style, i.e., $x{\downarrow}t$ for a stream tuple $x \in (M^{R+})^m$ denotes the tuple we get by applying ${\downarrow}t$ to each component of x.

The equivalent property in Scott's theory is monotonicity. From a theorem by Knaster and Tarski it is well-known that monotonic functions over complete partial orders have a least fixed point [Win93].

We model specifications by sets of (delayed) behaviors. They can be composed into networks of functions, which themselves behave as (delayed) behaviors. For this purpose, we will introduce three composition operators in the next section. For one of them, the feedback operator, the existence of a unique fixed point of the feedback loop is guaranteed only for *delayed* behaviors. To prove this formally we introduce a notion corresponding to delayed behaviors in metric space theory.

Definition 4 Lipschitz Functions. Let (D_1, d_1) and (D_2, d_2) be metric spaces and let $f : D_1 \to D_2$ be a function. We call f a *Lipschitz function* if there is a constant $c \geq 0$ such that the following condition is satisfied for all $x, y \in D_1$:

$$d_2(f(x), f(y)) \leq c \cdot d_1(x, y).$$

The Lipschitz constant $Lip(f)$ of a Lipschitz function f is denoted by the infimum of all c that fulfill the above mentioned inequation. If $Lip(f) \leq 1$ we call f *non-expansive*. If $Lip(f) < 1$ we call f *contractive*.

The following theorem relates the notions of behaviors and delayed behaviors to non-expansiveness and contractivity. Whereas the first ones have a operational justification, the latter ones represent their transfer to metric space theory and will be used as a requirement for Banach's fixed point theorem.

Theorem 5. *A stream processing function is a delayed behavior iff it is contractive with respect to the metric of stream tuples. A stream processing function is a behavior iff it is non-expansive with respect to the metric of stream tuples.*

Proof. We prove the first statement of the theorem. First, we prove the only-if-direction. Suppose that $d^{(m)}(x, y) = 2^{-t_0}$ and that f is a delayed behavior with delay δ. $d^{(m)}(x, y) = 2^{-t_0}$ implies that $x \downarrow t_0 = y \downarrow t_0$. Therefore, $f(x) \downarrow (t_0 + \delta) = f(y) \downarrow (t_0 + \delta)$. Finally, we get $inf\{2^{-t} \mid t \in \mathbb{R}_+ \land f(x) \downarrow t = f(y) \downarrow t\} \leq 2^{-(t_0 + \delta)} = 2^{-\delta} \cdot d^{(m)}(x, y)$. Since $2^{-\delta} < 1$ for all $\delta > 0$, f is contractive.

Now, we prove the if-direction. Suppose that $d^{(m)}(x, y) = 2^{-t_1}$, $d^{(n)}(f(x), f(y)) = 2^{-t_2}$, and that f is contractive, i.e., $\exists c < 1 : \forall x, y : d^{(n)}(f(x), f(y)) \leq c \cdot d^{(m)}(x, y)$. As $c < 1$ we can find a positive δ such that $2^{-\delta} = c$. Then $2^{t_1 - t_2} \leq c = 2^{-\delta}$. This implies because of the monotonicity of the logarithmic function that $t_1 + \delta \leq t_2$. As a consequence we get $x \downarrow t_1 = y \downarrow t_1 \Rightarrow f(x) \downarrow (t_1 + \delta) = f(y) \downarrow (t_1 + \delta)$ because $f(x) \downarrow t_2 = f(y) \downarrow t_2$ implies $f(x) \downarrow (t_1 + \delta) = f(y) \downarrow (t_1 + \delta)$. In other words, f is a delayed behavior. The second equivalence can be proven accordingly.

3 Composition Operators

The definition of networks is the main structuring principle on the functional specification level. There is no (semantical) difference in principle between a single component and a network of components. A network can be defined either by recursive equations or by special composition operators. We choose the second alternative and consider three basic composition operators, namely *sequential/parallel composition* and *feedback*.

In our functional specification technique, networks of components can be represented by directed graphs, where the nodes represent components and the edges represent point-to-point, directed communication channels (see, for instance, Fig. 2).

3.1 Sequential Composition

Sequential composition is simply defined by functional composition of two stream processing functions. The graphic representation of this composition is pictured in Fig. 2.

Definition 6 Sequential Composition. Let f and g be (m, n)-ary and (n, k)-ary stream processing functions, respectively. Then $f \circ g$ is the (m, k)-ary stream processing function defined by $(f \circ g)(x) = g(f(x))$.

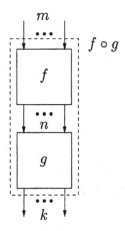

Fig. 2. Sequential Composition

The following theorem and corollary depict important properties of the sequential composition:

Theorem 7. *The sequential composition of two Lipschitz functions* $f : D_1 \to D_2$ *and* $g : D_2 \to D_3$ *is a Lipschitz function with constant* $Lip(f) \cdot Lip(g)$.

Proof.

$$d_3(g(f(x_1)), g(f(x_2))) \leq Lip(g) \cdot d_2(f(x_1), f(x_2)) \tag{1}$$
$$\leq Lip(g) \cdot Lip(f) \cdot d_1(x_1, x_2). \tag{2}$$

Corollary 8. *The sequential composition of two behaviors is a behavior. The sequential composition of two delayed behaviors with delays δ_1 and δ_2, respectively, is a delayed behavior with delay $\delta_1 + \delta_2$. The sequential composition of a behavior and a delayed behavior is a delayed behavior.*

Due to the above theorem, the proof of this corollary is obvious.

3.2 Parallel Composition

The parallel composition is defined intuitively. Sticking two components orthogonally together yields a component which input/output ports consists of all input/output ports of the composed components (see Fig. 3). Formally:

Definition 9 Parallel Composition. Let f and g be (m, n)-ary and (k, l)-ary stream processing functions. Then $f\|g$ is the $(m+k, n+l)$-ary stream processing function defined by

$$(f\|g)(x_1, \ldots, x_{m+k}) = (f(x_1, \ldots, x_m), g(x_{m+1}, \ldots, x_{m+k})).$$

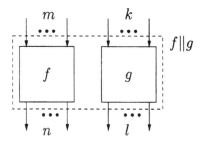

Fig. 3. Parallel Composition

As for the sequential composition, an equivalent property can also be formulated for the parallel composition:

Theorem 10. *The parallel composition of two behaviors is a behavior. The parallel composition of two delayed behaviors with delays δ_1 and δ_2, respectively, is a delayed behavior with delay $min(\delta_1, \delta_2)$. The parallel composition of a behavior and a delayed behavior is a behavior.*

Proof. We prove the second statement of the theorem. Let f be a (m, n)-ary delayed behavior with delay δ_1 and g be a (k, l)-ary delayed behavior with delay δ_2. Without loss of generality we assume that $\delta_1 < \delta_2$. Let $x, y \in (M^{\mathbb{R}_+})^k$, then $g(x)\!\downarrow\!(t + \delta_2) = g(y)\!\downarrow\!(t + \delta_2)$ implies that $g(x)\!\downarrow\!(t + \delta_1) = g(y)\!\downarrow\!(t + \delta_1)$. The other statements can be proven accordingly.

Note that the sequential composition of a behavior and a delayed behavior is a delayed behavior, whereas the parallel composition of a behavior and a delayed behavior is "only" a behavior.

3.3 Feedback Operator

Systems described by functional specifications may contain loops. In the graphic notation, this is denoted by circular graphs (Fig. 4). The feedback operator feeds k output channels back to k input channels of a $(m + k, n + k)$-ary delayed behavior.

Definition 11 Feedback Operator. Let $f : (M_1^{R+})^m \times (M^{R+})^k \to (M_2^{R+})^n \times (M^{R+})^k$ be a $(m + k, n + k)$-ary delayed behavior. Then $\mu^k f$ is a (m, n)-ary delayed behavior such that the value $(z_1, \ldots, z_n)$ of $(\mu^k f)(x_1, \ldots, x_m)$ is calculated as follows:

$$(z_1, \ldots, z_n, y_1, \ldots, y_k) = f(x_1, \ldots, x_m, y_1, \ldots, y_k)$$

where $(y_1, \ldots, y_k)$ is the solution of the equation

$$(y_1, \ldots, y_k) = g_{(x_1, \ldots, x_m)}(y_1, \ldots, y_k).$$

Here $g_{(x_1, \ldots, x_m)}$ is defined as a (k, k)-ary delayed behavior:

$$g_{(x_1, \ldots, x_m)}(y_1, \ldots, y_k) = \pi_{n+1, n+k}(f(x_1, \ldots, x_m, y_1, \ldots, y_k))$$

where $\pi_{n+1, n+k}$ denotes the projection on the last k ports.

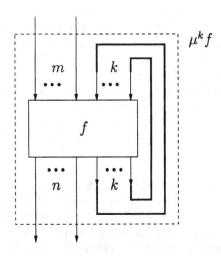

Fig. 4. Feedback Operator

The central issue of our contribution is that the fixed point operator is well-defined, i.e., that the unique solution of

$$(y_1, \ldots, y_k) = g_{(x_1, \ldots, x_m)}(y_1, \ldots, y_k)$$

exists. The existence of this fixed point is guaranteed by Banach's fixed point theorem:

Theorem 12 Banach's Fixed Point Theorem. *Let (D, d) be a complete metric space and $f : D \to D$ a contractive function. Then there exists an $x \in D$, such that the following holds:*

   (1) $x = f(x)$ *(x is a fixed point of f)*
   (2) $\forall y \in D : y = f(y) \Rightarrow y = x$ *(x is unique)*
   (3) $\forall z \in D : x = \lim_{n \to \infty} f^n(z)$ *where*
        $f^0(z) \quad = z$
        $f^{n+1}(z) = f(f^n(z))$

Proof. For instance, see [Sut75].

In the context of this paper, we can apply Banach's theorem in the following way. First of all, the metric space $((M^{I\!R+})^k, d^{(k)})$ is complete. Secondly, f is a $(m + k, n + k)$-ary delayed behavior and therefore contractive. Remember that f need not to be a basic stream processing function, but can also be a composed, delayed behavior. Moreover, also $g_{(x_1, \ldots, x_m)} : (M^{I\!R+})^k \to (M^{I\!R+})^k$ is by definition a contractive function. Altogether, all assumptions of Banach's fixed point theorem are fulfilled and the existence of a unique fixed point $(y_1, \ldots, y_k)$ of $g_{(x_1, \ldots, x_m)}$ is ensured. Hence, the feedback part of every delayed behavior has a unique fixed point.

Banach's fixed point theorem is the counterpart of Knaster/Tarski's fixed point theorem in the theory of metric spaces. However, note that Knaster/Tarski's theorem only guarantees the existence of a *least* fixed point, i.e., that potentially more than one fixed point can exist. In contrast, Banach's fixed point theorem guarantees the existence of a unique fixed point.

Again it is a straightforward proof to show that the feedback $\mu^k f$ is a delayed behavior, provided that f is a delayed behavior.

4 Example

In this section we give a functional specification of a thermostat, a simple hybrid system used as an introductory example in [ACH$^+$95]. The temperature of a room is controlled by a thermostat, which continuously senses the temperature and turns a heater on and off. The temperature is governed by differential equations. When the heater is off, the temperature $Temp$ of the environment, denoted by the dense stream x, decreases according to the function $x(t) = \Theta e^{-Kt}$ (see Example 1). When the heater is on, the temperature of the environment follows the function $x(t) = \Theta e^{-Kt} + h(1 - e^{-Kt})$, where h is a constant that depends on the power of the heater, Θ is the initial temperature of the room, and K is a constant determined by the environment. K can be considered to be direct proportional to the geometric size of the room. We wish to keep the temperature between min and max degrees and turn the heater on and off accordingly.

4.1 Thermostat as Open System

The controlling part of the resulting system for this informal description is shown in Fig. 5. The system consists of the two components *Control* and *Heater*. The first one is described by a function f_C of type

$$f_C : Temp^{I\!R+} \to \{on, off\}^{I\!R+}$$

that produces signals off or on, if the incoming stream of temperature signals overshoots max or undershoots min, respectively. These signals serve as an input stream for the *Heater* f_H:

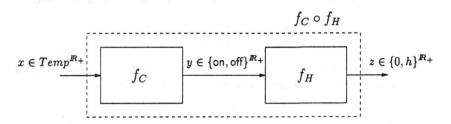

Fig. 5. Thermostat Modeled as Open System

$$f_H : \{on, off\}^{I\!R+} \to \{0, h\}^{I\!R+}$$

that produces the corresponding heating power, which can be 0 or h. Note that we model only the heating power of the heater, but not the resulting absolute temperature. The temperature of the room is regarded as part of the system's environment. This is different from [ACH+95], where the temperature is an inherent part of the system description. Therefore, the environment is there modeled as part of the system.

In fact, the model of hybrid automata does not emphasize on an interface concept to the environment, so that [ACH+95] describes merely closed systems without dividing the overall specification into system and environment. The advantage of our approach is its modularity, which allows us to separate the environment from the system specification. This is one of the essential issues of our approach. The application of our functional specification method to the thermostat example shows that indeed only the environment behaves continuously. The system itself, i.e., *Controller* and *Heater* behave as value-discrete components. They produce signals on, off, 0, and h. The environment, however, is characterized by the temperature, which is denoted by a real-valued $(Temp)$ stream. In the sequel, we give the precise specifications of the components *Control* and *Heater*. First of all, we define *Control*:

$$f_C(x) = y$$

where the output stream $y \in \{\text{on}, \text{off}\}^{R_+}$ is for all $t \in \mathbb{R}_+$ defined as follows:

$$
\begin{aligned}
x(t) \leq \text{min} &\quad \Rightarrow y(t + \delta_C) = \text{on} \\
x(t) \geq \text{max} &\quad \Rightarrow y(t + \delta_C) = \text{off} \\
\text{min} < x(t) < \text{max} &\Rightarrow y(t + \delta_C) = y(t).
\end{aligned}
$$

Here $\delta_C > 0$ denotes the delay of the component *Control*. However, this specification leaves the value $y(t)$ in the interval $[0, \delta_C)$ unspecified. We can abolish this under-specification by simply defining $y(t) = \text{off}$ in this interval. Now, we specify the *Heater*:

$$f_H(y) = z$$

where the output stream $z \in \{0, h\}^{R_+}$ is for all $t \in \mathbb{R}_+$ defined as follows:

$$
\begin{aligned}
y(t) = \text{off} &\Rightarrow z(t + \delta_H) = 0 \\
y(t) = \text{on} &\Rightarrow z(t + \delta_H) = h.
\end{aligned}
$$

Again, to avoid under-specification, we define $z(t) = 0$ for $t \in [0, \delta_H)$. The whole thermostat can then be described using the sequential composition

$$f_C \circ f_H.$$

This function has delay $\delta_C + \delta_H$ according to Corollary 1.

4.2 Thermostat as Closed System

To model the continuous part of the specification, we add the environment to it, yielding a closed system (Fig. 6):

$$f_E : \{0, h\}^{R_+} \rightarrow Temp^{R_+}$$

Env is specified as a component that cools the temperature down according to the exponential function Θe^{-Kt} (see also Example 1), if the *Heater* is off. When it is on, the temperature follows the function $\Theta e^{-Kt} + h(1 - e^{-Kt})$. We combine these two functions to one function $x(t) = \Theta e^{-Kt} + z(t) \cdot (1 - e^{-Kt})$ and get:

$$f_E(z) = x$$

where the output stream $x \in Temp^{R_+}$ is defined by the differential equation:

$$x'(t) = z(t) - K\Theta x(t)$$

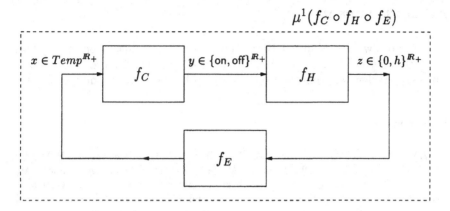

Fig. 6. Thermostat Modeled as Closed System

where $x'(t)$ denotes the first differentiation of $x(t)$. f_E and $f_C \circ f_H$ form a closed system in the shape of a feedback:

$$\mu^1(f_C \circ f_H \circ f_E).$$

This function is well-defined, as the occurring fixed point is uniquely determined according to our theory in Section 3: as $f_C \circ f_H$ is contractive with delay $\delta_C + \delta_H$, $f_C \circ f_H \circ f_E$ is contractive according to Corollary 1, even if f_E has no delay at all. Therefore Banach's fixed point theorem can be applied.

5 Conclusion and Further Work

We have shown that the specification formalism of discrete timed stream processing functions can easily be extended to deal with real-time and hybrid systems. We could give functional specifications with feedback a semantical foundation by introducing the concept of delayed behaviors that allows us to employ Banach's fixed point theorem. Characteristic of our approach is that our functional model naturally reflects the physical and conceptual structure of the system and its environment. In particular, it is possible to distinguish clearly between system and environment. In the thermostat example this structural clarity has been documented. Furthermore, we have the impression that the concept of well-known mathematical functions leads to a simple and clear specification style. Moreover, this generic approach fulfills the major requirement for any reasonable modeling formalism, namely modularity. In the discrete case a verification methodology by (structural, behavioral, and interface) refinements is well studied and understood. It is work in progress to carry over these results to our setting. It would also be interesting to analyze another type of streams as functions of type $I\!N \to M \times I\!R$, yielding a sampling semantics.

Acknowledgment

Thanks are owed to Manfred Broy who provided first ideas concerning both dense streams and behaviors. The authors have benefited from many discussion with Ketil Stølen and from the technical report by Radu Grosu and Ketil Stølen.

References

[ACH⁺95] R. Alur, C. Courcoubetis, N. Halbwachs, T.A. Henzinger, P.-H. Ho, X. Nicollin, A. Olivero, J. Sifakis, and S. Yovine. The algorithmic analysis of hybrid systems. *Theoretical Computer Science*, 138(1):3–34, 1995.

[AKNS95] P. Antsaklis, W. Kohn, A. Nerode, and S. Sastry. *Hybrid Systems II*, volume 999. Springer Verlag, 1995. Lecture Notes in Computer Science.

[BDD⁺93] M. Broy, F. Dederichs, C. Dendorfer, M. Fuchs, T. F. Gritzner, and R. Weber. The Design of Distributed Systems: An Introduction to Focus — Revised Version. Technical Report TUM-I9202-2, Technische Universität München, Fakultät für Informatik, 80290 München, Germany, 1993.

[Bro93] M. Broy. Interaction Refinement – The Easy Way. In M. Broy, editor, *Program Design Calculi*, volume 118 of *NATO ASI Series F: Computer and System Sciences*. Springer, 1993.

[Eng77] R. Engelking. *General Topology*. PWN - Polish Scientific Publishers, 1977.

[GNRR93] R.L. Grossman, A. Nerode, A.P. Ravn, and H. Rischel. *Hybrid Systems*, volume 736. Springer Verlag, 1993. Lecture Notes in Computer Science.

[GS96] R. Grosu and K. Stølen. A Model for Mobile Point-to-Point Dataflow Networks without Channel Sharing. In *Proc. of the 5th International Conference on Algebraic Methodology and Software Technology AMAST'96, Munich*, volume 1101 of *Lecture Notes in Computer Science*, pages 513–519, 1996. Also available as Technical Report TUM-I9527, Technische Universität München.

[Lam93] L. Lamport. Hybrid Systems in TLA+. In R.L. Grossman et al., editor, *[GNRR93]*, 1993.

[LSVW95] N. Lynch, R. Segala, F. Vaandrager, and H.B. Weinberg. Hybrid I/O automata. Technical Report CS-R9578, CWI, Computer Science Department, Amsterdam, 1995. Available under http://www.cs.kun.nl/~fvaan/.

[MP93] Z. Manna and A. Pnueli. Verifying Hybrid Systems. In Grossman et al., editor, *[GNRR93]*, 1993.

[MT75] M.D. Mesarovic and Y. Takahara. *General Systems Theory: Mathematical Foundations*, volume 113. Academic Press, 1975. Mathematics in Science and Engineering.

[SG90] D. Scott and C. Gunter. Semantic Domains and Denotational Semantics. In *Handbook of Theoretical Computer Science*, chapter 12, pages 633 – 674. Elsevier Science Publisher, 1990.

[Sut75] W. A. Sutherland. *Introduction to metric and topological spaces*. Claredon Press - Oxford, 1975.

[Win93] G. Winskel. *The Formal Semantics of Programming Languages*. The MIT Press, 1993.

Relating Time Progress and Deadlines in Hybrid Systems[*]

Sébastien Bornot[1] and Joseph Sifakis[1]

SPECTRE-VERIMAG[**],
Sebastien.Bornot@imag.fr and Joseph.Sifakis@imag.fr

Abstract. Time progress conditions in hybrid systems are usually specified in terms of *invariants*, predicates characterizing states where time can continuously progress or dually, *deadline conditions*, predicates characterizing states where time progress immediately stops. The aim of this work is the study of relationships between general time progress conditions and these generated by using state predicates. It is shown that using deadline conditions or invariants allows to characterize all practically interesting time progress conditions. The study is performed by using a Galois connection between the corresponding lattices. We provide conditions for the connection to be a homomorphism and apply the results to the compositional description of hybrid systems.

1 Introduction

Hybrid systems are systems that combine discrete and continuous dynamics. Their semantics is usually defined as a transition system on a set of states Q consisting of

- *transition relations* $\xrightarrow{a} \subseteq Q \times Q$ for $a \in A$ where A is a possibly infinite set of action names.
- *time progress relations* $\xrightarrow{t} \subseteq Q \times Q$ for $t \in \mathbf{R_+}$ such that

$$\forall q_1\, t_1\, t_2.\ \exists q_2\, q_3.\ q_1 \xrightarrow{t_1} q_2 \wedge q_2 \xrightarrow{t_3} q_3 \Leftrightarrow q_1 \xrightarrow{t_1+t_2} q_3 \quad \text{(additivity property)}.$$

The behavior of a hybrid system is characterized by the set of the execution sequences of the transition system. Additivity property guarantees that the set of states reached from a state within a given time is independent of the sequence of the time steps performed.

Usually, hybrid systems are modeled as hybrid automata (cf [ACH+95]), automata extended with a set of real valued variables. The variables can be tested and modified at transitions. Continuous state changes are specified by associating with automaton states evolution laws and constraints restricting the domain of variables.

[*] Partially supported by CNET Contract #95 7B.
[**] VERIMAG is a joint laboratory of CNRS, Institut National Polytechnique de Grenoble, Université J. Fourier and Vérilog SA associated with IMAG. VERIMAG Centre Equation, 2, av. de Vignate, 38610 Gières, France

Example 1. The following example represents the hybrid automaton for a thermostat. The variable θ represents the temperature which decreases (resp. increases) at states OFF and ON according to the laws $\theta \triangleright_{OFF} t$ (resp. $\theta \triangleright_{ON} t$). Furthermore, the conditions $m < \theta$ and $\theta < M$ are *invariants* restricting the values of θ between minimal and maximal values m and M respectively. Transitions occur when θ reaches limit values.

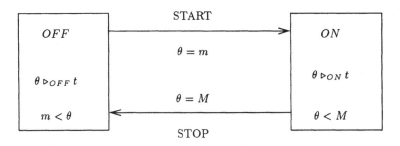

Fig. 1. the thermostat example

The hybrid automaton represents the transition system with $Q = \{ON, OFF\} \times \mathbf{R}$ and $A = \{START, STOP\}$ such that

- $q \xrightarrow{START} q' \Leftrightarrow q = (OFF, m) \land q' = (ON, m)$
  $q \xrightarrow{STOP} q' \Leftrightarrow q = (ON, M) \land q' = (OFF, M)$
- $(OFF, \theta) \xrightarrow{t} (OFF, \theta \triangleright_{OFF} t)$ if $\forall t'\ 0 \le t' < t\ .\ m < \theta \triangleright_{OFF} t'$
  $(ON, \theta) \xrightarrow{t} (ON, \theta \triangleright_{ON} t)$ if $\forall t'\ 0 \le t' < t\ .\ \theta \triangleright_{ON} t' < M$

Notice that the invariants $m < \theta$ and $\theta < M$ play an important role in this description as they do not allow temperature to progress beyond limit values, $\theta = m$ and $\theta = M$. Furthermore, when these values are reached time cannot progress by making the execution of the enabled transitions "urgent".

In this paper we consider hybrid systems represented as transition systems whose time progress relations are specified as a pair $(\triangleright, f)$ where $\triangleright$ is an *evolution law*, total function from $Q \times \mathbf{R}_+$ into Q and f is a *time progress function*, predicate on $Q \times \mathbf{R}_+$ such that $q \xrightarrow{t} q' \Leftrightarrow q' = q \triangleright t \land f(q, t)$.

Such a representation is common in hybrid automata where evolution laws are specified either explicitly or by a system of differential equations. Time progress function describes how from a given state time can progress by some amount. If for a given state it is false for any positive time, time cannot progress from this state. We call such a state *deadline state* because stopping time progress is used in practice to enforce a transition meeting a deadline.

Time progress functions are usually specified in terms of state predicates without mentioning time explicitly. These predicates characterize either the states where time can continuously progress (*invariants* in [ACH+95]) or dually, the states where time progress immediately stops (*deadline conditions* in [SY96]). Given such a predicate and an evolution law ▷, one can define progress functions: from a given state q time can progress by t if all the states encountered along the ▷-trajectory satisfy the invariant or dually do not satisfy the deadline condition. Invariants and deadline conditions are dual notions. In this paper we consider deadline conditions; the results can be adapted to invariants by dualization.

Formally, given a deadline condition $d(q)$, a time progress function $tp(d)(q,t)$ can be defined : $tp(d)(q,t) = \forall t'\ 0 < t' \leq t\ .\ \neg d(q \triangleright t')$

Conversely, from given a time progress function $f(q,t)$ one can define a deadline condition $dl(f)(q) : dl(f)(q) = \forall t > 0\ .\ \neg f(q,t)$. This simply means that the deadline condition corresponding to $f(q,t)$ is satisfied by all the states from which time cannot progress by any positive quantity.

If deadline conditions or invariants are useful for specification purposes, it is important to have available in some explicit form progress functions for simulation or analysis purposes. Explicit knowledge of progress function can help accelerating simulation by making it driven by deadline events.

The question arises about the nature of the correspondence between time progress functions and deadline conditions. Is it possible by using deadline conditions or invariants to characterize all time progress functions? The (obvious) answer is no. However, we show in section 2 that using deadlines allows to characterize some reasonably large class of progress functions. Formally speaking, we show that the pair of functions (dl, tp) is a Galois connection between the lattice of time progress functions and the lattice of deadline conditions.

In section 3, we investigate the relationships between the structures of the two lattices and provide conditions for tp and dl to be homomorphisms. Furthermore, we illustrate the use of the results for the compositional description of hybrid systems. We show how for modal formulas describing a global deadline condition in terms of local deadline conditions, a global progress function can be obtained in terms of local progress functions.

2 The correspondence between *TP* and *DL*

We study relations between time progress functions and deadline conditions for hybrid systems with set of states Q and evolution law $\triangleright : Q \times \mathbf{R}^+ \to Q$ that is additive and assumed fixed through the paper. Both time progress functions and deadline conditions are considered as predicates, that is, functions into the set $\{tt, ff\}$. We use standard notation $\vee, \wedge, \neg$ and $\Rightarrow$ to represent disjunction, conjunction, negation and implication. We represent by *true* and *false* respectively the functions $\lambda x.tt$ and $\lambda x.ff$.

2.1 The lattice of time progress functions TP

For a given evolution law $\triangleright$, a *time progress function* is a function f, $f : Q \times \mathbf{R}^+ \rightarrow \{tt, ff\}$ such that :

- $f(q, 0) = tt$
- $\forall t_1, t_2 \, . \, f(q, t_1 + t_2) = f(q, t_1) \wedge f(q \triangleright t_1, t_2)$ *(additivity)*

Example 2. If $q \triangleright t = q + t$ then

$$f_1(q, t) = q + t \leq 2 \vee (t = 0) \text{ and}$$
$$f_2(q, t) = 0 \leq q \wedge (q + t \leq 2) \vee (t = 0)$$

are time progress functions while

$$f_3(q, t) = 0 \leq q + t \leq 2 \vee (t = 0)$$

is not a time progress function as $f_3(-1, 2) = tt$, and $f_3(-1, t) = ff \; \forall t \in [0, 1)$.

Let TP be the set of time progress functions. TP is partially ordered by $\Rightarrow$ with bottom element the function $\lambda t \lambda q . t = 0$ and top element *true*.

We represent by $\sqcap$ and $\sqcup$ respectively, the greatest lower bound and least upper bound operations on TP. Notice that from the above definition, we have that if f_1, f_2 are time progress functions then $f_1 \wedge f_2$ is a time progress function. Consequently, $f_1 \sqcap f_2 = f_1 \wedge f_2$. However, $f_1 \vee f_2$ is not in general a time progress function. For instance, if $q \triangleright t = q + t$, the function

$$f_4(q, t) = 0 \leq q \wedge (q + t \leq 2) \vee 2 \leq q \wedge (q + t \leq 4) \vee (t = 0)$$

is the disjunction of two time progress functions but it is not a time progress function as $f_4(0, 4) = ff$ while $f_4(0, 2) = tt$ and $f_4(2, 2) = tt$. However, one can find

$$f_5(q, t) = 0 \leq q \wedge (q + t \leq 2) \vee (t = 0) \sqcup 2 \leq q \wedge (q + t \leq 4) \vee (t = 0)$$

which is equal to $0 \leq q \wedge (q + t \leq 4) \vee (t = 0)$ and is the least time progress function implied by both $0 \leq q \wedge (q + t \leq 2) \vee (t = 0)$ and $2 \leq q \wedge (q + t \leq 4) \vee (t = 0)$.

Proposition 1. $(TP, \Rightarrow, \sqcap, \sqcup)$ *is a distributive lattice with :*

$$f_1 \sqcap f_2 = f_1 \wedge f_2 \quad and \quad f_1 \sqcup f_2 = \bigvee_{i=1}^{\infty} f_1 \vee_i f_2$$

where :

$$f_1 \vee_1 f_2 = f_1 \vee f_2$$
$$f_1 \vee_{i+1} f_2(q, t) = \exists t' \; 0 \leq t' \leq t \, . \, (f_1 \vee_i f_2)(q, t') \wedge (f_1 \vee f_2)(q \triangleright t', t - t')$$

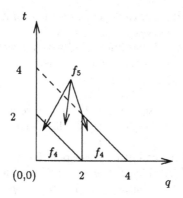

Fig. 2.

Proof.

- $f_1 \sqcap f_2 = f_1 \wedge f_2$ is immediate.
- For $f_1 \sqcup f_2 = \bigvee_{i=1}^{\infty} f_1 \vee_i f_2$:
  $f_j \Rightarrow f_1 \vee_1 f_2 \Rightarrow f_1 \sqcup f_2$ for $j \in \{1, 2\}$
  On the other hand, if for some arbitrary time progress function f, $f_j \Rightarrow f$
  for $j \in \{1, 2\}$, we will show by induction that $\forall i \in \mathbf{N} . f_1 \vee_i f_2 \Rightarrow f$ and
  therefore $f_1 \sqcup f_2 \Rightarrow f$:
  $f_1 \vee_1 f_2 \Rightarrow f$
  If $f_1 \vee_{i-1} f_2 \Rightarrow f$, then for all (q, t) such that $(f_1 \vee_i f_2)(q, t) = tt$ we have by
  definition :
  $\exists t' \ 0 \le t' \le t . (f_1 \vee_{i-1} f_2)(q, t') \wedge (f_1 \vee f_2)(q \triangleright t', t - t')$
  and then : $\exists t' \ 0 \le t' \le t . f(q, t') \wedge f(q \triangleright t', t - t')$
  by additivity of f: $f(q, t) = tt$.

2.2 The lattice of deadlines

Consider the set of state predicates DL whose elements d are unary predicates
on Q (functions from Q into $\{tt, ff\}$). We shall interpret the elements of DL
as deadline conditions. DL is a boolean lattice with the standard operations of
conjunction, disjunction and negation.

We define the pair of functions (tp, dl) relating DL and TP:
$tp : DL \to TP$ such that $tp(d)(q, t) = \forall t' \ 0 \le t' < t . \neg d(q \triangleright t')$
$dl : TP \to DL$ such that $dl(f)(q) = \forall t > 0 . \neg f(q, t)$

It is trivial to check that $tp(d)$ is a progress function. We call $tp(d)$ the
progress function corresponding to d and $dl(f)$ the deadline condition corresponding to f.

Notice that the definition of tp depends on the evolution law $\triangleright$ which can be considered as a family of curves parameterized with time in the space of variables. If a curve at a state q is parameterized with t_0 then the state $q \triangleright t$ reached by letting time pass by t, is on the curve parameterized by $t_0 + t$.

Example 3. For $q \triangleright t = q + t$ and $d(q) = 2 < q < 3$ we have,
$tp(d)(q,t) = \forall t'\ 0 \le t' < t .\ \neg 2 < q + t' < 3$ which gives
$tp(d)(q,t) = (t = 0) \vee q + t \le 2 \vee 3 \le q$.
If we compute the deadline condition corresponding to the latter time progress function we find: $dl(tp(d))(q) = 2 \le q < 3$ which differs from d in that it is left-closed. However, we have $tp(2 \le q < 3) = tp(2 < q < 3)$.

Consider now that $d = \neg(2 < q < 3)$ which means that time can progress only from states q such that $2 < q < 3$. We find

$$tp(d)(q,t) = \forall t'\ 0 \le t' < t .\ 2 < q + t' < 3$$

which is equivalent to

$$tp(d)(q,t) = (t = 0) \vee 2 < q \wedge q + t \le 3.$$

The deadline condition corresponding to the latter is again $d = \neg(2 < q < 3)$.

2.3 The Galois connection between TP and DL

Proposition 2. *For any deadline condition d, $d \Rightarrow dl\ tp(d)$*

Proof.

$$
\begin{aligned}
dl\ tp(d)(q) &= \forall t > 0 .\ \neg tp(d)(q,t) \\
&= \forall t > 0 .\ \neg \forall t'\ 0 \le t' < t .\ \neg d(q \triangleright t') \\
&= \forall t > 0 .\ \exists t'\ 0 \le t' < t .\ d(q \triangleright t')
\end{aligned}
$$

If $d(q) = tt$, by choosing $t' = 0$, we have $dl\ tp(d) = tt$.

Proposition 3. *For any progress function f, $f \Rightarrow tp\ dl(f)$*

Proof.

$$
\begin{aligned}
tp\ dl(f)(q,t) &= \forall t'\ 0 \le t' < t .\ \neg dl(f)(q \triangleright t') \\
&= \forall t'\ 0 \le t' < t .\ \neg(\forall t'' > 0 .\ \neg f(q \triangleright t', t'')) \\
&= \forall t'\ 0 \le t' < t .\ \exists t'' > 0 .\ f(q \triangleright t', t'')
\end{aligned}
$$

If $f(q,t) = tt$, choose $t'' = t - t'$, and by additivity, $tp\ dl(f)(q,t) = tt$.

A consequence of the above propositions and of the fact that tp and dl are anti-monotonic, is that the pair $(tp, \widetilde{dl})$ is a Galois connection (see for example [Ore44, San77]) between DL and $\widetilde{TP}$ where $\widetilde{dl}$ and $\widetilde{TP}$ are respectively the dual function of dl $(\widetilde{dl} = \lambda f.\neg dl(\neg f))$ and the dual lattice of TP.

The following properties result from the application of well-known results about Galois connections.

Properties:
$$tp(d_1 \vee d_2) = tp(d_1) \sqcap tp(d_2) = tp(d_1) \wedge tp(d_2)$$
$$dl(f_1 \sqcup f_2) = dl(f_1) \wedge dl(f_2)$$

Definition 4. Given a time predicate g $(g : \mathbf{R}_+ \to \{tt, ff\})$ we say that g is *left-closed* if
$\forall t_0 \, . \, \neg g(t_0) \Rightarrow \exists \epsilon > 0 \, . \, \forall \epsilon' \leq \epsilon \, . \, \neg g(t_0 + \epsilon')$ (cf figure 3). We say that g is *right-closed* if in the above definition $g(t_0 + \epsilon')$ is replaced by $g(t_0 - \epsilon')$.

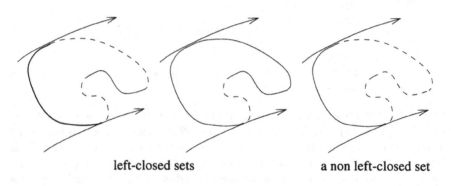

left-closed sets   a non left-closed set

Fig. 3. left-closure

Proposition 5.

- *The image of dl, $im(dl)$, contains only left-closed deadline conditions i.e., dealine conditions d such that for all q $\lambda t.d(q \rhd t)$ is left-closed.*
- *The image of tp, $im(tp)$, contains only right-closed time progress functions i.e., functions f such that for all q, $f(q, t)$ is right-closed.*
- *$im(dl)$ and $im(tp)$ are isomorphic via tp.*

Proof.

– For all f and q,

$$dl(f)(q) = \forall t > 0 . \neg f(q, t).$$

If $dl(f)(q) = ff$ then $\exists t > 0 . f(q, t)$. By additivity we have

$$\exists t > 0 . \forall t' \leq t . f(q \triangleright t', t - t').$$

Consequently :

$$\exists t > 0 . \forall t' \leq t . dl(f)(q \triangleright t') = ff.$$

– For all d, q and t,

$$tp(d)(q, t) = \forall t' < t . \neg d(q \triangleright t').$$

If $tp(d)(q, t) = ff$ then $\exists t' \, 0 \leq t' < t . d(q \triangleright t')$.
We can write this $\exists t' \, 0 < t' \leq t . d(q \triangleright t - t')$.
For all t'' such that $0 \leq t'' < t'$ we have

$$\exists t_0 = t - t' . 0 \leq t_0 < t - t'' \wedge d(q \triangleright t_0)$$

and then $tp(d)(q, t - t'') = ff$. Finally,

$$\exists \epsilon < t' . \forall \epsilon' \leq \epsilon . tp(d)(q, t - \epsilon') = ff.$$

$\lambda t.tp(d)(q, t)$ is right closed.
– Isomorphism of $im(dl)$ and $im(tp)$ via tp is a direct result from the fact that (tp, dl) is a Galois connection.

Notice that a consequence of the above propositions is that if d is left-closed then $d = dl \, tp(d)$ and if f is right-closed then $f = tp \, dl(f)$. This means that left-closed deadline conditions and right-closed time progress functions are in bijection. This implies that functions which are not right-closed such as $f(q, t) = q + t < 2$ for $q \triangleright t = q + t$ cannot be obtained as images of deadline conditions. Such functions can be considered as non well-defined because time can get arbitrarily close to a bound without reaching it, enforcing the existence of converging infinite time sequences. It can be shown that if f is not right-closed then $tp \, dl(f)$ is the right-closure of f. Dually, deadline conditions that are not left-closed have the same image via tp as their left-closure which means that they do not characterize all the states from which time cannot progress.

3 Translating deadline conditions into progress functions

3.1 Well-defined deadline conditions

The results of the previous section establish some strong correspondence between deadline conditions and time progress functions. However, in practice, deadline conditions or equivalently invariants of a hybrid system are obtained as a combination of deadline conditions of its components. In this section we provide results for the compositional computation of time progress functions. We investigate the conditions for the functions tp and dl to be lattice homomorphisms. Then, we provide results for translating modal deadline formulas into progress functions.

To have a lattice homomorphism it is necessary that
$tp(d_1 \wedge d_2) = tp(d_1) \sqcup tp(d_2)$.
First observe that in general this equality does not hold. Consider the deadline conditions d_1 and d_2 defined by :
$d_1 = \bigvee_i p_{2i}$, $d_2 = \bigvee_i p_{2i+1}$, where $p_i(q) = 1 - 2^{-i} \leq q < 1 - 2^{-(i+1)}$.
We have $d_1 \wedge d_2 = false$ and consequently, $tp(d_1 \wedge d_2) = true$. However, $\forall t > 1 . tp(d_1) \sqcup tp(d_2)(0, t) = ff$. Thus, in general, $tp(d_1) \sqcup tp(d_2) \Rightarrow tp(d_1 \wedge d_2)$ and the implication is strict. Notice that this is due to the fact there is an accumulation point of the alternations between dealine conditions which does not allow time progress beyond $t = 1$ (included). In fact, $tp(d_1) \sqcup tp(d_2)(0, 1) = ff$.

We call *well-defined* the deadline conditions d that are left-closed and such that the function $\lambda t.d(q \triangleright t)$ changes only a finite number of times in any finite interval. Notice that well-defined deadline conditions are closed under disjunction and conjunction and form a sub-lattice of DL.

Proposition 6. *The restriction tp to well-defined deadline conditions is a homomorphism.*

Proof. We have trivially $tp(d_1 \vee d_2) = tp(d_1) \sqcap tp(d_2)$ and
$tp(d_1) \sqcup tp(d_2) \Rightarrow tp(d_1 \wedge d_2)$, by definition of tp.
Let us prove that $tp(d_1 \wedge d_2) \Rightarrow tp(d_1) \sqcup tp(d_2)$ if the finite variability condition holds. Suppose $tp(d_1 \wedge d_2)(q, t) = tt$:

$$tp(d_1 \wedge d_2)(q, t) = \forall t' < t . \neg(d_1 \wedge d_2)(q \triangleright t')$$
$$= \forall t' < t . (\neg d_1(q \triangleright t')) \vee (\neg d_2(q \triangleright t'))$$

Since $\lambda \tau.d_1(q \triangleright \tau)$ and $\lambda \tau.d_2(q \triangleright \tau)$ have a finite set of points of discontinuity in $[0, t]$, we can divide this interval into a finite set of open subintervals $[t_0, t_1[, [t_1, t_2[, \ldots, [t_{n-1}, t_n[$ with $t_0 = 0$ and $t_n = t$, such that

$$\forall i < n . (\forall t' \in]t_i, t_{i+1}[. \neg d_1(q \triangleright t')) \vee (\forall t' \in]t_i, t_{i+1}[. \neg d_2(q \triangleright t')).$$

We show that one can find t_i''s such that : $t_0' = 0, t_{n'}' = t$ and

$$\forall i < n' . (\forall t' \in [t_i', t_{i+1}'[. \neg d_1(q \triangleright t')) \vee (\forall t' \in [t_i', t_{i+1}'[. \neg d_2(q \triangleright t')).$$

As d_1 and d_2 are left-closed, one can find t_i's such that

$$(\forall t' \in [t_i, t_{i+1}[. \neg d_1(q \triangleright t')) \vee (\forall t' \in [t_i, t_{i+1}[. \neg d_2(q \triangleright t')).$$

Suppose that for a given i we have $\forall t' \in]t_i, t_{i+1}[. \neg d_1(q \triangleright t')$, and $d_1(q \triangleright t_i) = tt$. Then $d_2(q \triangleright t_i) = ff$, since $tp(d_1 \wedge d_2)(q, t) = tt$, and $\forall t' \in]t_{i-1}, t_i[. \neg d_2(q \triangleright t')$, since d_2 is left-closed. It follows that there exists some ϵ such that $\forall t' \in]t_{i-1}, t_i + \epsilon[. \neg d_2(q \triangleright t')$, and $\forall t' \in [t_i + \epsilon, t_{i+1}[. \neg d_1(q \triangleright t')$. So it is sufficient to take $t_i' = t_i + \epsilon$ instead of t_i.

Thus we obtain $\forall i < n \, . \, tp(d_1)(q \triangleright t_i, t_{i+1} - t_i) \vee tp(d_2)(q \triangleright t_i, t_{i+1} - t_i)$ which is equivalent to $(tp(d_1) \sqcup tp(d_2))(q, t)$.

3.2 Translating modal deadline formulas - Application to compositional specification

In this section we present results for the compositional computation of time progress functions when deadline conditions are expressed as modal formulas.

In [SY96] is proposed a variant of timed automata where transitions are labeled with two kinds of conditions : *guards* (enabling conditions) that characterize states from which transitions can be executed and *deadline conditions* that characterize states from which transition execution is enforced by stopping time progress. In general, a deadline condition d depends on the corresponding guard g. To avoid time deadlocks it is necessary that $d \Rightarrow g$; when $d = g$ the transition is *eager* and when $d = false$ there is no constraint on time progress. Timed automata with deadline conditions have been used to show that extending compositionally an untimed (discrete) description into a timed one requires in general the use of modal formulas to express the guards of the composed system in terms of the guards of the components.

Example 4. To illustrate this thesis, consider a discrete (untimed) producer-consumer system with a one-space buffer (figure 4). It is composed of two processes, a producer and a consumer, whose parallel composition is a four state automaton. Suppose that the actions *produce, put, get* and *consume* are submitted to timing constraints expressed respectively with guards $g_1 = 2 \leq x \leq 5$, $g_2 = 1 \leq x \leq 2$, $g_3 = 2 \leq y \leq 4$, $g_4 = 1 \leq y \leq 4$, where x and y are clocks used to measure sojourn times at states of each process (reset at transitions of the associated process). There are at least two different practically interesting choices for the guard of the transition 23.

- For $g_{23} = 1 \leq x \leq 2 \wedge 2 \leq y \leq 4 = g_2 \wedge g_3$ the actions put and get terminate synchronously by respecting the lower and upper bounds of the guards of the components. It is easy to see that this kind of strong synchronization may be the cause of Zeno behavior [HNSY94] in the composed system even though the components are nonZeno.
- For $g_{23} = (1 \leq x \leq 2 \wedge 2 \leq y) \vee (2 \leq y \leq 4 \wedge 1 \leq x)$ a process may wait for his partner. Both lower bounds are respected but only one upper bound. This kind of synchronization with waiting is implicit in timed Petri nets [Sif77, SDdSS94] and can be defined so as to preserve nonZenoness by parallel composition. It is easy to see that expressing g_{23} in terms of g_2 and g_3 requires the use of modal operators : g_{23} is true if one of the two guards has been true and the other is currently true.

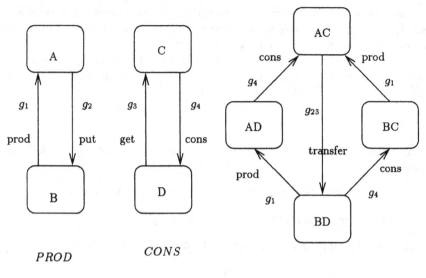

Fig. 4. the producer-consumer example

We assume that the language of the deadline conditions is defined by the syntax :

$d ::= g \mid d \downarrow \mid false$ where,

$g ::= true \mid false \mid c \in C \mid g \wedge g \mid g \cup g \mid \Box g \mid \Diamond g \mid \boxdot g \mid \diamondsuit g.$

C is a set of conditions representing atomic guards.

The following definitions express the semantics of this language as a function $\mid . \mid$ associating with a formula d a predicate $\mid d \mid$ on Q in terms of the meaning of the constants $\mid false \mid = false$, $\mid true \mid = true$ and by taking the meaning $\mid c \mid$ of c to be well-formed and closed predicates on Q.

$$\mid g_1 \wedge g_2 \mid = \mid g_1 \mid \wedge \mid g_2 \mid$$
$$\mid g_1 \vee g_2 \mid = \mid g_1 \mid \vee \mid g_2 \mid$$
$$\mid \Diamond g \mid (q) = \exists t \geq 0 . \mid g \mid (q \triangleright t)$$
$$\mid \Box g \mid (q) = \forall t \geq 0 . \mid g \mid (q \triangleright t)$$
$$\mid \diamondsuit g \mid (q) = \exists t \geq 0 . \exists q' . q = q' \triangleright t \wedge \mid g(q') \mid$$
$$\mid \boxdot g \mid (q) = \forall t \geq 0 . \forall q' . q = q' \triangleright t \Rightarrow \mid g(q') \mid$$
$$\mid d \downarrow \mid (q) = g(q) \wedge \exists t \geq 0 . \forall t' \leq t . \mid g \mid (q \triangleright t')$$

Notice that $\Box, \Diamond, \boxminus, \diamondsuit$ correspond to well-known modalities of temporal logic [MP91] meaning respectively always, eventually, always in the past, and once in the past. The operator $\downarrow$ is a falling edge operator.

We did not consider negation in order to preserve the property of closeness. However, we use in the sequel negated formulas with the usual meaning. This implies the following relations : $\neg true = false$, $\neg false = true$, $\neg(g_1 \wedge g_2) = \neg(g_1) \vee \neg(g_2)$, $\neg(g_1 \vee g_2) = \neg(g_1) \wedge \neg(g_2)$, $\neg \Box g = \Diamond \neg g$, $\neg \Diamond g = \Box \neg g$, $\neg \boxminus = \diamondsuit \neg g$ and $\neg \diamondsuit = \boxminus \neg g$.

Proposition 7. *Any deadline can be expressed as a formula of the following language :*

$$X ::= g \mid (g \vee g) \downarrow \mid (\Diamond g) \downarrow \mid (\boxminus g) \downarrow \mid c \downarrow \mid X \wedge X \mid X \vee X$$

In order to prove this we will need the following lemma :

Lemma 8. *For all guards g, g_1 and g_2, the following relations hold :*

$$true \downarrow = false = false \downarrow$$

$$(g_1 \wedge g_2) \downarrow = (g_1 \downarrow \wedge g_2) \vee (g_1 \wedge g_2 \downarrow)$$

$$(g_1 \vee g_2) \downarrow = (g_1 \downarrow \wedge (\neg g_2 \vee g_2 \downarrow)) \vee ((\neg g_1 \vee g_1 \downarrow) \wedge g_2 \downarrow)$$

$$(\Box g) \downarrow = false = (\diamondsuit g) \downarrow$$

Proof. We have $g \downarrow (q) = g(q) \wedge \exists t > 0 . \forall t'\ 0 < t' \le t . \neg g(q \triangleright t')$. So it is clear that $true \downarrow = false = false \downarrow = (\Box g) \downarrow = (\diamondsuit g) \downarrow$.

For the other cases we have :

- $(g_1 \wedge g_2) \downarrow$: the falling edges of $g_1 \wedge g_2$ are the falling edges of one of the guards while the other is true. $(g_1 \wedge g_2) \downarrow = (g_1 \downarrow \wedge g_2) \vee (g_1 \wedge g_2 \downarrow)$.
- $(g_1 \vee g_2) \downarrow$: the falling edges of $g_1 \vee g_2$ are the falling edges common to g_1 and g_2 and the falling edges of one guard when the other is false. $(g_1 \vee g_2) \downarrow = (g_1 \downarrow \wedge (\neg g_2 \vee g_2 \downarrow)) \vee ((\neg g_1 \vee g_1 \downarrow) \wedge g_2 \downarrow)$.

Proof. of proposition 7 : trivial if the deadline is not of the form $g \downarrow$; otherwise by induction on the structure of g.

Theorem 9. *For any deadline formula d, $tp(d)$ can be expressed as a formula of the following language :*

$$Y ::= \neg g(q) \vee (t = 0) \mid g(q) \vee t = 0 \mid \neg g(q \triangleright t) \vee (t = 0) \mid tp(c) \mid tp(\neg c) \mid Y \wedge Y \mid Y \sqcup Y$$

Sketch of proof : By induction on the structure of d :

- $tp(d_1 \wedge d_2) = tp(d_1) \sqcup tp(d_2)$
- $tp(d_1 \vee d_2) = tp(d_1) \wedge tp(d_2)$
- $tp(g)$: by induction on the structure of g :
  - $tp(true)$, $tp(false)$, $tp(c)$, $tp(g_1 \wedge g_2)$, $tp(g_1 \vee g_2)$ are easily reduced.

- If $\Diamond g$ or $\boxminus g$ are false at a state q, they remain false forever. Thus, it is sufficient to test their value at the current state q to know if time can pass : $tp(\Diamond g) = \neg \Diamond g(q) \vee (t = 0)$ and $tp(\boxminus g) = \neg \boxminus g(q) \vee (t = 0)$.

- If $\Box g$ or $\Diamond g$ are true at a state q they remain true forever. Following a similar reasoning as before one can prove :
$tp(\Box g) = (\neg \Box g(q \triangleright t) \vee t = 0) \sqcup tp(\neg g)$ and
$tp(\Diamond g) = (\neg \Diamond g(q \triangleright t) \vee t = 0) \sqcup tp(\neg g)$.

- $tp((g_1 \vee g_2) \downarrow)$: from the previous lemma we know that
$(g_1 \vee g_2) \downarrow = (g_1 \downarrow \wedge (\neg g_2 \vee g_2 \downarrow)) \vee ((\neg g_1 \vee g_1 \downarrow) \wedge g_2 \downarrow)$.
It is easy to check that $\neg g \vee g \downarrow$ is well-formed if g is well-formed and closed.
Then we can reduce $tp((g_1 \vee g_2) \downarrow)$ to
$[tp(g_1 \downarrow) \sqcup (tp(\neg g_2) \wedge tp(g_2 \downarrow))] \wedge [tp(g_2 \downarrow) \sqcup (tp(\neg g_1) \wedge tp(g_1 \downarrow))]$.
As $tp(\neg g) \Rightarrow tp(g \downarrow)$ we obtain :
$tp((g_1 \vee g_2) \downarrow) = (tp(g_1 \downarrow) \sqcup tp(\neg g_2)) \wedge (tp(g_2 \downarrow) \sqcup tp(\neg g_1))$.

- $tp(\neg g)$: the following reduction rules can be proven :

$$tp(\neg(g_1 \vee g_2)) = tp(\neg g_1) \sqcup tp(\neg g_2),$$
$$tp(\neg \Box g) = (\Box g)(q) \vee t = 0,$$
$$tp(\neg \Diamond g) = (\Diamond g)(q \triangleright t) \vee t = 0,$$
$$tp(\neg \boxminus g) = (\boxminus g)(q \triangleright t) \vee t = 0 \text{ and}$$
$$tp(\neg \Diamond g) = (\Diamond g)(q) \vee t = 0.$$

- $tp(c \downarrow) = tp(c) \sqcup tp(\neg c)$. This equivalence is illustrated for an example in figure 5. Consider q, t_1, t_2 as in figure 5. We have $tp(c)(q, t_1)$ and $tp(\neg c)(q \triangleright t_1, t_2 - t_1)$. Thus, $(tp(c) \sqcup tp(\neg c))(q, t_2)$ is true as is $tp(c \downarrow)(q, t_2)$. But for any $t > 0$ we have $\neg tp(c)(q \triangleright t_2, t)$ and $\neg tp(\neg c)(q \triangleright t_2, t)$. Thus $(tp(c) \sqcup tp(\neg c))(q, t_2 + t)$ is false as is $tp(c \downarrow)(q, t_2 + t)$.

- $tp((\Diamond g) \downarrow) = tp(\Diamond g) \sqcup tp(\neg \Diamond g) = (\neg \Diamond g(q) \vee t = 0) \sqcup (\Diamond g(q \triangleright t) \vee (t = 0))$

- $tp((\boxminus g) \downarrow) = tp(\boxminus g) \sqcup tp(\neg \boxminus g) = (\neg \boxminus g(q) \vee t = 0) \sqcup (\boxminus g(q \triangleright t) \vee (t = 0))$

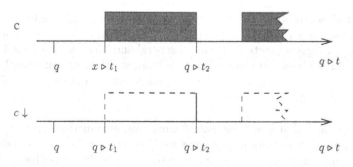

Fig. 5. $tp(c \downarrow)$

Example 4 (continued) Consider the consumer-producer example.

- If $g_{23} = g_2 \wedge g_3$ then one can take the corresponding deadline condition d_{23} :
  - either $d_{23} = g_{23}$ (eager transition), in which case,
    $tp(d_{23}) = tp(g_2) \sqcup tp(g_3)$.
  - or $d_{23} = g_{23} \downarrow = (g_2 \downarrow \wedge g_3) \vee (g_2 \wedge g_3 \downarrow)$ (delayable transition) which means that the time progress function is :

$$
\begin{aligned}
tp(d_{23}) &= tp((g_2 \downarrow \wedge g_3) \vee (g_2 \wedge g_3 \downarrow)) \\
&= tp(g_2 \downarrow \wedge g_3) \wedge tp(g_2 \wedge g_3 \downarrow) \\
&= (tp(g_2 \downarrow) \sqcup tp(g_3)) \wedge (tp(g_2) \sqcup tp(g_3 \downarrow)) \\
&= (tp(g_2) \sqcup tp(\neg g_2) \sqcup tp(g_3)) \wedge (tp(g_2) \sqcup tp(\neg g_3) \sqcup tp(g_3)) \\
&= tp(g_2) \sqcup tp(g_3) \sqcup (tp(\neg g_2) \wedge tp(\neg g_3))
\end{aligned}
$$

- If $g_{23} = (1 \leq x \leq 2 \wedge 2 \leq y) \vee (2 \leq y \leq 4 \wedge 1 \leq x) = (g_2 \wedge \Diamond g_3) \vee (\Diamond g_2 \wedge g_3)$ we have the case of synchronization with mutual waiting. One can take as deadline condition d_{23} :
  - either $d_{23} = g_{23}$ (eager transition) in which case

$$
\begin{aligned}
tp(d_{23}) &= tp((g_2 \wedge \Diamond g_3) \vee (\Diamond g_2 \wedge g_3)) \\
&= tp(g_2 \wedge \Diamond g_3) \wedge tp(\Diamond g_2 \wedge g_3) \\
&= (tp(g_2) \sqcup tp(\Diamond g_3)) \wedge (tp(\Diamond g_2) \sqcup tp(g_3)) \\
&= (tp(g_2) \wedge tp(\Diamond g_2)) \sqcup (tp(g_2) \wedge tp(g_3)) \sqcup \\
&\quad (tp(\Diamond g_3) \wedge tp(\Diamond g_2)) \sqcup (tp(\Diamond g_3) \wedge tp(g_3)) \\
&= tp(\Diamond g_2) \sqcup (tp(g_2) \wedge tp(g_3)) \sqcup (tp(\Diamond g_3) \wedge tp(\Diamond g_2)) \sqcup tp(\Diamond g_3) \\
&= tp(\Diamond g_2) \sqcup tp(\Diamond g_3) \sqcup (tp(g_2) \wedge tp(g_3))
\end{aligned}
$$

This can be simplified furthermore, by reducing $tp(\Diamond g_2)$ and $tp(\Diamond g_3)$.
  - or $d_{23} = g_{23} \downarrow$ (delayable transition). The reader can verify

$$
\begin{aligned}
tp(d_{23}) &= (tp(\neg g_1) \sqcup tp(\neg g_2) \sqcup tp(g_1) \sqcup tp(\Diamond g_2)) \\
&\quad \wedge (tp(\neg g_2) \sqcup tp(\neg g_1) \sqcup tp(g_2) \sqcup tp(\Diamond g_1))
\end{aligned}
$$

4 Discussion

The paper studies relationships between progress functions and deadline conditions or invariants used in hybrid systems to specify when continuous evolution can take place. Progress functions are more general and their explicit knowledge is important for analysis and simulation. Deadline conditions or equivalently invariants, are easier to specify as they express constraints on the states without explicitly mentioning time.

The results show that any "reasonable" time progress function can be generated by using deadline conditions or invariants. However, for this correspondence to be a homomorphism, it is necessary to restrict to deadline conditions with finite variability. In this case and under some closeness conditions, it is possible to compute compositionally progress functions corresponding to deadline conditions that are formulas with conjunction, disjunction and modal operators.

Apart from their theoretical interest, the results can find an application in a framework for the compositional specification of hybrid systems, currently under study.

Acknowledgement: We thank Sergio Yovine and Oded Maler for constructive critiques of the ideas developed in the paper.

References

[ACH+95] R. Alur, C. Courcoubetis, N. Halbwachs, T. Henzinger, P. Ho, X. Nicollin, A. Olivero, J. Sifakis, and S. Yovine. The algorithmic analysis of hybrid systems. *Theoretical Computer Science*, 138:3–34, 1995.

[HNSY94] T.A. Henzinger, X. Nicollin, J. Sifakis, and S. Yovine. Symbolic model checking for real-time systems. *Information and Computation*, 111(2):193–244, 1994.

[MP91] Z. Manna and A. Pnueli. *The Temporal Logic of Reactive and Concurrent Systems: Specification*. Springer-Verlag, New York, 1991.

[Ore44] O. Ore. Galois connections. *Trans. Amer. Math. Society*, 55:493–513, February 1944.

[San77] L.E. Sanchis. Data types as lattices : retractions, closures and projections. *RAIRO, Theoretical Computer Science*, 11, no 4:339–344, 1977.

[SDdSS94] P. Sénac, M. Diaz, and P. de Saqui-Sannes. Toward a formal specification of multimedia scenarios. *Annals of telecomunications*, 49(5-6):297–314, 1994.

[Sif77] J. Sifakis. Use of petri nets for performance evaluation. In H. Beilner and E. Gelenebe, editors, *Measuring, modelling and evaluating computer systems*, pages 75–93. North-Holland, 1977.

[SY96] J. Sifakis and S. Yovine. Compositional specification of timed systems. In *13th Annual Symposium on Theoretical Aspects of Computer Science, STACS'96*, pages 347–359, Grenoble, France, February 1996. Lecture Notes in Computer Science 1046, Spinger-Verlag.

Semantics and Verification of Extended Phase Transition Systems in Duration Calculus

Xu Qiwen*
International Institute for Software Technology
The United Nations University
P.O.Box 3058, Macau

Abstract. Hybrid systems contain both discrete and continuous components, and therefore it is desirable to have a method which decomposes the verification of a hybrid system into verification of its continuous and discrete components. In this paper, the system we consider may have non-trivial discrete components, such as computer algorithms with iteration structures. We take the view that a single discrete action is instantaneous, and a finite number of discrete actions happening at one time point is still instantaneous, but infinitely many discrete actions occurring at one time leads to losing control of the continuous components and this should in general be avoided in a design. We develop a variant of Duration Calculus in which we embed Hoare Logic used to verify discrete components. Several other rules are proposed in the calculus for proving temporal properties. The paper concludes with a simple example.

1 Introduction

Many practical systems are hybrid in that they contain both discrete and continuous components. At the semantical level, they are usually modelled by piecewise continuous functions [10]. Continuous evolution is accumulated over time, whereas discrete actions are usually considered instantaneous, since the durations of the latter are many order lower than the former. Such abstraction provides essential convenience in reasoning about hybrid systems.

A number of system models have been proposed for hybrid systems, along the lines of Manna & Pnueli's Phase Transition Systems [7]. In most of these models, discrete actions are considered atomic. To support the design of discrete controllers, we need to refine discrete actions into programs which not only are non-atomic but may also contain some forms of iterations. Although it is reasonable to consider a finite number of discrete actions happening at one time point instantaneous, infinitely many discrete actions occurring at one time, caused by for example the control program entering an infinite loop or the system being switched repeatedly between several control laws without spending any time in

---

* Email: qxu@iist.unu.edu

any phase, leads to losing control of the continuous components. We call this situation divergence and it should in general be avoided in a design.

As expected, it is desirable to have a method which decomposes the verification of a hybrid system into verification of its continuous and discrete components, and for the discrete component uses the good and old verification method for traditional programs. Therefore, when discrete segments are modelled as sequential programs, it seems the easiest to use Hoare Logic, which has proven to be effective for reasoning about sequential programs, to verify them. Of course, the results must be incorporated into a real time logic in which the continuous timing behaviours are reasoned.

It is clear that we need a logic capable of both handling super dense (i.e, containing several instantaneous actions at one time point) and infinite behaviours. In this paper, we choose a variant of Duration Calculus [2] as the real time logic. Based on the Interval Temporal Logic [8], Duration Calculus has been used in the specification and verification of real time systems, and a number of extensions have been proposed recently, including the Infinite Duration Calculus [3] and the Super Dense Duration Calculus [1]. Our calculus is basically an integration of the two calculi.

This paper is organised as follows. In section 2, we give a reformulation of the Super Dense Duration Calculus, in which program semantics and logic can be directly embedded. Section 3 extends the Super Dense Duration Calculus to infinite intervals. Based on Manna & Pnueli's Phase Transition System model of hybrid systems, we propose a language in which the usual programming notations and durational formulas can be mixed. Within this language, systems can be specified at a high level, and refined step by step into implementations. The syntax and semantics are defined in section 4. Hoare Logic for total correctness is embedded in section 5, and rules for proving temporal properties are studied in section 6. The paper concludes with a simple example in section 7 and a brief discussion in section 8.

2   Super Dense Duration Calculus

The traditional model underlying the Duration Calculus is a function mapping each time point to a state which gives valuation for various variables. To describe an instantaneous transition which involves two states, the usual approach in Duration Calculus is to use states in the left and right neighbourhood intervals of the time point. This works quite smoothly when there is only one transition at a time, since the states in the left and right neighbourhood intervals of it are stable and represent the states before and after the transition. To handle super dense computations which contain several instantaneous transitions at one time point, Zhou and Hansen proposed to adopt a two level time space and a time point in the grand time space can be enlarged to an interval in a finer time space. A calculus is developed in this way [1]. Although it has been used successfully to reason about super dense computations, it does not relate directly to the usual semantics of sequential programs.

In this paper, we suggest an alternative formulation which allows the program semantics to be directly embedded in Duration Calculus. Instead of mapping each time point to one state, we assign two states to it: one before and one after the transition (if there is a discrete transition at that). For uniformity, we still associate two states with a time point when there is not a discrete transition, and the two states are simply defined to be identical. Formally, an interpretation Π is a mapping

$$\Pi : \text{Time} \rightarrow \text{State} \times \text{State}$$

For a time point t, $\Pi(t)_1$ and $\Pi(t)_2$ denote the first and second states of $\Pi(t)$. In the following diagram, the upper and lower curves represent respectively $\Pi(t)_2$ and $\Pi(t)_1$ as functions of time. At continuous points, $\Pi(t)_2$ and $\Pi(t)_1$ are the same, but we draw them slightly apart for visual convenience.

Interpretation

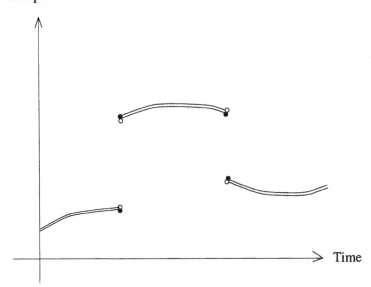

Fig. 1. Timing diagram of valuations

The chop operator is defined as follows: for any durational formulas A and B

$\Pi, [c, d] \models A \bullet B$
   iff $\exists c \leq m \leq d, \Pi', \Pi''$, such that
      $\Pi'(t) = \Pi(t)$ if $c \leq t < m$, and $\Pi'(m)_1 = \Pi(m)_1$
      $\wedge \Pi''(t) = \Pi(t)$ if $m < t \leq d$, and $\Pi''(m)_2 = \Pi(m)_2$
      $\wedge \Pi'(m)_2 = \Pi''(m)_1$
      $\wedge \Pi', [c, m] \models A$ and $\Pi'', [m, d] \models B$

Pre- and post-conditions of a transition can be specified as follows. For any interpretation Π, interval $[c, d]$ and state assertion r

$$\Pi, [c, d] \models \nwarrow r \qquad \text{iff} \qquad c = d \text{ and } \Pi(c)_1 \models r$$

$$\Pi, [c, d] \models \nearrow r \qquad \text{iff} \qquad c = d \text{ and } \Pi(c)_2 \models r$$

Super dense chop has the following properties

$$(\nwarrow r_1 \wedge \nearrow r_2) \bullet (\nwarrow r_3 \wedge \nearrow r_4) = \nwarrow r_1 \wedge \nearrow r_4 \qquad \text{if } r_2 \wedge r_3 \text{ is satisfiable}$$
$$(\nwarrow r_1 \wedge \nearrow r_2) \bullet (\nwarrow r_3 \wedge \nearrow r_4) = \text{false} \qquad \text{if } r_2 \wedge r_3 \text{ is not satisfiable}$$

An important concept in Duration Calculus is the duration that an assertion [2] over physical variables holds in the interval. Although in our framework, two states are associated with a time point, under the usual finite variability condition (which says that assertions are continuous almost everywhere, i.e., everywhere except for a finite number of points), the values of the assertion are the same over the two states almost everywhere. Therefore, for calculating the durations, we can choose any of the two states to interpret the assertion, and as a simple solution, we use the first state: the duration of r over an interval $[c, d]$ is

$$\int_c^d r(\Pi(t)_1) dt$$

As usual, the length of an interval is denoted by l, and r holds almost everywhere over a non-empty interval is defined

$$\lceil r \rceil \stackrel{\text{def}}{=} (\int r = l) \wedge l > 0$$

A point interval is characterised by $l = 0$, also denoted by $\lceil \rceil$; an interval is either a point or r holds almost everywhere over it is defined as

$$\lceil r \rceil^* \stackrel{\text{def}}{=} \lceil \rceil \vee \lceil r \rceil$$

As in the Extended Duration Calculus [4], $b.z$ and $e.z$ refer to the values of a term z at the beginning and end of interval $[c, d]$, but instead of the usual definition via limits, we can simply define them as

$$z(\Pi(c)_2) \qquad \text{and} \qquad z(\Pi(d)_1)$$

The usual axioms and rules of Interval Temporal Logic and Duration Calculi are all valid. Moreover, the following MT rule of the Extended Duration Calculus allows mathematical theories to be imported into the calculus:

Let $R(\lceil H(z) \rceil, b.z, e.z, l)$ be a Duration Calculus formula without chop.
If in mathematics $\forall c < d.R(\forall c \leq t \leq d.H(z(t)), z(c^+), z(d^-), d - c)$ then
$\lceil \rceil \vee R(\lceil H(z) \rceil, b.z, e.z, l)$

where $z(c^+)$ and $z(d^-)$ denote limits of z from the right of c and left of d respectively.

---

[2] In Duration Calculus literature, it is called a state, but in this paper, we use state to denote a mapping from variables to values as in traditional theory of programming.

3 Infinite Duration Calculus

In [3], Zhou, Hung and Li proposed a two level syntax approach to Duration Calculus with infinite intervals. A formula in the Infinite Duration Calculus is satisfied over an infinite interval, if and only if the formula is satisfied over all the finite prefixes of it in Duration Calculus. In this approach, calculating length of infinite intervals is avoided.

Formally, for any formula A of Duration Calculus, A^{fin} and A^{inf} are formulae of infinite Duration Calculus, and

$$\Pi, [c, d] \models A^{\text{fin}} \quad \text{iff} \quad d < \infty \text{ and } \Pi, [c, d] \models A$$
$$\Pi, [c, d] \models A^{\text{inf}} \quad \text{iff} \quad d = \infty \text{ and } \Pi, [c, h] \models A \quad \text{for any } h \geq c$$

Denote true^{fin} and true^{inf} by fin and inf and let

$$\text{True} \stackrel{\text{def}}{=} \text{fin} \vee \text{inf}$$
$$\text{False} \stackrel{\text{def}}{=} \neg \text{True}$$

As abbreviations, we define

$$A^{\text{fin}} \bullet B^{\text{fin}} \stackrel{\text{def}}{=} (A \bullet B)^{\text{fin}}$$
$$(A^{\text{fin}})^{\text{fin}} \stackrel{\text{def}}{=} A^{\text{fin}}$$
$$(A^{\text{fin}})^{\text{inf}} \stackrel{\text{def}}{=} A^{\text{inf}}$$
$$A^{\text{inf}} \bullet B \stackrel{\text{def}}{=} A^{\text{inf}}$$
$$A^{\text{fin}} \bullet B^{\text{inf}} \stackrel{\text{def}}{=} \exists x.(l < x \vee ((A \wedge l = x) \bullet B))^{\text{inf}}$$

From the definition of the last formula, we can see that as a price for avoiding calculation over infinity, some simple properties will have to be expressed in a more involved way. To simplify notations, we write A for A^{fin} when $A \Rightarrow \lceil \rceil$, and subsequently, $\nwarrow r$ and $\nearrow r$ also stand for $(\nwarrow r)^{\text{fin}}$ and $(\nearrow r)^{\text{fin}}$ respectively.

We have the following theorems

$$(A \bullet B)^{\text{inf}} = A \bullet B^{\text{inf}} \quad \text{if } A \Rightarrow \lceil \rceil$$
$$\text{false}^{\text{inf}} = \text{False}$$
$$(l < x)^{\text{inf}} = \text{False}$$
$$(\lceil q \rceil^*)^{\text{inf}} = \forall x \exists y.y \geq x \wedge (l < y \vee ((\lceil q \rceil^* \wedge l = y) \bullet \text{true}))^{\text{inf}}$$

4 Syntax and Semantics of Extended Phase Transition Systems

4.1 Syntax

We select a small set of basic commands

$$P ::= \text{skip} \mid \langle b \rangle \mid \bar{x} := \bar{e} \mid \ll C \gg \mid P_1;P_2 \mid P_1 \; [] \; P_2 \mid \text{while } b \text{ do } P \text{ od}$$

The interpretation of most statements is standard as in sequential programming: the statement skip does not change any variables and terminates instantly; $\langle b \rangle$ is an "assertion" statement which does nothing if b holds otherwise it behaves like a magic satisfying any requirement; $\bar{x} := \bar{e}$ is the usual assignment and is considered instantaneous; sequential composition is represented by $P_1; P_2$ and its meaning is that if P_1 terminates then P_2 is executed immediately afterwards; $[]$ is the nondeterministic choice. A phase in which continuous evolution occurs is denoted by $\ll C \gg$ where C is a durational formula specifying the dynamical laws that the continuous change follows. The boolean test of the iteration statement does not take time, but its body may and may not. If the body does not take time and the loop terminates after a finite number of iterations, the whole structure is considered instantaneous. If the body does not take time but the loop does not terminated, the whole structure is considered to be infinite.

Other commands can be defined from the basic ones. For example, the delay statement can be defined as

$$\text{delay } t \stackrel{\text{def}}{=} \ll l = t \gg$$

and the alternative command as

$$\text{if } b_1 \to P_1 \; [] \; \cdots \; [] \; b_n \to P_n \text{ fi} \stackrel{\text{def}}{=} \langle b_1 \rangle; P_1 \; [] \; \cdots \; [] \; \langle b_n \rangle; P_n$$

4.2 Semantics

The semantics of most statements are straightforward

$$[\![\text{skip}]\!] \stackrel{\text{def}}{=} \exists \bar{v}.(\nwarrow (\bar{x} = \bar{v}) \wedge \nearrow (\bar{x} = \bar{v}))$$
$$[\![\langle b \rangle]\!] \stackrel{\text{def}}{=} \exists \bar{v}.(b(\bar{v}) \wedge \nwarrow (\bar{x} = \bar{v}) \wedge \nearrow (\bar{x} = \bar{v}))$$
$$[\![\bar{x} := \bar{e}]\!] \stackrel{\text{def}}{=} \exists \bar{v}.(\nwarrow (\bar{x} = \bar{v}) \wedge \nearrow (\bar{x} = \bar{e}(\bar{v})))$$
$$[\![\ll C \gg]\!] \stackrel{\text{def}}{=} \exists \bar{u}, \bar{v}.(\nwarrow (\bar{x} = \bar{u}) \bullet (b.\bar{x} = \bar{u} \wedge C \wedge e.\bar{x} = \bar{v}) \bullet \nearrow (\bar{x} = \bar{v}))$$
$$[\![P_1; P_2]\!] \stackrel{\text{def}}{=} ((\text{fin} \wedge [\![P_1]\!]) \bullet [\![P_2]\!]) \vee (\text{inf} \wedge [\![P_1]\!])$$
$$[\![P_1 \; [] \; P_2]\!] \stackrel{\text{def}}{=} [\![P_1]\!] \vee [\![P_2]\!]$$

As an example, we prove $[\![x := x + 1; x := x + 2]\!] = [\![x := x + 3]\!]$

$$[\![x := x + 1; x := x + 2]\!]$$
$$= ((\text{fin} \wedge [\![x := x + 1]\!]) \bullet [\![x := x + 2]\!]) \vee (\text{inf} \wedge [\![x := x + 1]\!])$$
$$= [\![x := x + 1]\!] \bullet [\![x := x + 2]\!]$$
$$= (\exists v. \nwarrow (x = v) \wedge \nearrow (x = v + 1)) \bullet (\exists v. \nwarrow (x = v) \wedge \nearrow (x = v + 2))$$
$$= \exists v_1, v_2.(\nwarrow (x = v_1) \wedge \nearrow (x = v_1 + 1)) \bullet (\nwarrow (x = v_2) \wedge \nearrow (x = v_2 + 2))$$
$$= \exists v_1, v_2.(v_1 + 1 = v_2) \wedge \nwarrow (x = v_1) \wedge \nearrow (x = v_2 + 2)$$
$$= \exists v. \nwarrow (x = v) \wedge \nearrow (x = v + 3)$$
$$= [\![x := x + 3]\!]$$

The semantics of the iteration statement is somewhat complicated. Let $W \stackrel{\text{def}}{=}$ while b do P od, and

$$[\![W]\!](0) \stackrel{\text{def}}{=} [\![\text{skip}]\!]$$
$$[\![W]\!](i + 1) \stackrel{\text{def}}{=} [\![\langle b \rangle]\!] \bullet [\![P]\!] \bullet [\![W]\!](i)$$

The semantics of the statement is defined as

$$[\![W]\!] \stackrel{\text{def}}{=} (\exists i.[\![W]\!](i) \bullet [\![\langle\neg b\rangle]\!])^{\text{fin}}$$
$$\vee(\exists i.[\![W]\!](i)) \wedge \text{inf}$$
$$\vee\forall x \exists y, i.(y \geq x \wedge (l < y \vee (([\![W]\!](i) \wedge l = y) \bullet \text{true}))^{\text{inf}})$$
$$\vee((\exists i.[\![W]\!](i))^{\text{fin}} \bullet (\forall i.([\![W]\!](i) \wedge l = 0) \bullet \text{true})^{\text{inf}})$$

The four disjunctives model respectively the cases that, (1) the statement terminates, (2) there are finitely many iterations and the last one does not terminate, (3) there are infinitely many iterations and time is not bounded, (4) the statement repeats forever and no time is spent in any iteration after certain point. The last case arises when the program enters an infinite loop, or the system is switched repeatedly among several phases without spending any time in any of them. We consider this as a undesirable situation since the system is not subject to any control, and following the divergence semantics of CSP [5], any behaviours after that is considered to be possible. We have not considered the so-called zeno behaviour in which the statement repeats forever and although each iteration takes a positive amount of time, the total execution time is bounded.

As another example, we show $[\![W]\!] = ([\![\text{skip}]\!]\bullet\text{true})^{\text{inf}}$ where $W =$ while true do skip od.

$$[\![W]\!](0) = [\![\text{skip}]\!]$$
$$[\![W]\!](1) = [\![\langle\text{true}\rangle]\!] \bullet [\![\text{skip}]\!] \bullet [\![\text{skip}]\!]$$
$$= [\![\text{skip}]\!]$$

. . .

$$[\![W]\!](i) = [\![\text{skip}]\!]$$

Therefore, $[\![W]\!](i) \bullet [\![\neg\text{true}]\!] = \text{false}$ and $([\![W]\!](i) \wedge l = y) = \text{false}$ if $y > 0$.

$$[\![W]\!] = (\forall x \exists y, i.y \geq x \wedge (l < y)^{\text{inf}}) \vee ([\![\text{skip}]\!] \bullet ([\![\text{skip}]\!] \bullet \text{true})^{\text{inf}})$$
$$= ([\![\text{skip}]\!] \bullet [\![\text{skip}]\!] \bullet \text{true})^{\text{inf}}$$
$$= ([\![\text{skip}]\!] \bullet \text{true})^{\text{inf}}$$

4.3 Correctness Formulae

To effectively reason about phase transition systems, we would like to use the familiar notations of Hoare logic. Define

$\{p\} \; P \; \{q\}$	iff	$\nearrow p \bullet [\![P]\!] \Rightarrow \text{fin} \bullet \nearrow q$
$\{p\} \; P \; A$	iff	$\nearrow p \bullet [\![P]\!] \Rightarrow A$

where p and q are state assertions, and A is a durational formula.

5 Verification of Discrete Components

In this section, we investigate the verification of discrete components. Any discrete components without iteration are assumed to be instantaneous and we use

D, D_1 and D_2 to denote them. If an iteration terminates, we assume it is also instantaneous, but if it does not terminate, then it is divergent.

For discrete components, the Hoare triple $\{p\}\ D\ \{q\}$ of total correctness is equivalent to

$$\nearrow p \bullet [\,D\,] \Rightarrow \nearrow q$$

The proof system for sequential discrete programs can be considered as abbreviations of the Duration Calculus theorems.

$$\{p[\bar{e}/\bar{x}]\}\ \bar{x} := \bar{e}\ \{p\}$$
$$\nearrow p[\bar{e}/\bar{x}] \bullet (\exists \bar{v}.(\nwarrow (\bar{x} = \bar{v}) \wedge \nearrow (\bar{x} = \bar{e}(\bar{v})))) \Rightarrow \nearrow p(\bar{x})$$

$$\{p\}\ \langle b \rangle\ \{p \wedge b\} \qquad\qquad \nearrow p \bullet [\,\langle b \rangle\,] \Rightarrow \nearrow (p \wedge b)$$

$$\frac{\{p\}\ D_1\ \{r\} \qquad \{r\}\ D_2\ \{q\}}{\{p\}\ D_1; D_2\ \{q\}}$$
$$\frac{\nearrow p \bullet [\,D_1\,] \Rightarrow \nearrow r \qquad \nearrow r \bullet [\,D_2\,] \Rightarrow \nearrow q}{\nearrow p \bullet [\,D_1; D_2\,] \Rightarrow \nearrow q}$$

$$\frac{\{p\}\ D_1\ \{q\} \qquad \{p\}\ D_2\ \{q\}}{\{p\}\ D_1\ [\!]\ D_2\ \{q\}}$$
$$\frac{\nearrow p \bullet [\,D_1\,] \Rightarrow \nearrow q \qquad \nearrow p \bullet [\,D_2\,] \Rightarrow \nearrow q}{\nearrow p \bullet [\,D_1\ [\!]\ D_2\,] \Rightarrow \nearrow q}$$

$$\frac{r(\alpha) \wedge \alpha > 0 \Rightarrow b \qquad r(0) \Rightarrow \neg b}{\{r(\alpha) \wedge \alpha > 0\}\ D\ \{\exists \beta < \alpha.r(\beta)\}}$$
$$\overline{\{\exists \alpha.r(\alpha)\}\ \text{while } b \text{ do } D \text{ od}\ \{r(0)\}}$$
$$\frac{r(\alpha) \wedge \alpha > 0 \Rightarrow b \qquad r(0) \Rightarrow \neg b}{\nearrow (r(\alpha) \wedge \alpha > 0) \bullet [\,D\,] \Rightarrow \nearrow (\exists \beta < \alpha.r(\beta))}$$
$$\overline{\nearrow (\exists \alpha.r(\alpha)) \bullet [\,\text{while } b \text{ do } D \text{ od}\,] \Rightarrow \nearrow r(0)}$$

$$\frac{p \Rightarrow p_1 \quad \{p_1\}\ D\ \{q_1\} \quad q_1 \Rightarrow q}{\{p\}\ D\ \{q\}}$$
$$\frac{p \Rightarrow p_1 \quad \nearrow p_1 \bullet [\,D\,] \Rightarrow \nearrow q_1 \quad q_1 \Rightarrow q}{\nearrow p \bullet [\,D\,] \Rightarrow \nearrow q}$$

The soundness of the rules follow from the correctness of the corresponding theorems, of which only the one concerning iteration is a little involved.

Lemma 1 *Given the assumptions in the iteration theorem, it is true that*

$$\nearrow (\exists \alpha.r(\alpha)) \bullet [\,W\,](i) \Rightarrow \nearrow (\exists \alpha.r(\alpha))$$

Proof: by induction on i.
Base $i = 0$:

$$\nearrow (\exists \alpha.r(\alpha)) \bullet [\,\text{skip}\,]$$
$$\Rightarrow \nearrow (\exists \alpha.r(\alpha))$$

Induction step: assume $\nearrow (\exists \alpha.r(\alpha)) \bullet [\![W]\!](i) \Rightarrow \nearrow (\exists \alpha.r(\alpha))$, then

$$\nearrow (\exists \alpha.r(\alpha)) \bullet [\![W]\!](i+1)$$
$$\Rightarrow \nearrow (\exists \alpha.r(\alpha)) \bullet [\, \langle b \rangle \,] \bullet [\, D \,] \bullet [\![W]\!](i)$$
$$\Rightarrow \nearrow (\exists \alpha.r(\alpha) \wedge \alpha = 0) \bullet [\, \langle b \rangle \,] \bullet [\, D \,] \bullet [\![W]\!](i)$$
$$\vee \nearrow (\exists \alpha.r(\alpha) \wedge \alpha > 0) \bullet [\, \langle b \rangle \,] \bullet [\, D \,] \bullet [\![W]\!](i)$$
$$\Rightarrow \nearrow (\exists \alpha.r(\alpha)) \bullet [\![W]\!](i)$$
$$\Rightarrow \nearrow (\exists \alpha.r(\alpha))$$

Lemma 2 *Given the assumptions in the iteration theorem, it is true that*

$$\nearrow (\exists \alpha.r(\alpha)) \bullet [\![W]\!](i) \bullet [\, \langle \neg b \rangle \,] \Rightarrow \nearrow r(0)$$
$$\nearrow (\exists \alpha.r(\alpha)) \bullet (\exists i.[\![W]\!](i))^{\mathsf{fin}} \bullet (\forall i.([\![W]\!](i) \wedge l = 0) \bullet \mathsf{true}))^{\mathsf{inf}} = \mathsf{False}$$

Proof: The first one follows immediately from lemma 1

$$\nearrow (\exists \alpha.r(\alpha)) \bullet [\![W]\!](i) \bullet [\, \langle \neg b \rangle \,]$$
$$\Rightarrow \nearrow (\exists \alpha.r(\alpha)) \bullet [\, \langle \neg b \rangle \,]$$
$$\Rightarrow \nearrow (\exists \alpha.r(\alpha) \wedge \neg b)$$
$$\Rightarrow \nearrow r(0)$$

For the second one, it follows from lemma 1 that

$$\nearrow (\exists \alpha.r(\alpha)) \bullet (\exists i.[\![W]\!](i))^{\mathsf{fin}} \Rightarrow \nearrow (\exists \alpha.r(\alpha))$$

so it remains to show

$$\nearrow (\exists \alpha.r(\alpha)) \bullet (\forall i.([\![W]\!](i) \wedge l = 0) \bullet \mathsf{true}))^{\mathsf{inf}} = \mathsf{False}$$

and the proof is as follows

$$\nearrow (\exists \alpha.r(\alpha)) \bullet (\forall i.([\![W]\!](i) \wedge l = 0) \bullet \mathsf{true})^{\mathsf{inf}}$$
$$= (\nearrow (\exists \alpha.r(\alpha)) \bullet (\forall i.([\![W]\!](i) \wedge l = 0) \bullet \mathsf{true}))^{\mathsf{inf}}$$
$$= (\exists \alpha. \nearrow r(\alpha) \bullet (\forall i.([\![W]\!](i) \wedge l = 0) \bullet \mathsf{true}))^{\mathsf{inf}}$$
$$\Rightarrow (\exists \alpha. \nearrow r(\alpha) \bullet ([\![W]\!](\alpha + 1) \wedge l = 0) \bullet \mathsf{true})^{\mathsf{inf}}$$
$$\Rightarrow (\exists \alpha.\exists \beta > 0. \nearrow r(0) \bullet ([\![W]\!](\beta) \wedge l = 0) \bullet \mathsf{true})^{\mathsf{inf}}$$
$$\Rightarrow (\exists \alpha.\exists \beta > 0. \nearrow r(0) \bullet [\, \langle b \rangle \,] \bullet [\, D \,] \bullet ([\![W]\!](\beta - 1) \wedge l = 0) \bullet \mathsf{true})^{\mathsf{inf}}$$
$$\Rightarrow (\exists \alpha.\exists \beta > 0.\mathsf{false} \bullet [\![W]\!] \bullet ([\![W]\!](\beta - 1) \wedge l = 0) \bullet \mathsf{true})^{\mathsf{inf}}$$
$$= \mathsf{false}^{\mathsf{inf}}$$
$$= \mathsf{False}$$

Lemma 3 *If $[\![W]\!](i) \Rightarrow \bigsqcap$ for any i, then $\forall x \exists y, i.(y \geq x \wedge (l < y \vee (([\![W]\!](i) \wedge l = y) \bullet \mathsf{true}))^{\mathsf{inf}}) = \mathsf{False}$*

Proof:

$$\forall x \exists y, i.(y \geq x \wedge (l < y \vee (([\![W]\!](i) \wedge l = y) \bullet \mathsf{true}))^{\mathsf{inf}})$$
$$\Rightarrow \exists y, i.(y \geq 2 \wedge (l < y \vee (([\![W]\!](i) \wedge l = y) \bullet \mathsf{true}))^{\mathsf{inf}})$$
$$= \exists y, i.(y \geq 2 \wedge (l < y \vee \mathsf{false} \bullet \mathsf{true})^{\mathsf{inf}})$$
$$= \exists y, i.(y \geq 2 \wedge (l < y)^{\mathsf{inf}})$$
$$= \mathsf{False}$$

The soundness of the iteration rule then follows from lemmas 1, 2 and 3.

6 Verification of Temporal Properties

In this section, we propose a number of rules which can be used to verify properties that are not restricted to pre-conditions and post-conditions. In the following, P, P_1 and P_2 may contain continuous components.

$$\{p\} \ \langle b \rangle \ \sqcap$$
$$\{p\} \ \bar{x} := \bar{e} \ \sqcap$$

$$\frac{\{p\} \ P_1 \ A \quad \{p\} \ P_2 \ A}{\{p\} \ P_1 \ [] \ P_2 \ A}$$

$$\frac{\{p\} \ P_1 \ \{r\} \quad \{p\} \ P_1 \ A \quad \{r\} \ P_2 \ A \quad A \bullet A \Rightarrow A}{\{p\} \ P_1; P_2 \ A}$$

$$\frac{\{p\} \ P_1 \ A^{\text{inf}}}{\{p\} \ P_1; P_2 \ A^{\text{inf}}}$$

One particularly interesting property is invariant, expressed as $(\lceil S \rceil^*)^{\text{fin}} \vee (\lceil S \rceil^*)^{\text{inf}}$. The following rule says that an iteration statement satisfies an invariant if the body always takes some time (this ensures that there are no divergence) and satisfies the invariant.

$$\frac{\{b \wedge r\} \ P \ ((\lceil S \rceil^* \wedge l > 0)^{\text{fin}} \bullet \nearrow r) \vee (\lceil S \rceil^*)^{\text{inf}}}{\{r\} \ \text{while} \ b \ \text{do} \ P \ \text{od} \ (\lceil S \rceil^*)^{\text{fin}} \vee (\lceil S \rceil^*)^{\text{inf}}}$$

The soundness of this rule follows from the lemmas below.

Lemma 4 $\nearrow r \bullet [W](i) \Rightarrow ((\lceil S \rceil^* \wedge l > 0)^{\text{fin}} \bullet \nearrow r) \vee (\lceil S \rceil^*)^{\text{inf}}$

Proof: by induction over i.
Base $i = 0$: it is obvious that $\nearrow r \bullet [\text{skip}] \Rightarrow \nearrow r$.
Induction step: assume $\nearrow r \bullet [W](i) \Rightarrow ((\lceil S \rceil^* \wedge l > 0)^{\text{fin}} \bullet \nearrow r) \vee (\lceil S \rceil^*)^{\text{inf}}$,

$$\begin{aligned}
& \nearrow r \bullet [W](i+1) \\
=\ & \nearrow r \bullet [\langle b \rangle] \bullet [P] \bullet [W](i) \\
\Rightarrow\ & \nearrow (r \wedge b) \bullet [P] \bullet [W](i) \\
\Rightarrow\ & ((((\lceil S \rceil^* \wedge l > 0)^{\text{fin}} \bullet \nearrow r) \vee (\lceil S \rceil^*)^{\text{inf}}) \bullet [W](i) \\
\Rightarrow\ & ((((\lceil S \rceil^* \wedge l > 0)^{\text{fin}} \bullet \nearrow r) \bullet [W](i)) \vee (\lceil S \rceil^*)^{\text{inf}} \\
\Rightarrow\ & ((((\lceil S \rceil^* \wedge l > 0)^{\text{fin}} \bullet ((((\lceil S \rceil^* \wedge l > 0)^{\text{fin}} \bullet \nearrow r) \vee (\lceil S \rceil^*)^{\text{inf}})) \vee (\lceil S \rceil^*)^{\text{inf}} \\
\Rightarrow\ & ((\lceil S \rceil^* \wedge l > 0)^{\text{fin}} \bullet \nearrow r) \vee (\lceil S \rceil^*)^{\text{inf}}
\end{aligned}$$

Lemma 5 $\nearrow r \bullet \forall x \exists y, i.(y \geq x \wedge (l < y \vee (([W](i) \wedge l = y) \bullet \text{true}))^{\text{inf}}) \Rightarrow$
$(\lceil S \rceil^*)^{\text{inf}}$

Proof:

$$\nearrow r \bullet \forall x \exists y, i.(y \geq x \wedge (l < y \vee (([\![W]\!](i) \wedge l = y) \bullet \text{true}))^{\text{inf}})$$
$$\Rightarrow \forall x \exists y, i.(y \geq x \wedge (\nearrow r \bullet (l < y \vee (([\![W]\!](i) \wedge l = y) \bullet \text{true})))^{\text{inf}})$$
$$\Rightarrow \forall x \exists y, i.(y \geq x \wedge ((\nearrow r \bullet l < y) \vee (\nearrow r \bullet ([\![W]\!](i) \wedge l = y) \bullet \text{true}))^{\text{inf}})$$
$$\Rightarrow \forall x \exists y, i.(y \geq x \wedge (l < y \vee (((((\lceil S \rceil^* \wedge l > 0)^{\text{fin}} \bullet \nearrow r)$$
$$\vee (\lceil S \rceil^*)^{\text{inf}}) \wedge l = y) \bullet \text{true}))^{\text{inf}})$$
$$\Rightarrow \forall x \exists y, i.(y \geq x \wedge (l < y \vee ((\lceil S \rceil^*)^{\text{fin}} \wedge l = y) \bullet \text{true}))^{\text{inf}})$$
$$= (\lceil S \rceil^*)^{\text{inf}}$$

Lemma 6 $(\exists i.[\![W]\!](i))^{\text{fin}} \bullet (\forall i.([\![W]\!](i) \wedge l = 0) \bullet \text{true})^{\text{inf}} = \text{False}$

Proof:

$$(\exists i.[\![W]\!](i))^{\text{fin}} \bullet (\forall i.([\![W]\!](i) \wedge l = 0) \bullet \text{true})^{\text{inf}}$$
$$\Rightarrow (\exists i.[\![W]\!](i))^{\text{fin}} \bullet (\forall i.\text{false} \bullet \text{true})^{\text{inf}}$$
$$= (\exists i.[\![W]\!](i))^{\text{fin}} \bullet \text{False}$$
$$\Rightarrow \text{False}$$

7 A Simple Example

We want to control the speed V of a car by a computer. Consider a simple control strategy. At each unit time point, the computer samples the current speed of the car and decides the rate a at which the speed should be changed in the next time unit. Assume that sampling time can be neglected, the action can be modelled by an assignment $v := V$ where v is a variable that the computer can directly access. Depending on whether the speed is more than 80 or less than 70, a will be assigned a positive or negative value. Due to hardware restrictions, the actual rate of speed change $\dot{V}$ may not be exactly the same as a, but we can assume that the difference between them is not more than 3, and in the accelerating mode $\dot{V} \geq 1$, while in the decelerating mode, $\dot{V} \leq -1$.

One requirement is that the speed should always be lower than 100. The system and the requirement can be denoted as

```
V := 0;
alarm := off;
while true do
    v := V;
    if  v < 80 → A; ≪ ⌈|V̇ − a| ≤ 3 ∧ V̇ ≥ 1⌉ ∧ l = 1 ≫
       [] v ≥ 70 → D; ≪ ⌈|V̇ − a| ≤ 3 ∧ V̇ ≤ −1⌉ ∧ l = 1 ≫
       [] v > 100 → alarm := on
    fi
od
```

$$(\lceil V < 100 \rceil^*)^{\text{fin}} \vee (\lceil V < 100 \rceil^*)^{\text{inf}}$$

where the formula is marked with a shaded box. The design task is to construct algorithms A and D so that the requirement is satisfied. Due to the compositionality of our method, we can verify the correctness of the system based on the specifications of A and D:

$V := 0;$
$\{V = 0\}$
alarm := off;
$\{V < 100\}$
while true do
    $\{V < 100\}$
    $v := V;$
    $\{V < 100 \wedge v = V\}$
    if $v < 80 \rightarrow A;$ $\{V < 80 \wedge a < 15\}$
                    $\ll \lceil |\dot{V} - a| \leq 3 \wedge \dot{V} \geq 1\rceil \wedge l = 1 \gg$
                    $(\lceil V < 100 \rceil^* \wedge l > 0)^{\mathsf{fin}} \bullet \nearrow (V < 100)$
    $[]\ v \geq 70 \rightarrow D;$ $\{V < 100\}$
                    $\ll \lceil |\dot{V} - a| \leq 3 \wedge \dot{V} \leq -1\rceil \wedge l = 1 \gg$
                    $(\lceil V < 100 \rceil^* \wedge l > 0)^{\mathsf{fin}} \bullet \nearrow (V < 100)$
    $[]\ v > 100 \rightarrow$ $\{\mathsf{false}\}$
                    alarm := on
                    $(\lceil V < 100 \rceil^* \wedge l > 0)^{\mathsf{fin}} \bullet \nearrow (V < 100)$
    fi
od
$(\lceil V < 100 \rceil^*)^{\mathsf{fin}} \vee (\lceil V < 100 \rceil^*)^{\mathsf{inf}}$

The proof outline indicates that to guarantee the simple invariant requirement, it is sufficient that algorithm A does not assign a value bigger than 15 to a, or formally

$\{\mathsf{true}\}\ A\ \{a \leq 15\}$

It is preferable to accelerate fast when the speed is low and accelerate slowly when the speed is high. Therefore, we calculate the value of a based on function $0.1 * e^{100/(v+20)}$. To this end, we let

$\{\mathsf{true}\}\ A\ \{0.1 * e^{100/(v+20)} - 1 \leq a \leq 0.1 * e^{100/(v+20)}\}$

Correctness of the system is maintained, because

$0.1 * e^{100/(v+20)} \leq 15$ for any $v \geq 0$

Finally, the control algorithm is developed

```
procedure      A
c := 100/(v + 20);
k := 1;
a := 0.1;
r := 0.6 * 5;
```

$$\{a = 0.1 * \sum_{n=0}^{k-1} \frac{1}{n!}(\frac{100}{v+20})^n \wedge r = 0.6 * 5^k/k!\}$$

```
while k ≤ 5 ∨ r ≥ 1 do
    a := a + 0.1 * c;
    k := k + 1;
    c := (100 * c)/((v + 20) * k);
    r := (5 * r)/k
od
```

$$\{a = 0.1 * \sum_{n=0}^{k-1} \frac{1}{n!}(\frac{100}{v+20})^n \wedge r = 0.6 * 5^k/k! \wedge k > 6 \wedge r < 1\}$$

$$\{0.1 * e^{100/(v+20)} - r \leq a \leq 0.1 * e^{100/(v+20)}\}$$

$$\{0.1 * e^{100/(v+20)} - 1 \leq a \leq 15\}$$

8   Discussion

In this paper, we have investigated integrating program logic into real time specification logic. We believe this provides one of the most effective way to verify hybrid systems. Our method is compositional with respect to usual sequential constructs and therefore also supports derivation of discrete controller.

We take the view that finite iterations of instantaneous instructions are still instantaneous. In practice, it may happen that only a small number of iterations can be considered timeless. In our framework, we cannot simply represent such programs as discrete components, which are either instantaneous or infinite, and instead we must model them as timed systems. For example, if the execution time of less than 10 iterations of $x := x + 1$ can be omitted, but every 10 iterations should be considered as taking one time unit, then the program should be represented as

```
while x < 100
    do
        x := x + 1;
        if (x mod 10) then delay 1
    od
```

In this paper, we have defined semantics using a variant of the Duration Calculus with infinite disjunctions. This is somewhat ad-hoc. It is worthwhile to investigate if the semantics can be defined in a Duration Calculus with fixed-points, such as [9].

Our model of Super Dense Duration Calculus is related to the "higher order" time domain used by a number of researchers, e.g., [6]. As an alternative, instead of associating two states to one time point, we could define the time domain as

$$\text{NTime} = \text{Time} \times \{1, 2\}$$

and use the states associated with $(t, 1)$ and $(t, 2)$ to denote states before and after the transition at t.

Our next step is to study concurrent systems. In fact, this work is motivated by and subsequently provides a basis for a case study which involves concurrency [11], but not knowing a way to decompose the verification of the system into verification of its concurrent components, we had to first transform the concurrent system into a sequential one. To take into account of the interactions among concurrent components, it is necessary to record the intermediate states of discrete actions. We expect that verification techniques for concurrent programs can also be incorporated into Duration Calculus.

Acknowledgement I thank a number of researchers, especially Burghard von Karger, Yassine Lakhnech and Willem-Paul de Roever of Kiel University where the work was reported in a seminar, for comments which led me to write down the reformulation of the Super Dense Duration Calculus. I also thank Zhou Chaochen, Dang Van Hung, Paritosh Pandya, Yang Zhenyu and anonymous referees for comments and related discussions.

References

1. Zhou Chaochen and M. R. Hansen. Chopping a point. In J-F. He, editor, *Proc. BCS FACS 7th Refinement Workshop:Theory and Practice of System Design*, Bath, U.K., July 1996. Springer–Verlag.
2. Zhou Chaochen, C.A.R. Hoare, and A.P. Ravn. A calculus of durations. *Information Processing Letters*, 40(5):269–276, 1991.
3. Zhou Chaochen, Dang Van Hung, and Li Xiaoshan. A duration calculus with infinite intervals. In *Fundamentals of Computation Theory, Horst Reichel (Ed.)*, pages 16–41. LNCS 965, Springer-Verlag, 1995.
4. Zhou Chaochen, A.P. Ravn, and M.R. Hansen. An extended duration calculus for hybrid systems. In *Hybrid Systems, R.L. Grossman, A. Nerode, A.P. Ravn, H. Rischel (Eds.)*, pages 36–59. LNCS 736, Springer-Verlag, 1993.
5. C. A. R. Hoare. *Communicating Sequential Processes*. Prentice–Hall, 1985.
6. R. Koymans. *Specifying Message Passing and Time-Critical Systems with Temporal Logic*. LNCS 651, Springer–Verlag, 1992.
7. Z. Manna and A. Pnueli. Verifying hybrid systems. In *Hybrid Systems, R.L. Grossman, A. Nerode, A.P. Ravn, H. Rischel (Eds.)*, pages 36–59. LNCS 736, Springer-Verlag, 1993.
8. B. Moszkowski. A temporal logic for multilevel reasoning about hardware. *IEEE Computer*, 18(2):10–19, 1985.

9. P.K. Pandya. A recursive duration calculus. Technical report, CS-95/3, Computer Science Group, TIFR, Bombay, 1995.

10. A. Pnueli. Development of hybrid systems. In H. Langmack, W.-P. de Roever, and J. Vytopil, editors, *Formal Techniques in Real-Time and Fault-Tolerant Systems*, pages 77–85. LNCS 863, Springer-Verlag, 1994.

11. Xu Qiwen and Yang Zengyu. Derivation of control programs: a heating system. In *Presented at the 4th International Conference on Hybrid Systems*, Ithaca, NY, USA, 1996.

Weak Refinement for Modal Hybrid Systems

Carsten Weise and Dirk Lenzkes

Lehrstuhl für Informatik I,
Aachen University of Technology, Germany
{carsten|dlen}informatik.rwth-aachen.de

Abstract. Timed modal specifications have proven useful for the verification of real time systems. In the paper, we show how the notion of
modal transitions systems, the basic formalism of timed modal specifications, can be generalized to *modal hybrid systems*. This allows verification problems for hybrid systems to be formulated using the notion
of (weak) refinement, a generalization of bisimulation. We also explain
how to adopt an improved algorithm for the decision of weak refinement
for timed modal specifications to modal linear hybrid systems, which offers the possibility of automatic verification. The algorithm has already
been implemented for parametrized modal timed automata, a subclass
of modal linear hybrid systems.

1 Introduction

Recently, we have put forward a more space efficient decision algorithm for bisimulation of timed automata [WL97]. The same basic algorithm can be used to
decide weak refinement for timed modal specifications ([CGL93]). In this paper, we extend the notion of timed modal specification to *modal timed automata*
and *modal linear hybrid systems*. We show that our algorithm can be extended
straightforwardly to an algorithm for weak refinement for modal hybrid systems.
As (weak) refinement is the more general notion, this also yields algorithms for
(weak) bisimulation and simulation. However, while the algorithm always terminates for timed automata, in the case of linear hybrid systems it only yields a
semi-decision procedure. We will demonstrate the usefulness of weak refinement
by examples with *parametrized modal transition systems* and *modal linear hybrid
systems*.

We start with recalling the definition of modal transition systems, weak refinement and hybrid systems. In section 3 we extend hybrid systems to modal
hybrid systems. In section 4 we explain informally our improved algorithm for
bisimulation of timed automata. This algorithm can be extended to modal linear
hybrid systems. In the following sections we give application examples of modal
hybrid systems.

2 Modal Transition Systems

Labeled transition systems are a well-established notion in the field of Formal
Methods, used as a direct graphical specification language and as the semantic

model for specification formalisms. Refinement relations serve mainly two purposes: either as an implementation relation, used to verify if an implementation meets a specification, or in the stepwise development of applications, going from coarser to finer versions of the specification, eventually leading to a concrete implementation.

Modal transition systems [LT88] are labeled transition systems with a may- and a must-modality: must-transitions describe required properties of the specified system, while may-transitions describe admitted properties. Formally they are defined by:

Definition 1 (Modal Transition System). A *modal transition system* is a structure $S = (\Sigma, \sigma, A, \underset{\Box}{\rightarrow}, \underset{\Diamond}{\rightarrow})$, where Σ is a set of states, $\sigma \subseteq \Sigma$ is a set of initial states, A is a set of actions and $\underset{\Box}{\rightarrow}, \underset{\Diamond}{\rightarrow} \subseteq \Sigma \times A \times \Sigma$ are transition relations. For the transition relations, $\underset{\Box}{\rightarrow} \subseteq \underset{\Diamond}{\rightarrow}$ must hold.

The requirement $\underset{\Box}{\rightarrow} \subseteq \underset{\Diamond}{\rightarrow}$ expresses consistency (anything required should be allowed). In the case $\underset{\Box}{\rightarrow} = \underset{\Diamond}{\rightarrow}$, the above reduces to the traditional notion of labelled transition systems. Modal transition systems come equipped with a notion of refinement, which matches the given intuition for may- and must-transitions:

Definition 2 (Refinement). Let $S_i = (\Sigma_i, \sigma_i, A, \underset{\Box}{\rightarrow}^i, \underset{\Diamond}{\rightarrow}^i)(i \in \{1,2\})$ be two modal transition systems. A refinement $\mathcal{R}$ is a binary relation in $\Sigma_1 \times \Sigma_2$ such that whenever $s_1 \mathcal{R} s_2$ and $a \in A$ then the following holds:

1. Whenever $s_1 \underset{\Diamond}{\overset{a}{\rightarrow}}^1 s_1'$, then $s_2 \underset{\Diamond}{\overset{a}{\rightarrow}}^2 s_2'$ for some s_2' with $s_1' \mathcal{R} s_2'$,
2. Whenever $s_2 \underset{\Box}{\overset{a}{\rightarrow}}^2 s_2'$, then $s_1 \underset{\Box}{\overset{a}{\rightarrow}}^1 s_1'$ for some s_1' with $s_1' \mathcal{R} s_2'$.

The state s_1 is said to be a *refinement* of s_2 in case (s_1, s_2) is contained in some refinement $\mathcal{R}$, written $s_1 \lhd s_2$. S_1 is a refinement of S_2 if $s_1 \lhd s_2$ holds for all their start states $s_1 \in \sigma_1, s_2 \in \sigma_2$.

If $\underset{\Box}{\rightarrow}^i = \underset{\Diamond}{\rightarrow}^i$, we obtain the well–known notion of bisimulation [Mil89], while $\underset{\Box}{\rightarrow}^i = \emptyset$ yields forward simulation [LV91].

In practice, one is often interested in hiding details of the implementation or inter-process communication. This is done by the usual method from process algebras: a special symbol τ denotes the *silent action*. The refinement relation is then defined on the derived *weak transition systems* of P and Q, which is basically the τ-closure of the transition relation. This notion of refinement is called *weak refinement*. We postpone the formal definition of the weak transition system, as we use a special version suitable for timed transition systems.

The parallel operator for timed modal transition systems typically requires *maximal progress*: whenever two components must engage in a synchronous action, they will do so immediately.

Definition 3 (Maximal Progress). A given modal transition system $S = (\Sigma, \sigma, A, \underset{\Box}{\rightarrow}, \underset{\Diamond}{\rightarrow})$ has the *maximal progress property* if for all $s \in \Sigma$:

$$s \underset{\Box}{\overset{\tau}{\rightarrow}} \text{ impl. } \forall d \in \mathbb{R}^{>0}. \, s \underset{\Box}{\overset{d}{\nrightarrow}}$$

Note that this is a requirement concerning must-transitions only. The parallel operator for modal transition systems can be defined so that the resulting system always has the maximal progress property.

3 Modal Hybrid Systems

Modal hybrid systems are the generalization of hybrid systems in the same sense as modal transition systems are generalizations of transition systems. We assume familiarity with hybrid systems (see e.g. [AC+95]), but give a formal definition on which we will rely for the generalization. We use the following terminology:

Definition 4 (Valuation, Activity). For a given set Var of variables, the *set of (continuous) valuations* is defined by $V := \{v : \text{Var} \rightarrow \mathbb{R}\}$. An *activity* is a mapping $\mathbb{R}^{\geq 0} \rightarrow V$. The *set of all activities* (of Var) is written act. For an activity f and a variable $x \in \text{Var}$, the function f^x is defined by $f^x(t) = v(x)$, where $v = f(t)$. Let f be an activity and $t \in \mathbb{R}^{\geq 0}$, then $f+t$ is the activity defined by $\forall d \in \mathbb{R}^{\geq 0}. \, (f+t)(d) = f(d+t)$. A set S of activities is called *time-invariant* if for all $f \in S$ and $t \in \mathbb{R}^{\geq 0}$ always $f+t \in S$ holds.

hybrid systems are defined as:

Definition 5 (Hybrid System). A *hybrid system* is a structure (Loc, init, Lab, $\rightarrow$, Var, Act, Inv) where Loc is a finite set of *(control) locations*, init $\subseteq$ Loc $\times V$ is the set of initial states, Lab is a finite set of *synchronization labels*, $\rightarrow \subseteq$ Loc $\times$ Lab $\times 2^{(V^2)} \times$ Loc is the transition relation, Act : Loc $\rightarrow 2^{\text{act}}$ is a mapping of time invariant activity sets to control locations, Inv : Loc $\rightarrow 2^V$ is a mapping of *invariants* to control locations.

The semantics of a hybrid system are given as usual by a labeled transition system whose nodes are pairs of locations and valuations, and whose labels are from Lab $\cup \mathbb{R}^{\geq 0}$. We use SOS rules to define the semantics of a hybrid system:

Definition 6 (Valuation Graph). A given hybrid system (Loc, init, Lab, $\rightarrow_H$, Var, Act, Inv) defines a valuation graph (Loc $\times V$, init, Lab $\cup \mathbb{R}^{\geq 0}$, $\rightarrow$), where the transition relation is defined by:

$$(A) \quad \frac{\ell \xrightarrow{a,\mu}_H \ell', (v,v') \in \mu, \quad v \in \text{Inv}(\ell), v' \in \text{Inv}(\ell')}{(\ell,v) \xrightarrow{a} (\ell',v')}$$

$$(T) \quad \frac{f \in \text{Act}(\ell), f(0) = v, \quad \forall 0 \leq t' \leq t. \, f(t') \in \text{Inv}(\ell)}{(\ell,v) \xrightarrow{t} (\ell, f(t))}$$

States of the valuation graph – i.e. pairs of location and valuation – are called *points*, as the state set $\mathsf{Loc} \times V$ is isomorphic to the space $\mathsf{Loc} \times \mathbb{R}^n$ (if $|V| = n$). We call a transition $e = \ell \xrightarrow{a,\mu} \ell'$ enabled in the state (ℓ, v) if there is v' such that (v, v') is in μ. The set of all valuations v for which e is enabled in (ℓ, v) is called the *guard* of the transition e.

There are two important subclasses of hybrid systems: *linear hybrid systems* and *timed automata*. The former have only activities which are piecewise linear functions, while the latter have only activities with derivative 1, and all variables (called *clocks*) either keep their value along a transition or are reset to zero. Note that the guard of a transition can be described by closed formulae for these sub-classes: for linear hybrid systems, the formulae are boolean combinations of linear inequalities over the variables (called *linear formulae*), while for timed automata, the formulae are conjunctions over inequalities comparing variables or differences of two variables against constants (which we will call *simple linear formulae*).

For timed automata and linear hybrid systems, the set of initial states consists of one point (ℓ_0, v_0) only. We call ℓ_0 the *initial location*. For timed automata, v_0 is the valuation $\bigwedge_{C \in V\text{ar}} C = 0$ with all clocks set to Zero, while for linear hybrid systems, all variables must have integral values in v_0. A *modal hybrid system* is a generalization of hybrid systems as modal transition systems are a generalization of transition systems:

Definition 7 (Modal Hybrid System). A *modal hybrid systems* is a structure $(\mathsf{Loc}, \mathsf{init}, \mathsf{Lab}, \underset{\square}{\rightarrow}, \underset{\diamond}{\rightarrow}, \mathsf{Var}, \mathsf{Act}_\square, \mathsf{Act}_\diamond, \mathsf{Inv}_\square, \mathsf{Inv}_\diamond)$ such that $(\mathsf{Loc}, \mathsf{init}, \mathsf{Lab}, \underset{\square}{\rightarrow}, \mathsf{Var}, \mathsf{Act}_\square, \mathsf{Inv}_\square)$ and $(\mathsf{Loc}, \mathsf{init}, \mathsf{Lab}, \underset{\diamond}{\rightarrow}, \mathsf{Var}, \mathsf{Act}_\diamond, \mathsf{Inv}_\diamond)$ are hybrid systems, and for their valuation graphs $(\mathsf{Loc} \times V, \mathsf{init}, \mathsf{Lab} \cup \mathbb{R}^{\geq 0}, \underset{\square}{\rightarrow})$ and $(\mathsf{Loc} \times V, \mathsf{init}, \mathsf{Lab} \cup \mathbb{R}^{\geq 0}, \underset{\diamond}{\rightarrow})$, the structure $(\mathsf{Loc} \times V, \mathsf{init}, \mathsf{Lab} \cup \mathbb{R}^{\geq 0}, \underset{\square}{\rightarrow}, \underset{\diamond}{\rightarrow})$ is a modal transition system.

For a given modal hybrid system, we will call $(\mathsf{Loc} \times V, \mathsf{init}, \mathsf{Lab} \cup \mathbb{R}^{\geq 0}, \underset{\square}{\rightarrow}, \underset{\diamond}{\rightarrow})$ its *semantics* or *(modal) valuation graph*. The notion of refinement can now be applied to modal hybrid systems:

Definition 8 ($P \triangleleft Q$). Let P and Q be two modal hybrid systems, then P *refines* Q, written $P \triangleleft Q$, if for their modal valuation graphs S_P, S_Q we have $S_P \triangleleft S_Q$.

As pointed out before, in practice weak refinement is the important notion. The formal definition of weak refinement uses the weak form of a timed transition relation:

Definition 9 (Weak Transition Relation). A *timed transition relation* is a transition relation with labels L where $\mathbb{R}^{\geq 0} \cap L \neq \emptyset$.

Given a timed transition relation $\to$ and a special silent action $\tau \in L$, the *weak timed transition relation* $\Rightarrow$ is defined by:

$$(W1) \quad \frac{s \xrightarrow{\tau^n} \xrightarrow{\ell} \xrightarrow{\tau^m} s', \ell \in L, n, m \in \mathbb{N}}{s \xRightarrow{\ell} s'}$$

$$(W2) \quad \frac{s \xrightarrow{\tau^n} s', n \in \mathbb{N}}{s \xRightarrow{0} s'} \qquad (W3) \quad \frac{s \xRightarrow{d} s', s' \xRightarrow{d'} s'', d, d' \in \mathbb{R}^{\geq 0}}{s \xRightarrow{d+d'} s''}$$

For two modal transition systems S and T we say that S *weakly refines* T (written $S \unlhd T$) if the refinement relation holds for the systems where the relations $\xrightarrow[\square]{}$ and $\xrightarrow[\diamond]{}$ are replaced by their weak counterparts $\Rightarrow_\square$ and $\Rightarrow_\diamond$. Analogously, for two modal hybrid systems $P \unlhd Q$ if $S_P \unlhd S_Q$ holds for its modal valuation graph.

There is a certain problem with modal hybrid systems and maximal progress: generally, maximal progress cannot be imposed on parallel products of arbitrary hybrid systems. This is only possible if for a location ℓ, every zone (ℓ, ϕ) which has outgoing τ-transitions is left-closed, i.e. ϕ has a minimal element v. If this is the case, the zone (ℓ, ϕ) must be replaced by (ℓ, v). Note that this is a semantical restrictions which cannot easily be formulated syntactically.

Thus we have extended the definition of modal transition systems and weak refinement to hybrid systems. In the following we will only deal with modal linear hybrid systems, and modal (parametrized) timed automata.

4 Fast Decision Algorithm for Weak Refinement

Recently we have put forward a decision algorithm for (weak) bisimulation for timed automata [WL97], which is more space efficient than the original algorithm of Čerāns ([Čer92]). Due to the close relation between (weak) bisimulation and (weak) refinement, the algorithm can also be used to decide (weak) refinement for modal transition systems.

A precise formal description of our algorithm for timed automata can be found elsewhere (see [WL97]). In this section we will only give an informal sketch of the decision procedure, but in contrast to our presentation in [WL97] for timed automata, here we give a description within the setting of linear hybrid systems.

4.1 Product and Zone Graphs

Our algorithm follows the idea of the original algorithm of Čerāns and uses the product graph of two systems H_1 and H_2 to decide bisimilarity. Thus we define the notion of a *product hybrid system*:

Definition 10 (Product Hybrid System). Given two hybrid systems $H_i = (\mathsf{Loc}_i, \mathsf{init}_i, \mathsf{Lab}, \to_i, \mathsf{Var}_i, \mathsf{Act}_i, \mathsf{Inv}_i)(i \in \{1, 2\})$, where $\mathsf{Var}_1 \cap \mathsf{Var}_2 = \emptyset$, their strong product $H_1 \times H_2$ is the hybrid system $(\mathsf{Loc}_1 \times \mathsf{Loc}_2, \mathsf{init}_1 \times \mathsf{init}_2, \mathsf{Lab}, \to$

, $\mathsf{Var}_1 \cup \mathsf{Var}_2, \mathsf{Act}, \mathsf{Inv})$ where for all $\ell_i \in \mathsf{Loc}_i$: $\mathsf{Act}(\ell_1, \ell_2) := \mathsf{Act}(\ell_1) \cap \mathsf{Act}(\ell_2)$ and $\mathsf{Inv}(\ell_1, \ell_2) := \mathsf{Inv}(\ell_1) \cap \mathsf{Inv}(\ell_2)$, and the transition relation $\rightarrow$ is defined by

$$\frac{n_1 \xrightarrow{a,\mu_1}_1 n_1', n_2 \xrightarrow{a,\mu_2}_2 n_2'}{(n_1, n_2) \xrightarrow{a,\mu_1 \cap \mu_2} (n_1', n_2')}$$

The *weak product* $H_1 \times_w H_2$ is defined by two additional rules for the transition relation:

$$\frac{n_1 \xrightarrow{\tau,\mu_1}_1 n_1'}{(n_1, n_2) \xrightarrow{\tau,\mu_1} (n_1', n_2)} \qquad \frac{n_2 \xrightarrow{\tau,\mu_2}_2 n_2'}{(n_1, n_2) \xrightarrow{\tau,\mu_2} (n_1, n_2')}$$

This definition extends to modal hybrid systems in the usual way:

Definition 11 (Modal Product). For two modal hybrid systems $H_i = (\mathsf{Loc}_i, \mathsf{init}_i, \mathsf{Lab}, \xrightarrow{}_\Box^i, \xrightarrow{}_\Diamond^i, \mathsf{Var}_i, \mathsf{Act}_\Box^i, \mathsf{Act}_\Diamond^i, \mathsf{Inv}_\Box^i, \mathsf{Inv}_\Diamond^i)(i \in \{1, 2\})$, where $\mathsf{Var}_1 \cap \mathsf{Var}_2 = \emptyset$, their *strong resp. weak product* $H_1 \times H_2$ is the modal hybrid system $(\mathsf{Loc}_1 \times \mathsf{Loc}_2, \mathsf{init}_1 \times \mathsf{init}_2, \mathsf{Lab}, \xrightarrow{}_\Box, \xrightarrow{}_\Diamond, \mathsf{Var}_1 \cup \mathsf{Var}_2, \mathsf{Act}_\Box, \mathsf{Act}_\Diamond, \mathsf{Inv}_\Box, \mathsf{Inv}_\Diamond)$ where $(\mathsf{Loc}_1 \times \mathsf{Loc}_2, \mathsf{init}_1 \times \mathsf{init}_2, \mathsf{Lab}, \xrightarrow{}_\Box, \mathsf{Var}_1 \cup \mathsf{Var}_2, \mathsf{Act}_\Box, \mathsf{Inv}_\Box)$ and $(\mathsf{Loc}_1 \times \mathsf{Loc}_2, \mathsf{init}_1 \times \mathsf{init}_2, \mathsf{Lab}, \xrightarrow{}_\Diamond, \mathsf{Var}_1 \cup \mathsf{Var}_2, \mathsf{Act}_\Diamond, \mathsf{Inv}_\Diamond)$ are the strong resp. weak product of the hybrid systems for allowed and required transitions contained in the H_i.

A point (s, v) of the product graph is a pair of points $(s_1, v_1), (s_2, v_2)$ of the systems H_1 and H_2, such that $s = (s_1, s_2)$ and v_i is v restricted to the variables of H_i. We will use the suggestive notation $(s_1 \times s_2, v_1 \times v_2)$ for such points. If $(s_1' \times s_2', v_1' \times v_2')$ is reachable by a path from $(s_1 \times s_2, v_1 \times v_2)$ in the product graph, then (s_i', v_i') is reachable from (s_i, v_i) by a path with the same labeling in H_i for $i = 1, 2$.

Our decision algorithm works on *zone graphs*. In general, a zone is a set of points of the valuation graph. For a zone Z, let $\phi(Z)$ be the set of all valuations appearing in the zone's states, i.e. $\phi(Z) := \{v \mid (\ell, v) \in Z\}$. Further let $\mathcal{L}(Z) := \{\ell \mid (\ell, v) \in Z\}$ be set of locations of Z. In the case of linear hybrid systems and timed automata, Z is a zone only if $\mathcal{L}(Z)$ is a singleton set and $\phi(Z)$ is describable by a linear resp. a simple linear formula. A zone graph is a graph whose nodes are zones. We use the following operations on zones Z, most of which appear e.g. in [AC+95]. All operations are interpreted w.r.t. to a given valuation graph with transition relation $\rightarrow$. Let V, V' be the sets of valuations of variables from $\mathsf{Var}, \mathsf{Var}'$ resp.:

- time forward closure $Z^\uparrow := \{s' \mid \exists s \in Z, d \in \mathbb{R}^{\geq 0} . s \xrightarrow{d} s'\}$: the set of all points reachable from a zone by time progress,
- time backward closure $Z^\downarrow := \{s' \mid \exists s \in Z, d \in \mathbb{R}^{\geq 0} . s' \xrightarrow{d} s\}$: the set of all points from which the zone is reachable by time progress,
- postcondition $\mathsf{post}_e(Z) := \{s' \mid \exists s = (\ell, v) \in Z, v' \in V . (v, v') \in \mu, s \xrightarrow{a} s' = (\ell', v')\}$ of a transition $e = (\ell, a, \mu, \ell')$: the set of all points reachable by taking the transition e from Z,

- precondition $\mathsf{pre}_e(Z) := \{s \mid \exists s' = (\ell', v') \in Z, v \in V. (v, v') \in \mu, s = (\ell, v) \xrightarrow{a} s'\}$ of a transition $e = (\ell, a, \mu, \ell')$: the set of all points from which the zone is reachable by taking the transition e,
- restriction $Z|_{\mathsf{Var'}} := \{(\ell, v') \mid (\ell, v) \in Z, v' \in V', \forall x \in \mathsf{Var'}. v'(x) = v(x)\}$ to $\mathsf{Var'} \subseteq \mathsf{Var}$: the set of all points of a zone where we restrict the valuations to the variables of $\mathsf{Var'}$,
- embeddding $Z|^{\mathsf{Var}} := \{(\ell, v) \mid \{(\ell, v)\}|_{\mathsf{Var'}} \subseteq Z\}$ of a zone $Z \subseteq \mathsf{Loc} \times V'$ from $\mathsf{Var'} \subseteq \mathsf{Var}$ into Var: all points over $\mathsf{Loc} \times V$ which are identical to the given points when the valuation is restricted to $\mathsf{Var'}$.

In the case of timed automata and linear hybrid systems, all these operations are effectively computable and map zones onto zones. This is an important property to lift our algorithm from timed automata to linear hybrid systems.

4.2 Deciding Refinement

In [Čer92], the first algorithm for deciding timed bisimulation was given. This algorithm was generalized to timed refinement in [CGL93]. Note that a decision algorithm for timed refinement is implicitly an algorithm for timed bisimulation and timed forward simulation, as these are special cases of refinement (as seen on p. 2). This subsection starts with recollecting Čerāns' ideas for timed bisimulation.

Deciding bisimulation between two system H_1, H_2 means to either find a bisimulation relation on their states or to show that no such relation can exist. As the start states are required to be in the bisimulation relation, it follows that for any pair (n_1, n_2) in the bisimulation, the states n_i must be reachable by paths with the same labeling in the systems H_i. Thus only the points in the reachable part of the product graph of H_1, H_2 are candidates for elements of the bisimulation. Čerāns algorithm constructs a subgraph of $H_1 \times H_2$'s valuation graph which represents the bisimulation relation. If the construction fails there exists no bisimulation relation. The algorithm uses the *region graph* of the product. Regions are special zones: the zone (ℓ, Z) is a region if Z is the smallest valuation set representable by a simple linear formula[1], where all constants in the formula are less than a given constant c (dependent on H_1 and H_2). The important property of a region is that if a point of a region is a pair of bisimilar states, then all points of the region are pairs of bisimilar states. Therefore it is sufficient to check the pertinent property for the regions only. An example of the region graph computed by Čerāns' algorithm can be found in Fig. 1.

The main problem of this region technique is that the number of regions grows like $n!$ for n variables. In many cases it is better to keep a large zone instead of splitting it into many regions. This is especially true as convex zones (i.e. polyhedra) can be represented within the same amount of space as a region. We refer to the idea of keeping zones as large as possible in the analysis of a hybrid system as the *zone technique*. Examples of the power of the zone technique can be found e.g. in [AC+95].

---

[1] remember that Čerāns' algorithm is formulated for timed automata

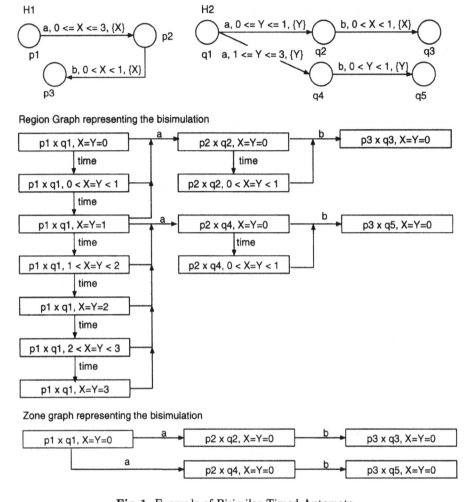

Fig. 1. Example of Bisimilar Timed Automata

Our algorithm starts from a very coarse representation of the product graph, which we call a *full backward stable zone graph* (short: FBS-graph).

Definition 12 (FBS-graph). Given a linear hybrid system $H = (\text{Loc}, \text{init}, \text{Lab}, \rightarrow, \text{Var}, \text{Act}, \text{Inv})$, its *full backward stable zone graph* is the structure $(N, n_0, \text{Lab}, \rightarrow)$ where N is the set of all zones, n_0 is the zone of all initial states, $\rightarrow \subseteq N \times \text{Lab} \times N$ is the transition relation such that for every edge $e = \ell \xrightarrow{a, \mu} \ell'$ of H and zone $n \in N$ with $\mathcal{L}(n) = \{\ell\}$ there is a transition $n \xrightarrow{a} \text{post}_e(n^\uparrow)$.

The graph is *backward stable* as for a transition $n \xrightarrow{a} n'$, every point $s' \in n'$ is reachable from n, i.e. there is $d \in \mathbb{R}^{\geq 0}$ and $s \in n$ such that $s \xrightarrow{d} \xrightarrow{a} s'$ in the underlying valuation graph. Note that the graph represents only points of the valuation graph which are reachable by an action transition. The information on time steps can only be found by additional knowledge of the hybrid system from which the FBS-graph was constructed. In general, FBS-graphs are not finite, not even for timed automata. In the case of timed automata however zones can be replaced by their closure up to regions, yielding finite zone graphs. For a zone Z, its closure $cl(Z)$ up to regions is the union of all regions R such that $R \cap Z \neq \emptyset$.

From the FBS-graph of $H_1 \times H_2$ (resp. $H_1 \times_w H_2$) our algorithm successively deletes all points $(s_1 \times s_2, v_1 \times v_2)$ where (s_1, v_1) does not refine (s_2, v_2). A point is deleted if either of its components cannot match a required resp. allowed transition of the other. After a point is deleted, all its predecessors must be re-examined. The algorithm terminates if either all points are examined or a pair of start states is removed. In the former case, the resulting zone graph represents a refinement relation, while in the latter case no refinement exists.

In addition to the region graph computed by Čerāns' algorithm, Fig. 1 gives also the zone graph computed by our algorithm, demonstrating that our algorithm computes a much smaller graph. In the worst case, our algorithm cannot be better than Čerāns', but in many practical cases our algorithm will compute a graph of a much smaller size than Čerāns' algorithm. Further our algorithm is *scaling-invariant*: multiplicating every formula in a timed automata by an arbitrary constant does not change the principal behaviour of the automaton, but is indeed a change of the time scale. While the region graph blows up exponentially with the scaling factor, our algorithm yields graphs which are the same size for the orginal and the re-scaled automaton.

The algorithm has been implemented and tested in a tool called the *Real Time Explorer*. This tool is the real-time component for MetaFrame. The implementation uses *difference bound matrices* ([Dil89]) (short: dbm) for the representation of the zones. Note that the current implementation handles modal timed automata only, and that dbm's can only represent zones describable by simple linear formulae. For detailed information on the algorithm, see [WL97,WL96].

As our algorithm uses only the operations on zones presented on page 6, it can straightforwardly be applied to linear hybrid systems. The Real Time Explorer is an extendable tool. We plan adding support for modal linear hybrid systems. In the next section we will explain how the Real Time Explorer was extended to deal with parametrized modal timed automata.

5 Applications: Parametrized Modal Timed Automata

In [LSW95] we have presented the *constraint oriented methodology*: this methodoloy allows the reduction of real-time verification problems with a large state space to a large number of small problems, where the small problems are subject to automatic verification. In [LSW95] we have shown how to reduce the verification of Fischer's protocol ([AL93]) with an arbitrary number of processes to a small verification problem on parametrized modal timed automata. Fig. 2 gives the timed automata involved in the verification: P_r, P_s are the processes with process identification r and s resp., U is an abstraction of all other processes different from $P(r), P(s)$, V is the shared variable of the processes, and $EX(r, s)$ is a specification of mutual exclusion of the processes $P(r), P(s)$. In the figure, must-transitions are straight lines, while may-transitions are dotted lines. Of course, for every must- there has to be the appropriate may-transition, but we do not picture them explicitly. The labeling of the transitions is the usual for timed automata: synchronization label, condition and the set of clocks to be reset. The proof of the correctness of Fischer's protocol is then reduced to showing

$$(b_r < c_s \Rightarrow (P(r) \mid P(s) \mid U \mid V) \setminus L \trianglelefteq EX(r, s)) \qquad (1)$$

where L is the set of actions not in $EX(r, s)$. In [LSW95], the correctness proof was done by hand, as no tool coping with weak refinement for parametrized modal timed automata was available at that time. The implementation of the improved decision algorithm for weak refinement presented in [WL97] is also not able to deal with parameters, as dbm's cannot handle parameters in timed automata.

However it is straightforward to extend dbm's to deal with parameters in simple linear formulae. A dbm is a matrix of size $(n + 1) \times (n + 1)$ in the case of n variables (i.e. clocks, as we stay within timed automata). Let $C_1, \ldots, C_n$ be the clocks of the automaton, then the entry $d_{i,j}$ of a dbm gives a (strict or non-strict) upper bound on the difference between C_i and C_j. Thus a dbm describes the zone where the values of the clocks all lie within the bounds $d_{i,j}$. The major advantages of dbm's are that all operations on zones are easily implemented and that there is a canonical form which allows comparison of zones. The "clock" C_0 in a dbm is special: it is used to give the absolute values of the clocks C_i by comparing them against Zero. Therefore the clock C_0 has always the value Zero. So in contrast to all other clocks, C_0 never changes its value, and must be treated differently in the operations. This is exactly the behaviour of a parameter: a parameter can be seen as a variable which never changes its value (cf. [AC+95]). Thus dbm's are extended to parameters by adding "clocks" for every parameter of the timed Automaton which are treated in the same way as C_0.

Checking bisimilarity changes slightly: instead of starting with the zone where all clocks are zero, we now start with the zone where all real clocks are zero and parameters can have arbitrary positive values. We only allow positive values for parameters, but we do not think that this is a severe restriction. The two automata are bisimilar if the initial zone is not removed entirely. Even better, if on termination a non-empty sub-zone of the initial zone is left, this sub-zone gives a sufficient condition on parameters for bisimilarity of the automata. This analogous to the parameter synthesis as found in the linear hybrid systems tool HyTech ([HHT95]).

We have implemented this extension of the original algorithm and tested it on Fischer's protocol. As expected, it terminates with the sufficient condition for Fischer's protocol as given in (1).

However the algorithm will not terminate in general. From [AHV93] it follows that weak bisimilarity is undecidable for parametrized modal timed automata. Thus we only get a semi-decision procedure for the case of parametrized modal timed automata.

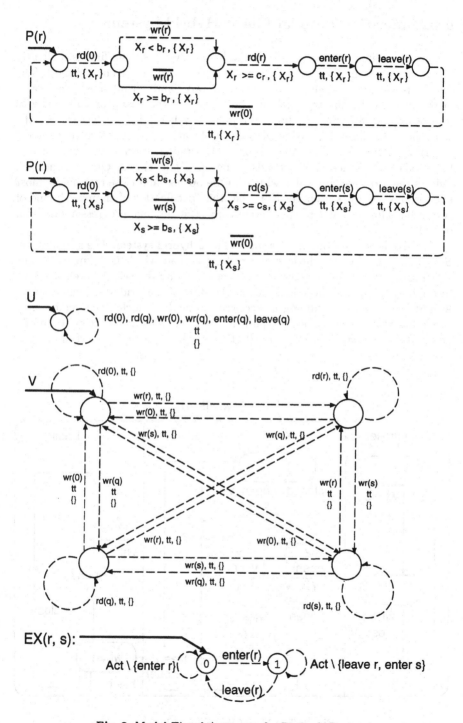

Fig. 2. Modal Timed Automata for Fischer's Protocol

6 Applications: Modal Linear Hybrid Systems

We give two simple examples for modal hybrid systems. The first is a slightly different version of the well-known water level monitor, given in Fig. 3. The system is composed from a water basin, a valve and a controller. The water in the basin sinks due to a hole in the ground of the basin, and may rise as fresh water can flow into the basin through the valve which is controlled by the controller. The water level wl of the basin sinks at 2m/sec if the valve is closed, and rises at 1m/sec if the valve is open. The original specification was correct for a valve which needed 2 seconds to be open and closed. Our specification reflects the fact that any valve which can be opened and closed within at most two seconds will work properly. Note that this specification of an upper bound of two seconds for the valve to open or close uses the maximal progress assumption of TMS.

We give a simple example of a modal linear hybrid system. The specification S in Fig. 4 models a vehicle moving along a road which is 5 km long. The specification uses two variables: an exact clock X and the vehicle's velocity VV. In the beginning the vehicle is placed at the start of the road with velocity zero. A start-event marks beginning of acceleration of the vehicle. The vehicle may accelerate within certain bounds until it reaches a velocity between 30 and 50 km/h. If the velocity is below 40 km/h, the vehicle may run for between 425 and

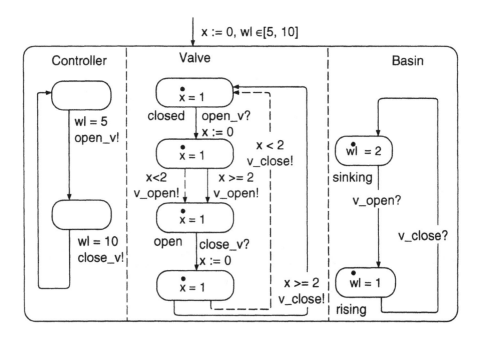

Fig. 3. The Modal Water Level Monitor

435 seconds until it has to slow down, while for a velocity between 40 and 50 km/h, the vehicle must restrain speed after 335 to 345 seconds. This guarantees that the vehicle will not crash against the wall which is at the end of the road.

This admitted behaviour is modelled by may-transitions. Note that we also require that every vehicle is at least able to reach a velocity of 40 km/h, modeled by a must-transition.

Fig. 4 gives the modal linear hybrid systems in a standard way: each location has a name and the activities for the variables, each transition has a synchronization label, a guard and the assignment to the variables (if they change their value). May- and must-transitions are straight and dotted lines as before. We do not give the invariants for the locations: each location despite stopped must be left as long as it is possible, it is not allowed to stay in a location if there is no future time point at which the location can be left by an outgoing transition.

Fig. 4 also gives a possible realization I of such a vehicle: this vehicle accelerates until it reaches a velocity between 38 and 40 km/h and starts slowing down after 430 seconds. The velocity is modelled by a variable WW, and the variable Y is another exact clock. It is easily verified that I is a refinement of S.

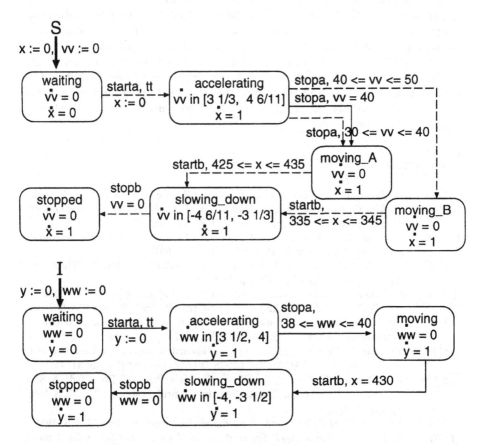

Fig. 4. The Modal Vehicle

7 Conclusion

We have presented how to generalize the idea of timed modal specifications to linear hybrid systems. We have also shown that our new decision algorithm for weak refinement for modal timed automata can be extended to an algorithm for modal linear hybrid systems. For the case of parametrized modal timed automata the algorithm is already implemented in the *Real Time Explorer*. Thus the *Real Time Explorer* can be used as the backend of the constraint oriented methodology ([LSW95]). Details of the algorithm can be found in [WL97,WL96]. We demonstrated the usefulness of the notion of modal hybrid systems with some simple application examples: the verification of Fischer's protocol involved the checking of refinement for parametrized modal timed automata, and we gave two simple examples of how to use modal linear hybrid systems.

There are two main problems with the generalization of timed modal specifications to hybrid systems. The first is that in the case of the vehicle, establishing the refinement does not imply that the implementation will reach the correct velocity, as the refinement relation only relates the time steps, but not the evolution of the variables. The second problem is that the presented refinement relation cannot be used to express properties like this: an action must occur within a given interval, but it may occur at any instant of the interval. We will investigate these questions in the future. For the first problem, it seems that modal refinement must be combined with the simulation relation for hybrid I/O automata as defined in [LSVW96]. Note however that the advantage of using our definition of refinement is that a decision algorithm and a tool are already present and usable.

References

[AL93]   M. Abadi, L. Lamport. An Old-Fashioned Recipe for Real Time. LNCS 600.

[AC+95]   R. Alur,C. Courcoubetis, N. Halbwachs, T.A. Henzinger, et al. The algorithmic analysis of hybrid systems. TCS, February 1995.

[AHV93]   R. Alur, T.A. Henzinger, M.Y. Vardi. Parametric real-time reasoning. Proc. 25th STOC, ACM Press 1993, pp. 592–601.

[Čer92]   K. Čerāns. Decidability of Bisimulation Equivalences for Parallel Timer Processes. Proceedings CAV '92, pp. 289 – 300.

[CGL93]   K. Čerāns, J.C. Godsken, K.G. Larsen. Timed Modal Specification - Theory and Tools. in: CAV '93, LNCS 697, Springer Berlin 1993, pp. 253–267.

[Dil89]   D.L. Dill. Timing Assumptions and Verification of Finite-State Concurrent Systems. in: LNCS 407, Springer Berlin 1989, pp. 197-212.

[HHT95]   Thomas A. Henzinger, Pei-Hsin Ho, Howard Wong-Toi. A User Guide to HyTech. in: Proc. TACAS '95, LNCS 1019, Springer 1995, pp. 41–71.

[LSW95]   K.G. Larsen, B. Steffen, C. Weise. Fischer's Protocol Revisited: A Simple Proof Using Modal Constraints. in: Hybrid Systems III, LNCS 1066, Springer 1996.

[LT88]   K. Larsen and B. Thomsen. A modal process logic. In: Proc. LICS, 1988.

[LV91]   N. Lynch, F. Vaandrager. Forward and Backward Simulations for Timing-Based Systems. in: REX Workshop, LNCS 600, pp. 397–446, 1991.

[LSVW96] N. Lynch, R. Segala, F. Vaandrager, H.B. Weinberg. Hybrid I/O automata. in: Hybrid Systems III, LNCS 1066, Springer 1996.

[Mil89]   R. Milner. Communication and Concurrency. Prentice-Hall, 1989.

[WL96]   C. Weise, D. Lenzkes. A Fast Decision Algorithm for Timed Refinement. AIB 96-11, Technical Report University of Tech. Aachen, 1996.

[WL97]   C. Weise, D. Lenzkes. Efficient Scaling Invariant Checking of Timed Bisimulation. To appear Proc. STACS '97

Robust Timed Automata*

Vineet Gupta[1]   Thomas A. Henzinger[2]   Radha Jagadeesan[3]

[1] Xerox PARC, 3333 Coyote Hill Road, Palo Alto, CA 94304; vgupta@parc.xerox.com
[2] EECS Department, University of California, Berkeley, CA 94720; tah@eecs.berkeley.edu
[3] Mathematical Sciences Department, Loyola Univ.-Lake Shore Campus, Chicago, IL 60626; radha@math.luc.edu

Abstract. We define *robust timed automata*, which are timed automata that accept all trajectories "robustly": if a robust timed automaton accepts a trajectory, then it must accept neighboring trajectories also; and if a robust timed automaton rejects a trajectory, then it must reject neighboring trajectories also. We show that the emptiness problem for robust timed automata is still decidable, by modifying the region construction for timed automata. We then show that, like timed automata, robust timed automata cannot be determinized. This result is somewhat unexpected, given that in temporal logic, the removal of real-time equality constraints is known to lead to a decidable theory that is closed under all boolean operations.

1 Introduction

The formalism of timed automata [AD94] has become a standard model for real-time systems, and its extension to hybrid automata [ACHH93, ACH+95, Hen96] has become a standard model for mixed discrete-continuous systems. Yet it may be argued that the precision inherent in the formalism of timed and hybrid automata gives too much expressive power to the system designer. For example, while there is a timed automaton A that issues an event a at the exact real-numbered time t, such a system cannot be realized physically. This is because for every physical realization R_A of A there is a positive real ϵ, however small, so that one can guarantee at most that R_A issues the event a in the time interval $(t - \epsilon, t + \epsilon)$. The discretization of time into units of size ϵ, on the other hand, may not allow a sufficiently abstract representation of A. Among the reasons for leaving the precision ϵ parametric are the following: the actual value of ϵ may be unknown; future realizations of A may achieve a precision smaller than ϵ; if A is an open system, it may be composed with systems whose precision is smaller than ϵ; a small ϵ may cause a dramatic increase in the state space.

Similarly, consider two hybrid automata $B_{\leq}$ and $B_{<}$ for modeling the controller of a chemical plant. The two automata are identical except that $B_{\leq}$ activates a furnace iff the plant temperature falls to T degrees, and $B_{<}$ activates the furnace iff the plant temperature falls *below* T degrees. These two formal objects differ, and may have entirely different mathematical properties; for example, some plant transition may be possible only at temperatures less than T, thus causing any number of states to be reachable

* The first and third author were supported in part by grants from ARPA and ONR. The second author was supported in part by the ONR YIP award N00014-95-1-0520, by the NSF CAREER award CCR-9501708, by the NSF grant CCR-9504469, by the AFOSR contract F49620-93-1-0056, by the ARO MURI grant DAAH-04-96-1-0341, by the ARPA grant NAG2-892, and by the SRC contract 95-DC-324.036. The third author was also supported by the NSF.

in $B_<$ but not in $B_\le$. Yet the difference between the two automata cannot be realized physically, because every physical thermometer has a positive error ϵ, however small, and cannot reliably distinguish between T and $T - \epsilon$ degrees. Again, the discretization of temperature into ϵ-units may not be adequate for reasons given above.

We remove the "excessive" expressive power of timed and hybrid automata without discretization, by having automata define (i.e., generate or accept) not individual trajectories, but bundles of closely related trajectories. A bundle of very similar trajectories is called a *tube*. For example, while a single trajectory τ may have event a at time t, every tube containing τ also contains some trajectories with event a very close to, but not exactly at time t. Formally, we suggest several metrics on the trajectories of timed and hybrid automata, and define a tube to be an *open* set of trajectories. This definition is shown to be independent of the choice of metric, because the "reasonable" metrics all induce the same topology. Then, a tube is accepted by a timed or hybrid automaton iff the accepted trajectories form a *dense* subset of the tube. Accordingly, while "isolated" accepted trajectories do not belong to any accepted tube, isolated rejected trajectories are added to accepted tubes, as motivated by the observation that an automaton ought not be able to accept or reject individual trajectories.

Timed and hybrid automata with tube acceptance are called *robust*, because they are insensitive against small input perturbations (and they may produce small output perturbations). In this paper, we look at some theoretical implications of robustness. First, we solve the emptiness problem for robust timed automata: given a timed automaton A, does A accept any tube? Our emptiness check for tube acceptance is derived from the region method of [AD94] for trajectory acceptance, but is somewhat more efficient, because only open regions need be considered. The emptiness check leads, in the usual way, to algorithms for verifying requirements of robust timed automata that are specified in a linear-time logic such as MITL [AFH96], in a branching-time logic such as TCTL [ACD93], or by event-clock automata [AFH94].

Second, we study the complementation problem for robust timed automata. Complementation is instrumental for using automata as a requirements specification language: abstract requirements of trajectories are often specified naturally using nondeterministic automata; then, in order to check that all trajectories that are generated by an implementation automaton A are accepted by the specification automaton B, the latter needs to be complemented (before checking the product of A and $\neg B$ for emptiness). While timed automata with trajectory acceptance are not closed under complement [AD94] (i.e., there is a timed automaton whose rejected trajectories are not the accepted trajectories of any other timed automaton), one may harbor some hope that robust timed automata can be complemented (i.e., for every timed automaton B there may be a timed automaton $\neg B$ that accepts precisely the tubes which are disjoint from the tubes accepted by B). This hope stems from the following observations:

1. In the case of linear-time temporal logic, the removal of all timing constraints that enforce exact real-numbered time differences between events leads to a decidable theory, called MITL, which is closed under all boolean operations [AFH96]. It is therefore not unreasonable to expect that in the case of timed automata, the removal of individual trajectories, which express exact real-numbered time differences between events, leads likewise to a decidable and boolean-closed theory.

2. The impossibility of complementation for timed automata follows from the fact that while the emptiness problem is decidable, the universality problem (i.e., given a timed automaton, does it accept all trajectories?) is not [AD94]. Undecidabil-

ity proofs for real-time problems, however, typically depend on an encoding of Turing-machine computations which uses the exact real-numbered times available in individual trajectories. These proofs do not straight-forwardly extend to tubes.

3. Since the complement of an open set is closed, the definition of complementation for timed automata with tube acceptance does not coincide with the definition of complementation for timed automata with trajectory acceptance.

We considerably dampen the hope that robust timed automata can be complemented by proving that, like ordinary timed automata, robust timed automata cannot be determinized (which is the usual first step in complementation). Indeed, the theory of ordinary timed automata turns out to be remarkably *robust* (pun intended) against perturbations in the definition of the automata: our results show that neither the syntactic removal of equality from timing constraints (open timing constraints only) nor the semantic removal of equality (tube acceptance) alter the theory of timed automata qualitatively.

2 Trajectories and Tubes

In this paper, we consider finite trajectories only. A *trajectory* over an alphabet Σ is an element of the language $(\Sigma \times \mathbb{R}^+)^*$, where $\mathbb{R}^+$ stands for the set of positive reals excluding 0. Thus, a trajectory is a finite sequence of pairs from $\Sigma \times \mathbb{R}^+$. We call the first element of each pair an *event*, and the second element the *time-gap* of the event. The time-gap of an event represents the amount of time that has elapsed since the previous event of the trajectory (the first time-gap can be thought of representing the amount of time that has elapsed since the "beginning of time"). For a trajectory τ, we denote its length (i.e., the number of pairs in τ) by $\text{len}(\tau)$, and its projection onto Σ^* (i.e., the sequence of events that results from removing the time-gaps) by $\text{untime}(\tau)$. For $1 \leq i \leq \text{len}(\tau)$, we denote the i-th event of τ by $a_\tau(i)$, and the i-th time-gap by $\delta_\tau(i)$. We also assign time-stamps to the events of a trajectory: for the i-th event of τ, the *time-stamp* is defined to be $t_\tau(i) = \sum_{1 \leq j \leq i} \delta_\tau(j)$.

Metrics on trajectories

Let the set of all trajectories be denoted **Traj**. Assuming that trajectories cannot be generated and recorded with infinite precision, in order to get an estimate of the amount of error in the data that represents a trajectory, we need a metric on **Traj**. We will not choose a specific metric, but give some examples of "reasonable" metrics, and then state a condition on "reasonableness" that will be sufficient for all later results.

For all metrics d we consider, given two trajectories τ and τ', we define $d(\tau, \tau') = \infty$ if $\text{untime}(\tau) \neq \text{untime}(\tau')$. Thus, only two trajectories with the same sequence of events have a finite distance, and finite errors may occur only in measuring time. In the following examples, assume that $\text{untime}(\tau) = \text{untime}(\tau')$.

Example 1. Define

$$d_{max}(\tau, \tau') = \max\{|t_\tau(i) - t_{\tau'}(i)| : 1 \leq i \leq \text{len}(\tau)\}.$$

This metric measures the maximal difference in the time-stamps of any two corresponding events: two timed words are close to each other if they have the same events in the same order, and the times at which these events occur are not very different. For instance, for $\tau_1 = (a, 1)(a, 1)(a, 1)$ and $\tau_2 = (a, 0.9)(a, 1.2)(a, 1.2)$, we have $d_{max}(\tau_1, \tau_2) = 0.3$. □

Example 2. The following metric considers the sum of all differences in the time-stamps:

$$d_{sum}(\tau, \tau') = \sum\{|t_\tau(i) - t_{\tau'}(i)| : 1 \le i \le \text{len}(\tau)\}.$$

For instance, $d_{sum}(\tau_1, \tau_2) = 0.5$. □

Example 3. Another metric considers the pairwise time-differences between any two events of a trajectory:

$$d_{allpair}(\tau, \tau') = \max\{|\sum_{i < k \le j} (\delta_\tau(k) - \delta_{\tau'}(k))| : 0 \le i < j \le \text{len}(\tau)\}.$$

This metric is based on the intuition that a clock may constrain or measure the distance between any two events in a trajectory. For instance, $d_{allpair}(\tau_1, \tau_2) = 0.4$. □

Example 4. An alternate metric measures only the differences in the time-gaps between consecutive events:

$$d_{sucpair}(\tau, \tau') = \max\{|\delta_\tau(i) - \delta_{\tau'}(i)| : 1 \le i \le \text{len}(\tau)\}.$$

For instance, $d_{sucpair}(\tau_1, \tau_2) = 0.2$. □

Example 5. The drift metric assumes that the clocks for measuring time-stamps may drift, and their deviation from a correct clock is a percentage of the total elapsed time. For example, if x is a correct clock, and x' is a clock with maximal drift 0.1, then always $x/1.1 \le x' \le 1.1x$. Thus we can define the distance between two trajectories as

$$d_{drift}(\tau, \tau') = \max\{\max\left(\frac{t_{\tau'}(i)}{t_\tau(i)}, \frac{t_\tau(i)}{t_{\tau'}(i)}\right) : 1 \le i \le \text{len}(\tau)\} - 1.$$

It follows that $d_{drift}(\tau, \tau') = \epsilon$ iff $t_{\tau'}(i)/(1 + \epsilon) \le t_\tau(i) \le (1 + \epsilon)t_{\tau'}(i)$ for all $1 \le i \le \text{len}(\tau)$. For instance, $d_{drift}(\tau_1, \tau_2) = 0.111\ldots$ Note that if all time-gaps in both τ and τ' were doubled, the distance $d_{drift}(\tau, \tau')$ would remain the same. This is not true of the other metrics. It is also possible to define sum, all-pairs, and successive-pairs versions of the drift metric. □

Example 6. Finally, we have the discrete metric, with $d_{disc}(\tau, \tau') = 1$ if $\tau \ne \tau'$. This metric with the tube acceptance conditions (defined below) would give us ordinary timed automata (with trajectory acceptance). We will not use this metric any further. □

Given a metric, we use the standard definition of open sets. The listed metrics, with the exception of the discrete metric, all define the same topology on trajectories. Formally, for a metric d, a trajectory τ, and a positive real $\epsilon \in \mathbb{R}^+$, define the d-tube around τ of diameter ϵ to be the set $T_d(\tau, \epsilon) = \{\tau' : d(\tau, \tau') < \epsilon\}$ of all trajectories at a d-distance less than ϵ from τ. A d-open set O, called a d-tube, is any subset of **Traj** such that for all trajectories $\tau \in O$, there is a positive real $\epsilon \in \mathbb{R}^+$ with $T_d(\tau, \epsilon) \subseteq O$. Thus, if a d-tube contains a trajectory τ, then it also contains all trajectories in some neighborhood of τ. Let the set of all d-tubes be denoted **Tube**(d).

Proposition 1. *The five metrics $d = d_{max}, d_{sum}, d_{allpair}, d_{sucpair}, d_{drift}$ all define the same set $\mathbf{Tube}(d)$ of tubes.*

Proof. We show $\mathbf{Tube}(d_{max}) \subseteq \mathbf{Tube}(d_{sum}) \subseteq \mathbf{Tube}(d_{allpair}) \subseteq \mathbf{Tube}(d_{sucpair}) \subseteq \mathbf{Tube}(d_{max}) \subseteq \mathbf{Tube}(d_{drift}) \subseteq \mathbf{Tube}(d_{max})$.

Consider a tube $O \in \mathbf{Tube}(d_{max})$ and a trajectory $\tau \in O$. Then there is a positive real $\epsilon \in \mathbb{R}^+$ such that $T_{d_{max}}(\tau, \epsilon) \subseteq O$. Since $d_{sum}(\tau, \tau') \leq \operatorname{len}(\tau) \times d_{max}(\tau, \tau')$, it follows that $T_{d_{sum}}(\tau, \epsilon/\operatorname{len}(\tau)) \subseteq T_{d_{max}}(\tau, \epsilon) \subseteq O$. Hence $O \in \mathbf{Tube}(d_{sum})$.

The other proofs are similar. In each case we need to relate two metrics. The following relations suffice, and can be easily proved from the definitions:

$$d_{allpair}(\tau, \tau') \leq d_{sum}(\tau, \tau')$$
$$d_{sucpair}(\tau, \tau') \leq d_{allpair}(\tau, \tau')$$
$$d_{max}(\tau, \tau') \leq \operatorname{len}(\tau) \times d_{sucpair}(\tau, \tau')$$
$$d_{drift}(\tau, \tau') \leq d_{max}(\tau, \tau')/t_\tau(1)$$
$$d_{max}(\tau, \tau') \leq d_{drift}(\tau, \tau') \times t_\tau(\operatorname{len}(\tau)) \qquad \square$$

We say that a metric d on trajectories is *reasonable* if $\mathbf{Tube}(d)$ is equal to $\mathbf{Tube}(d_{max})$. Henceforth we assume to be given a reasonable metric d. A d-tube will be simply called a tube, and the set of tubes will be denoted $\mathbf{Tube}$.

From trajectory languages to tube languages

A *trajectory language* is any subset of $\mathbf{Traj}$; a *tube language* is any subset of $\mathbf{Tube}$. Every trajectory language L induces a tube language $[L]$, which represents a "fuzzy" rendering of L. In $[L]$ we wish to include a tube iff sufficiently many of its trajectories are contained in L. We define "sufficiently many" as any dense subset, in the topological sense.

For this purpose we review some simple definitions from topology. A set S of trajectories is closed if its complement $S^c = \mathbf{Traj} - S$ is open. The closure $\overline{S}$ of a set S of trajectories is the least closed set containing S, and the interior S^{int} is the greatest open set contained in S. The set S' of trajectories is dense in S iff $S \subseteq \overline{S'}$.

Formally, given a trajectory language L, the corresponding tube language is defined as

$$[L] = \{O \in \mathbf{Tube} : O \subseteq \overline{L}\}.$$

Thus, a tube O is in $[L]$ if for each trajectory $\tau \in O$ there is a sequence of trajectories with limit τ such that all elements of this sequence are in L. Equivalently, L must be dense in O; that is, for every trajectory $\tau \in O$ and for every positive real $\epsilon \in \mathbb{R}^+$, there is a trajectory $\tau' \in L$ such that $d(\tau, \tau') < \epsilon$. Since the tubes in $[L]$ are closed under subsets and union, the tube language $[L]$ can be identified with the maximal tube in $[L]$, which is the interior $\overline{L}^{int}$ of the closure of L.

We will define the semantics of a robust timed automaton with trajectory set L to be the tube set $[L]$. This has the effect that a robust timed automaton cannot generate (or accept) a particular trajectory when it refuses to generate (rejects) sufficiently many surrounding trajectories. Neither can the automaton refuse to generate a particular trajectory when it may generate sufficiently many surrounding trajectories. Our definition of "sufficiently many" as "dense subset" does not seem all that strong, because every tube O, while uncountable, has dense subsets that are countable (such as the set of trajectories in O all of whose time-gaps are rationals). However, when we define timed automata below, we will see that the syntax of timed automata will not allow us to specify very strange trajectory languages L. In particular, we will not be able to specify a trajectory language L such that both L and L^c are dense in some tube O. Thus, for timed automata, a tube will be accepted iff all but finitely many of its trajectories are accepted, and it will be rejected iff all but finitely many of its trajectories are rejected.

3 Robust Timed Automata

We define a variant of Alur-Dill timed automata [AD94]. While the variant makes several aspects of our presentation easier, we will show that it is equivalent in expressive power to Alur-Dill timed automata.

A *timed automaton* is a 6-tuple $A = \langle \Sigma, Q, Q_0, Q_f, C, E \rangle$:

- Σ is a finite alphabet of events;
- Q is a finite set of locations;
- $Q_0 \subseteq Q$ is a set of start locations;
- $Q_f \subseteq Q$ is a set of accepting locations;
- C is a finite set of real-valued clock variables;
- $E \subseteq Q \times Q \times \Sigma \times \Phi(C) \times 2^C \times \Phi(C)$ is a finite set of transitions. Each transition $(q, q', a, \phi, \rho, \phi')$ consists of a source location q, a target location q', an event a, a constraint ϕ on the clocks in C, a set ρ of clock variables, and a constraint ϕ' on the clocks in ρ. The *clock constraints* $\Phi(C')$ on a set C' of clock variables are generated by the grammar

$$\phi ::= x \leq c \mid x \geq c \mid \neg\phi \mid \phi \wedge \phi,$$

where c is a rational number and $x \in C'$ is a clock variable. We call ϕ the *precondition*, ρ the *update set*, and ϕ' the *postcondition* of the transition. Intuitively, the transition is enabled when the clock values satisfy the precondition. When the transition is taken, the clock variables in the update set are assigned new values so that the postcondition is satisfied, and all other clock values remain unchanged.

Instead of precondition, update set, and postcondition, we often write guarded commands. For example, if x and y are clock variables, then the nondeterministic guarded command $x > 1 \rightarrow y := (0, 1)$ corresponds to the precondition $x > 1$, the update set $\{y\}$, and the postcondition $0 < y < 1$. It is often convenient to annotate locations with clock constraints, so-called *invariant conditions* [HNSY94]. Our results extend straight-forwardly to timed automata with invariant conditions.

A *clock-valuation function* $\gamma \colon C \rightarrow \mathbb{R}_0^+$ assigns to each clock variable a nonnegative real in $\mathbb{R}_0^+ = \mathbb{R}^+ \cup \{0\}$. The clock-valuation function γ satisfies the clock constraint ϕ iff ϕ evaluates to true when each clock x is replaced by the value $\gamma(x)$. For a positive real δ, the clock-valuation function $\gamma + \delta$ assigns to each clock x the value $\gamma(x) + \delta$.

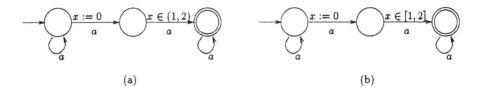

Fig. 1. The timed automata A_1 and A_2

Trajectory acceptance

A trajectory τ is *accepted* by the timed automaton A iff there is a sequence $r = \langle q_i, \gamma_i \rangle_{0 \leq i \leq \mathrm{len}(\tau)}$ of locations $q_i \in Q$ and clock-valuation functions $\gamma_i : C \to \mathbb{R}_0^+$ such that

1. Initialization: $q_0 \in Q_0$.
2. Consecution: for all $1 \leq i \leq \mathrm{len}(\tau)$, there is a transition in E of the form $(q_{i-1}, q_i, a_\tau(i), \phi_i, \rho_i, \phi_i')$ such that $\gamma_{i-1} + \delta_\tau(i)$ satisfies ϕ_i, $\gamma_i(x) = \gamma_{i-1}(x) + \delta_\tau(i)$ for all $x \in C - \rho_i$, and γ_i satisfies ϕ_i'.
3. Acceptance: $q_k \in Q_f$ for $k = \mathrm{len}(\tau)$.

The sequence r is called a *run* of τ through the timed automaton A, and τ is said to be accepted along the *path* $\langle q_i \rangle_{0 \leq i \leq \mathrm{len}(\tau)}$ of locations. We write $L(A)$ for the set of trajectories accepted by A.

In an *Alur-Dill timed automaton*, if ρ is the update set of a transition, then the corresponding postcondition must have the form $\bigwedge_{x \in \rho} x = 0$; that is, the clock variables in the update set are always reset to 0.

Proposition 2. *For each timed automaton there is an Alur-Dill timed automaton that accepts the same set of trajectories.*

Proof. Every time a clock variable is updated and assigned, nondeterministically, a new value in the interval I_1, and later tested against the interval I_2, we replace the update with a reset to 0, and the test with a test against the interval $I_2 - I_1 = \{\delta_2 - \delta_1 : \delta_1 \in I_1, \delta_2 \in I_2\}$. If a clock variable is updated into different intervals on different transitions, these updates are handled using new clocks. □

Tube acceptance

The timed automaton A accepts the set $[L(A)]$ of tubes. That is, a tube O is accepted by A iff there is a set $O' \subseteq O$ of trajectories such that O' is dense in O and all trajectories in O' are accepted by A.

The following examples illustrate tube acceptance. First, consider the timed automaton A_1 of Figure 1(a). This automaton accepts all trajectories over the unary alphabet $\{a\}$ which contain two consecutive a events with a time-gap in the open interval $(1, 2)$. This property is invariant under sufficiently small perturbations of the time-stamps. Hence the automaton A_1 accepts precisely those tubes that consist of

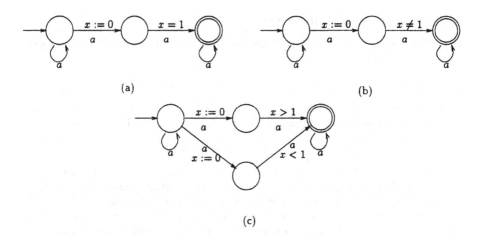

(a)

(b)

(c)

Fig. 2. The timed automata A_3, A_4, and A_5

trajectories in $L(A_1)$, and the maximal accepted tube is $L(A_1)$ itself. In the timed automaton A_2 of Figure 1(b), the open interval $(1, 2)$ is replaced by the closed interval $[1, 2]$. This changes the set of accepted trajectories but not the set of accepted tubes: $L(A_1) \subset L(A_2)$ but $[L(A_1)] = [L(A_2)]$. Notice that the "boundary trajectories" accepted by A_2, with two consecutive a's at a time-gap of 1 or 2 but no consecutive a's at a time-gap strictly between 1 and 2, are not accepted robustly, because there are arbitrarily small perturbations that are not acceptable.

Next, consider the timed automaton A_3 of Figure 2(a). This automaton accepts all trajectories of a's in which some a is followed by the subsequent a exactly 1 time unit later. The automaton A_3 accepts no tubes, because by perturbing the pair of a's that causes a trajectory to be accepted, however slightly, the trajectory becomes unacceptable. By contrast, the timed automaton A_4 of Figure 2(b) accepts all trajectories of a's such that there is a pair of consecutive a's that are not 1 time unit apart. Thus, by perturbing an unacceptable trajectory slightly, it will be accepted. As a result, the automaton A_4 accepts all tubes. The timed automaton A_5 of Figure 2(c) accepts the same trajectories as A_4, but along different paths of locations. Since our definition of tube acceptance depends only on the accepted trajectories, and not on the structure of the automaton, the automaton A_5 still accepts all tubes. Note, however, that some tubes, such as the tube of all trajectories, are not accepted along any single path through the automaton.

4   Verification of Robust Timed Automata

The basic verification problem for automata is the emptiness problem. In this section, we give an algorithm for checking if a timed automaton accepts any tube. For this purpose, we relate tube acceptance and trajectory acceptance by considering syntactic subclasses of timed automata.

Closed and open timed automata

A timed automaton A is *closed* iff all preconditions and postconditions of A are generated by the grammar

$$\phi ::= x \leq c \mid x \geq c \mid \phi \wedge \phi \mid \phi \vee \phi,$$

where c is a rational number and c is a clock variable. Dually, the timed automaton A is *open* iff all preconditions and postconditions of A are generated by the grammar

$$\phi ::= x < c \mid x > c \mid \phi \wedge \phi \mid \phi \vee \phi.$$

For each timed automaton A, we define the closure automaton and the interior automaton as follows. First, write every precondition and postcondition of A in negation-free form, using both logical-and and logical-or and both strict and nonstrict comparison operators. Then, the *closure automaton* $\overline{A}$ is the timed automaton that results from replacing each strict comparison operator by the corresponding nonstrict operator (replace $<$ by $\leq$, and replace $>$ by $\geq$). For example, $\overline{A_1} = A_2$ for the timed automata of Figure 1.

The interior automaton A^{int} cannot be formed by the reverse of the above process, as some of the postconditions would be replaced by "false." Instead, we relax the postconditions and tighten up the preconditions. For example, we obtain the interior automaton of the timed automaton A_2 of Figure 1(b) by replacing the constraint $x := 0$ by $x :\in (0, 0.5)$, and replacing the constraint $x \in [1, 2]$ by $(1.5, 2)$. This construction, however, is not sufficient if more than one test is performed. Consider the automaton shown on the left in Figure 3. Its interior is shown on the right in Figure 3. The two clocks x_1 and x_2 are necessary to capture the maximal and minimal possible values of the clock x. Then, any number of tests can be satisfied.

Formally, we assume that the clock constraints of the timed automaton A contain only integer constants. This can be ensured by multiplying all constants with their least common denominator, and once the interior automaton has been constructed, dividing all constants again by the same amount. For each clock variable x of A, the *interior automaton* A^{int} uses two clock variables, x_1 and x_2. In A^{int}, each clock constraint in a postcondition of A is replaced as follows:

$$x \geq c \qquad\qquad x_1 > c \wedge x_2 > c$$
$$x > c \qquad\qquad x_1 > c \wedge x_2 > c$$
$$x \leq c \qquad\qquad x_1 < c + 0.5 \wedge x_2 < c + 0.5$$
$$x < c \qquad\qquad x_1 < c + 0.5 \wedge x_2 < c + 0.5$$

In addition, each clock constraint in a precondition of A is replaced as follows:

$$x \geq c \qquad\qquad x_1 > c + 0.5 \vee x_2 > c + 0.5$$
$$x > c \qquad\qquad x_1 > c + 0.5 \vee x_2 > c + 0.5$$
$$x \leq c \qquad\qquad x_1 < c \vee x_2 < c$$
$$x < c \qquad\qquad x_1 < c \vee x_2 < c$$

Clearly, $\overline{A}$ is a closed automaton and A^{int} is open. Moreover, if A is closed, then $\overline{A} = A$, and if A is open, it is easy to check that $L(A^{int}) = L(A)$. The following proposition shows that as far as tube languages are concerned, we can either restrict our attention to closed automata or to open automata.

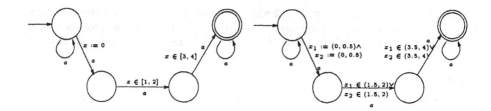

Fig. 3. A timed automaton and its interior automaton

Proposition 3. *For every timed automaton A, we have $L(\overline{A}) = \overline{L(A)}$ and $[L(A)] = [L(\overline{A})] = [L(A^{int})]$.*

Proof. Consider a trajectory $\tau \in L(\overline{A})$. By the construction of $\overline{A}$, in every ϵ-neighborhood of τ, there must exist a trajectory that is accepted by A along the same path. Hence $L(\overline{A}) \subseteq \overline{L(A)}$. Conversely, since $L(\overline{A})$ is closed, $\overline{L(A)} \subseteq L(\overline{A})$. Since $\overline{L(\overline{A})} = \overline{L(A)}$, we have $[L(\overline{A})] = [L(A)]$.

We now show that $L(A^{int}) \subseteq L(A)$, from which it follows that $[L(A^{int})] \subseteq [L(A)]$. By the construction of A^{int}, each postcondition $x \in (a, b)$ is changed to $x_1 \in (a, b + 0.5) \wedge x_2 \in (a, b+0.5)$, and each precondition $x \in (c, d)$ is changed to $x_1 \in (c+0.5, d) \vee x_2 \in (c+0.5, d)$. Thus the amount of time that may elapse between the setting and the testing of any of the clocks x, x_1, and x_2 is characterized by the interval $(c - b, d - a)$.

Conversely, consider a tube $O \in [L(A)]$ and a trajectory $\tau \in O$. Then there exists a positive real $\epsilon > 0$ such that $T(\tau, \epsilon) \subseteq O$. Since $T(\tau, \epsilon) \subseteq \overline{L(A)}$, there exists a trajectory $\tau' \in T(\tau, \epsilon)$ such that τ' is accepted strictly by A; that is, τ' is accepted along a path of A such that none of the nonstrict preconditions are satisfied at the boundary. This is because if $L_N(A)$ is the set of trajectories in $L(A)$ that are not accepted strictly by A, then the interior of $\overline{L_N(A)}$ is empty. Now τ' is accepted along the same path in A^{int}: when the clock x is set to t, let $x_1 = t$ and let $x_2 = t + 1/2$. Thus we can construct a sequence of trajectories accepted by A^{int} with limit τ, for each $\tau \in O$. Hence $O \in [L(A^{int})]$. $\square$

The following proposition shows that for open timed automata, tube emptiness coincides with trajectory emptiness.

Proposition 4. *For every open timed automaton A and every trajectory τ, if τ is accepted by A along some path, then there is a positive real $\epsilon \in \mathbb{R}^+$ such that all trajectories in the tube $T(\tau, \epsilon)$ are accepted by A along the same path.*

Proof. It suffices to consider the metric d_{max}, because any d_{max}-tube contains a d-tube for every reasonable metric d. Consider a run $r = \langle q_i, \gamma_i \rangle_{0 \le i \le \mathrm{len}(\tau)}$ of A that accepts the trajectory τ. Since all clock constraints are open, for each $0 \le i \le \mathrm{len}(\tau)$, there is a real $\epsilon_i > 0$ such that substituting $\gamma_i + \epsilon_i$ or $\gamma_i - \epsilon_i$ for γ_i in r still gives a run through A. Now let $\epsilon = \min\{\epsilon_i : 0 \le i \le \mathrm{len}(\tau)\}$. $\square$

From the proof it follows that in the case of $d = d_{max}$, if a trajectory τ is accepted by an open timed automaton A whose clock-constraint constants are all integers, then τ

belogs to a d-tube of diameter $\epsilon = 1/2$ which is accepted by A. It should also be noticed that for every closed timed automaton A, we have $\bigcup[L(A)] \subseteq L(A)$, and for every open timed automaton B, we have $L(B) \subseteq \bigcup[L(B)]$. The latter follows from Proposition 4.

Checking emptiness

By Proposition 3 we can reduce the problem of checking if a timed automaton A accepts any tube to the problem of checking if the interior automaton A^{int} accepts any tube. Moreover, by Proposition 4, the open automaton A^{int} accepts any tube iff it accepts any trajectory. The latter problem can be solved using the region construction of [AD94]. In fact, for checking the emptiness of open timed automata such as A^{int}, only open regions need be considered.

Theorem 5. *The problem of deciding whether a timed automaton accepts any tube is complete for PSPACE.*

Proof. Given an open timed automaton A, we construct an open-region automaton $reg(A)$, which is a finite-state machine that accepts a string s iff A accepts a trajectory τ with untime$(\tau) = s$. First, we multiply all clock-constraint constants in A with their least common denominator, so that all resulting constants are integers. Let c_{max} be the largest of these integers. An *open clock region* is a satisfiable conjunction of formulas that contains

- for each clock x of A, either the conjunct $x > c_{max}$ or a conjunct of the form $c < x < c + 1$, for an integer $0 \le c \le c_{max}$, and
- for each pair of clocks x and y of A, either the conjunct $x - \lfloor x \rfloor < y - \lfloor y \rfloor$ or the conjunct $y - \lfloor y \rfloor < x - \lfloor x \rfloor$.

For two open clock regions R and R', the region R' is a *successor region* of R iff there is a clock-valuation function γ and a real $\delta \in \mathbb{R}^+$ such that R satisfies γ and R' satisfies $\gamma + \delta$.

The input alphabet Σ of the finite-state machine $reg(A)$ is the same as for A. Each state of $reg(A)$ is a pair $\langle q, R \rangle$ that consists of a location q of A and an open clock region R. The state $\langle q, R \rangle$ is a start state of $reg(A)$ iff q is a start location of A, and $\langle q, R \rangle$ is an accepting state of $reg(A)$ iff q is an accepting location of A. For each $a \in \Sigma$, there is an a-transition from the state $\langle q, R \rangle$ to the state $\langle q', R' \rangle$ in $reg(A)$ iff A has a transition of the form $(q, q', a, \phi, \rho, \phi')$ such that R implies ϕ and R' implies both ϕ' and ϕ'', which results from ϕ by replacing with "true" all comparisons that involve clocks in ρ. In addition, there is an ε-transition from the state $\langle q, R \rangle$ to the state $\langle q', R' \rangle$ in $reg(A)$ iff $q = q'$ and R' is a successor region of R.

PSPACE-completeness follows from the corresponding proof in [AD94]. □

5 Nondeterminizability of Robust Timed Automata

The previous section shows that timed automata yield a decidable theory of tubes. In this section, we present evidence that the resulting theory of tubes is not closed under all boolean operations. Using trajectory-based methods, for every two timed automata A and B, we can construct a product automaton C with $[L(C)] = [L(A)] \cap [L(B)]$[4]

---

[4] From the results of this section it will follow that $[L(A)] \cap [L(B)] = [L(A) \cap L(B)]$.

and a union automaton D with $[L(D)] = ([L(A)] \cup [L(B)])^* = [L(A) \cup L(B)]$, where $\mathcal{L}^*$ denotes the closure of a tube language $\mathcal{L}$ under union (i.e., $\mathcal{L}^*$ is the least set of tubes containing $\mathcal{L}$ which is closed under union). As in ordinary timed automata, however, complementation presents a problem.

Complementation

The timed automaton B is a *trajectory complement* of the timed automaton A iff B accepts precisely the trajectories that are not accepted by A; that is, $L(B) = L(A)^c$. Before defining the tube complements of a timed automaton, we observe an important property of the trajectory languages that can be defined by timed automata.

Proposition 6. *For every timed automaton A, there is no tube O such that both $L(A)$ and $L(A)^c$ are both dense in O.*

Proof. Suppose that $L(A)$ is dense in O. Then $O \subseteq \overline{L(A)}$, and $O \in [L(A)] = [L(A^{int})]$. By Proposition 4, for each trajectory $\tau \in O$ there is a positive real $\epsilon \in \mathbb{R}^+$ such that $T(\tau, \epsilon) \subseteq L(A^{int}) \subseteq L(A)$. Hence $L(A)^c$ is not dense in O. □

It follows that a tube cannot be accepted by both a timed automaton A and a trajectory complement of A. This observation will allow us to relate the tube complements of a timed automaton to its trajectory complements.

For defining the tube complements of a timed automaton A, it is not useful to consider the boolean complement **Tube** $-[L(A)]$ of the tube language $[L(A)]$. For $[L(A)]$ is closed under subsets and union. Therefore, unless $[L(A)] = \emptyset$ or $[L(A)] = $ **Tube**, the boolean complement **Tube** $-[L(A)]$ cannot be induced by any trajectory language and, hence, cannot be accepted by any timed automaton. Thus, for every tube language $\mathcal{L} \subseteq $ **Tube**, we define the *tube complement* of $\mathcal{L}$ to be the set

$$\mathcal{L}^c = \{O \in \textbf{Tube} : O \cap \bigcup \mathcal{L} = \emptyset\}$$

of tubes that are disjoint from the tubes in $\mathcal{L}$. The following proposition shows that for every timed automaton A, the tube complement $[L(A)]^c$ is induced by the trajectory complement $L(A)^c$; that is, $[L(A)^c] = [L(A)]^c$.

Proposition 7. *If L is a trajectory language and there is no tube O such that both L and L^c are dense in O, then $[L]^c = [L^c]$.*

Proof. Let O be a tube in $[L]$, and let τ be a trajectory in O. Then there is an $\epsilon > 0$ and a tube around τ of diameter ϵ whose trajectories are all in O. Suppose that $\tau \in O'$ for some tube $O' \in [L^c]$. Then there is an $\epsilon' > 0$ and a tube around τ of diameter ϵ' whose trajectories are all in O'. Without loss of generality, assume that $\epsilon \leq \epsilon'$. Thus the trajectories of the ϵ-tube around τ are contained in both $\overline{L}$ and $\overline{L^c}$. This contradicts the fact that there is no tube in which both L and L^c are dense. Hence the tubes in $[L^c]$ are pairwise disjoint from the tubes in $[L]$. □

For two timed automata A and B, we say that B is a *tube complement* of A iff B accepts precisely the tubes that do not intersect any tube accepted by A; that is, $[L(B)] = [L(A)]^c$. From Propositions 6 and 7, it follows that every trajectory complement of a timed automaton is also a tube complement (the converse is generally not true). Since

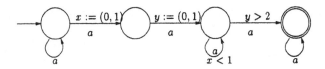

Fig. 4. A nondeterminizable open timed automaton

$[L(A)]^c = [L(A^{int})]^c = [L(A^{int})^c]$, in order to construct tube complements, it would suffice to construct trajectory complements of open timed automata.[5] This, however, does not seem feasible, because we now show that open timed automata cannot be determinized, which is the usual first step in automaton complementation.

Nondeterminizability of open timed automata

A timed automaton A is *tube-determinizable* iff there is a deterministic Alur-Dill timed automaton that accepts the set $[L(A)]$ of tubes. An Alur-Dill timed automaton is *deterministic* iff for all locations, every two outgoing transitions contain either different events or mutually exclusive preconditions. Note that trajectory-determinizability implies tube-determinizability, but not vice versa.

Theorem 8. *The open timed automaton A of Figure 4 is not tube-determinizable.*

Proof. The automaton A accepts a trajectory over the unary alphabet $\{a\}$ iff there is some consecutive pair of a's with time-stamps t and t' such that there are no a's with time-stamps in the interval $[t+1, t'+1]$. Every such trajectory is accepted robustly, as part of an accepted tube; that is, $L(A) = \bigcup[L(A)]$. To accept any tube in $[L(A)]$ with sufficiently small diameter deterministically, an automaton would have to remember the time-stamps of all a's within the last 1 time unit, which is not possible with a finite number of clock variables.

Formally, suppose there is a deterministic Alur-Dill timed automaton B with n clock variables and $[L(B)] = [L(A)]$. For simplicity, assume that all clock-constraint constants of B are integers, and assume the metric $d_{sucpair}$ on trajectories. For $\delta_0 = 0$, consider a trajectory τ of the form $(a, \delta_1) \ldots (a, \delta_{n+2})(a, (\delta_0 + \delta_1)/2) \ldots (a, (\delta_{n+1} + \delta_{n+2})/2)$ with $t_\tau(n+2) = 1$. The trajectory τ is rejected robustly by A, as part of a rejected tube. Hence we can choose a positive real $\epsilon \in \mathbb{R}^+$ such that $O = T(\tau, \epsilon) \in [L(A)]^c$ and $\epsilon < \delta_i/8$ for all $1 \leq i \leq n+2$. Then $O \in [L(B)]^c = [L(B)^c]$. Since B is deterministic and has at most n clock variables, for each trajectory $\tau' \in O \cap L(B)^c$, there is at most one run of B over τ'. After reading the first $n+2$ events of this run, there is at least one $1 \leq i \leq n+1$ such that no clock variable of B has a value in the interval $(1 - t_\tau(i) - \epsilon, 1 - t_\tau(i) + \epsilon)$. We partition the trajectories in $O \cap L(B)^c$ into $n+1$ sets, corresponding to the possible values for i. At least one of these sets must be dense in $O \cap L(B)^c$. Let this be the k-th set. Now consider the set O' of trajectories obtained from $O \cap L(B)^c$ by reducing the time-gap of each $(n+k+3)$-rd event by $\delta_{k+1}/2 + \delta_k/4$,

---

[5] Similarly, since $[L(A)]^c = [L(\overline{A})]^c = [L(\overline{A})^c]$, it would suffice to construct trajectory complements of closed timed automata. This, however, is known to be impossible [AD94].

increasing the time-gap of each $(n+k+4)$-th event by the same amount, and truncating each sequence after $n + k + 4$ events. All trajectories in O' are rejected by B, because they follow the same paths as the corresponding trajectories in O, which, if truncated after $n + k + 4$ events, are also rejected. But all trajectories in O' are accepted by A. Since O' is dense in some tube, $[L(B)] \neq [L(A)]$. $\square$

We suspect that the open timed automaton A has no tube complement. For, a tube complement of A would have to accept all trajectories of a's such that every consecutive pair of a's with time-stamps t and t' is followed by another a with a time-stamp in the interval $(t + 1, t' + 1)$. For this purpose, the automaton would have to remember the time-stamps of an unbounded number of a's, which does not seem possible (however, we know of no formal proof, as the above proof depends on the determinism of B).

6 Robust Hybrid Automata

The definitions of tube acceptance can be extended to hybrid automata. If **HTraj** is the set of hybrid trajectories, then each hybrid automaton accepts a subset of **HTraj** [ACH+95]. Given a metric on **HTraj**, we again define tubes as the open sets of the corresponding topology. Now, following our definition for timed automata, a tube is accepted by a hybrid automaton iff a dense subset of trajectories in the tube are accepted by the automaton. Several metrics on hybrid trajectories can be defined similar to the corresponding metrics for timed trajectories (Section 2). Here we propose three additional metrics.

A *hybrid trajectory* σ over a given set of real-valued variables V is a piecewise smooth function $\sigma : I \to (V \to \mathbb{R})$ from a bounded interval $I \subset \mathbb{R}_0^+$ of the nonnegative real line to valuation functions for the variables in V. By piecewise smooth we mean that the domain of σ can be partitioned into a finite sequence $I = I_1 \cup \ldots \cup I_m$ of intervals such that for each $1 \leq j \leq m$ and each variable $x \in V$, the real-valued function $\sigma(x)$ restricted to the domain I_j is infinitely differentiable (i.e., $(\sigma \restriction I_j)(x) \in \mathbf{C}^\infty$).

Suppose that $V = \{x_1, \ldots, x_n\}$. Each valuation function for V is a point in $\mathbb{R}^n$. For two points p and p' in $\mathbb{R}^n$, let $d_{euc}(p, p')$ be the euclidean distance between p and p':

$$d_{euc}(p, p') = \sqrt{(p_1 - p_1')^2 + \cdots + (p_n - p_n')^2}.$$

The timed metric on hybrid trajectories compares the values of all variables at each point in time: if two hybrid trajectories σ and σ' have different domains, $d_{time}(\sigma, \sigma') = \infty$; otherwise, if $\mathrm{dom}(\sigma)$ is the domain of both σ and σ', then

$$d_{time}(\sigma, \sigma') = \sup\{d_{euc}(\sigma(t), \sigma(t)) : t \in \mathrm{dom}(\sigma)\}.$$

Alternatively, each hybrid trajectory σ can be regarded as a subset of the $(n + 1)$-dimensional real space $\mathbb{R}^{n+1}$, with one component representing time, and the other components representing values for the variables in V: let $(p_0, p_1, \ldots, p_n) \in \sigma$ iff $p_0 \in \mathrm{dom}(\sigma)$ and $\sigma(p_0)(x_i) = p_i$ for all $1 \leq i \leq n$. Then the distance of a point $p \in \mathbb{R}^{n+1}$ from a hybrid trajectory $\sigma' \subset \mathbb{R}^{n+1}$ can be defined using the euclidean metric on $\mathbb{R}^{n+1}$:

$$d_{euc}(p, \sigma') = \inf\{d_{euc}(p, p') : p' \in \sigma'\}.$$

Now the distance between two hybrid trajectories σ and σ' can be defined as

$$d_{euc}(\sigma, \sigma') = \max(\sup\{d(p, \sigma') : p \in \sigma\}, \sup\{d(p', \sigma) : p' \in \sigma'\}).$$

While the metric d_{euc} treats time as a data variable, one can also project away the time component and look at hybrid trajectories as subsets of the phase space $\mathbb{R}^n$. This gives us the metric d_{phase}.

Consider, for example, the two functions f_1 and f_2 with $f_1(t)(x) = t$ and $f_1(t)(x) = t+2$ for all $t \in \mathbb{R}_0^+$. For the two hybrid trajectories $\sigma_1 = (f_1 \upharpoonright [0,5])$ and $\sigma_2 = (f_2 \upharpoonright [0,5])$, we have $d_{time}(\sigma_1, \sigma_2) = 2$. For the two hybrid trajectories $\sigma_1 = (f_1 \upharpoonright [1,6])$ and $\sigma_2 = (f_2 \upharpoonright [0,5])$, we have $d_{euc}(\sigma_1, \sigma_2) = \sqrt{2}$. For the two hybrid trajectories $\sigma_1 = (f_1 \upharpoonright [2,7])$ and $\sigma_2 = (f_2 \upharpoonright [0,5])$, we have $d_{phase}(\sigma_1, \sigma_2) = 0$ and, for a hybrid extension of the metric d_{max} from Section 2, $d_{\sup}(\sigma_1, \sigma_2) = 1$.

Linear hybrid automata can be analyzed for tube acceptance as in the case of trajectory acceptance [AHH96], but only open regions (open polyhedral sets in $\mathbb{R}^n$) are needed during the computation. This significantly simplifies the algorithms that have been implemented in tools such as HyTech [HHWT95]. We conclude by posing an important open question: are there interesting classes of hybrid automata whose emptiness is undecidable under trajectory acceptance but decidable under tube acceptance?

References

[ACD93] R. Alur, C. Courcoubetis, and D.L. Dill. Model checking in dense real time. *Information and Computation*, 104(1):2–34, 1993.

[ACH+95] R. Alur, C. Courcoubetis, N. Halbwachs, T.A. Henzinger, P.-H. Ho, X. Nicollin, A. Olivero, J. Sifakis, and S. Yovine. The algorithmic analysis of hybrid systems. *Theoretical Computer Science*, 138:3–34, 1995.

[ACHH93] R. Alur, C. Courcoubetis, T.A. Henzinger, and P.-H. Ho. Hybrid automata: an algorithmic approach to the specification and verification of hybrid systems. In R.L. Grossman, A. Nerode, A.P. Ravn, and H. Rischel, editors, *Hybrid Systems I*, Lecture Notes in Computer Science 736, pages 209–229. Springer-Verlag, 1993.

[AD94] R. Alur and D.L. Dill. A theory of timed automata. *Theoretical Computer Science*, 126:183–235, 1994.

[AFH94] R. Alur, L. Fix, and T.A. Henzinger. A determinizable class of timed automata. In D.L. Dill, editor, *CAV 94: Computer-aided Verification*, Lecture Notes in Computer Science 818, pages 1–13. Springer-Verlag, 1994.

[AFH96] R. Alur, T. Feder, and T.A. Henzinger. The benefits of relaxing punctuality. *Journal of the ACM*, 43(1):116–146, 1996.

[AHH96] R. Alur, T.A. Henzinger, and P.-H. Ho. Automatic symbolic verification of embedded systems. *IEEE Transactions on Software Engineering*, 22(3):181–201, 1996.

[Hen96] T.A. Henzinger. The theory of hybrid automata. In *Proceedings of the 11th Annual Symposium on Logic in Computer Science*, pages 278–292. IEEE Computer Society Press, 1996. Invited tutorial.

[HHWT95] T.A. Henzinger, P.-H. Ho, and H. Wong-Toi. HyTech: the next generation. In *Proceedings of the 16th Annual Real-time Systems Symposium*, pages 56–65. IEEE Computer Society Press, 1995.

[HNSY94] T.A. Henzinger, X. Nicollin, J. Sifakis, and S. Yovine. Symbolic model checking for real-time systems. *Information and Computation*, 111(2):193–244, 1994. Special issue for LICS 92.

Data-Structures for the Verification of Timed Automata*

Eugene Asarin[3] Marius Bozga[1] Alain Kerbrat[1]
Oded Maler[1] Amir Pnueli[2] Anne Rasse[1]

[1] VERIMAG, Centre Equation, 2, av. de Vignate, 38610 Gières, France,
Oded.Maler@imag.fr
[2] Dept. of Computer Science, Weizmann Inst. Rehovot 76100, Israel,
amir@wisdom.weizmann.ac.il
[3] Institute for Information Transmission Problems, 19 Bol. Karetnyi per., 101447
Moscow, Russia, asarin@ippi.ac.msk.su

Abstract. In this paper we suggest *numerical decision diagrams*, a BDD-based data-structure for representing certain subsets of the Euclidean space, namely those encountered in verification of timed automata. Unlike other representation schemes, NDD's are *canonical* and provide for all the necessary operations needed in the verification and synthesis of timed automata. We report some preliminary experimental results.

1 Introduction

Consider a transition system $A = (Q, \delta)$, where Q is the set of *states* and $\delta : Q \mapsto 2^Q$ is a *transition function*, mapping each state $q \in Q$ into the set of q-*successors* $\delta(q) \subseteq Q$.

The problem of calculating or characterizing all the states reachable from a subset $F \subseteq Q$ of the state-space is one of the central problems in verification. The basic algorithm to calculate this set of states is the following:

$$F_0 := F$$
$$\textbf{for } i = 0, 1, \ldots, \textbf{ repeat}$$
$$F_{i+1} := F_i \cup \delta(F_i)$$
$$\textbf{until } F_{i+1} = F_i$$

where $\delta(F_i) = \bigcup_{q \in F_i} \delta(q)$.

Symbolic methods [BCM+93], [McM93] have proved to be a very useful tool in the analysis of large discrete transition systems composed of many interacting components. Instead of transforming the description of the system into an enormous "flat" transition table over[4] $I\!\!B^m$, on which reachability analysis is practically impossible, these methods represent the transition relation as a formula over the state variables. Given such a formula $\mathcal{T}$ and a formula P describing

---

* This research was supported in part by the European Community projects HYBRID EC-US-043 and INTAS-94-697. VERIMAG is a joint laboratory of CNRS and UJF.
[4] We use $I\!\!B$ for $\{0, 1\}$ and $I\!\!R$ for the non-negative reals.

the subset F of the state-space one can calculate a new formula P' characterizing the set $\delta(F)$ of immediate successors of F. Iterating the procedure until a fixed-point is reached yields a formula P^* characterizing the set of all states reachable from F. When δ is expressed by a formula $\mathcal{T}(X, X')$, and F by a formula $P(X)$. the above algorithm can be reformulated as:

$$P_0(X) := P(X)$$
for $i = 0, 1, \ldots,$ **repeat**
$$P_{i+1}(X) := P_i(X) \vee \exists Y \, (P_i(Y) \wedge \mathcal{T}(Y, X))$$
until $P_i(X) = P_{i+1}(X)$

The essence of any symbolic method is a data-structure for representing sets (equivalently, the formulas characterizing them) on which the above operations can be performed, in particular the forward (or backward) projection (line 3), boolean operations and equivalence testing. Binary decision diagrams (BDD's) [Bry86] are such a data-structure for boolean domains. The calculation of the forward projection is relatively-easy on large practical problems and the space requirements for the representations are reasonable. Given an ordering of the variables, BDD's also have the canonicity property: all equivalent formulas lead to the same BDD and equivalence testing is thus trivial.

The verification of timed automata introduces an additional ingredient, that is, a set of continuous variables (clocks) ranging over non-countable domains. The dynamics of the passage of time cannot be captured by a "next-state" transition relation, and symbolic methods are unavoidable as states and trajectories cannot be enumerated. The sets encountered in reachability analysis of timed automata are thus certain subsets of $\mathbb{B}^m \times \mathbb{R}^d$. While the discrete part is standard, the subsets of $\mathbb{R}^d$ that need to be represented and manipulated are what we call *k-polyhedral sets*, namely sets definable by a boolean combination of basic inequalities of the form $x_i < c$, $x_i \leq c$, $x_i - x_j < c$ and $x_i - x_j \leq c$, for $i, j \in \{1, \ldots, d\}$ and $c \in \{0, \ldots, k\}$. Such polyhedral sets have been called *regions* in [AD94].

As long as these polyhedra are convex (i.e., definable by conjunctions of basic inequalities and their negations), there exists a canonical representation, the difference bounds matrix (DBM, see for example [Dil89]). This is a $(d+1) \times (d+1)$ matrix with entries taken from $\{0, \ldots, k\}$ denoting the constants in a non-redundant set of inequalities whose intersection forms the region. For this representation, the intersection is done very easily via min and max operations. The forward and backward projections via elimination of the time quantifier are also done very efficiently on DBM's. Things however get complicated when we have arbitrary unions of convex polyhedra. In this case there is no unique representation and most tools represent such sets as a list of DBM's. The more "non-convex" the set becomes, more matrices are required in order to represent it and this makes equivalence testing and redundancy elimination difficult. Moreover, it is not clear how this representation is to be combined with a symbolic representation of the discrete part.

In this paper, we suggest an alternative BDD-based data-structure, *Numerical Decision Diagram* (NDD) that has a caonicity property: *given an ordering of the clock variables, every k-polyhedral set has a unique minimal representation.* For this data-structure we have boolean set-theoretic operations and equivalence testing for free.[5] We present an algorithm to calculate forward and backward projection in time for this data-structure and thus have all the ingredients needed in order to do reachability analysis for timed automata. Since this representation is BDD-based it can be combined naturally with symbolic methods for the discrete part of the system.

The rest of the paper is organized as follows. In section 2 we present timed automata and define the components of their reachability analysis algorithms. In section 3 we define NDD's and their forward projection algorithm for the *discrete-time* interpretation of timed automata. In section 4 we show how a discretization scheme, first reported in [GPV94], can be used to extend the scope of NDD's to the *dense-time* interpretation. Finally we present some experimental results.

2 Timed Automata

First, some notations. We use bold-face letters to denote points in $\mathbb{R}^d$. Thus, $\mathbf{v}$ stands for $(v_1, \ldots, v_d)$, where $v_i \in \mathbb{R}$, for every $i = 1, \ldots, d$. For points $\mathbf{u}, \mathbf{v} \in \mathbb{R}^d$, we write $\mathbf{u} \leq \mathbf{v}$ to denote that $u_i \leq v_i$, for every $i = 1, \ldots, d$. A subset $S \subseteq \mathbb{R}^d$ is called *monotonic* if $\mathbf{v} \in S$ implies $\mathbf{u} \in S$, for every $\mathbf{u} \in \mathbb{R}^d$ satisfying $\mathbf{u} \leq \mathbf{v}$.

For the sake of (the few) readers not familiar with timed automata we start with an informal illustration of the behavior of these creatures. Consider the timed automaton of figure 1. It has two states and two clocks z_1 and z_2. Suppose it starts operating in the configuration $(q_1, 0, 0)$ (the two last coordinates denote the values of the clocks). Then it can stay at q_1 as long as the staying condition for q_1 is true, namely $z_1 \leq 2$. Meanwhile the values of the clocks grow and the set of all configurations reachable from $(q_1, 0, 0)$ without leaving q_1 is $\{(q_1, t, t) : 0 \leq t \leq 2\}$. However, after one second, the condition $z_1 \geq 1$ (the guard of the transition from q_1 to q_2) is satisfied and the automaton can move to q_2 while setting z_2 to 0. Hence the additional reachable configurations are $\{(q_2, t, 0) : 1 \leq t \leq 2\}$. Having entered q_2 in one of these configurations, the automaton can either stay there as long as $z_1 \leq 5 \wedge z_2 \leq 3$ or can unconditionally move to $(q_1, 0, 0)$, etc.

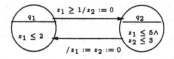

Fig. 1. A timed automaton.

---

[5] That is, for the same price as for BDD's in general.

Since the state-space of timed automata contains real values, we have an infinite-state automaton and an enumerative approach, where all states and transitions are enumerated, is impossible. We will use notation such as $G_{qq'}$ to denote the set of values in the clock space that satisfy the condition ("guard") for the transition from q to $q' \neq q$. Similarly, G_{qq} denotes the set of clock values allowing the automaton to stay in q ("staying conditions"). In timed automata such sets are restricted to be k-polyhedral subsets of $\mathbb{R}^d$, that is, the class of sets obtainable by applying set-theoretic operations to half-spaces of the form $\{\mathbf{v} : v_i \leq c\}$, $\{\mathbf{v} : v_i < c\}$, $\{\mathbf{v} : v_i - v_j \leq c\}$ or $\{\mathbf{v} : v_i - v_j < c\}$ for some integer $c \in \{0, \ldots, k\}$, where k is some positive integer.[6] These sets constitute the finite *region graph* [AD94] whose properties underlie all analysis methods for timed automata.

A function from $\mathbb{R}^d$ to itself is a *reset function* if it sets some of its arguments to 0 and leaves the others intact. We will use $R_{qq'}$ to denote the reset function associated with every pair of states (we take R_{qq} to be the identity function).

We will make the following simplifying assumptions concerning the timed automata that we consider: 1) There is only one transition associated with every pair of states. 2) The values of the clocks are bounded by k. Hence the clock space is $[0, k)^d$. 3) $G_{qq'}$ is convex for every $q, q' \in Q$, and 4) G_{qq} is monotonic for every $q \in Q$. The readers can convince themselves that it costs few states to convert any timed automaton into one satisfying these properties.

We let K denote the interval $[0, k)$ in the dense-time interpretation or the set $\{0, \ldots, k-1\}$ in the discrete-time interpretation. For every $\mathbf{z} \in \mathbb{R}^d$ we use $\mathbf{z} + t$ to denote $\mathbf{z} + t \cdot \mathbf{1}$ where $\mathbf{1} = (1, 1, \ldots, 1)$ is a d-dimensional unit vector.

Definition 1 (Timed Automaton). *A timed automaton is $\mathcal{A} = (Q, Z, \delta)$ such that, Q is a discrete set, $Z = K^d$ is the clock space ($Q \times Z$ is the configuration space) and $\delta : Q \times Z \mapsto 2^{Q \times Z}$ is the transition relation. It is required that δ admits the following decomposition: For every $q, q' \in Q$, let $G_{qq'} \subseteq Z$ be a k-polyhedral monotonic set and let $R_{qq'} : Z \mapsto Z$ be a reset function. Then, for every $(q, z) \in Q \times Z$*

$$\delta(q, z) = \left\{ (q', z') \; : \; \exists t \in K \begin{pmatrix} z + t \in G_{qq} \cap G_{qq'} \; \wedge \\ z' = R_{qq'}(z + t) \end{pmatrix} \right\} \tag{1}$$

The meaning of $\delta(q, \mathbf{z})$ is the set of $Q \times Z$ configurations the automaton can reach starting at $(q, \mathbf{z})$ by waiting t time (possibly zero), and then taking *at most* one transition.[7]

Every subset of $Q \times Z$ encountered in the analysis of timed automata can be decomposed into a finite union of sets of the form $\{q\} \times P$ where P is k-polyhedral. We will write such sets as (q, P). We will extend functions on ele-

---

[6] In fact, we can use $c \in \{0, r, 2r \ldots, kr\}$ for some positive rational r.

[7] In the treatment here, we assume that all sets of the form G_{qq} are definable by a positive boolean combination of inequalities of the form $x_i \leq c_i$ and $x_i - x_j \leq c_{ij}$. All the techniques presented here can be generalized to apply to the more general case that some of the inequalities defining G_{qq} are strict.

ments to functions on sets in the natural way, e.g. $\delta(q, P) = \bigcup_{\mathbf{z} \in P} \delta(q, \mathbf{z})$ and $R_{q,q'}(P) = \bigcup_{\mathbf{z} \in P} R_{qq'}(\mathbf{z})$.

Next, we define a function $\Phi : 2^Z \mapsto 2^Z$ (time forward projection) as:

$$\Phi(P) = \{\mathbf{z} + t : \mathbf{z} \in P, t \in K\} \cap Z.$$

It is not hard to see that the immediate successors of a set of configuration (q, P) can be written as

$$\delta((q, P)) = (q, \widehat{P}) \cup \bigcup_{q' \neq q} (q', P_{q'})$$

where $\widehat{P} = \Phi(P) \cap G_{qq}$ and for every q', $P_{q'} = R_{qq'}(\widehat{P} \cap G_{qq'})$. This concludes the motivation for the paper as we see that the additional machinery needed to analyze timed automata consists of calculations of boolean operations, $R(P)$, and $\Phi(P)$ on k-polyhedral sets.

3 Numerical Decision Diagrams: Discrete Time

3.1 Representation

The idea of NDD's is elementary. When we consider K^d under the discrete interpretation, we have nothing but subsets of a finite set. Obviously, every element of K can be coded in binary using $b = \lceil \log k \rceil$ bits, where $\lceil \log k \rceil$ is the smallest integer not smaller than $\log k$. Consequently, we can represent every subset of K^d as a boolean function of $d \cdot b$ boolean variables. This function can be represented by a BDD in the usual way. We will use standard positional encoding, i.e., every number $n \in K$ is represented by a set of values $x_0, \ldots x_{b-1}$ such that $n = \sum_{i=0}^{b-1} x_i \cdot 2^i$.

The first question concerning the implementation is the ordering of the bits of every number. Although, especially for sets of the form $x \leq c$, putting the most significant bit first might lead to smaller BDD's, we prefer to put the least significant bit first, because it facilitates the calculation of Φ. Examples of sets and their NDD representation for $d = 1$ and $k = 8$ appear in figure 2. When there are more than one clock variables, there are various ways to order their bits, for example, $x_0, x_1, \ldots, y_0, y_1, \ldots$.

In order to represent decision trees and BDD's textually we will use the expression $bdd(x_i, L, R)$ to denote a tree that tests x_i, branches to the subtree L on zero and to the subtree R on one. For example, the tree obtained from the BDD for $x < 5$ in figure 2 is written as:

$$bdd(x_0, bdd(x_1, \quad 1, \quad bdd(x_2, 1, 0)),$$
$$bdd(x_1, bdd(x_2, 1, 0), \quad 0 \quad))$$

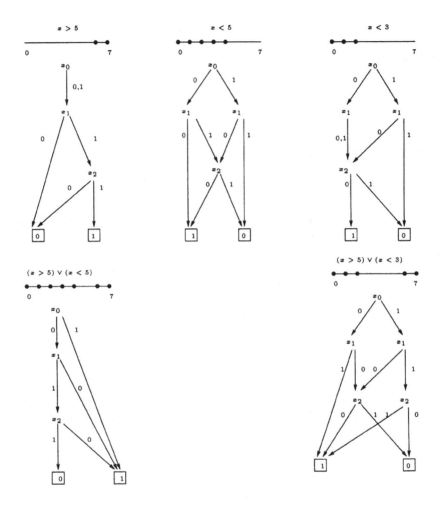

Fig. 2. Some 8-polyhedral sets in one-dimension and their corresponding NDD's.

3.2 Operations

Beside set-theoretic operations that we have for free, the reset operation is also elementary: in order to calculate $R(P)$ for a reset function R that resets, say, the variable x, you build a BDD for the set $x = 0$ and intersect it with the BDD for $\exists x\, P$. What remains to show is how to calculate $\Phi(P)$, which we will first demonstrate on the semantic level.

Given $P \in K^d$, $\Phi(P)$ can be written as $\{z : \exists t \in K \ s.t. \ z - t \in P\}$. Before applying the existential quantifier we have a set $P' \in K^{d+1}$ representing all the tuples (t, z) such that $z - t \in P$. We will present a procedure that converts a $b \cdot d$-variable NDD for P into a $b \cdot (d + 1)$-variable NDD for P' (with t as an additional K-variable, encoded using the boolean variables $t_0, t_1, \ldots, t_{b-1}$).

Eliminating the existential quantifier for t from P', we obtain the BDD for $\Phi(P)$. The procedure will initially create the NDD for the set $P_1 = \{(t, z) : z \ominus t \in P\}$ where $\ominus$ stands for subtraction modulo k. Then, by intersecting P_1 with the set $P_2 = \{(t, z) : z \geq t \cdot 1\}$ we get P' (see figure 3).

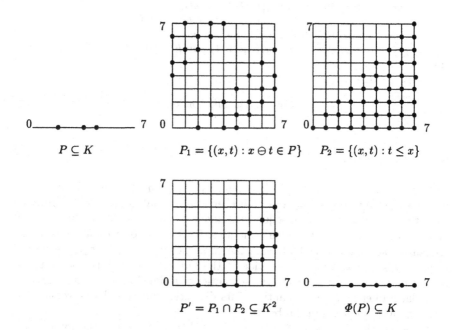

$$P \subseteq K \qquad P_1 = \{(x, t) : x \ominus t \in P\} \qquad P_2 = \{(x, t) : t \leq x\}$$

$$P' = P_1 \cap P_2 \subseteq K^2 \qquad \Phi(P) \subseteq K$$

Fig. 3. Calculating $\Phi(P)$ via moving to $d+1$ dimensions (P') and then projecting away the time. We first make the subtraction modulo k (creating P_1) and then intersect with P_2 to get rid from overflow.

To illustrate the construction of P_1, we consider first the case that $d = 1$, i.e. only one clock. The recursive function $sub(B, borrow)$ presented in table 1 takes an NDD B for $P \subseteq K$ and produces an NDD for $P_1 \subseteq K^2$ as described above. The parameter $borrow$ represents the "borrow" bit which is propagated from right to left on performing binary subtraction. The external invocation of this function is done with $borrow = 0$. For simplicity of presentation, we assume that B has nodes for all variables, with entries of the form $bdd(x_i, L, L)$ in case the function is independent of x_i. An optimized version can be derived for the more general case of skipped variable tests.[8]

The effect of applying the function to an arbitrary decision tree over $\{0, \ldots, 3\}$ is depicted in figure 4. The extension to $d > 1$ is rather straightforward.

---

[8] As usual in BDD applications, all calls are hashed so that repeated calls with the same arguments will not repeatedly traverse the complete subtrees.

```
function sub(B, borrow)
begin
if B is a leaf
  then return(B);
  else let B = bdd(x_i, L, R)
    if borrow = 0
      then return(bdd(t_i, bdd(x_i, (sub(L, 0), sub(R, 0))),
                            bdd(x_i, (sub(R, 1), sub(L, 0))))));
      else  return(bdd(t_i, bdd(x_i, (sub(R, 1), sub(L, 0)))
                            bdd(x_i, (sub(L, 1), sub(R, 1))))));
end
```

Table 1. The function sub.

4 Dense Time

The above construction is sufficient for analyzing timed automata under the discrete-time interpretation. It is however known that some timed automata can produce behaviors (state-sequences) under a dense semantics which are not possible under any discrete-time semantics. In this section we introduce a discretization scheme [GPV94][9] having the two following important properties: 1) It preserves the qualitative behavior of the automaton, that is, for every sequence of discrete transitions in the semantics of a timed automaton $\mathcal{A}$, there is a similar sequence in the semantics of its discretization $\widetilde{\mathcal{A}}$ and vice versa. 2) It is amenable to representation by NDD's.

For each clock value z_i, $i = 1, \ldots, d$, let I_i and f_i denote the integer and fractional parts of z_i, respectively. Two clock valuations $\mathbf{z} = (I_1 + f_1, \ldots, I_d + f_d)$ and $\mathbf{z}' = (I_1' + f_1', \ldots, I_d' + f_d')$ are defined to be *region equivalent*, written $\mathbf{z} \sim \mathbf{z}'$, if

$$\bigwedge_{i=1}^{d} \Big((I_i = I_i') \wedge (f_i > 0 \leftrightarrow f_i' > 0) \Big) \quad \wedge \bigwedge_{i,j \in \{1,\ldots,d\}} (f_i > f_j \leftrightarrow f_i' > f_j').$$

We consider automata with $Z = K^d$. We will use a discretization step $\Delta = 1/(2d)$ and let $\widetilde{K} = \{n\Delta : 0 \leq n < 2kd\}$. In other words, we cut every unit interval into $2d$ equal segments and pick the endpoints. The discretized clock space (that is, the domain over which discretized clocks range) is

$$\widetilde{Z} = \widetilde{K}^d \cap \{(z_1, \ldots, z_d) : \forall i, j \, |z_i - z_j| = 2m\Delta\}.$$

Note that we take from $\widetilde{K}^d$ only points such that the difference between any pair of clock valuations is an *even* multiple of Δ (see figure 5). For any polyhedral

---

[9] Discovered independently by the authors but a year later.

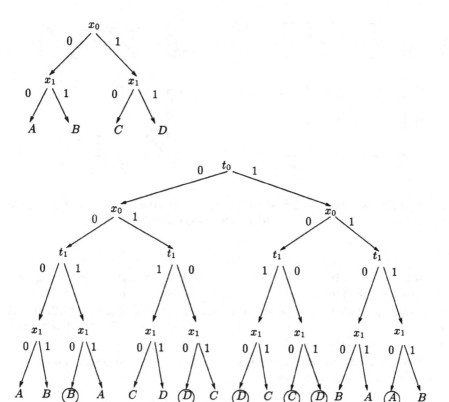

Fig. 4. Applying *sub* to an arbitrary decision tree over $\{0,1,2,3\}$. The circled leaves will become zero when we intersect with $\{(x,t) : x \geq t\}$.

set P, we let its discretization be $\widetilde{P} = P \cap \widetilde{Z}$. It is not hard to see that, for every k-polyhedral set P, we have $P \neq \emptyset$ iff $\widetilde{P} \neq \emptyset$. Another important property of this scheme is the following:

Claim 1. Let $z = \widetilde{z} + \varepsilon$ for some $\widetilde{z} \in \widetilde{Z}$, $|\varepsilon| < \Delta$. Then

$$z \in P \Rightarrow (\widetilde{z} \in P \vee \widetilde{z} + \Delta \in P)$$

(and hence at least one of them belongs also to $\widetilde{P}$).

Proof: If $\widetilde{z} = (z_1, \ldots, z_d) \in P$ we are done, otherwise there is one or more inequalities of the form $z_i > c_i$ satisfied by $\widetilde{z} + \varepsilon$ but not by $\widetilde{z}$ (which implies that $z_i = 2dn\Delta$ for some integer n). These inequalities must be satisfied by $\widetilde{z} + \Delta$ as well. On the other hand, if there is an inequality of the form $z_j < c_j$ satisfied by $\widetilde{z} + \varepsilon$ but not by $\widetilde{z} + \Delta$, we have $z_j = (2dn - 1)\Delta$, which contradicts the

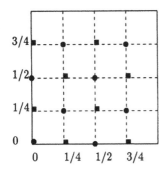 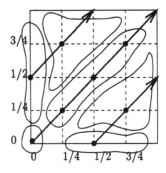

Fig. 5. Left: discretizing $[0,1)^2$: the circled points are the elements of $\widetilde{Z}$ while the squared points belong to $\widetilde{K}^d - \widetilde{Z}$. Right: illustration of claims 1 and 2.

assumption $\widetilde{\mathbf{z}} \in \widetilde{Z}$. In addition, $\widetilde{\mathbf{z}} + \varepsilon$ and $\widetilde{\mathbf{z}} + \Delta$ satisfy together every diagonal inequality (of the form $z_i - z_j < c$) and we can conclude that $\widetilde{\mathbf{z}} + \Delta \in P$. ◢

Note that this fails to be true for points outside $\widetilde{Z}$. Consider $\mathbf{z} = (0, 3/4)$ and $P = (0 < x < 1) \wedge (0 < y < 1) \wedge (y > x)$. Here $\mathbf{z} + \varepsilon \in P$ but neither $\mathbf{z}$ nor $\mathbf{z} + \Delta = (1/4, 1)$ belong to P.

The discretized forward projection $\widetilde{\Phi} : 2^{\widetilde{Z}} \to 2^{\widetilde{Z}}$ is the restriction of Φ to points in $\widetilde{Z}$ and time values in $\widetilde{K}$:

$$\widetilde{\Phi}(\widetilde{P}) = \{\widetilde{\mathbf{z}}' \in \widetilde{Z} : \exists \widetilde{\mathbf{z}} \in \widetilde{P}\, \exists \widetilde{t} \in \widetilde{K}\ s.t.\ \widetilde{\mathbf{z}}' = \widetilde{\mathbf{z}} + \widetilde{t}\}.$$

Claim 2 Discretization Preserves Forward Projection.   *For every k-polyhedral set P and P' such that $P' = \Phi(P)$*

$$\widetilde{\Phi}(\widetilde{P}) = \widetilde{P'}.$$

Proof: One direction, $\widetilde{\Phi}(\widetilde{P}) \subseteq \widetilde{P'}$ is obvious because $\widetilde{P} \subseteq P$ and $\widetilde{K} \subseteq K$. For the other direction, suppose some $\widetilde{\mathbf{z}}' \in \Phi(P) \cap \widetilde{Z}$, implying that $\widetilde{\mathbf{z}}'$ can be written as $(n_1 \Delta, \ldots, n_d \Delta)$, and that for some $\mathbf{z} \in Z$, $t \in K$, $\mathbf{z} + t = \widetilde{\mathbf{z}}'$ hence $\mathbf{z} = (n_1 \Delta - t, \ldots, n_d \Delta - t)$. Let $t = m\Delta + t'$ for some $t' < \Delta$. Then $\mathbf{z} = ((n_1 - m)\Delta + t', \ldots, (n_d - m)\Delta + t')$. According to the previous claim either $\widetilde{\mathbf{z}}$ or $\widetilde{\mathbf{z}} + \Delta$ is in $\widetilde{P}$ and their temporal successor $\widetilde{\mathbf{z}}'$ is in $\widetilde{\Phi}(\widetilde{P})$. ◢

Having shown that forward projection (as well as boolean operations) on Z and K can be imitated by discretized operations on $\widetilde{Z}$ and $\widetilde{K}$, the only remaining problem is concerned with the reset operator. The problem is that $\widetilde{Z}$ is not closed under reset functions – for example, resetting the first coordinate of $(\Delta, \Delta) \in \widetilde{Z}$ we obtain $(0, \Delta) \in \widetilde{K}^d - \widetilde{Z}$ (because the difference between the points is not an even multiple of Δ). This is important because claim 2 does not hold on $\widetilde{K}^d$ but only on $\widetilde{Z}$. In order to calculate successors on the discretization we need an "adjustment" operator, which, after applying a reset, will delete points that went out of $\widetilde{Z}$ and replace each of them by one or more region-equivalent points in $\widetilde{Z}$. This extra operator can be viewed as the price we pay for dense reasoning.

For each $m \in \{0, \ldots, d-1\}$, let us define a function $\alpha_m : \widetilde{K}^d \to 2^{\widetilde{Z}}$ as follows:

$$\alpha_m(z) = \{z' : \bigwedge_{i=1}^{d} \left((I_i = I'_i) \wedge \left(\begin{array}{c} f_i = 0 \wedge f'_i = 0 \\ \vee \\ f_i = (2l+1)\Delta \wedge l < m \wedge f'_i = (2l+2)\Delta \\ \vee \\ f_i = (2l+1)\Delta \wedge l > m \wedge f'_i = (2l)\Delta \end{array} \right) \right) \}$$

The function α_m returns a non-empty set if and only if all the non-zero fractional parts of z are odd multiples of Δ, and none of them falls in the interval $[2m\Delta, (2m+2)\Delta]$. Its effect is to add Δ to all fractional parts f_i satisfying $0 < f_i < 2m\Delta$ and subtract Δ from all fractional parts satisfying $(2m+2)\Delta < f_i$. Zero fractional parts are left unchanged. One can see that if z satisfies these conditions then $\alpha_m(z) = \{z'\}$ and $z' \sim z$. This operator is illustrated in figure 6. Based on this family of functions we define $\alpha : \widetilde{K}^d \to 2^{\widetilde{Z}}$ as

$$\alpha(z) = \begin{cases} z & \text{if } z \in \widetilde{Z} \\ \bigcup_{m=0}^{d-1} \alpha_m(z) & \text{otherwise} \end{cases}$$

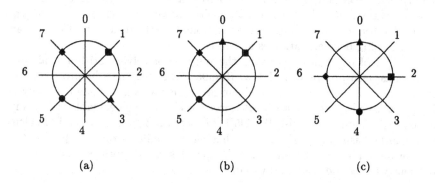

(a)   (b)   (c)

Fig. 6. The effect of the adjustment operator α_1 for $K = [0, 1)^4$ and $\Delta = 1/8$ (all points are multiples of Δ): (a) A point $z_1 = (1, 3, 5, 7) \in \widetilde{Z}$. (b) After restting the second clock we obtain $z_2 = (1, 0, 5, 7) \notin \widetilde{Z}$. (c) Applying α_1 we push the non-zero clocks toward the "hole" around 3 and obtain $z_3 = (2, 0, 4, 6)$ which is region-equivalent to z_2. Note that in this graphical representation the passage of time is via clock-wise rotation.

It is not hard to see that the application of α to any $P \subseteq \widetilde{K}^d$ yields a subset $P' \subseteq \widetilde{Z}$ such that, for every $z \in P$, there is at least one $z' \in P'$ satisfying $z \sim z'$. Based on this function, we can define for every reset function $R : Z \to Z$ a discretized reset function $\widetilde{R} : \widetilde{Z} \to 2^{\widetilde{Z}}$ as $\widetilde{R}(\widetilde{z}) = \alpha(R(\widetilde{z}))$. It follows that, for every

P, $\widetilde{R}(\widetilde{P})$ has elements of every region equivalence class which are represented in $R(P)$.

This is all we need: we just add to the NDD solution for integer time is $d \cdot \log(2d)$ bits to represent the finer grid and to replace every reset function $R_{qq'}$ by its adjusted version $\widetilde{R_{qq'}}$. The same arithmetical calculation of time successors described in section 3 will work, when we add the fractional bits to the clock and time variables. The adjustment operation seems to be the hard part of the calculation (we have only implemented the discrete-time representation so far) but at least this operation is performed only on the fractional bits (and hence is sensitive to the number of clocks but not to their ranges). In fact there is a trade-off between two discretization schemes (see [GPV94]), one with $\Delta = 1/(d+1)$, where resets behave normally but the evolution of time is distorted and loses some of its arithmetical content, and the other one we describe here, were time evolution remains arithmetic while resetting is more involved.

5 Concluding Remarks

5.1 Related Work

Various tools for the analysis of timed and hybrid automata have been developed recently, e.g. KRONOS [DOY94], UPPAL [BGK+] and Hy-Tech [AHH93]. The first two represent polyhedral sets by DBMs. An alternative approaches is to transform the timed automaton into a huge discrete automaton (the region graph) and than encode its using boolean variables and BDD's.[10]

The idea of extending BDD's for the purpose of solving arithmetical constraints has been proposed by Rauzy [Rau95]. The structures he proposes are, however, not canonical. Our method can be applied as well outside the analysis of timed automata, e.g., as a decision procedure for some decidable theories in bounded arithmetics (see also [WB95]). In fact, the forward projection calculation can be easily adapted to clocks having non-uniform rates in $\{0, -1, +1\}$ and can be applied to the analysis of larger classes of hybrid systems and to programs with bounded integer variables.

5.2 Experimental Results

For experimentation we have used a system developed at VERIMAG for representing and manipulating communicating automata augmeneted with bounded variables [BFK96]. This system takes such automata and translates them into BDDs using one of several publicly-available BDD packages – we have used the CUDD package [S95] of Colorado University. We have incorporated a discrete-time version of the NDD representation into that system and tested its performance on various timed automata corresponding to digital circuits with delays

---

[10] In [CC95] this approach has been applied to the degenerate case of one-clock automata.

(the exact definitions and the translation procedure from circuits to timed automata are described in [MP95]). We will report here the results obtained with two generic families of automata, for which we tried to calculate all the reachable configurations starting from an initial state.

The first family consists of one-state automaton having n clocks and n transitions. The automaton (see figure 7-a) can let time progress as long as none of the clocks has reached its upper bound u_i. Whenever a clock C_i reaches the lower bound $l_i < u_i$, a self-looping transition which resets C_i can be taken. These automata allow us to isolate the complexity of representing and manipulating polyhedral sets from that of treating the discrete state-space.

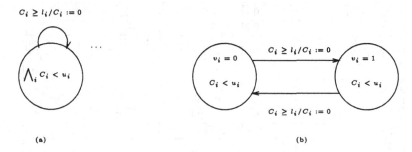

(a)   (b)

Fig. 7. The automata used for experiments: (a) A one-state automaton with n clocks and n transitions. (b) A basic two-state automaton with one clock.

The second family consists of a product of n two-state automata of the type appearing in (figure 7-b). Such a product is the natural way to model n independent non-deterministic input oscillators and it is a necessary ingredient in any attempt to do exhaustive timing analysis of asynchronous circuits.

The main difference between the two examples is that in the second we have 2^n discrete states. While the set of reachable configuration for this example will be of the form $\{(q, P_1)\}_{q \in \{0,1\}^n}$, the set of reachable clock configurations of the first automaton will be $\bigcup_q P_q$, which seems in general to be simpler to represent. In fact, for the constants we have chosen all the clock space is eventually reachable, but some very "hard" sets are encountered in the intermediate stages of the fixed-point calculations.

We have taken clock values in the range 0..15, let $l_i = 9$, $u_i = 12$ and compared the results with those of KRONOS which uses DBMs. For this set of examples the NDD results were much better. It should be noted, however, that DBMs implement the richer dense semantics and are *not sensitive to the range of clock values* as long as they do not cross the maximal integer value. In contrast, the performance of NDDs depends critically on the number of bits used to encode clock values. Moreover, the examples were chosen so that they generate intermediate polyhedral sets which are very "non-convex", which makes life hard for the DBM implementation (but also for NDDs). Finally the current forward simulation algorithm of KRONOS keeps all reachable regions in a form of a simulation graph[11] and this turns out to be inefficient for these examples – we believe that changing this implementation detail will allow KRONOS to treat larger examples. The results are summarized in table 2. They were obtained on a SUN Ultra-Sparc 1 with 256MB of memory.

no	one state				many states			
	DBM		NDD		DBM		NDD	
	time	regions	time	max ndd	time	regions	time	max ndd
2	0.0	17	0.1	51	0.1	35	0.2	107
3	0.4	175	0.3	235	1.5	783	2.1	837
4	53.9	2561	1.0	699	9:01.4	28974	9.9	3226
5	5:43:29.0	48499	2.7	1647	-	-	35.1	10617
6	-	-	5.9	3311	-	-	2:15.2	35640
7	-	-	10.9	5741	-	-	12:35.6	85863
8	-	-	18.8	10559	-	-	54:09.0	225057
9	-	-	29.3	19102	-	-	5:21:13.8	539092
10	-	-	48.6	29524	-	-	-	-
11	-	-	1:32.1	43947	-	-	-	-
12	-	-	2:58.8	62433	-	-	-	-
13	-	-	5:58.9	84696	-	-	-	-
14	-	-	13:09.3	114210	-	-	-	-
15	-	-	31:41.3	147016	-	-	-	-
16	-	-	1:23:38.6	190976	-	-	-	-
17	-	-	4:00:04.6	237484	-	-	-	-
18	-	-	11:59:32.1	299762	-	-	-	-

Table 2. Comparative results of the NDD and DBM implementations.

Our intermediate conclusion is that the analysis of timed automata with *many clocks* is not yet feasible. We have managed to handle additional non-trivial examples (such as an interconnected chain of XOR gates) with 10-13 clocks, but a closer investigation of polyhedral sets and their various representation schemes is needed in order to push performance limitations forward.

Acknowledgment: We thank S. Yovine, K. Daws and S. Tripakis for useful discussions concerning verification of timed automata in general and the KRONOS implementation in particular, and P. Raymond for some help in BDDs.

---

[11] This is the number appearing in the "regions" column in the table.

References

[AD94]   R. Alur and D.L. Dill, A Theory of Timed Automata, *Theoretical Computer Science* 126, 183–235, 1994.

[AHH93]   R. Alur, T.A. Henzinger, and P.-H. Ho. Automatic symbolic verification of embedded systems. In *IEEE Real-time Systems Symposium*, pages 2–11, 1993.

[BGK+]   J. Bengtsson, W.O.D. Griffioen, K.J. Kristoffersen, K. G. Larsen, F. Larsson, P. Pettersson, and W. Yi. Verification of an audio protocol with bus collision using UPPAAL. In R. Alur and T.A. Henzinger (Eds. *Proc. CAV'96*, LNCS 1102, 244–256, Springer, 1996.

[Bry86]   R.E. Bryant, Graph-based Algorithms for Boolean Function Manipulation, *IEEE Trans. on Computers* C-35, 677-691, 1986.

[BCM+93]   J.R. Burch, E.M. Clarke, K.L. McMillan, D.L. Dill, and L.J. Hwang, Symbolic Model-Checking: 10^{20} States and Beyond, *Proc. LICS'90*, Philadelphia, 1990.

[CC95]   S.V. Campos and E.M. Clarke. Real-time symbolic model checking for discrete time models, in T. Rus and C. Rattray (Eds.), *AMAST Series in Computing: Theories and Experiences for Real-Time System Development*, World Scientific, 1995.

[DOY94]   C. Daws, A. Olivero and S. Yovine, Verifying ET-LOTOS Programs with KRONOS, *Proc. FORTE'94*, Bern, 1994.

[Dil89]   D.L. Dill, Timing Assumptions and Verification of Finite-State Concurrent Systems, in J. Sifakis (Ed.), *Automatic Verification Methods for Finite State Systems*, LNCS 407, Springer, 1989.

[BFK96]   M. Bozga, J.-C. Fernandez, A. Kerbrat, A Symbolic μ-calculus Model Checker for Automata with Variables, Unpublished Manuscript, VERIMAG, 1996.

[GPV94]   A. Göllü, A. Puri and P. Varaiya, Discretization of Timed Automata, *Proc. 33rd CDC*, 1994.

[HNSY94]   T. Henzinger, X. Nicollin, J. Sifakis, and S. Yovine, Symbolic Model-checking for Real-time Systems, *Information and Computation* 111, 193–244, 1994.

[MP95]   O. Maler and A. Pnueli, Timing Analysis of Asynchronous Circuits using Timed Automata, in P.E. Camurati and H. Eveking (Eds.), *Proc. CHARME'95*, 189-205, LNCS 987, Springer, 1995.

[McM93]   K.L. McMillan, *Symbolic Model-Checking: an Approach to the State-Explosion problem*, Kluwer, 1993.

[S95]   F. Somenzi, CUDD: CU Decision Diagram Package, 1995.

[Rau95]   A. Rauzy, Toupie = μ-calculus + constraints. In P. Wolper, editor, *Proc. CAV'95*, LNCS 939, 114–126, Springer, 1995.

[WB95]   P. Wolper and B. Boigelot. An automata-theoretic approach to presburger arithmetic constraints. In *Proc. Static Analysis Symposium*, LNCS 983, 21–32, Springer, 1995.

Synthesizing Controllers for Hybrid Systems*

Deepak Kapur[1] and R.K. Shyamasundar[2]

[1] Department of Computer Science, State University of New York, Albany, NY,
USA, kapur@cs.albany.edu
[2] Computer Science Group, Tata Institute of Fundamental Research, Bombay 400
005, India, shyam@tcs.tifr.res.in

Abstract. A methodology for synthesizing control laws of hybrid systems is proposed using the hybrid automaton framework. The objective is to synthesize guards for making phase transitions to ensure that the system satisfies the global invariance over the whole state space. Classical analysis is used to derive a controller given the phases, transitions and global invariance of a hybrid system. The methodology seems to be general in the sense that control laws can be synthesized even if the hybrid system is modeled by an automaton more general than bounded-drift linear hybrid automaton model frequently discussed in the literature. The main requirement is that it should be possible to generate a closed form expression for state variables in every phase, as a function of time elapsed in that phase. Conditions on system parameters can be identified for which the control laws can be synthesized. Optimality criteria can be incorporated for selecting among different control strategies. The methodology is illustrated using three examples. This work is in contrast to most of the work on hybrid systems in which the focus has been on the *analysis* problem.

1 Introduction

Hybrid systems combine discrete and continuous computations. A hybrid system model describes activities that modify their variables continuously over intervals of positive duration in addition to the familiar transitions that change the values of variables instantaneously, representing the discrete components. Many systems that interact with a physical environment such as a digital module controlling a process or a manufacturing plant, a digital-analog guidance of transport systems, control of a robot, flexible manufacturing systems, etc., can benefit from the study of hybrid models.

Design, analysis (verification) and synthesis (control) of hybrid systems have been an important area of active research [GNRR93, AKNS95, AHS96]. In verification, an objective is to show that a given hybrid system will satisfy a certain (temporal) property and will never enter any *unsafe* state. In contrast, in synthesis, the focus is on designing a *controller* that ensures the given system will never enter any unsafe state.

---

* The work was partially supported by an NSF Indo-US grant INT-9416687. The first author was partially supported by an NSF grant nos. CCR-9303394, CCR-9308016, and CCR-9404930.

The study of hybrid systems is interdisciplinary, lying at the junction of computer science and control. Automata-theoretic and logical approaches have been primarily used for verification and synthesis. Hybrid automata [ACHHH95, De94] are used for modeling hybrid systems. Following [De94], informally, a hybrid transition system has finitely many discrete states, called *phases* (or *locations*), and infinitely many continuous states (henceforth just called states). A phase is used for modeling a dynamic subsystem specified by a set of differential equations. Transitions between phases are made based on conditions on continuous states. Since the behavior of a hybrid automaton can be complex, much research has been devoted to identifying subclasses of hybrid automata for which verification problems are decidable. Model checking techniques have been studied for verification [Ho95].

The controller synthesis problem discussed in this paper is that of identifying conditions on transitions among phases such that the continuous state of a given hybrid automaton always satisfies a given global invariance. We propose an approach using the classical techniques. The crux lies in deriving the *switching* points from one phase to another using some policy/optimality criterion. The method involves the following steps for each phase:

1. Solve the associated differential equations to obtain expressions for continuous state variables in terms of the time elapsed in the phase,
2. Determine how the value of each continuous variable evolves (constant, increasing, decreasing, or both increasing and decreasing). And,
3. Identify whether the phase is *control-critical* for a continuous variable in the sense that the value of a continuous variable constrained by the global invariant evolves in the direction opposite to its evolution in the previous phase. In that case, deduce possible constraints on system parameters so that the global invariant remains satisfied.

The methodology is illustrated using three examples. A *linear* water-level controller (in which the rates of change of water level are constants in different phases) [Ho95] is used as a running example as the methodology is presented. A slightly nontrivial version of a *nonlinear* water-level controller (in which the rates of change of water level are nonlinear functions as well as depend upon a system parameter) [SM94] is discussed later. This is followed by a discussion of a more general version of nuclear reactor temperature control [ACHHH95, Ho95]. It is illustrated that adopting different policies can lead to different controllers.

The work on the synthesis of controllers for discrete and timed systems has been elucidated in [AMP95]. The synthesis problem in a nondeterministic automaton and a timed automaton is reduced to the problem of deriving a winning strategy between two players game. Under certain classes of acceptance conditions of an automaton, the problem is shown to be decidable by showing that another automaton can be constructed that accepts the set of all strings corresponding to good (acceptable) behaviors of the given automaton (see [RW89]). The approach is based on a result by Buchi and Landweber for ω-automata. An algorithm is given that iteratively finds the subset of predecessors always leading to a designated subset of states, thus arriving at a selection (or control) predicate at each transition. The backward propagation is to be continued till a fixed point is reached. For timed systems, "time" becomes a third player which can interfere in favor of the other players. It then becomes essential to ensure that players are not allowed to win through *Zenonism* tricks (i.e., by preventing time from progressing). In the synthesis of a controller for discrete and timed systems, one thus tries to augment the conditions for the transitions to achieve the desired property from the designated final states. In contrast, the main objective in hybrid control is to derive a control strategy from the continuous dynamics of the system and the plausible transitions among phases such that the system state always satisfies the global invariant.

2 Hybrid Automaton Model

We adopt the framework in [De94], and reproduce relevant definitions below.

Definition 1. A *hybrid transition system* H is a quintuple $(Q, R^n, \Sigma, E, \mathcal{D})$, where

1. Q is a finite set of discrete *phases*, also called *locations*.
2. R^n constitutes the set of continuous states, where R is the set of real numbers. Let V_c be a finite set of continuous variables, so the size of V_c is n.
3. Σ is a finite set of discrete events.
4. $E \subset Q \times \mathcal{P}(R^n) \times \Sigma \times \{R^n \to R^n\} \times Q$ is a finite set of edges, where $\mathcal{P}$ is the power set operator. Edges model the discrete event dynamics. An edge $e \in E$ denoted by $(q_e, X_e, V_e, r_e, q'_e)$ is interpreted as:
   - The transition e is enabled in phase q_e if the continuous state is in X_e. X_e characterizes the switching points from q_e to q'_e, and is specified by a formula, called a *guard*.
   - If the transition e is selected by the system (since the transition system may be nondeterministic, there may be many transitions enabled out of which one may be selected), the continuous state is reset using the function r_e before the phase q'_e is entered.
5. $\mathcal{D}$ associates a finite set of differential equations to each phase expressing its dynamics. Let D_q be the differential equations associated with the phase $q \in Q$. Each differential equation is assumed to be of the form:

$$\dot{x} = E,$$

where $x \in V_c$ is a continuous state variable, and E is an expression over V, where V includes V_c, other discrete variables and parameters, if any.

As in [De94], a hybrid transition system executes in a sequence of phases, say $q_0 \, e_0 \, q_1 \, e_1 \cdots q_i \, e_i \, q_{i+1} \cdots$, where e_i is a transition from phase q_i to q_{i+1}. If the transitions (events) in the sequence are obvious, then they can be omitted. In each phase, the system evolves continuously by allowing time to pass. It then makes a discrete transition to the next phase instantaneously and evolves in the next phase. The evolution of the hybrid system can be described by the sequence of intervals over R^+: $\tau = [\tau'_0, \tau_1], [\tau'_1, \tau_2], [\tau'_2, \tau_3], \cdots$, where $\tau'_0 = 0$ and $\forall i\ (\tau_i = \tau'_i$ and $\tau_{i+1} \geq \tau'_i)$. The system evolves continuously in the interval $[\tau'_i, \tau_{i+1}]$ while being in some phase q_i, and makes an instantaneous transition e_i at τ_{i+1} from q_i to q_{i+1}. If $\tau'_i = \tau_{i+1}$, then it makes a transition at τ_{i+1} without any continuous evolution in q_i.

The system behavior within a phase q_i is governed by the associated differential equations D_{q_i}. We assume that a closed-form differentiable function $f_x(t)$ of time t can be found for each continuous variable $x \in V_c$ such that over the interval $[\tau'_i, \tau_{i+1}]$, $\sigma(\dot{x}) = \sigma(E)$ for every differential equation $\dot{x} = E$ in D_{q_i}, where σ is a substitution replacing each $x \in V_c$ by the corresponding $f_x(t)$ and $\dot{x}$ by the first derivative of $f_x(t)$. In other words, $\{f_x(t) \mid x \in V_c\}$, a finite set of functions, is a solution of D_{q_i}.

An edge $e = (q_e, X_e, V_e, r_e, q'_e)$ in the transition system plays a crucial role. X_e specifies the set of continuous states in which a transition from q_e to q'_e is enabled. Different X_e's correspond to different control strategies. In the sequel, for ease of reference, we use the term *phase transition system* to refer to hybrid transition systems without the switching states (or with switching states specified with some of the transitions, usually related to timers).

2.1 Controller

The specification of a hybrid system is usually augmented by a *global invariant* on continuous states of the system irrespective of the phase it is in. Examples of global invariants might be physical requirements like maximum tolerable temperature, required range of water level, the desired behavior of a plant, etc. A global invariant constrains the values of continuous variables in each phase and switching sets X_e's on transitions among phases.

The problem studied in this paper is to design a controller so that the system never evolves into a state that does not satisfy the global invariant. The controller synthesis problem is synthesizing any unknown X_e's for transitions based on a control policy. In order for a system to go on indefinitely, it is preferred that the controller be *viable* in the sense of [De94, DV95]. In this paper, a methodology is given for deriving X_e's. We also illustrate how one can arrive at different control strategies to achieve the global invariant using other criteria such as cost.

3 A Methodology for Synthesizing Controllers

Given a phase transition system without the switching points X_e's on (some of) the transitions, we give a methodology for synthesizing controllers when

1. the global invariance is specified as a conjunction of lower and upper bounds on a subset of continuous state variables, i.e., $\mathbf{l} \leq \mathbf{V_c} \geq \mathbf{h}$, where $\mathbf{l}$, $\mathbf{h}$ are elements in $(R \cup \{-\infty, \infty\})^n$, and
2. the differential equations associated with each phase can be solved to generate a closed-form differentiable expression for each continuous state variable as a function of time.

These requirements are less restrictive than the requirement that the state change in a phase has to be governed by differential equations in which variables are bounded in a fixed range (called bounded-drift linear hybrid automata in [Ho95]).

We explain and illustrate the main steps of the methodology using a water-level controller, an example discussed extensively in the literature [HRP93, ACHHH95, Ho95].

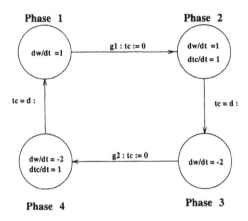

Figure 1: Water Level Controller

Water Level Controller : The controller continuously senses the water level in a water tank and regulates the level by opening and closing an outflow valve. There is a delay between the issuing of a valve command and carrying out of the command. The goal of the controller is to keep the water level in the tank between 1 and 12 ft. The phase transition system is given in Figure 1.

The dynamics of the system is given below. The water level in the tank is represented by w. The variable tc is a timer which is used to model the delay between the issuing of valve commands and their taking effect. The global invariant is $1 \leq w \leq 12$. The synthesis problem for this controller is to derive the guards g_1 (for the transition from phase 1 to phase 2) and g_2 (for the transition from phase 3 to phase 4).

- *Phase 1:* **Valve closed** : $dw/dt = 1$; $tc := 0$.
- *Phase 2:* **Valve opening:** $dw/dt = 1$; $dtc/dt = 1$ and the complete opening of the valve takes d seconds.
- *Phase 3:* **Valve open:** $dw/dt = -2$; $tc := 0$.
- *Phase 4:* **Valve closing:** $dw/dt = -2$; $dtc/dt = 1$, and the complete valve closing takes d seconds.[3]

---

[3] The analysis made extends even if the delays associated with valve opening and closing are different.

3.1   Closed-form Expressions for Continuous State Component

The first step is to generate closed-form expressions for the continuous variables in each phase as a function of time elapsed in that phase. These expressions and the differential equations are used to determine the direction of variables constrained by the global invariants, i.e., increasing, decreasing or both.

For the water-level controller example, the differential equations for each phase are solved. Consider a sequence of phases: 1, 2, 3, 4. Below, w_i is used to denote the value of w in phase i, t_i is the time elapsed since the system last entered the i-th phase. Let $w_{i,0}$ be the initial value of w when the system last entered the i-th phase; $w_{i,f}$ is the final value of w when the system last exited phase i; $t_{i,f}$ is the time spent in the i-th phase. Let $w_{i,t}$ be the value of the variable w in phase i after time $t(\le t_{i,f})$ has elapsed since the system entered i-th phase last.

- *Phase 1*: $w_1 = t_1 + w_{1,0}$.
- *Phase 2*: $w_2 = t_2 + w_{2,0}$. Also, $tc_2 = t_2$, and is bounded by d.
- *Phase 3*: $w_3 = -2\,t_3 + w_{3,0}$
- *Phase 4*: $w_4 = -2\,t_4 + w_{4,0}$, $tc = t_4$, and hence $t_{4,f} = d$.

Since the reset is the identity function, $w_{2,0} = w_{1,f}$, $w_{3,0} = w_{2,f}$, $w_{4,0} = w_{3,f}$. The reader should note that $w_{1,0} = w_{4,f}$ is not included because $w_{1,0}$ here refers to the value of w when the system enters phase 1 in the next cycle.

3.2   Identifying Control-Critical Phases

Each continuous state variable constrained by the global invariant is checked in every phase as to whether its value is constant, increasing, decreasing or both. Often, this can be inferred directly from the associated differential equations. The reset function for each transition is also analyzed for this information. A maximal sequence of phases is identified such that the variable continues to evolve in the same direction (nonincreasing or nondecreasing) if the system sequences through these phases. The last phase in such a sequence thus has a transition to another phase in which the variable stops evolving in the same direction. If the variable was increasing, then the upper bound specified on the variable in the global invariant can be used for synthesizing guards. Similarly, if the variable was decreasing, then the lower bound on the variable is used.

Let x be a continuous state variable constrained to take the maximum value h by the global invariant. Further, let $q_1\, q_2\, \cdots q_i$ be the longest sequence of phases in which x is increasing. Thus, every transition from q_i results in a phase in which x decreases. Consider an edge e from q_i. If the reset function r_e can result in x decreasing, then x can at most be h in q_i. If r_e is nondecreasing for x, then we need to consider how x evolves in phase q_{i+1} when the transition associated with e is made. If x is decreasing, then again x can at most be h after r_e is applied. If x first increases and then decreases, then x can become h in q_{i+1}. If q_{i+1} is both increasing as well as decreasing for x, it is called *control-critical*; otherwise if q_{i+1} is decreasing for x, then q_i is control-critical. A control-critical phase is so called since it is critical in determining the possible violation of the global invariant on x.

Similar analysis is done if the global invariant specifies a lower bound for x. Once control-critical phases are identified, bounds specified by a global invariant on the constrained variables can be used to determine the maximum time spent by the system in such phases. These constraints are then backward-propagated to other phases.

After such an analysis is performed for each state variable, constraints are solved to identify X_e's based on a desired control policy. Guards associated with transition edges are then synthesized.

Phases may have been introduced in a phase transition model of a hybrid system to model delays involved in certain actions, e.g., opening and closing of valves in the running example, or pre-emptive signals at regular intervals using variables modeling clocks, e.g., a variation of water-tank example in chapter 5 in [Ho95]. Clock or timer is modeled using a clock variable whose derivative is 1, e. g., the variable tc in the running example. Such constraints also enforce the exact time or maximum time that could be spent in a phase.

The analysis so far has focussed on considering a single state variable constrained by the global invariant. It extends to several variables constrained by the global invariant.

In the water-level controller example, the global invariant constrains w from below as well as from above. In phases 1 and 2, w is increasing (since dw/dt is +ve in these phases), whereas w is decreasing in phases 3 and 4 (since dw/dt is −ve in these phases). Phases 2 and 4 are control-critical since w will reach a maximum value in phase 2 and a minimum value in phase 4.

Since w is increasing in phase 2, w gets to the maximum value on exit from phase 2. $w_{max} = w_{2,f} = t_{2,f} + w_{2,0}$, where $w_{2,0} = w_{1,f} = t_{1,f} + w_{1,0}$. Since $t_{2,f} = d$, $w_{max} = w_{2,f} = d + t_{1,f} + w_{1,0} \leq 12$. This gives $w_{1,f} \leq 12 - d$.

Similarly, w gets to the minimum value in phase 4 when it is to exit phase 4. $w_{min} = w_{4,f} = -2 * t_{4,f} + w_{4,0}$, where $w_{4,0} = w_{3,f} = t_{3,f} + w_{3,0}$. Since $t_{4,f} = d$, $w_{min} = w_{4,f} = -2d + t_{3,f} + w_{3,0} \geq 1$. This gives $w_{3,f} \geq 1 + 2d$.

As stated above, there can be critical-control phases in which a state variable may be increasing as well as decreasing (see example 2 in the next section). In such a phase, from its first and second derivatives, global minimum or maximum value of this variable can be determined, and the appropriate constraints from the global invariant can be deduced.

3.3 Different Control Strategies

Constraints deduced from the above analysis can be used to develop different control strategies that ensure that the system never violates the global invariant. One possible strategy is to use the constraints on variables as the guards themselves. We will call such a controller to be *generic* since it can be further refined to implement a specific policy. With a generic controller, there is no guarantee that the system can go on forever.

For the water-level controller example, a generic controller is: $g_1 = (1 \leq w \leq 12 - d)$, and $g_2 = (1 + 2 * d \leq w \leq 12)$. For $d = 2$, the respective guards are

$1 \leq w \leq 10$ and $5 \leq w \leq 12$. Guards for other transitions are known as they are based on the timer value, namely, $tc = d$ for transition from phase 2 to 3, and $tc = d$ for transition from phase 4 to 1.

Another possible control strategy is to exit a phase *as late as possible* (*ALAP*) policy to mean: the system exits a phase only if further evolution in that phase would lead to the global invariant being violated (i.e., do not do anything unless it is really necessary). The associated guards for this case are obtained by converting inequalities in the deduced constraints to equalities.

Using the ALAP controller policy, the guard g_1 is $w = 12 - d_1$, and g_2 is $w = 1 + 2 * d_2$. For $d_1 = d_2 = 2$, we get the same guards assumed in the definition used in [Ho95].

It is easy to see that the controllers can be synthesized even if lower and upper bounds on water level, say L and U, respectively, in the global invariant are given as parameters. Using the *ALAP* policy, from phase 1 to 2, the transition is made when the water level reaches $U - d$, and from phase 3 to 4, the transition is made when the water level reaches $L + 2 * d$.

The above approach is also helpful in identifying conditions when a controller can be synthesized. For the water level controller example, it must be the case that $U \geq L + 2d$ as well as $U - d \geq L$. For instance, if the pump was very slow in responding to commands to open and close the valve, say $d = 6$ and $L = 1, U = 12$, then a controller cannot be synthesized to satisfy the global invariant since the first constraint above ($U \geq L + 2d$) cannot be satisfied.

3.4 Periodicity

Cycles in a phase transition system can give rise to additional constraints on continuous variables and parameters, depending upon the nature of a controller being synthesized. It is possible to synthesize a *viable* controller (in the sense of [De94, DV95]) which will enable the system to run forever. If guards on transitions can be synthesized that would lead to a *repeatable cycle* of phases being executed without the global invariant being violated, then the resulting controller can be shown to be viable. A repeatable cycle of phases can be used to generate an infinite sequence of phases.

For the water-level controller example, there is a cycle involving all the phases. In order for the system to traverse an infinite sequence of phases using this cycle, the system must have the property that when it enters the first phase in the i-th cycle, the water level is exactly the same as it exits the fourth phase in the i-th cycle, where $i > 0$. In that case, if after the first cycle, the hybrid system enters the first phase, then every phase can be visited infinitely often.

This requirement makes $w_{1,0} = w_{4,f}$, from which the relation $2d + 2t_{3,f} = d + t_{1,f}$ can be deduced. A viable controller can be synthesized only if there exist values of $t_{1,f}, t_{3,f}$ satisfying the above equation for a given value of d. And, $t_{1,f}$ and $t_{3,f}$ can be related to lower and upper bounds on w if given as parameters. Otherwise, a controller cannot be synthesized.

For synthesizing viable controllers, repeatable cycles in a phase transition system can be analyzed. Different repeatable cycles can lead to different control policies that generate infinite sequences of phases.

4 Two Nontrivial Examples

In the above example, the system dynamics was specified by simple differential equations in which the first derivatives of variables are constants. Below, we discuss two slightly nontrivial examples in which the system dynamics exhibits nonlinear behavior. The second example, taken from [SM94], is a variation of the first example. The rate of net inflow into the tank is a function of the extent to which the valve is open and a system parameter. The third example is a variation on the temperature controller for a nuclear reactor discussed in [Ho95]. This example illustrates how the proposed methodology can be used to evaluate different control laws, and an optimal control strategy based upon some other criterion such as cost can be designed.

4.1 Nonlinear Water Controller

The phase transition system for this example is similar to that shown in Figure 1. Equations governing the water level are different; there is no explicit timer either. In every phase, the water-level w is governed by

$$dw/dt = 8\,f - 16\,v,$$

where $0 \leq f \leq 1$ is a system parameter, and v is another continuous variable between 0 and 1, measuring the extent to which the valve is open. If the valve is fully closed, $v = 0$; when it is fully open, $v = 1$. The controller is required to work no matter what f is. Equations for v in each phase are given below.

- *Phase 1:* **Valve closed:** $dv/dt = 0, v = 0$.
- *Phase 2:* **Valve opening:** $dv/dt = 1, v \leq 1$.
- *Phase 3:* **Valve open:** $dv/dt = 0, v = 1$.
- *Phase 4:* **Valve closing:** $dv/dt = -1, v \geq 0$.

The global invariant is $60 \leq w \leq 76$. The objective is to synthesize g_1 and g_2, as before.

Just as in example 1, for each phase, closed-form expressions are obtained by solving equations in terms of the time elapsed in the phase.

- *Phase 1:* $w_1 = 8\,f\,t_1 + w_{1,0}, v_1 = 0$.
- *Phase 2:* $v_2 = t_2 + v_{2,0}, v_{2,0} = 0, 0 \leq v_2 \leq 1, w_2 = 8\,f\,t_2 - 8\,t_2^2 + w_{2,0}$.
- *Phase 3:* $v_3 = 1, w_3 = 8\,f\,t_3 - 16\,t_3 + w_{3,0}$.
- *Phase 4:* $v_4 = -t_4 + v_{4,0}, v_{4,0} = 1$, and $w_4 = 8\,f\,t_4 - 16\,t_4 + 8\,t_4^2 + w_{4,0}$.

Further, $w_{2,0} = w_{1,f}, w_{3,0} = w_{2,f}, w_{4,0} = w_{3,f}$.

Next, control-critical phases are identified. The rate of change of water-level dw_1/dt is positive for nonzero f in phase 1. Phase 2 is interesting; dw_2/dt can be positive to begin with depending upon the value of f, and then it goes from positive to negative. Phase 2 is thus control-critical; w attains a maximum value in this phase. When $dw_2/dt = 0$, $8f = 16v_2$. Since w_2 increases with f and the controller is required to work for every f between 0 and 1, the maximum value for f is assumed. So, for $f = 1$, when v is $1/2$, the water level is maximum (this is checked from the fact that the second derivative of w_2 is negative for values of $v > 1/2$). And, this happens when $t_2 = 1/2$. Thus,

$$w_{2,max} = 4 - 2 + w_{1,f} = 76,$$

and so $w_{1,f} = 74$.

In phase 3, dw_3/dt is negative. In phase 4, much like phase 2, dw_4/dt starts being negative and it becomes nonnegative. Phase 4 is control-critical, and w attains a minimum value. Equating $dw_4/dt = 0$, we have $v = f/2$. Since w increases with f, the minimum water level will be reached the fastest when $f = 0$, so $v = 0$, i.e., when the valve is fully open, and this happens when $t_4 = 1$.

$$w_{4,min} = -16 + 8 + w_{3,f} = 60,$$

and so $w_{3,f} = 68$.

A generic controller would have the guards, $w \leq 74$ on the transition from phase 1 to phase 2, and $w \geq 68$ on the transition from phase 3 to 4. For the ALAP policy, the guard for transition from phase 1 to phase 2 is $w = 74$, and for transition from phase 3 to 4, it is $w = 68$.

The control law can be synthesized even if lower and upper bounds on water level in the global invariant are given as parameters. It is possible to determine conditions on L, U for which a controller cannot be synthesized.

4.2 Nuclear Reactor Temperature Control

In [ACHHH95, Ho95], variations of a nuclear reactor temperature controller are discussed. The system has a reactor core and two rods for cooling the reactor core at different rates. Once a rod is immersed in the core for cooling and removed, it cannot be used for a certain period of time. In the discussion in [ACHHH95, Ho95], the controller is straightforward: put one of the rods in if the temperature reaches U, and take the rod out if temperature reaches L. If the time for which rods have to be outside is beyond a certain period, then rods 1 and 2 are alternated. We generalize the formulation so that it is possible to consider different possible control strategies based on cost criteria for cooling. The generalization illustrates the power of the methodology.

The phase transition system is shown in Figure 2. Let T stand for the minimum time for which rods should not be immersed after they have been taken out (for simplicity, it is assumed to be the same for both rods even though the analysis extends to the case if this time parameter is different for the two rods). The temperature of the reactor is assumed to be governed by:

$$dx/dt = x/f - (t_c + k_1\,p_1 + k_2\,p_2),$$

where p_i is the fraction of i-th rod immersed in the reactor, k_i is the cooling factor of the i-th rod. It is assumed, for simplicity, that at most one rod is immersed at any time, i.e., at least one of p_1 and p_2 is 0.

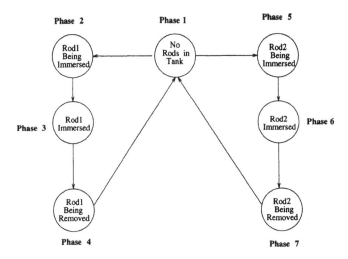

Figure 2. Temperature Controller of a Nuclear Reactor

Let c_i be the cost rate for keeping the i-th rod inside the reactor. For instance, if only rod 1 is used and it is inside the reactor (starting from when it is immersed inside) for t_1 time in any given cycle of time period t_m, then it costs $c_1 t_1$ units to keep the reactor in the prescribed temperature range during that cycle, and average cost is $c_1 t_1/t_m$ units per second.

The objective is to design a controller so as to ensure that (i) the reactor temperature is in between L and U, and (ii) rods are outside for at least T seconds after use. In case there are many control strategies possible, an optimality requirement is introduced: a cooling cycle should not cost more than K per unit per second on average.

To simplify the analysis, some of the above parameters – c_1, c_2, k_1, k_2, L and U are fixed. It is assumed that it takes 2 seconds to immerse a rod or to take out a rod from the reactor. And, $f = 1/10, t_c = 50, k_1 = 6$, and $k_2 = 10$, $L = 510, U = 550$, giving us a system similar to the one in [ACHHH95, Ho95].

Let o_j be the time j-th rod is outside after it was taken out of the reactor last time. Let i_j be the time j-th rod is inside the reactor.

The equation for the temperature of the nuclear reactor in each phase is:

$$dx/dt = x/10 - 50 - 6p_1 - 10p_2.$$

The differential equations for some of the phases are given below.

Phase 1: **No rod inside:** $do_1/dt = 1, do_2/dt = 1$. Further $di_1/dt = di_2/dt = 0$.

Phase 2: **Rod 1 being immersed:** $di_1/dt = 1, do_2/dt = 1, do_1/dt = 0, di_2/dt = 0$. The transition from phase 1 to phase 2 initializes i_1 to 0. And, $dp_1/dt = 1/2$, the rate at which rod 1 is being immersed, with $p_2 = 0$.

Phase 3: **Rod 1 immersed:** $di_1/dt = 1, do_2/dt = 1, do_1/dt = 0, di_2/dt = 0$. And, $dp_1/dt = 0$ with $p_1 = 1, p_2 = 0$.

Phase 4: **Rod 1 being removed:** $di_1/dt = 1, do_2/dt = 1, do_1/dt = 0, di_2/dt = 0$, and $dp_1/dt = -1/2$, the rate at which rod 1 is being removed, with $p_2 = 0$.

The transition from phase 4 to phase 1 resets the counter for 1st rod outside, i.e., $o_1 = 0$, and records time during which it is inside. Similarly, corresponding to the immersion and removal of rod 2 in the reactor, there are phases 5, 6 and 7 with similar differential equations.

In phase 1, dx/dt is positive if $x > 500$, implying that the temperature is increasing. In phase 2, $dx/dt = 0$ when x attains the maximum value of 550, implying that $p_1 = 5/6$, i.e., the first rod is immersed 5/6th into the reactor. Before this point, dx/dt is positive, and it becomes negative. In phase 3, dx/dt is $-ve$ as long as $x < 560$ implying that the temperature is decreasing. In phase 4, dx/dt goes from $-ve$ to $+ve$. At $dx/dt = 0$, it attains the minimum value – 510, when $p_1 = 1/6$, meaning that the first rod is still 1/6 immersed as it is being pulled out. Similar analysis can be done for phases 5, 6 and 7 for the second rod.

There can be at least three possible control strategies: (i) Only the first rod is used for cooling, (ii) only the second rod is used for cooling, and (iii) both rods are used in alternation. Below, a detailed analysis for the case when only the first rod is used, is given. Details are omitted for the second and third control strategies.

Strategy I **Using Rod 1 only**: The equation for the first phase is $dx/dt = x/10 - 50$ whose solution is:

$$x_1 = 500 + (x_{1,0} - 500) \times e^{t_1/10}. \tag{1}$$

For the second phase, the equation is, $dx_2/dt = x_2/10 - 50 - 6t_2$, whose solution is (obtained by substituting $x_2/10 - 6t$ by z) is:

$$x_2/10 = 110 + (x_{2,0}/10 - 110) \times e^{t_2/10} + 6 \times t_2, \tag{2}$$

where t_2 is the time the system has been in the second phase. Solving for x_2 at $t_2 = 5/6$, $x_{2,5/6} = 550$, we get $x_{2,0} = 547.97 = x_{1,f}$.

Similarly for the third phase,

$$x_3 = 560 + (x_{3,0} - 560) \times e^{t_3/10}. \tag{3}$$

For the fourth phase,

$$x_4/10 = -4 + (x_{3,f}/10 + 4) \times e^{t_4/10} - 6 \times t_4. \tag{4}$$

Solving for x_4 at $t_4 = 5/6$ and $x_{4,5/6} = 510$, we get $x_{3,f} = 512.03$.

Using these values, time elapsed in each phase when rod 1 is used for cooling can be calculated. Solving equation 4 for phase 4 at $t_4 = 1$, $x_{4,f} = 510.09$ which is also $x_{1,0}$ of the next cycle in the steady-state. Substituting these in 1,

$$e^{t_{1,f}/10} = (x_{1,f} - 500)/(x_{1,0} - 500) = 4.754,$$

from which $t_{1,f} = 15.59$ seconds.

Time spent in phase 3 is obtained by substituting in equation 3:

$$x_{3,f} - 560 = (x_{3,0} - 560) \times e^{t_{3,f}/10}.$$

Since $x_{3,0} = x_{2,f}$, $x_{2,f} = 549.91$ by substituting $t_{2,f} = 1$ in 2. Using these values in above equation for x_3, $t_{3,f} = 15.59$ seconds as well.

An analysis similar to that of strategy I can be done in case only rod 2 is used. In the following, we only highlight the results and omit the details for the sake of brevity.

Strategy II **Using Rod 2 only**: The governing equation is $dx/dt = x/10 - 50 - 10p_2$. When the temperature reaches the maximum during the immersion of the second rod, it is half-immersed since when $dx_5/dt = 0$; $x_5 = 550$, $p_2 = 1/2$. Similarly, when the temperature reaches the minimum when rod 2 is being removed, it is still $1/10$ immersed since $dx_7/dt = 0$ and $x_7 = 510$, $p_2 = 1/10$. Equations can be solved for different phases, as in the case of the first rod. Time spent in phase 6 is $t_{6,f} = 14.08$ seconds, and time spent in phase 1 in the steady-state when rod 2 is only being used is 15.795 seconds.

Strategy III **Using Both Rods**: The third control strategy is to alternate between the use of rod 1 and rod 2. Each of these strategies can be analyzed for meeting the requirements as well as cost.

1. **Only rod 1 is to be used**: The time T before rod 1 can be reused is less than or equal to 15.59 seconds. And, the cost incurred per unit time for this cycle is $c_1(2 + t_{3,f})/(t_{1,f} + 2 + t_{3,f}) = 0.5301c_1$, since $2 + t_{3,f}$ is the time for which rod 1 is inside, and $2 + t_{1,f} + t_{3,f}$ is the total cycle time.

2. **Only rod 2 is to be used**: Time T before rod 2 can be reused is less than or equal to 15.795 seconds, slightly longer than in the first case as to be expected since the cooling rate of rod 2 is greater. In this case, the cost incurred per unit time for this cycle is $c_2(2 + t_{6,f})/(t_{1,f} + 2 + t_{6,f}) = 0.5045c_2$. It does not appear to be very advantageous to use rod 2 with a higher cooling factor if c_2 is significantly higher than c_1.

3. **Rods are used together**: We could alternate between the use of rod 1 and rod2, then the time before either rod can be used has to be less than or equal to 47.465 seconds (minimum of $15.795 + 2 + 14.08 + 15.59 = 47.465$ and $15.795 + 2 + 15.59 + 15.59 = 48.975$). These calculations are approximate. And, the cost incurred per unit time is given by

$$(c_1(2+t_{3,f})+c_2(2+t_{6,f}))/(t_{1,f}+2+t_{3,f}+2+t_{5,f}+t'_{1,f}) = .2704c_1+.2472c_2.$$

Based on T that can be tolerated, different control strategies may be possible, and the best among them may be chosen based on the cost. If T is lower than 15.59 seconds, then either of the three strategies can be chosen, and the cost criterion can be brought into play to determine which one is optimal. If T is lower than 15.795 seconds but more than 15.59 seconds, then only two control strategies are feasible. If T is more than 15.795 but lower than 47.465 seconds, then a controller can be synthesized but rods 1 and 2 must alternate. If T is more than 47.465 then a controller cannot be synthesized.

It is possible to perform a parametric analysis and relate different control strategies in a way that T depends upon U, L, k_1, k_2 etc.

5 Concluding Remarks

We have proposed a methodology for synthesizing guards of a hybrid system for which a phase transition system and a global invariant are given. As illustrated, the proposed methodology can be used to synthesize generic controllers which could be further refined to viable controllers based on the ALAP policy. It is possible to identify conditions on system parameters under which a controller may not be realizable as the global invariant cannot be guaranteed. Even though the methodology is based on simple concepts, it appears to be quite powerful, and is applicable to the typical examples discussed in the hybrid system literature. The main assumption about the continuous dynamics of a hybrid system is that the differential equations associated with each phase are solvable resulting in closed-form expressions; it may be noted that this is also the assumption used in the study of hybrid automata. Further, the global invariant is restricted to be a conjunction of upper and/or lower bounds on continuous variables. The proposed methodology can handle hybrid systems that cannot be modeled by a bounded-drift linear hybrid automaton, as suggested by the two examples discussed in the previous section.

We believe that the methodology is mechanizable using symbolic and linear algebra algorithms supported in computer algebra systems such as Mathematica and Maple. This aspect is being explored. A detailed formal analysis as well as possible generalizations of the methodology based on concepts in [RW89] are planned.

Acknowledgments: The authors thank Akash Deshpande, Paliath Narendran, Michael Kourjanski, Christian Harris, and the referees for comments on earlier drafts of this paper.

References

[ACHH93] R. Alur, C. Courcoubetis, T. A. Henzinger and P.-H. Ho, *Hybrid Automata: An Algorithmic Approach to the Specification and Verification of Hybrid Systems*, LNCS, 736, Springer-Verlag, 1993, 209-229, enhanced version is in [ACHHH95].

[ACHHH95] R. Alur, C. Courcoubetis, N. Halbwachs, T. A. Henzinger and P.-H. Ho, X. Nicollin , A. Olivero, J. Sifakis, and S. Yovine, *The Algorithmic Analysis of Hybrid Systems*, Theoretical Computer Science, 138, 1995, 3-34.

[AHS96] R. Alur, T.A. Henzinger, and E.D. Sontag, editors, Hybrid Systems III: *Verification and Control*, LNCS, 1066, Springer-Verlag, 1996.

[AKNS95] P. Antsaklis, W. Kohn, A. Nerode and S. Shastry (Eds.), *Hybrid Systems II*, LNCS, 999, Springer Verlag, 1995.

[AMP95] E. Asarin, O. Maler, and A. Pnueli, *Symbolic Controller Synthesis for Discrete and Timed Systems*, Proc. of Hybrid Systems II, LNCS 999, 1995.

[De94] A. Deshpande, *Control of Hybrid Systems.* Ph.D. Thesis, University of California at Berkeley, 1994.

[DV95] A. Deshpande and P. Varaiya, *Viability in Hybrid Systems*, in [AKNS95], 128-147.

[GNRR93] R. L. Grossman, A. Nerode, A. P. Ravn, and H. Rischel (Eds.), *Hybrid Systems I*, LNCS, 736, Springer-Verlag, 1993.

[HRP93] N. Halbwachs, P. Raymond, Y.-E. Proy. *Verifying Linear Hybrid Systems by means of Convex Approximations*, Proc. SAS 94, LNCS 864, Sept. 1994.

[HKPV95] T. A. Henzinger, W. Kopke, A. Puri, P. Varaiya, *What's Decidable about Hybrid Automata*, Proc. of the 27th Annual ACM Symposium on Theory of Computing STOC 1995.

[Ho95] P.-H. Ho, *The Algorithmic Analysis of Hybrid Systems.* Ph.D Thesis, Cornell University, 1995.

[SM94] H.B. Sipma and Z. Manna, *Specification and Verification of Controlled Systems*, Proc. FTRTFT, LNCS, 863, 641-659, Sept. 1994.

[RW89] P.J.G. Ramadge, and W.M. Wonham, *The Control of Discrete Event Systems* , Proc. IEEE, Vol. 77, No. 1, Jan 1989, 81-98.

Control Synthesis for a Class of Hybrid Systems Subject to Configuration-Based Safety Constraints*

Michael Heymann[1]   Feng Lin[2]   George Meyer[3]

Abstract

We examine a class of hybrid systems called *Composite Hybrid Machines* (CHMs), that consist of the concurrent (and partially synchronized) operation of *Elementary Hybrid Machines* (EHMs).

Legal behavior is specified by a set of *illegal* configurations that the CHM may not enter, and is to be achieved by the concurrent operation of the CHM with a suitably designed *legal* controller. A legal controller is *minimally restrictive* if, when composed to operate concurrently with another legal controller, it will never interfere with the operation of the other controller. We focus attention on the problem of synthesizing a minimally restrictive legal controller, whenever a legal controller exists.

We present an algorithm for the synthesis of minimally restrictive legal controllers for CHMs with rate-limited dynamics, where legal guards are conjunctions or disjunctions of atomic formulas in the dynamic variables (of the type $x \leq x_0$ or $x \geq x_0$).

We demonstrate our approach by synthesizing a minimally restrictive controller for a steam boiler (the verification of which recently received a great deal of attention).
1 2 3

1 Introduction

Hybrid systems are dynamic systems in which discrete and continuous behaviors coexist and interact [3] [6] [14] [17] and changes occur both in response to events that take place discretely and instantaneously, and in response to dynamics as described by differential or difference equations of time.

*This research is supported in part by the National Science Foundation under grant ECS-9315344 and NASA under grant NAG2-1043 and in part by the Technion Fund for Promotion of Research.

[1] Department of Computer Science, Technion, Israel Institute of Technology, Haifa 32000, Israel, e-mail: heymann@cs.technion.ac.il. The work by this author was completed while he was a Senior NRC Research Associate at NASA Ames Research Center, Moffett Field, CA 94035.

[2] Department of Electrical and Computer Engineering, Wayne State University, Detroit, MI 48202, e-mail: flin@ece.eng.wayne.edu.

[3] NASA Ames Research Center, Moffett Field, CA 94035, e-mail: meyer@tarski.arc.nasa.gov.

In recent years interest in hybrid systems has grown rapidly both in the computer-science community and in the control theory community. In the computer-science community interest focused mainly on modeling, analysis, formal specification and verification (see, e.g. [3] [7] [16] [18] [19]), and more recently also on controller synthesis (see e.g. [5] [15]) and evolved progressively from logical systems, through "logically-timed" temporal systems to real-time systems modeled as timed-automata [2] and, most recently, to a restricted class of hybrid systems called *hybrid automata* [3]. In the control theory community, the growing realization that neither the purely discrete (discrete-event) [20] [13] nor the purely continuous frameworks are adequate for describing many physical systems [4] [6], led to focusing attention on hybrid system models. Contrary to the computer science viewpoint, the control-theory viewpoint focused attention mainly on issues of modeling and design.

Typical hybrid systems interact with the environment both by sharing signals (i.e., by transmission of input/output data), and by event-synchronization. Control of hybrid systems can therefore be achieved by employing both interaction mechanisms simultaneously. Yet, while this flexibility adds significantly to the potential control capabilities, it clearly makes the problem of design much more difficult.

In the present paper we examine the control problem for a restricted class of hybrid systems that we call *composite hybrid machines* (CHMs). We confine our attention to bounded-rate CHMs, in which the dynamic rates are bounded by lower and upper constant bounds. Control is confined to event-synchronization where the controller can affect the system's behavior only by discrete commands. These hybrid systems are a generalization of timed automata which, in turn, generalize discrete event systems by introducing real-time constraints. For such systems it is natural to specify the cotrol objective in terms of safety and liveness constraints, much in the spirit of the control of discrete-event systems. Indeed, this generalization is on one hand simple enough to be computationally tractable, and on the other hand, complex enough to provide some substantial new insights and a sense of new research direction.

Intuitively, a controller for legal behavior of a hybrid system is minimally restrictive if it never takes action unless constraint violation becomes imminent. When the latter happens, the controller is expected to do no more than prevent the system from becoming "illegal". This is a familiar setting in the discrete-event control literature, where the role of the controller has traditionally been that of a *supervisor* that can only intervene in the system's activity by event disablement [20]. Thus, a supervisor of a discrete-event system is minimally restrictive if it only disables events whenever filure to do so might permit the system to violate the specification.

A natural candidate for a minimally restrictive controller, is a system whose range of possible behaviors precisely coincides with the set of behaviors permitted by the specification. The concurrent execution of the controlled system and such a controller, in the sense that events are permitted to occur in the controlled system whenever they are possible in the controller, would constrain the system to satisfy the specification exactly. If all the events that are possible in the system but not permitted by the candidate controller can actually be disabled, we say that the specification is *implementable* or (when the specification is given as a legal language) *controllable* [20]. Generally, a specification may not be implementable because not all the events can be disabled.

The standard approach to supervisory-controller synthesis is an iterative procedure where, starting with the specification as a candidate implementation, the specification of legal behavior is progressively tightened, by excluding behaviors that cannot be prevented, by instantaneous event disablement, from becoming illegal [10] [11]. The modified specification thus obtained, is then used as a new candidate implementation. When the procedure converges in a finite number of steps (a fact guaranteed in case the system is a finite-automaton and the specification a regular-language), the result is either an empty specification (i.e., a legal supervisor does not exist) or a minimally restrictive implementable specification.

In the present paper we shall employ the same design philosophy for the synthesis of minimally restrictive controllers of hybrid systems and consider the problem where the controller is required to prevent the system from entering a specified set of *illegal* configurations. It will be shown elsewhere that a wide class of specifications can be transformed into the setting considered here. The reader is also referred to [5] [15] where a different (but somewhat related) hybrid controller synthesis problem is discussed.

We shall restrict our attention to *bounded-rate* hybrid systems, in which the rates of the dynamic variables are bounded by finite constants.

2 Hybrid Machines

We first introduce a modeling formalism for a class of hybrid systems which we call *hybrid machines* and which are a special case of *hierarchical hybrid machines* to be discussed elsewhere [12]. Hybrid machines are similar in spirit to hybrid automata as introduced in [3].

2.1 Elementary hybrid machines

An elementary hybrid machine is denoted by

$$EHM = (Q, \Sigma, D, I, E, (q_0, x_0)).$$

The elements of EHM are as follows.

- Q is a finite set of vertices.

- Σ is a finite set of event labels. An event is an input event, denoted by $\underline{\sigma}$ (underline), if it is received by the EHM from its environment; and an output event, denoted by $\overline{\sigma}$ (overline), if it is generated by the EHM and transmitted to the environment.

- $D = \{d_q = (x_q, y_q, u_q, f_q, h_q) : q \in Q\}$ is the dynamics of the EHM, where d_q, the dynamics at the vertex q, is given by:

$$\dot{x}_q = f_q(x_q, u_q),$$
$$y_q = h_q(x_q, u_q),$$

with x_q, u_q, and y_q, respectively, the state, input, and output variables of appropriate dimensions. f_q is a Lipschitz continuous function and h_q a continuous function. (A vertex need not have dynamics associated with it, that is $d_q = \emptyset$, in which case we say that the vertex is *static*.)

- $I = \{I_q : q \in Q\}$ is a set of invariants. I_q represents conditions under which the EHM is permitted to reside at q. A formal definition of I_q will be given in the next subsection.

- $E = \{(q, G \wedge \underline{\sigma} \to \overline{\sigma'}, q', x_{q'}^0) : q, q' \in Q\}$ is a set of edges (transition-paths), where q is the exiting vertex, q' the entering vertex, $\underline{\sigma}$ the input-event, $\overline{\sigma'}$ the output-event, G the guard to be formally defined in the next subsection, and $x_{q'}^0$ the initialization value for $x_{q'}$ upon entry to q'.

  $(q, G \wedge \underline{\sigma} \to \overline{\sigma'}, q', x_{q'}^0)$ is interpreted as follows: If G is true and the event $\underline{\sigma}$ is received as an input, then the transition to q' takes place with the assignment of the initial condition $x_{q'}(t_0) = x_{q'}^0$ (here t_0 denotes the time at which the vertex q' is entered). The output-event $\overline{\sigma'}$ is transmitted at the same time. If $\underline{\sigma}$ is absent, then the transition takes place immediately upon G become true; if $\overline{\sigma'}$ is absent, then no output-event is transmitted; if G is absent, the guard is always true and the transition will be triggered by the input-event $\underline{\sigma}$; and if $x_{q'}^0$ is absent, then the initial condition is inherited from x_q (assuming x_q and $x_{q'}$ represent the same physical object and hence are of the same dimension).

- (q_0, x_0) denote the initialization condition: q_0 is the initial vertex and $x_{q_0}(t_0) = x_0$.

For the EHM to be well-defined, we require that the vertices be completely guarded with each possible invariant violation. That is, every invariant violation implies that some guard becomes true and the associated transition is input-event-free in the sense that it has the form $(q, G \to \overline{\sigma'}, q', x_{q'}^0)$. (It is, in principle, permitted that more than one guard become true at the same instant. In this case the transition that will actually take place is resolved nondeterministically.) Note that we do not require the converse to be true. That is, a transition can be triggered even if the invariant is not violated. We do require that, upon entry to q', the invariant $I_{q'}$ not be violated. It is however possible that, upon entry to q', one of the guards at q' is already true. In this case, the EHM will immediately exit q' and go to the vertex specified by the guards. Such a transition is considered instantaneous. Naturally, we only allow finite chains of such instantaneous transitions. That is, the guards must be such that no sequence of instantaneous transitions will form a loop.

In this paper we will study a restrictive class of hybrid machines by making the following assumption.

Assumption 1 *The dynamics described by f_q and h_q has the following properties: (1) $h_q(x_q, u_q)$ is a linear function; and (2) $f_q(x_q, u_q)$ is bounded by a lower limit k_q^L and an upper limit k_q^U, that is, $f_q(x_q, u_q) \in [k_q^L, k_q^U]$.*

An execution of the EHM is a sequence

$$q_0 \xrightarrow{e_1, t_1} q_1 \xrightarrow{e_2, t_2} q_2 \xrightarrow{e_3, t_3} \ldots$$

where e_i is the ith transition and t_i is the time when the ith transition takes place.

2.2 Composite hybrid machine

A composite hybrid machine consists of several elementary hybrid machines running in parallel:

$$CHM = EHM^1 \| EHM^2 \| \ldots \| EHM^n.$$

Interaction between EHMs is achieved by means of signal transmission (shared variables) and input/output-event synchronization (message passing) as described below.

Shared variables consist of output signals from all EHMs as well as signals received from the environment. They are shared by all EHMs in the sense that they are accessible to all EHMs. A shared variable can be the output of at most one EHM. If the EHM of the output variable does not update the variable, its value will remain unchanged. The set of shared variables defines a signal space $S = [S_1, S_2, \ldots, S_m]$.

Transitions are synchronized by an input/output synchronization formalism. That is, if an output-event $\overline{\sigma}$ is either generated by one of the EHMs or received from the environment, then all EHMs for which σ is an active transition label (i.e., σ is defined at the current vertex with a true guard) will execute σ (and its associated transition) concurrently with the occurrence of $\overline{\sigma}$. An output-event can be generated by at most one EHM. Notice that input-events do not synchronize among themselves. Notice further that this formalism is a special case of the prioritized synchronous composition formalism [9], where each event is in the priority set of at most one parallel component.

By introducing the shared variables S, we can now define invariants and guards formally as boolean combinations of inequalities of the form (called *atomic formulas*)

$$S_i > C_i \quad \text{or} \quad S_i < C_i,$$

where S_i is a shared variable and C_i is a real constant.

To describe the behavior of

$$CHM = EHM^1 \| EHM^2 \| \ldots \| EHM^n,$$

we define a *configuration* of the CHM to be

$$q = < q_{i_1}^1, q_{i_2}^2, \ldots, q_{i_n}^n > \in Q^1 \times Q^2 \times \ldots \times Q^n$$

where Q^j is the set of vertices of EHM^j (components of the EHMs are superscripted).

When all the elements of q are specified, we call q a *full* configuration. When only some of the elements of q are specified, we call q a *partial* configuration and we mean that an unspecified element can be any possible vertex of the respective EHM. For example, $< , q_{i_2}^2, ..., q_{i_n}^n >$ is interpreted as the set

$$< q_{i_2}^2, ..., q_{i_n}^n >= \{< q_{i_1}^1, q_{i_2}^2, ..., q_{i_n}^n >: q_{i_1}^1 \in Q^1\}$$

of full configurations. Thus, a partial configuration is a compact description of a set of (full) configurations.

A transition

$$< q_{i_1}^1, q_{i_2}^2, ..., q_{i_n}^n > \xrightarrow{\quad l \quad} < q_{i'_1}^1, q_{i'_2}^2, ..., q_{i'_n}^n >$$

of a CHM is a triple where $< q_{i_1}^1, q_{i_2}^2, ..., q_{i_n}^n >$ is the source configuration, $< q_{i'_1}^1, q_{i'_2}^2, ..., q_{i'_n}^n >$ the target configuration, and l the label that triggers the transition. l can be either an event or a guard (becoming true). Thus, if $l = \sigma$ is an event (generated by the environment), then either $q_{i'_j}^j = q_{i_j}^j$ if $\underline{\sigma}$ is not active at $q_{i_j}^j$, or $q_{i'_j}^j$ is such that $(q_{i_j}^j, \underline{\sigma} \to \overline{\sigma'}, q_{i'_j}^j, x_{q_{i'_j}^j}^0)$ is a transition in E^j.

On the other hand, if $l = G$ is a guard, then there must exists a transition $(q_{i_m}^m, G \to \overline{\sigma'}, q_{i'_m}^m, x_{q_{i'_m}^m}^0)$ in some EHM^m and for $j \neq m$, either $q_{i'_j}^j = q_{i_j}^j$ if $\underline{\sigma'}$ is not defined at $q_{i_j}^j$, or $q_{i'_j}^j$ is such that $(q_{i_j}^j, \underline{\sigma'} \to \overline{\sigma''}, q_{i'_j}^j, x_{q_{i'_j}^j}^0)$ is a transition in E^j.

Recall that our model also allow guarded event transitions of the form

$$q \xrightarrow{G \wedge \sigma} q'.$$

However, since for the transition to take place the guard must be true when the event is triggered, a guarded event transition can be decomposed into

$$q^1 \underset{\neg G}{\overset{G}{\underset{\longleftarrow}{\longrightarrow}}} q^2 \xrightarrow{\sigma} q',$$

where q has been partitioned into q_1 and q_2, with $I_{q^1} = I_q \wedge \neg G$ and $I_{q^2} = I_q \wedge G$. It follows that a guarded event transition can be treated as a combination of a dynamic and an event transition.

Thus, transitions in CHMs can be classified into two types: (1) dynamic transitions, that are labeled by guards only, and (2) event transitions, that are labeled by events.

The transitions are considered to occur instantaneously and concurrent vertex changes in parallel components occur exactly at the same instant (even when constituting a logically triggered finite chain of transitions).

Based on the above definition, a CHM can be viewed as the same object as an EHM:

$$CHM = (Q, \Sigma, D, I, E, (q_0, x_0)).$$

Therefore, we can define an execution of a CHM in the same way as that of an EHM.

3 Control

3.1 Specifications

As stated in the previous section, a CHM can interact with its environment in two ways: (1) by signal transmission (shared variables), and (2) by input/output-event synchronization. Formally, a *Controller* of a CHM is a hybrid machine C that runs in parallel with the CHM. The resultant system

$$CHM \| C$$

is called the *controlled* or *closed loop* system. The objective of control is to force the controlled system to satisfy a prescribed set of behavioral specifications.

For conventional (continuous) dynamical systems, control specification might consist of the requirement of stability, robustness, disturbance rejection, optimality and the like. For discrete-event systems, specifications of required behavior are typically given as *safety* specifications, where a prescribed set of unwanted behaviors or configurations is to be avoided, or *liveness* specifications, where a prescribed set of termination conditions is to be met, or both.

For general hybrid systems, specifications can, in principle, be of a very complex nature incorporating both dynamic requirements and the logical (discrete) aspects.

In the present paper we consider only safety specifications given as a set of *illegal* configurations

$$Q_b = \{q = <q_{i_1}^1, q_{i_2}^2, ..., q_{i_n}^n> \in Q^1 \times Q^2 \times ... \times Q^n : q \text{ is illegal}\}$$

that the system is not permitted to visit.

Our goal is to synthesize a controller that guarantees satisfaction of the above stated configuration-based safety requirement. A controller that achieves the specification is then said to be *legal*.

In this paper, we shall consider only restricted interaction between the controller and the CHM by permitting the controller to communicate with the CHM only through input/output-event synchronization. Thus, we make the following assumption.

Assumption 2 *C can only control the CHM by means of input/output-event synchronization. That is, C can only control event transitions in the CHM.*

Thus, the controller is assumed not to generate any output signals that may affect the CHM.

We shall assume further that C can control all the event transitions in the CHM. That is, all the (externally triggered) event transitions are available to the controller. This leads to no essential loss of generality because, when some of the events are *uncontrollable*, we can use the methods developed in supervisory control of discrete-event systems [20] to deal with uncontrollable event transitions. We shall elaborate on this issue elsewhere.

A legal controller C is said to be *less restrictive* than another legal controller C' if every execution permitted by C' is also permitted by C (a formal definition will be given in the next subsection). A legal controller is said to be *minimally restrictive* if it is less restrictive than any legal controller.

3.2 Control synthesis

As stated, our control objective is to ensure that the system **CHM** never enter the set of illegal configurations Q_b. Such entry can occur either via an event transition or via a dynamic transition. Since all event transitions are at the disposal of the controller, prevention of entry to the illegal set via event transitions is a trivial matter (they simply must not be triggered). Therefore, in our control synthesis we shall focus our attention on dynamic transitions. Intuitively, the minimally restrictive legal controller must take action, by forcing the **CHM** from the current configuration to some other legal configuration, just in time (but as late as possible) to prevent a dynamic transition from leading the system to an illegal configuration. Clearly, entry to a configuration which is legal but at which an inescapable (unpreventable) dynamic transition to an illegal configuration is possible, must itself be deemed technically illegal and avoided by the controller. Thus the controller synthesis algorithm that we present below, will iterate through the (still) legal configurations and examine whether it is possible to prevent a dynamic transition from leading to an illegal configuration. In doing so, it will frequently be necessary to "split" configurations by partitioning their invariants into their *legal* and *illegal* parts.

For each legal configuration, we classify the transitions leaving q into four types: (1) Legal event transitions that lead to legal configurations: $ET_g(q, Q_b) = \{(q, \underline{\sigma}, q') : q \xrightarrow{\sigma} q' \wedge q' \notin Q_b\}$. (2) Illegal event transitions that lead to illegal configurations: $ET_b(q, Q_b) = \{(q, \underline{\sigma}, q') : q \xrightarrow{\sigma} q' \wedge q' \in Q_b\}$. (3) Legal dynamic transitions that lead to legal configurations: $DT_g(q, Q_b) = \{(q, G, q') : q \xrightarrow{G} q' \wedge q' \notin Q_b\}$. (4) Illegal dynamic transitions that lead to illegal configurations: $DT_b(q, Q_b) = \{(q, G, q') : q \xrightarrow{G} q' \wedge q' \in Q_b\}$.

To formally present our algorithm, let us define $wp(q, \underline{\sigma}, q')$ to be the weakest precondition under which the transition $(q, \underline{\sigma}, q')$ will not violate the invariant $I_{q'}$ upon entry to q'. Since some of the shared variables that appear in $I_{q'}$ are possibly (re-)initialized upon entering q', the condition $wp(q, \underline{\sigma}, q')$ can be computed from $I_{q'}$ by substituting into $I_{q'}$ the appropriate initial (entry) values of all the variables that are also output variables of q'. That is, if y_j is the jth output variable of q' and $S_i = y_j$ is a shared variable that appears in $I_{q'}$, then the value of S_i must be set to

$$S_i = h_j(x_{q'}^0, u_{q'}).$$

Also, for a predicate $P = (S_i \leq C_i)$, we define

$$critical(P) = \begin{cases} (S_i \geq C_i) & \text{if } r_i{}^U > 0 \\ \text{false} & \text{otherwise,} \end{cases}$$

Similarly, we can define $critical(P)$ for $P = (S_i \geq C_i)$. For conjunction of two predicates $P = P_1 \wedge P_2$, $critical(P) = critical(P_1) \vee critical(P_2)$, and for disjunction of two predicates $P = P_1 \vee P_2$, $critical(P) = critical(P_1) \wedge critical(P_2)$.

Let the interval of time that will elapse before P can become true be bounded by the minimum value $T_{min}(true(P))$ and the maximum value $T_{max}(true(P))$.

Then we can define the following preemptive condition[4]

$$pc(q, G, q') = (T_{min}(true(G \wedge wp(q, G, q'))) > T_{max}(false(I_q))).$$

Now, we can present the following

Algorithm 1 *(Control Synthesis)*
Input

- *The model of the system $CHM = (Q, \Sigma, D, I, E, (q_0, x_0))$.*

- *The set of illegal configurations $Q_b \subseteq Q$.*

Output

- *The controller $C = (Q^c, \Sigma^c, D^c, I^c, E^c, (q_0^c, x_0^c))$.*

Initialization

1. *Set of bad configurations $BC := Q_b$;*

2. *Set of pending configurations $PC := Q - Q_b$;*

3. *New set of pending configurations $NPC := \emptyset$;*

4. *For each $q \in PC$ set its* configuration origin as $CO(q) = q$;

Iteration

5. *For all $q \in PC$ do*

$$\begin{aligned}
I_{q_1} &:= \quad I_q \wedge ((\wedge_{(q,G,q') \in DT_b(q,BC)} pc(q, G, q')) \\
&\qquad \vee (\vee_{(q,\underline{\sigma},q') \in ET_g(q,BC)} wp(q, \underline{\sigma}, q'))); \\
I_{q_2} &:= \quad I_q \wedge (\neg(\wedge_{(q,G,q') \in DT_b(q,BC)} pc(q, G, q')) \\
&\qquad \wedge \neg(\vee_{(q,\underline{\sigma},q') \in ET_g(q,BC)} wp(q, \underline{\sigma}, q'))); .
\end{aligned}$$

If $I_{q_1} \neq false$, then

$$NPC := NPC \cup \{q_1\}; \quad CO(q_1) := CO(q);$$

If $I_{q_2} \neq false$, then

$$BC := BC \cup \{q_2\};$$

6. *If $PC = NPC$, go to 8.*

7. *Set*

$$PC := NPC; \quad NPC := \emptyset;$$

Go to 5;

Construction of C

---

[4] We take the convention that if $T_{min}(true(G \wedge wp(q, G, q'))) = \infty$, then $pc(q, G, q') = true$ even if $T_{max}(false(I_q)) = \infty$.

8. Define vertices, events and dynamics:

$$Q^c := PC; \quad \Sigma^c := \Sigma \cup \{\tilde{\sigma} : \sigma \in \Sigma\}; \quad D^c := \emptyset;$$

9. Define transitions:

$$
\begin{aligned}
E^c := \quad & \{(q, critical(I_q) \wedge wp(q, \underline{\sigma}, q') \\
& \wedge (\neg (\wedge_{(q,G,q'') \in DT_b(q,BC)} pc(q, G, q''))) \rightarrow \overline{\sigma}, q') : \\
& q, q' \in Q^c \wedge (CO(q), \underline{\sigma}, CO(q')) \in E\}; \\
E^c := \quad & E^c \cup \{(q, wp(q, \underline{\sigma}, q') \wedge \tilde{\underline{\sigma}} \rightarrow \overline{\sigma}, q') : \\
& q, q' \in Q^c \wedge (CO(q), \underline{\sigma}, CO(q')) \in E\};
\end{aligned}
$$

10. End.

Another controller D can be embeded into C as follows. First, all the output-events $\overline{\sigma}$ in D are replaced by $\overline{\tilde{\sigma}}$ to obtain $\tilde{D}$. Then the embeded control system is given by

$$CHM\|C\|\tilde{D}.$$

We can now prove the following

Theorem 1 *If Algorithm 1 terminates in a finite number of steps and no sequence of instantaneous transitions forms a loop, then the controller synthesized is the minimally restrictive legal controller in the following sense. (1) For any controller D, an execution in $CHM\|C\|\tilde{D}$ will never visit illegal configurations Q_b. (2) For any legal controller D, an execution is possible in $CHM\|D$ if and only if it is possible in $CHM\|C\|\tilde{D}$.*

4 Steam Boiler Example

In this section, we shall illustrate application of the control synthesis algorithm developed in the previous section by synthesizing a controller for the familiar steam boiler example that was proposed in [1] as a benchmark problem for modeling and verification of hybrid systems (see also e.g. [8]). This example was proposed as a benchmark problem because it has many essential properties that are found in some commonly used industrial processes, such as chemical reactors, oil refineries, etc.

We use a simplified model of the steam boiler described in [1]. Some parameters are set at the same values as in [8]. This simplified model captures the essence of the control problem addressed in this paper.

The steam boiler consists of a water tank (boiler) equipped with two pumps (instead of four pumps as in [1]). Each pump can supply water to the boiler at the rate of 4 liter/sec. The pump can be switched on (event *start_i*) and off (event *stop_i*) by a controller. Due to the fact that the pump cannot balance the presure inside the boiler instantaneously, there is a five-second delay before water starts pouring into the boiler after the pump is switched on.

Steam is generated by an unmodeled mechanism. The rate at which steam is generated is therefore nondeterministic. But we do know that the rate is bounded between 0 liter/sec. and 6 liter/sec.

The control objective is to maintain the water level L in the boiler between the minimal level of 5 liters and the maximal level of 220 liters. This is achieved by turning the two pumps on and off. Since we are interested in synthesizing the minimally restrictive controller, our controller will accept (that is, permit) all behaviors (turning pumps on and off) that do not imply possible violation of the level constraints and will intervene by forcing the pumps (on or off) only whenever it is absolutely necessary to do so in order to guarantee constraint satisfaction.

The controller can sample the water level in the boiler only every five seconds. Since this implies sampled decision making, there is no loss in generality in assuming that control (turning the pumps on and off) can only be applied at the sampling instants.

In summary, the steam boiler to be controlled is modeled by the CHM in Figure 1.

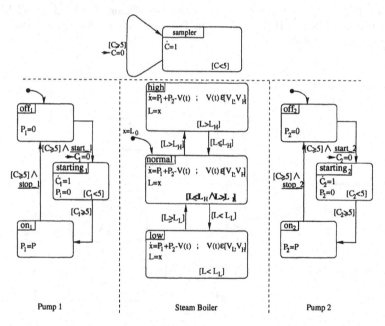

Figure 1: Steam Boiler System

As stated above, the parameters are given by

$$P_1 = 4, \quad P_2 = 4, \quad V_L = 0, \quad V_H = 6, \quad L_L = 5, \quad L_H = 220.$$

Without changing the nature of the problem but to avoid nondeterminism in the controller, we shall assume that Pump 1 will be turned on before Pump 2 can be turned on; and Pump 1 cannot be turned off before Pump 2 is turned off.

Thus, the configurations of the CHM to be controlled can be denoted by the legal configurations

$q^1 =< off_1, off_2, normal >,$ $q^2 =< starting_1, off_2, normal >,$
$q^3 =< on_1, off_2, normal >,$ $q^4 =< starting_1, starting_2, normal >,$
$q^5 =< on_1, starting_2, normal >,$ $q^6 =< on_1, on_2, normal >,$

and illegal configurations where $normal$ ($[L \geq 5] \wedge [L \leq 220]$) is replaced by $high$ ($[L > 220]$), or low ($[L < 5]$). That is,

$Q_b =< high > \cup < low > .$

Because of the delays in turning the pumps on and the delays caused by sampling, there are configurations in $< normal >$ from which unavoidable dynamic transitions may lead to illegal configurations in Q_b. Therefore, we must partition $< normal >$ properly using the synthesis algorithm.

Before applying the algorithm, we first replace the guarded event transitions by dynamic and event transitions. Also note that since $C_1 = C_2 = C$ whenever they are not equal to 0 or 5, only one clock is sufficient (to be denoted by C). Thus, the equivalent CHM is shown in Figure 2, where, for clarity, the illegal configurations are not drawn.

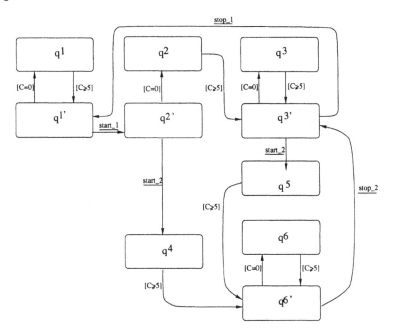

Figure 2: CHM

We will only illustrate how the algorithm performs on q^6 and $q^{6'}$, where

$I_{q^6} = [L \geq 5] \wedge [L \leq 220] \wedge [C < 5],$
$I_{q^{6'}} = [L \geq 5] \wedge [L \leq 220] \wedge [C \geq 5].$

	initial	1st iteration	2nd iteration	3rd iteration
q^1	$[L≥5]∧[L≤220]∧[C<5]$	$[L>35-6C]∧[L≤220]∧[C<5]$	$[L>35-6C]∧[L≤220]∧[C<5]$	$[L>65-6C]∧[L≤220]∧[C<5]$
$q^{1'}$	$[L≥5]∧[L≤220]∧[C≥5]$	$[L≥5]∧[L≤220]∧[C≥5]$	$[L>35]∧[L≤220]∧[C≥5]$	$[L>35]∧[L≤220]∧[C≥5]$
q^2	$[L≥5]∧[L≤220]∧[C<5]$	$[L>35-6C]∧[L≤220]∧[C<5]$	$[L>35-6C]∧[L≤220]∧[C<5]$	$[L>45-6C]∧[L≤220]∧[C<5]$
$q^{2'}$	$[L≥5]∧[L≤220]∧[C≥5]$	$[L≥5]∧[L≤220]∧[C≥5]$	$[L>35]∧[L≤220]∧[C≥5]$	$[L>35]∧[L≤220]∧[C≥5]$
q^3	$[L≥5]∧[L≤220]∧[C<5]$	$[L>15-2C]∧[L≤200+4C]∧[C<5]$	$[L>15-2C]∧[L≤200+4C]∧[C<5]$	$[L>25-2C]∧[L≤200+4C]∧[C<5]$
$q^{3'}$	$[L≥5]∧[L≤220]∧[C≥5]$	$[L≥5]∧[L≤220]∧[C≥5]$	$[L>15]∧[L≤220]∧[C≥5]$	$[L>15]∧[L≤220]∧[C≥5]$
q^4	$[L≥5]∧[L≤220]∧[C<5]$	$[L>35-6C]∧[L≤220]∧[C<5]$	$[L>35-6C]∧[L≤220]∧[C<5]$	$[L>35-6C]∧[L≤220]∧[C<5]$
q^5	$[L≥5]∧[L≤220]∧[C<5]$	$[L>15-2C]∧[L<200+4C]∧[C<5]$	$[L>15-2C]∧[L<200+4C]∧[C<5]$	$[L>15-2C]∧[L<200+4C]∧[C<5]$
q^6	$[L≥5]∧[L≤220]∧[C<5]$	$[L≥5]∧[L<180+8C]∧[C<5]$	$[L≥5]∧[L<180+8C]∧[C<5]$	$[L≥5]∧[L<180+8C]∧[C<5]$
$q^{6'}$	$[L≥5]∧[L≤220]∧[C≥5]$	$[L≥5]∧[L≤220]∧[C≥5]$	$[L≥5]∧[L≤220]∧[C≥5]$	$[L≥5]∧[L≤220]∧[C≥5]$

Table 1: Steam Boiler Controller Synthesis

By our algorithm,

$$wp(q^{6'}, \underline{stop\_2}, q^{3'}) = [L \geq 5] \wedge [L \leq 220].$$

Therefore, $q^{6'}$ will not be split. On the other hand, q^6 will be split as follows (note that at q_6, $\dot{L} \in [2, 8]$).

$$pc(q^6, [L > 220], < illegal >)$$
$$= (T_{min}([L > 220]) > T_{max}([L < 5] \vee [L > 220] \vee [C \geq 5]))$$
$$= ((220 - L)/8 > min\{\infty, (220 - L)/8, 5 - C\})$$
$$= ((220 - L)/8 > (5 - C)) = (L < 180 + 8C)$$

Similarly,

$$pc(q^6, [L < 5], < illegal >)$$
$$= (T_{min}([L < 5]) > T_{max}([L < 5] \vee [L > 220] \vee [C \geq 5]))$$
$$= (\infty > T_{max}([L < 5] \vee [L > 220] \vee [C \geq 5]))$$
$$= true.$$

Therefore, q^6 will be split into q_1^6 and q_2^6 with invariants

$$I_{q_1^6} = I_{q^6} \wedge pc(q^6, [L > 220], < illegal >) \wedge pc(q^6, [L < 5], < illegal >)$$
$$= [L \geq 5] \wedge [L \leq 220] \wedge [C < 5] \wedge [L < 180 + 8C]$$
$$= [L \geq 5] \wedge [C < 5] \wedge [L < 180 + 8C],$$
$$I_{q_2^6} = [L \geq 5] \wedge [L \leq 220] \wedge [C < 5] \wedge [L \geq 180 + 8C].$$

The results of each iteration using Algorithm 1 is summarized in Table 1, and the minimally restrictive controller is synthesized and shown in Figure 3.

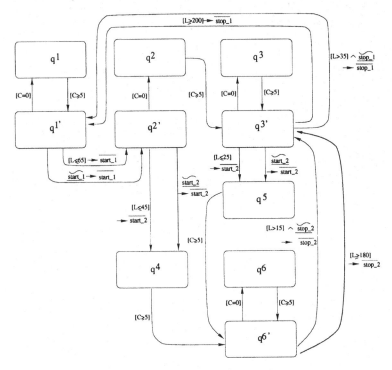

Figure 3: Steam Boiler Controller

References

[1] J.-R. Abrial, 1995. Steam-boiler control specification problem. *Dagstuhl Meeting: Method for Semantics and Specification.*

[2] R. Alur and D. Dill, 1990. Automata for modeling real-time systems. *Proc. of the 17th International Colloquium on Automata, Languages and Programming*, pp. 322-336.

[3] R. Alur, C. Courcoubetis, N. Halbwachs, T. A. Henzinger, P.-H. Ho, X. Nicollin, A. Olivero, J. Sifakis, and S. Yovine, 1995. The algorithmic analysis of hybrid systems. *Theoretical Computer Science, 138*, pp. 3-34.

[4] P.J. Antsaklis, J.A. Stiver, and M. Lemmon, 1993. Hybrid system modeling and autonomous control systems. *Hybrid Systems, Lecture Notes in Computer Science, 736*, Springer-Verlag, pp. 366-392.

[5] Asarin, O. Maler and A. Pnueli, 1995. Symbolic Controller Synthesis for Discrete and Timed systems, Hybrid Systems II.

[6] M. S. Branicky, 1995. Universal computation and other capabilities of hybrid and continuous dynamical systems. *Theoretical Computer Science, 138*, pp. 67-100.

[7] T. Henzinger, P. Kopke, A. Puri and P. Varaiya, 1995. What's decidable about hybrid automata, *Proc. of the 27th Annual ACM Symposium on the Theory of Computing.*

[8] T. A. Henzinger and H. Wong-Toi, 1996. Using HYTECH to synthesize control parameters for a steam boiler. *Preprint.*

[9] M. Heymann 1990. Concurrency and discrete event control, *IEEE Control Systems Magazine, Vol. 10, No.4*, pp 103-112.

[10] M. Heymann and F. Lin, 1994. On-line control of partially observed discrete event systems. *Discrete Event Dynamic Systems: Theory and Applications, 4(3)*, pp. 221-236.

[11] M. Heymann and F. Lin, 1996. Discrete event control of nondeterministic systems. control of nondeterministic systems, *CIS Report 9601*, Technion, Israel.

[12] M. Heymann and F. Lin, 1996. Hierarchical hybrid machines. To appear.

[13] F. Lin and W. M. Wonham, 1988. On observability of discrete event systems. *Information Sciences, 44(3)*, pp. 173-198.

[14] O. Maler, Z. Manna and A. Pnueli, 1991. From timed to hybrid systems. In *Real Time: Theory in Practice*, Lecture Notes in Computer Science 600, pp. 447-484. Springer Verlag.

[15] O. Maler, A. Pnueli, and J. Sifakis, 1995. On the Synthesis of Discrete Controllers for Timed Systems, STACS'95.

[16] Z. Manna and A. Pnueli, 1993. Verifying hybrid systems. *Hybrid Systems, Lecture Notes in Computer Science, 736*, Springer-Verlag, pp. 4-35.

[17] A. Nerode and W. Kohn, 1993. Models for hybrid systems: automata, topologies, controllability, observability. *Hybrid Systems, Lecture Notes in Computer Science, 736*, Springer-Verlag, pp. 317-356.

[18] X. Nicollin, A. Olivero, J. Sifakis, and S. Yovine, 1993. Am approach to the description and analysis of hybrid systems. *Hybrid Systems, Lecture Notes in Computer Science, 736*, Springer-Verlag, pp. 149-178.

[19] X. Nicollin, J. Sifakis, and S. Yovine, 1991. From ATP to timed graphs and hybrid systems. In *Real Time: Theory in Practice*, Lecture Notes in Computer Science 600, Springer-Verlag, pp. 549-572.

[20] R. J. Ramadge and W. M. Wonham, 1987. Supervisory control of a class of discrete event processes. *SIAM J. Control and Optimization, 25(1)*, pp. 206-230.

Hybrid Dynamic Programming

Wolf Kohn and Jeffrey B. Remmel*[1]

Sagent Corporation
11201 SE 8th Street #J140
Bellevue, Washington 98004
e-mail: wk@sagent.com jremmel@sagent.com

Abstract. In this paper we outline the derivation of a hybrid Hamilton-Jacobi-Bellman equation for our Multiple Agent Hybrid Control Architecture (MACHA) [5, 4] and a hybrid system dynamic progamming methodology.

1 Introduction

The classical discrete optimization problem is the following. Given x_0, find $min_{u_i \in U} \; \psi(X_N, n) + \sum_{k=0}^{N-1} \phi(x_k, u_k)$ where $x_{k+1} = f(x_k, u_k)$ for all k. Here U is the space of controls, X is the state space, $f : X \times U \to X$, $\psi : X \times N \to R$, and $\phi : X \times U \to R$ where R and N denote the set of real number and natural numbers respectively. One then defines the usual cost-to-go function

$$V(y, t) = min_{u_i \in U}[\psi(X_N, n) + \phi(y, u_t) + \sum_{k=t+1}^{N-1} \phi(x_k, u_k)]. \qquad (1)$$

Then $V(x_0, 0)$ is the desired optimal value and $V(y, t)$ satisfies Bellman's equation

$$V(y, t) = min_{u_t \in U} \; \phi(y, u_t) + V(f(y, u_t), t + 1) \qquad (2)$$

where $V(y, N) = \psi(y, N)$.

In our multiple agent hybrid control architechture we can use chattering to derive a hybrid Bellman equation. That is, corresponding to an automous problem, $\dot{x} = F(x, t)$, we construct a non-negative Lagrangian $L(x, \dot{x})$ where we have included time as an extra variable $t = x_{n+1}$ with the constraint that $\dot{x}_{n+1} = 1$. We assume that the Lagrangian $L(x, \dot{x})$ is positive definite in $\dot{x}$ along an extremal tube. That is,

$$(g_{ij}(x, \dot{x})) \mid_* = \left(\frac{\partial^2 (L(x, \dot{x}))}{\partial \dot{x}_j \dot{x}_i} \right) \mid_*$$

is a positive definite matrix for each fixed x where
(*) restricted to an extremal tube of trajectories which are solutions to $\dot{x} = F(x, t)$.

Under suitable convexity assumptions on L, we can then define a metric ground form ds on the underlying manifold M such that $ds^2 = \sum_{ij} g_{ij}(x, \dot{x}) \mid_* dx_i dx_j$. A key result of the theory is the fact that the geodesics under the metric ground form induced by L are the extremals of the corresponding parametric calculus of

---

* Research supported by Dept. of Commerce Agreement 70-NANB5H1164.

variations problem which is to find an admissible curve on M which minimizes $\int_{t_0}^{t_1} L(x(t), \dot{x}(t)) dt$ satisfying given endpoint conditions. That is, the geodesics are solutions to the Euler-Lagrange equation $\dfrac{d}{dt} \dfrac{\partial L(x, \dot{x})}{\partial \dot{x}_i} - \dfrac{\partial L(x, \dot{x})}{\partial x_i} = 0$.

In our multiple-agent hybrid control architecture, see section 1, each agent wants to minimize the intergral of a Lagrangian over trajectories on the carrier manifold M. More precisely, fix a goal x_1 and an element y_1 in the tangent space T_{x_1} Then for any given initial condition $x(0) = x_0$, each agent wants to force the plant to follow a trajectory $x(\cdot)$ such that $\int_0^T L(x(t), \dot{x}(t), t) dt$ is with ϵ of

$$min_{x(\cdot) \in M} \int_0^T L(x(t), \dot{x}(t), t) dt. \qquad (3)$$

Define a counting variable Z_i by $Z_i = \sum_{n=0}^i \int_{n\Delta}^{(n+1)\Delta} L(x(t), \dot{x}(t), t) dt$ where $(N+1)\Delta = T$ so that

$$Z_{i+1} = Z_i + \int_{(i+1)\Delta}^{(i+2)\Delta} L(x(t), \dot{x}(t), t) dt. \qquad (4)$$

If we replace the integral on the righthand side of (4) by its chattering approximation, see [3], we can define a new sequence of counting variables $\tilde{Z}_i$ by

$$\tilde{Z}_{i+1} = \tilde{Z}_i + \sum_{k=0}^{s} L\big(x((i+1)\Delta), v_{i+1}^k, (i+1)\Delta\big)\alpha_{i+1}^k \Delta. \qquad (5)$$

where for each i (a) v_i^k for $k = 0, \ldots, s$ is in the field of extremals of the original problem (3) at $x(i\Delta)$ and (b) α_i^k are nonnegative reals such that $\sum_{k=0}^s \alpha_i^k = 1$.

Note the desired minumum of the original problem (3) is a geodesic on a manifold M so that we can obtain $x((i+1)\Delta)$ and v_{i+1}^k for $k = 0, \ldots, s$ by parallel transport from $x(i\Delta)$ and v_i^k for $k = 0, \ldots, s$. That is, assume for any point x_0 in the manifold, there is a unique goedesic starting at x_0 and ending at x_1 with tangent vector y_1 at x_1. The idea is to compute with the Levi-Civita connection ∇ along an integral curve or geodesic $\alpha : [t_0, t_1] \to M$ such that $x_0 = \alpha(t_0)$ and $x_1 = \alpha(t_1)$, see [6]. For example, suppose that α lies entirely within some chart (U, ψ) where U is an open set of M and ψ is a homeomorphism of U to a region of Euclidean space. Assume that in the local coordinates of the chart, the curve α is given by $\psi(\alpha(t)) = (\alpha_1(t), \ldots, \alpha_n(t))$. Let TM_x denote the tangent space of M at the point x. Now suppose that we are given a tangent vector $y_1 \in TM_{x_1}$ and we want to compute the parallel transport of the tangent vector y_1 along the geodesic α to get a tangent vector y_0. In such a case, we will write that $x_1 = \overrightarrow{x_0}(t_1)$ and that $y_1 = \overrightarrow{y_0}(t_1)$. This requires transporting y_1 to a vector $h(t) = \overrightarrow{y_0}(t)$ tangent to the surface at the point $\alpha(t) = \overrightarrow{x_0}(t)$ for all intermediate t. In local coordinates $h(t)$ will be of the form $h(t) = \sum h_i(t)\frac{\partial}{\partial x_i}$. The condition that $h(t)$ is parallel along $\alpha(t)$ is that the covariant derivative be zero along that path. This says that $h(t)$ satisfies

$$0 = \nabla h = \sum_{ijk} \left[\frac{dh_k}{dt} + h_i \frac{d\alpha_j}{dt} \Gamma_{ij}^k \right] \frac{\partial}{\partial x_k} \qquad (6)$$

where the Γ_{ij}^k's are the Christoffel symbols.

Thus in (5), we have that $x((i+1)\Delta) = \overrightarrow{x(i\Delta)}(\Delta)$ and $v_{i+1}^k = \overrightarrow{v_i^k}(\Delta)$ for $k = 0, \ldots, s$. We can then show that the original problem (3) which is to find $min_{x(\cdot) \in M} Z_{N+1}$ can be replaced by the relaxed problem which is to find $inf_{\alpha_i^j} \tilde{Z}_{N+1}$ where $min_{x(\cdot) \in M} Z_{N+1} + O(\Delta^2) \geq inf_{\alpha_i^j} \tilde{Z}_{N+1}$.

This formulation allows us to derive a dynamic programming equation for the relaxed problem. That is, define

$$V(\tilde{Z}_i, x, i\Delta) = inf_{\alpha_i^k} \tilde{Z}_i + \sum_{t=i+1}^{N} \sum_{k=0}^{s} L(x(t\Delta), v_t^k, t\Delta)\alpha_t^k \Delta. \tag{7}$$

where $x = x(i\Delta)$ and for each $r > i$,

(i) $x(r\Delta) = \overrightarrow{x((r-1)\Delta)}(\Delta)$ and $v_r^k = \overrightarrow{v_{r-1}^k}(\Delta)$ for $k = 0, \ldots, s$,
(ii)) v_r^k for $k = 0, \ldots, s$ is in the field of extremals of the original problem (3) at $x(r\Delta)$, and
(iii) α_t^r are nonnegative real numbers such that $\sum_{k=0}^{s} \alpha_r^k = 1$. Then by Bellman's Principle of Optimality,

$$V(\tilde{Z}, x, i\Delta) = inf_{\alpha_i^k} V(\tilde{Z}_{x,(i+1)\Delta}, x((i+1)\Delta), (i+1)\Delta) \tag{8}$$

where $\tilde{Z}_{x,(i+1)\Delta} = \tilde{Z} + \sum_{k=0}^{s} L(\overrightarrow{x}(\Delta), \overrightarrow{v_x^k}(\Delta), (i+1)\Delta)\alpha_{i+1}^k \Delta$ and we have the boundary condition that $V(\tilde{Z}, x, N+1) = \tilde{Z}$.

We showed in [7] that V satisfies

$$-\frac{\partial_c V}{\partial_c t}(\tilde{Z}, x, i\Delta) = \tag{9}$$

$$inf_{\alpha_i^k} \left[\begin{array}{l} \left(\frac{\partial_c V}{\partial_c Z}(\tilde{Z}, x, i\Delta) \cdot (\sum_{k=0}^{s} L(\overrightarrow{x}(\Delta), \overrightarrow{v_x^k}(\Delta), (i+1)\Delta)\alpha_{i+1}^k) \right) \\ + \frac{\partial_c V}{\partial_c x}(\tilde{Z}, x, i\Delta) \cdot \frac{(\overrightarrow{x}(\Delta) - x)}{\Delta} + \frac{O(\Delta^2)}{\Delta} \end{array} \right].$$

Here the notation $\frac{c\partial}{c\partial u}$ stands for the covariant partial derivative with respect to u.

We can compute the terms $\overrightarrow{x}(\Delta)$ and $\overrightarrow{v_x^k}(\Delta)$ which appear on the righthand side of (9) by the following procedure. We can embed the manifold M into a Euclidean space E^n for sufficiently large n. In E^n, the geodesics are straight lines and the second derivatives along the geodesics are 0. Thus we can take the explicit equations of the geodesics and transfer them back to M. In local coordinates, we then get the following equations for the geodesics.

$$\frac{d_c^2 x^j}{d_c t^2} = -\sum_{h=1}^{n} \sum_{k=1}^{n} \Gamma_{h,k}^j \frac{d_c x^h}{d_c t} \frac{d_c x^k}{d_c t} \tag{10}$$

where the $\Gamma^j_{k,l}$ are the Christoffel coefficients for our affine connection ∇ on M and $\frac{d_c}{d_c u}$ denotes the total covariant derivative. Thus if we write $y^j = \frac{d_c x^j}{d_c t}$, then we can show that $d_c x^j = y^j \Delta + x^j$ and $d_c y^j = -(\sum_{h=1}^{n} \sum_{k=1}^{n} \Gamma^j_{h,k} y^h y^k) \Delta + y^j$ from which we can recover $\overrightarrow{x(\Delta)} - x$ via integration. Similary we can show that the j-th coordinate of the k-th element of the field of extremals at x, $v^{k,j}_x$ satisfies the equation

$$v^{k,j}_{\overrightarrow{x}(\Delta)} = -(\sum_{s=1}^{n} \sum_{l=1}^{n} \Gamma^j_{s,l} v^{k,s}_x (y^l + x^l)) \Delta. \tag{11}$$

It follows that we can use these equations to give an explict expression for $\frac{\partial_c V}{\partial_c t}$ and thus derive an explicit analogue of Jacobi-Bellman equation for Hybrid Systems.

2 Hybrid Dynamic Programming and Automata

Our solution to the hybrid dynamic programming equation described in the previous section is carried out by each agent in our Multiple Agent Hybrid Control Architecture (MAHCA). MAHCA is implemented as a distributed system composed of agents and a communication network which we call the logic communication network. The architecture realizing this system operates as an on-line distributed theorem prover. At any update time, each active agent generates control actions as side effects of proving an existentially quantified subtheorem (lemma) which encodes the model of the plant as viewed by the agent. The conjunction of lemmas at each instant of time, encodes the desired behavior of the entire network. In this section, we shall briefly describe the functionality of an agent's basic modules and the basic computational elements which is employed by an agent to solve its dynamic programming problem.

The basic architecture of an MAHCA agent is pictured Figure 1.

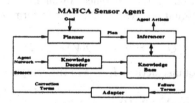

Fig. 1. MAHCA Agent Architecture

The agent consists of 5 modules with the following functionality:
Planner: The Planner constructs and repairs the agent optimization criteria, i.e. it constructs the agent's Lagrangian. The Planner, in fact, generates an existentially quantified logic expression which encapsulates the agent's model and optimization problem which is called the Behavior Statement.
Inferencer: The Inferencer determines whether there is a near optimal solution for

the agent's relaxed variational control problem. If there is such a solution, the agent infers a near optimal solution and sends data to the other agents. Othewise it infers failure terms and a new state for the agent and reports the failure to the other agents. In particular, the Inferencer determines whether the Behavior Statement is a theorem in the theory currently active in the Knowledge Base. If the Behavior Statement logically follows from the current status of the Knowledge Base, the inferencer generates, as a side effect of proving this Behavior Statement to be true, the current control for the plant. If the Control Statement does not logically follow from the current status of the Knowledge Base, that is, if the desired behavior is not realizable, the inferencer transmits the failed terms to the Adapter module for replacement or modification.

Adapter: The Adapter repairs failure terms and constructs correction terms.

Knowledge Base: The Knowledge Base stores and updates the agent's plant model and constraints. The Knowledge Base also stores the requirements of operations or processes within the scope of the agent's control problem. It also encodes system constraints, interagent protocols and constraints, sensory data, operational and logic principles and a set of primitive inference operations defined in the domain of equational terms.

Knowledge Decoder: The Knowledge Decoder receives and translates the other agent's data.

The agent solves the dynamic programming equation by constructing, on-line, a procedure, termed the inference automaton, that generates the solution. A solution consists of the control actions to be sent to the process under control and the current state of the agent. The solution is generated in three steps. First the dynamic programming equation is transformed into two subproblems: goal backward propagation and current interval optimization [6]. The coordinated resolution of these two subproblems is generated by a two-level finite state machine (each level computes a solution to one of the two subproblems) referred to as the inference automaton associated with the current plan. The event at the end of this step is a canonical equation for the inference automaton.

The canonical equation of the inference automaton is always version of the Kleene-Schutzenberger Equation (KSE). The generic form of a KSE is given below in (12).

$$Q(V) = E(V) \cdot Q(V) + T(X) \tag{12}$$

where V is in the space of instructions I, X is in the computation space CS, $Q(V) = \begin{bmatrix} Q_1(V) \\ \vdots \\ Q_n(V) \end{bmatrix}$ is a vector of rules, $E(V) = \begin{bmatrix} E_{1,1}(V) \ldots E_{1,n}(V) \\ \vdots \quad \vdots \ \vdots \\ E_{n,1}(V) \ldots E_{n,n}(V) \end{bmatrix}$ is the matrix of inference operators of the procedure, and $T(X) = \begin{bmatrix} T_1(X) \\ \vdots \\ T_n(X) \end{bmatrix}$ is the vector of goals.

In (12), each entry of the matrix E is a rational form constructed from the basis of inference operators and T is a vector of equational forms from the Knowledge Base. If the (i,j)-th entry of the matrix E is a non-empty entry, this represents an edge in the finite state machine from state j to state i, see figure 2. The binary

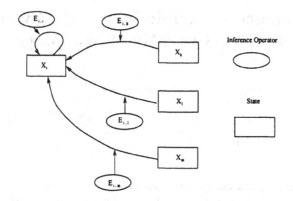

Fig. 2. State Graph of Inference Automaton

operator · between $E(V)$ and $Q(V)$ represents "apply inference to" operator. This operator is also called unification and is implemented as an evaluation routine.

In the second step, the canonical equation is operated on by the deduction, evolution and output relation inference operators to construct the inference automaton. The canonical equation is solvable if it is Lyapunov stable and its domain has quasi-regular convergence.

Finally, in the third step, the inference automaton is executed by first decomposing the inference automaton into an equivalent (i.e. same behavior) series-parallel network of simpler inference automata. This is carried out from the canonical equation by the unitary, prefix, loop decomposition, and trimmer inference operators. As the inference automaton executes, it accesses clauses in the Knowledge Base which are used to determine the control actions to be executed and the next state of the agent.

References

1. R. Bellman, *Dynamic Programming*, Princeton U. Press, Princeton, NJ, (1957).
2. R. W. R. Darling: *Differential Forms and Connections*, Cambridge U. Press, (1995).
3. X. Ge, W. Kohn, A. Nerode, and J. B. Remmel: Hybrid Systems: Chattering Approximations to Relaxed Controls. *Hybrid Systems III*, Lecture Notes in Computer Science 1066 , Springer-Verlag (1996), 76-100.
4. W. Kohn and A. Nerode: A Multiple agent Hybrid Control Arcitecture In *Logical Methods* (J. Crossley, J. B. Remmel, R. Shore, M. Sweedler, eds.), Birkhauser, (1993) 593-623.
5. W. Kohn and A. Nerode: Models for hybrid systems: automata, topologies, controllability and observability, Hybrid Systems, Lecture notes in Computer Science 736, Springer-Verlag, (1993), 317-356.
6. W. Kohn, A. Nerode, and J. B. Remmel, Hybrid Systems as Finsler Manifolds: Finite State Control as Approximation to Connections, *Hybrid Systems II*, Lecture Notes in Computer Science vol. 999, Springer-Verlag (1995), 294-321.
7. W. Kohn, A. Nerode, and J.B. Remmel: Feedback Derivations: Near Optimal Controls for Hybrid Systems, Proc. of CESA'96 IMACS Multiconference vol. 2, 517-521

Invariance Principle in Hybrid Systems Modeled by Mixed Mappings

Toshimitsu Ushio

Department of Electronic Engineering, Osaka University
2-1 Yamadaoka, Suita, Osaka 565 Japan
E-mail : *ushio@ele.eng.osaka-u.ac.jp*

Abstract. Mixed mapping systems have been introduced in order to analyze an effect of quantization in digital computers on controlled behaviors of digital control systems. We consider a mixed mapping in which discrete variables are governed by set-valued mappings. We propose a new concept of stability called w-asymptotical stability, and show an invariance principle and sufficient conditions of stability of invariant sets using Lyapunov functions.

1 Introduction

Mixed mapping systems are discrete-time systems whose state consists of both real-valued and integer-valued variables, and have been originally proposed, as an extension of a cell-to-cell mapping[1], in order to analyze effects of quantization in digital computers on controlled behaviors of digital control systems[2, 3]. Behaviors of physical plants and digital controllers with finite precision are described by continuous and digital variables respectively. Output signals are measured and input signals are changed at every sampling period. So behaviors of digital control systems at every sampling period can be modeled by discrete-time systems, that is, mixed mapping systems. In some discrete event systems such as Petri nets, states are described by integer vectors. So discrete-time hybrid systems whose discrete variables are integers are modeled by mixed mappings.

Stability problems are very important issues in system theory. Lyapunov function approaches are very popular in control systems and applied to various classes of systems. Related to hybrid systems, Passino *et al.* studied Lyapunov functions in discrete event systems, and applied to manufacturing systems and self-stabilizing systems[4]. Multiple Lyapunov functions are introduced in order to analyze switching effects in hybrid systems[5], and stabilization methods by hybrid controllers based on Lyapunov functions are discussed [6, 7, 8]. Ye *et al.* introduced several concepts of stability and proved converse theorems in hybrid systems[9].

In this paper, we consider mixed mapping systems as a model of hybrid systems. We introduce a new concept of stability taking topological feature of continuous variables and discrete variables into account explicitly. Next, we extend an invariance principle[10] in mixed mapping systems, and show sufficient conditions for stability of fixed points.

2 Mixed mappings

Let $\mathcal{M}^{(n,\,\ell)} = \mathcal{R}^n \times \mathcal{Z}^\ell$ be the set of mixed states, where $\mathcal{R}$ is the set of real numbers and $\mathcal{Z}$ is the set of integers.

A mixed mapping system whose discrete component is governed by a set-valued mapping is described by

$$x(k+1) = G(x(k),\ z(k))$$
$$z(k+1) \in C(x(k),\ z(k)) \tag{1}$$

or simply

$$\zeta(k+1) \in M(\zeta(k)) \tag{2}$$

where $x(k) \in \mathcal{R}^n$, $z(k) \in \mathcal{Z}^\ell$, $\zeta(k) := [x(k)^T,\ z(k)^T]^T$, respectively, and $G : \mathcal{R}^n \times \mathcal{Z}^\ell \to \mathcal{R}^n$, $C : \mathcal{R}^n \times \mathcal{Z}^\ell \to 2^{\mathcal{Z}^\ell}$, and $M : \mathcal{M}^{(n,\,\ell)} \to \mathcal{R}^n \times 2^{\mathcal{Z}^\ell}$. For a mixed state $\zeta = [x^T,\ z^T]^T \in \mathcal{M}^{(n,\,\ell)}$, x and z are called the continuous and discrete part of ζ, respectively.

A function $d_z(\cdot,\ \cdot) : \mathcal{M}^{(n,\,\ell)} \times \mathcal{M}^{(n,\,\ell)} \to \mathcal{R}$ is called a z-semidistance if for any $\zeta_i = [x_i^T,\ z_i^T]^T \in \mathcal{M}^{(n,\,\ell)} (i = 1,\ 2,\ 3)$,

(i) $d_z(\zeta_1,\ \zeta_2) \geq 0$, and $d_z(\zeta_1,\ \zeta_2) = 0$ iff $z_1 = z_2$,
(ii) $d_z(\zeta_1,\ \zeta_2) = d_z(\zeta_2,\ \zeta_1)$, and
(iii) $d_z(\zeta_1,\ \zeta_2) + d_z(\zeta_2,\ \zeta_3) \geq d_z(\zeta_1,\ \zeta_3)$.

A function $d_x(\cdot,\ \cdot) : \mathcal{M}^{(n,\,\ell)} \times \mathcal{M}^{(n,\,\ell)} \to \mathcal{R}$ is called an x-semidistance if for any $\zeta_i = [x_i^T,\ z_i^T]^T \in \mathcal{M}^{(n,\,\ell)} (i = 1,\ 2,\ 3)$,

(i) $d_x(\zeta_1,\ \zeta_2) \geq 0$, and $d_x(\zeta_1,\ \zeta_2) = 0$ iff $x_1 = x_2$,
(ii) $d_x(\zeta_1,\ \zeta_2) = d_x(\zeta_2,\ \zeta_1)$, and
(iii) $d_x(\zeta_1,\ \zeta_2) + d_x(\zeta_2,\ \zeta_3) \geq d_x(\zeta_1,\ \zeta_3)$.

A function $d(\cdot,\ \cdot) := d_x(\cdot,\ \cdot) + d_z(\cdot,\ \cdot)$ is a distance. For any $\zeta \in \mathcal{M}^{(n,\,\ell)}$ and any set $S \subseteq \mathcal{M}^{(n,\,\ell)}$, the distance $d(\zeta,\ S)$ of ζ and S is defined by

$$d(\zeta,\ S) := \inf\{\, d(\zeta,\ \zeta')\,;\, \zeta' \in S \,\}$$

In order to make discussion simple, we will use an ℓ^p-norm for definitions of functions $d_z(\cdot,\ \cdot)$ and $d_x(\cdot,\ \cdot)$ in this paper.

$$d_z(\zeta_1,\ \zeta_2) = \|\, z_1 - z_2\, \|_p, \quad d_x(\zeta_1,\ \zeta_2) = \|\, x_1 - x_2\, \|_p$$

where $\|\cdot\|_p$ denotes the ℓ^p-norm. Thus, for any ℓ^p-norm, we have

$$\min_{\substack{\zeta_1,\zeta_2 \in \mathcal{M}^{(n,\,\ell)} \\ z_1 \neq z_2}} d_z(\zeta_1,\ \zeta_2) = 1$$

The closure of S, denoted by $\bar{S}$, is defined by $\bar{S} := \{\, \zeta\,;\, d(\zeta,\ S) = 0 \,\}$. S is said to be closed if $S = \bar{S}$, and open if its complement is closed.

A set of all mixed states whose discrete parts are equal to z is called a real hyperplane at z, or simply a real hyperplane. A set included in a real hyperplane is called a hyperplane set. A set U_x is called an x-neighborhood of a mixed state ζ if

(i) U_x is a hyperplane set, and

(ii) there exists an open set V such that $\zeta \in V \subseteq U_x$.

Two mixed states ζ_1 and ζ_2 are said to be adjoining if $d_z(\zeta_1, \zeta_2) = 1$ and $d_x(\zeta_1, \zeta_2) = 0$. A set U is called a neighborhood of a mixed state ζ if

(i) there exists an x-neighborhood U_x of ζ such that $U_x \subseteq U$, and

(ii) for each adjoining mixed state ζ' of ζ, there exists an x-neighborhood $U_x(\zeta')$ of ζ' such that $U_x(\zeta') \subseteq U$.

We can also define an x-neighborhood and a neighborhood of a set $S \subseteq \mathcal{M}^{(n, \ell)}$ in a similar way.

A set S is said to be positively invariant if $M(S) \subseteq S$, and invariant if $M(S) = S$.

An infinite sequence $\{\zeta(k)\}_{k=0}^{\infty}$ of mixed state is called an orbit (starting from $\zeta(0)$) of Eq.(2) if $\zeta(k+1) \in M(\zeta(k))$ for each nonnegative integer k, and a set of orbits starting from ζ, denoted by $O(\zeta)$, is called an orbit set of ζ. A mixed state $\bar{\zeta}$ is a limit state of ζ under a mixed mapping M if there exists an orbit $\{\zeta(k)\}$ with $\zeta(0) = \zeta$ and an infinite sequence $\{k_i\}$ of positive integers with $k_i < k_{i+1}$ such that

$$\lim_{i \to \infty} d(\bar{\zeta}, \zeta(k_i)) = 0$$

and the limit set $\Omega(\zeta)$ of $\mathcal{M}^{(n, \ell)}$ under M is the set of all the limit mixed states of ζ under M.

Note that, for each limit mixed state $\bar{\zeta} = [\bar{x}^T, \bar{z}^T]^T$ of ζ, there exists an orbit $\{\zeta(k)\}$ with $\zeta(0) = \zeta$ and an infinite sequence $\{k'_i\}$ of positive integers with $k'_i < k'_{i+1}$ satisfying $\bar{z} \in \zeta(k'_i)$.

Lemma 1. *The limit set $\Omega(\zeta)$ is closed.*

Lemma 2. *The limit set $\Omega(\zeta)$ of ζ is positively invariant if there exists an x-neighborhood $U_x(\Omega(\zeta))$ of $\Omega(\zeta)$ such that M is continuous in $U_x(\Omega(\zeta))$. If, in addition, $M^k(\zeta)$ is bounded for all nonnegative integers k, then $\Omega(\zeta)$ is nonempty and invariant.*

3   Definitions of stability

We introduce several concepts of stability for mixed mapping systems.

Definition 3. A set $H \subseteq \mathcal{M}^{(n, \ell)}$ is said to be an attractor if there exists a neighborhood W of H such that, for any $\zeta \in W$ and any $\{\zeta(k)\} \in O(\zeta)$,

$$\lim_{k \to \infty} d(\zeta(k), H) = 0$$

Definition 4. A set $H \subseteq \mathcal{M}^{(n, \ell)}$ is said to be stable if, for any neighborhood U of H, there exists a neighborhood W of H such that $M^k(W) \subseteq U$ for any nonnegative integer k. H is said to be asymptotically stable if H is stable and an attractor.

Proposition 5. *Let ζ^* be a stable fixed mixed state of Eq. (2) and let S be the set of all adjoining mixed states of ζ^*. Then, $S \cup \{\zeta^*\}$ is positively invariant.*

The concept of stability defined above is a straightforward extension of that in discrete-time systems over $\mathcal{R}^n$, but is very restrictive in a mixed mapping systems as shown in the above proposition. So we will introduce a weaker concept of stability.

Definition 6. A set $H \subseteq \mathcal{M}^{(n,\,\ell)}$ is said to be a weak attractor or a w-attractor if there exists an x-neighborhood W_x of H such that, for any $\zeta \in W_x$ and any $\{\zeta(k)\} \in O(\zeta)$,

$$\lim_{k \to \infty} d(M^k(\zeta),\ H) = 0$$

Definition 7. A set $H \subseteq \mathcal{M}^{(n,\,\ell)}$ is said to be w-stable if, for any x-neighborhood U_x, there exists an x-neighborhood W_x such that $M^k(W_x) \subseteq U_x$ for all nonnegative integer k. H is said to be asymptotically w-stable(resp. w-asymptotically w-stable) if it is w-stable and an attractor(resp. a w-attractor).

4 Lyapunov function and invariance principle

We introduce a Lyapunov function for mixed mapping systems and prove an invariance principle.

Let V be a function from $\mathcal{M}^{(n,\,\ell)}$ to $\mathcal{R}$, and

$$\Delta V(x,\ z) = \{V(G(x,\ z),\ z') - V(x,\ z)\,|\,z' \in C(x,\ z)\}$$

For any set $H \subseteq \mathcal{M}^{(n,\,\ell)}$, we define

$$E_H = \{(x,\ z)\,|\,\Delta V(x,\ z) = \{0\}\}$$

The largest invariant set in E_H under M is denoted by L_H. We will write $\Delta V(x,\ z) \le 0$(resp. < 0) if we have $a \le 0$(resp. < 0) for any $a \in \Delta V(x,\ z)$.

Definition 8. Let H be a set in $\mathcal{M}^{(n,\,\ell)}$. A function V is said to be a Lyapunov function of Eq. (2) on H if

(i) for any $\zeta \in H$, $V(\zeta)$ is finite,
(ii) $\Delta V(\zeta) \le 0$, and
(iii) V is x-continuous in $\bar{H}$.

We will show an invariance principle in mixed mapping systems.

Theorem 9. *Suppose that*

(i) V is a Lyapunov function on H,
(ii) there exists an x-neighborhood U_x of E_H such that M is continuous in U_x, and
(iii) for any nonnegative integer k, $M^k(\zeta_0)$ is bounded and $M^k(\zeta_0) \subseteq H$.

Then, for each orbit $\{\zeta(k)\} \in O(\zeta_0)$ with $\zeta_0 \in H$, there exists a real number r such that

$$\lim_{k \to \infty} d(\zeta(k),\ L_H \cap V^{-1}(r)) = 0$$

We will show stability conditions for a fixed state of Eq. (2) using a Lyapunov function.

Theorem 10. *Let ζ^* be a fixed point of Eq. (2), and assume that M is continuous in some x-neighborhood U_x of ζ^*. If there exists an open set H with $\zeta^* \in H$ such that V is a Lyapunov function on H and $V(\zeta) > V(\zeta^*)$ for any $\zeta \in H - \{\zeta^*\}$, then ζ^* is w-stable. If $\Delta V(\zeta) < 0$ for any $\zeta \in H - \{\zeta^*\}$, then ζ^* is w-asymptotically w-stable. If, in addition, there exists a neighborhood U of ζ^* with $U \subseteq H$ such that $M(U) \subseteq U$, then ζ^* is asymptotically stable.*

As an example, we consider the following system:

$$x(k + 1) = 0.5x(k) + f(x(k),\ z(k))$$
$$z(k + 1) \in C(x(k),\ z(k)) \tag{3}$$

where

$$f(x,\ z) = \begin{cases} \dfrac{1}{2z} & \text{if } z \neq 0 \\ 0 & \text{otherwise} \end{cases} \tag{4}$$

$$C(x,\ z) = \begin{cases} \{\,z - 1,\ z - 2\,\} & \text{if } |x| \geq 1 \text{ and } z \geq 2 \\ \{\,z - 1\,\} & \text{if } |x| \geq 1 \text{ and } z = 1 \\ \{\,z + 1,\ z + 2\,\} & \text{if } |x| \geq 1 \text{ and } z \leq -2 \\ \{\,z + 1\,\} & \text{if } |x| \geq 1 \text{ and } z = -1 \\ \{\,0\,\} & \text{otherwise} \end{cases} \tag{5}$$

We consider the following Lyapunov function $V(x,\ z)$.

$$V(x,\ z) = |x| + |z| \tag{6}$$

Then, we have

$$\Delta V(x,\ z) \leq \begin{cases} -0.5|x| - 0.5 & \text{if } |x| \geq 1 \text{ and } z \neq 0 \\ -0.5|x| - |z| & \text{otherwise} \end{cases} \tag{7}$$

Thus, the mixed state $\zeta^* = [0,\ 0]^T$ is a w-asymptotically w-stable fixed point. However, it is easily shown that it is not asymptotically stable.

We replace Eq. (4) by the following mapping.

$$f(x,\ z) = \begin{cases} \dfrac{1}{2z} & \text{if } z \neq 0 \text{ and } |x| \geq 1 \\ 0 & \text{otherwise} \end{cases} \tag{8}$$

Then the Lyapunov function $V(x,\ z)$ also satisfies Eq. (7), and it is shown that the fixed point ζ^* is asymptotically stable.

5 Conclusion

We deal with stability theory of mixed mapping systems based on Lyapunov functions. A mixed mapping based approach is a good tool for analyzing digital control systems where plants are modeled by difference equations and controllers are represented by integer-valued models such as Petri nets.

References

1. C. S. Hsu, *Cell-to-Cell Mapping*, Springer-Verlag, 1987.
2. T. Ushio and C. S. Hsu, "A stability theory of mixed mapping systems and its applications to digital control systems," *Memoirs of the Faculty of Engineering, Kobe University*, no. 33, pp. 1–14, 1986.
3. T. Ushio and C. S. Hsu, "Chaotic rounding error in digital control systems," *IEEE Trans. Circuits & Syst.*, vol. CAS-34, no. 2, pp. 133–139, Feb. 1987.
4. K. M. Passino, A. N. Michel, and P. J. Antsaklis, "Lyapunov stability of a class of discrete event systems," *IEEE Trans. Automatic Control*, vol. 39, no. 2, pp. 269–279, Feb. 1994.
5. M. S. Branicky, " Stability of switched and hybrid systems," *Proc. 33rd Conference on Decision and Control*, Lake Buena Vista, FL, pp. 3498–3503, Dec. 1994.
6. M. A. Wicks and P. Peleties, "Construction of piecewise Lyapunov functions for stabilizing switched systems," *Proc. 33rd Conference on Decision and Control*, Lake Buena Vista, FL, pp. 3492–3497, Dec. 1994.
7. M. Doğruel and Ü. Özgüner, "Modeling and stability issues in hybrid systems," *Hybrid Systems II*, Lecture Notes in Computer Science, vol. 999, Springer-Verlag, pp. 148–165, 1995.
8. M. Doğruel, Ü. Özgüner, and S. Drakunov, "Sliding-mode control in discrete-state and hybrid systems," *IEEE Trans. Automatic Control*, vol. 41, no. 3, pp. 414–419, March, 1996.
9. H. Ye, A. N. Michel, and L. Hou, "Stability theory for hybrid dynamical systems," *Proc. 34th Conference on Decision and Control*, New Orleans, LA, pp. 2679–2684, Dec. 1995.
10. J. P. Lasalle, *The Stability of Dynamical Systems*, SIAM Publications, 1976.

Hybrid Systems Described by the Complementarity Formalism

A.J. van der Schaft *   and   J.M. Schumacher,[†]

In recognition of the fact that many systems contain both continuous and discrete aspects, considerable study has been devoted recently to "hybrid systems." The formulation of equations of motion for hybrid systems in explicit form, including the condition/event rules and the description of the continuous dynamics for every possible mode, is in many cases a formidable task, and there is a clear need for devices that enable the modeler to work in what might be called a "high-level language." A formalism that can be used for this purpose is the so-called *complementarity formalism* [5, 8, 9]. The formalism is applicable to a broad class of physical hybrid systems, as well as to hybrid systems described by an underlying dynamics subject to piecewise-linear constraints. This paper surveys some of the key issues treated in [8, 9], and discusses some possible extensions.

Complementary-slackness hybrid systems

We consider systems that are described by general differential-algebraic equations (DAE's)

$$G(z(t), \dot{z}(t)) = 0, \quad z \in \mathbb{R}^N, \tag{1}$$

together with a "complementary" set of inequality constraints defined as follows. Let

$$e = E(z(t)), \quad e \in \mathbb{R}^k$$
$$f = F(z(t)), \quad f \in \mathbb{R}^k \tag{2}$$

be two mappings, and consider the "complementary-slackness conditions" (the terminology stems from optimization theory)

$$e(t) \geq 0, \quad f(t) \geq 0, \quad e^T(t)f(t) = 0 \tag{3}$$

where the inequalities are understood componentwise. The conditions on $e(t)$ and $f(t)$ imply that for each index i in the index set $K := \{1, \cdots, k\}$ and each

---

*Systems and Control Group, Department of Applied Mathematics, University of Twente, P.O. Box 217, 7500 AE Enschede, The Netherlands, and CWI, P.O. Box 94079, 1090 GB Amsterdam, the Netherlands, e-mail: a.j.vanderschaft@math.utwente.nl

†CWI, P.O. Box 94079, 1090 GB Amsterdam, the Netherlands, and Tilburg University, CentER and Department of Economics, P.O. Box 90153, 5000 LE Tilburg, the Netherlands, e-mail: Hans.Schumacher@cwi.nl

time instant t we must have either $e_i(t) = 0$ and $f_i(t) \geq 0$, or $f_i(t) = 0$ and $e_i(t) \geq 0$. Thus for every subset $I \subset K$ we obtain a different set of DAE's

$$G(z(t), \dot{z}(t)) \quad = \quad 0$$

$$E_i(z(t)) \quad = \quad 0, \quad i \in I \tag{4}$$

$$F_i(z(t)) \quad = \quad 0, \quad i \in K \backslash I$$

together with feasibility conditions

$$E_i(z(t)) \quad \geq \quad 0, \quad i \in K \backslash I$$

$$F_i(z(t)) \quad \geq \quad 0, \quad i \in I \tag{5}$$

The dynamics described by (4) will be called a *mode* of the system, and the mode corresponding to a subset $I \subset K$ will be simply denoted as "mode I." Thus we have obtained a multi-mode (or hybrid) system with, in principle, 2^k different modes, which each have to satisfy a set of additional feasibility conditions (5). This special class of hybrid systems has been introduced in [8] as "*complementary-slackness systems*", and analysed in [8, 9]. Note that, loosely speaking, the mappings E and F defined in (2) can be regarded as some kind of "guards" or "system invariants", as they are often appearing in the literature on hybrid systems. However a main difference is that the mappings E and F (as well as the underlying dynamics $G(z(t), \dot{z}(t)) = 0$) are "globally" defined, that is, do not depend on the particular mode, and in fact *define* the different modes as in (4). Note that in fact for every I the functions E_i, $i \in I$, and F_i, $i \in K \backslash I$ define the mode I as in (4), with "complementary" guards E_i, $i \in K \backslash I$, and F_i, $i \in I$, given by (5).

The basic motivation for studying complementary-slackness hybrid systems in [8, 9] is two-fold. *First*, a rich class of physical hybrid systems can be directly modelled as complementary-slackness systems: e.g. electrical circuits with diodes, mechanical systems with stops, hydraulic systems with one-way valves. We refer to [11] for related developments in modeling physical systems with switches. Furthermore, with a little bit more effort also discontinuous physical phenomena as (ideal) Coulomb friction and backlash, and control elements as relays, can be modelled this way [9]. *Secondly*, the class of complementary-slackness systems has an appealing mathematical structure which suggests some natural rules for mode-switching and re-initialization, and which admits the derivation of strong theorems concerning the resulting hybrid dynamics. A major mathematical tool in this is the *Linear Complementarity Problem* (LCP): Given a vector $q \in \mathbb{R}^k$ and an $k \times k$ matrix M, find k-vectors e and f such that

$$\text{LCP:} \quad e = q + Mf, \quad e \geq 0, \quad f \geq 0, \quad e^T f = 0 \tag{6}$$

The LCP has been studied extensively, and a wealth of theoretical results and computational methods is available, see e.g. [2]. A basic result is that the LCP has a unique solution e, f if the principal minors of M are all positive. The strong relation of complementary-slackness systems with the LCP becomes more clear by considering as a special case of (1), (2) "input-output" systems

$$\dot{x}(t) = g(x(t), f(t))$$
$$e(t) = h(x(t), f(t))$$

(7)

subject to the complementary-slackness conditions (3). One may regard (7), (3) as a "dynamical" (nonlinear) version of the LCP. The recognition of the close connection of complementary-slackness systems with the LCP also sheds light on the fundamental question which class of hybrid systems can be modelled as complementary-slackness systems. In fact, it has been shown in [3] that any piecewise linear n-dimensional set of equations (under a mild "nonsingularity" assumption) is equivalent to a certain LCP (whose order k is typically larger than n). This implies that generally any (linear or nonlinear) dynamics subject to piecewise linear constraints can be modelled, in principle, as a complementary-slackness system. (See also [10] for a discussion on the relation of hybrid and piecewise-linear systems.) The modelling of Coulomb friction and relay elements by complementary-slackness conditions in [9] can be understood in this way. The power of the LCP for the modeling and analysis of mechanical systems with inequality constraints and Coulomb friction has been already advocated by Lötstedt [5], and for the modeling of static electrical circuits in [6].

Mode-selection and re-initialization in the complementarity framework

Let us now discuss how the complementarity framework may be used for suggesting natural rules for mode-selection and re-initialization, and for proving strong theorems concerning the resulting hybrid dynamics.

First we discuss the *mode-selection problem* for complementary-slackness systems of the form (7), (3). To simplify discussion we assume that every mode I is *autonomous* in the sense that from each continuous state that is consistent for mode I there is a unique solution of the dynamics of the mode I (not necessarily satisfying the feasibility conditions (5)) on a time-interval of positive length. We say that *smooth continuation* from a continuous state is possible *in mode I* if also the feasibility conditions (5) are met on some time-interval $[0, \epsilon]$, $\epsilon > 0$. We consider an initial continuous state for which smooth continuation is possible in at least one of the modes; so no re-initialization is necessary. By successively differentiating the "output" equations $e = h(x, f)$ along the dynamics $\dot{x} = g(x, f)$ one obtains sets of equations (linear in the highest derivatives of e and f) which can be brought inductively into the format of an LCP (in the unknowns $e, \dot{e}, \ddot{e}, \cdots, f, \dot{f}, \ddot{f}, \cdots$.) This allows the derivation of strong theoretical results concerning *uniqueness* of smooth continuation of complementary slackness hybrid systems, such as mechanical systems subject to multiple (independent) geometric inequality constraints, and passive electrical circuits containing diodes. At the same time it gives, via the LCP, a numerical recipe for *computing* the unique mode of smooth continuation, which is clearly of much importance e.g. for simulation purposes. All this is detailed in [9]. Note that in the above analysis of unique smooth continuation the discrete part of the state (that is, the mode the system is currently in) is taken to be "sub-ordinated" to the continuous state. This has been done on the basis of physical considerations. For instance, we believe that it would not be reasonable from a physical point of view to include in the initial conditions for an electrical circuit with diodes any information as to which diodes are voltage- or current-blocking; one should be able to derive this information from the continuous state components (assuming no hysteretic effects). This is in some contrast with a more standard formulation of hybrid systems in which the system "knows" which mode it is in.

If a consistent continuous initial condition does not admit smooth continuation in any of the modes then (and only then) the system has to be re-initialized to another state (in accordance with the Principle of Constraints formulated in [4, p.79]: "Constraints shall be maintained by forces, so long as this is possible; otherwise, and only otherwise, by impulses"). This *re-initialization* can be split into two parts: (a) a *switch rule* which determines to which mode the system will be re-initialized, (b) a *jump rule* which defines the new continuous state, consistent with the mode just determined.

A *jump rule* was proposed in [8] for linear dynamics and some special nonlinear dynamics (such as mechanical systems), based on geometric considerations. Indeed, with any linear set of *autonomous* DAE's

$$E\dot{z} = Az, \quad z \in \mathbb{R}^N \tag{8}$$

one can associate two *complementary* subspaces V and T in $\mathbb{R}^N$, where V is the set of consistent points, and T is directly related to the impulsive behavior of the system. Suppose now that the switch rule has determined a mode, whose dynamics is described by (8). Then the *jump rule* is simply to project the initial continuous state z_0 *along* T to a point $z_0' \in V$. (Applied to mechanical systems this corresponds to the application of impulsive forces to the system.)

With regard to *switch rules* the following options are available. In [8] the following switch rule was proposed. Let z_0 be a consistent point for some mode I. Consider the unique solution from z_0 in mode I, and detect *all* the feasibility conditions (5) that will be immediately violated. (Since by assumption there is no smooth continuation in mode I at least *one* feasibility condition is going to be violated.) Let Γ_1 be the subset of $K \backslash I$ for which the first set of feasibility conditions in (5) are going to be violated, and let Γ_2 be the subset of I for which the second set of feasibility conditions in (5) are going to be violated. Then determine the new mode J as

$$J := (I \backslash \Gamma_2) \cup \Gamma_1 \tag{9}$$

For the bimodal case (i.e. $k = 1$) the resulting hybrid dynamics was analyzed in detail in [8]. For $k > 1$ there are some alternatives to the above switch rule, which are partly motivated by the LCP as well as by the theory of (inelastic) mechanical collisions (see e.g. [1]). Indeed, let us consider a mechanical system with n degrees of freedom $q = (q_1, \cdots, q_n)$ having kinetic energy $\frac{1}{2}\dot{q}^T M(q)\dot{q}$;

$M(q) > 0$ being the generalized mass matrix. Suppose the system is subject to k geometric inequality constraints

$$e_i = E_i(q) \geq 0, \quad i \in K = \{1, \cdots, k\} \tag{10}$$

If the i-th inequality constraint is *active*, that is $E_i(q) = 0$, then the system will experience a constraint force of the form $\frac{\partial E_i}{\partial q}(q)\lambda_i$, with $\frac{\partial E_i}{\partial q}(q)$ the column-vector of partial derivatives of E_i and λ_i a Lagrangian multiplier. Let us now consider an initial continuous state $(q, \dot{q})$ of the system, from which no smooth continuation is possible in any of the possible modes (corresponding to some of the inequality constraints being active and the resulting constraint forces). Define the set of active indices $\bar{K} = \{i \in K \mid E_i(q) = 0\}$ and the sub-vector of generalized velocities

$$y := C(q)\dot{q} \tag{11}$$

where the i-th row of $C(q)$ is the gradient vector of E_i, $i \in \bar{K}$. In order to describe the inelastic collision we consider the LCP (in the unknowns y^+, λ)

$$y^+ = y + C(q)M^{-1}(q)C^T(q)\lambda$$
$$y^+ \geq 0, \ \lambda \geq 0, \ (y^+)^T\lambda = 0 \tag{12}$$

Here λ can be interpreted as the vector of *impulsive forces*. Since $C(q)M^{-1}(q)C^T(q) > 0$ this LCP has a unique solution, which determines a new mode as

$$J_c := \{i \in \bar{K} \mid \lambda_i > 0\} \tag{13}$$

In general the new mode J_c will be *different* from the mode J obtained in (10). On the other hand, given the new mode J_c the *jump rule* for these mechanical systems

$$(q, \dot{q}) \mapsto (q, \dot{q}^+ := \dot{q} + M^{-1}(q)C^T(q)\lambda), \tag{14}$$

coincides with the jump rule discussed before, consisting of projection along T onto V (with now T and V corresponding to the DAE's describing mode J_c). Interestingly enough, the new velocity vector $\dot{q}^+$ may be equivalently characterized as the solution of the quadratic programming problem

$$\min_{\{\dot{q}^+ \mid C(q)\dot{q}^+ \geq 0\}} \frac{1}{2}(\dot{q}^+ - \dot{q})^T M(q)(\dot{q}^+ - \dot{q}), \tag{15}$$

which is sometimes taken as the starting point for describing multiple inelastic collisions, see [1, 7]. An appealing feature of both switch rules (9) and (13), together with the resulting jump (14), is that the energy of the mechanical system will always decrease at the switching instant. This is a promising starting point for stability analysis. The extension of the "LCP-based" switching rule (13) to general complementary-slackness systems is currently under investigation.

Conclusion

We have argued that a sizeable class of hybrid (physical) systems can be modelled as complementary-slackness hybrid systems. We have indicated that the complementarity-formalism suggests some natural rules for mode-selection and re-initialization, which provide a "high-level language" for describing the full hybrid dynamics of such systems, and in this way simplify their specification. We have argued that the LCP is a powerful tool for proving existence and uniqueness of solutions of the hybrid dynamics and for actually *computing* the mode of continuation and the re-initialization.

References

[1] B. Brogliato, *Nonsmooth Impact Mechanics Models, Dynamics and Control*, Lect. Notes Contr. Inf. Sci. 220, Springer, Berlin, 1996.

[2] R.W. Cottle, J.-S. Pang, R.E. Stone, *The Linear Complementarily Problem*, Academic Press, Boston, 1992.

[3] B.C. Eaves, C.E. Lemke, "Equivalence of LCP and PLS", Mathematics of Operations Research, 6: 475-484, 1981.

[4] C.W. Kilmister, J.E. Reeve, *Rational Mechanics*, Longmans, London, 1966.

[5] P. Lötstedt, "Mechanical systems of rigid bodies subject to unilateral constraints", SIAM Journal of Applied Mathematics, 42: 281-296, 1982.

[6] L. Vandenberghe, B.L. de Moor, J. Vandewalle, "The generalized linear complementarity problem applied to the complete analysis of resistive piecewise-linear circuits, IEEE Trans. Circuits Syst., CS-36, pp. 1382-1391, 1989.

[7] J.J. Moreau, "Liaisons unilatérales sans frottement et chocs inélastiques", C.R. Acad. Sc. Paris, t.296, pp. 1473-1476, 1983.

[8] A.J. van der Schaft, J.M. Schumacher, "The complementary-slackness class of hybrid systems", Report BS-R9529, CWI, Amsterdam, 1995. To appear in Math. Contr. Sign. Syst.

[9] A.J. van der Schaft, J.M. Schumacher, "Complementarity modeling of hybrid systems", July 1996, submitted to IEEE-AC, Special Issue on Hybrid Systems.

[10] E. Sontag, "Interconnected automata and linear systems: A theoretical framework in discrete time", *Hybrid Systems: Verification and Control* (R. Alur, T. Henzinger, E.D. Sontag, eds.), pp. 436-448, Lect. Notes Comp. Sci., Vol. 1066, Springer Berlin 1996.

[11] J.-E. Strömberg, *A Mode Switching Modelling Philosophy*, PhD thesis, Linköping University, Dept. of Electrical Eng., 1994.

Generalized Linear Complementarity Problems and the Analysis of Continuously Variable Systems and Discrete Event Systems

Bart De Schutter* and Bart De Moor**

ESAT/SISTA, K.U.Leuven, Kardinaal Mercierlaan 94, B-3001 Leuven, Belgium
`bart.deschutter@esat.kuleuven.ac.be, bart.demoor@esat.kuleuven.ac.be`

Abstract. We present an overview of our research on the use of generalized linear complementarity problems (LCPs) for analysis of continuously variable systems and discrete event systems. We indicate how the Generalized LCP can be used to analyze piecewise-linear resistive electrical circuits. Next we discuss how the Extended LCP can be used to solve some fundamental problems that arise in max-algebraic system theory for discrete event systems. This shows that generalized LCPs appear in the analysis and modeling of certain continuously variable systems and discrete event systems. Since hybrid systems exhibit characteristics of both continuously variable systems and discrete event systems, this leads to the question as to whether generalized LCPs can also play a role in the modeling and analysis of certain classes of hybrid systems.

1 Introduction

In our research we have developed extensions of the linear complementarity problem (LCP), which is one of the basic problems in mathematical programming. We have used one extension in the analysis of electrical circuits with piecewise-linear characteristics [5, 13], which can be considered as examples of continuously variable systems (CVSs). Another extension of the LCP has been used in the analysis of a class of discrete event systems (DESs) that can be described by a state space model that is linear in the max-plus algebra [8, 9]. In this paper we present an overview of this research. Since hybrid systems exhibit characteristics of both CVSs and DESs, this suggests that extensions of the LCP will probably also be useful in the analysis of hybrid systems.

2 Generalized and Extended Linear Complementarity Problems

One of the possible formulations of the Linear Complementarity Problem (LCP) is the following [3]:

---

* Senior research assistant with the F.W.O. (Fund for Scientific Research – Flanders)
** Senior research associate with the F.W.O.

Given $M \in \mathbb{R}^{n \times n}$ and $q \in \mathbb{R}^n$, find $w, z \in \mathbb{R}^n$ such that $w \geq 0$, $z \geq 0$, $w = q + Mz$ and $z^T w = 0$.

The LCP has numerous applications such as quadratic programming problems, determination of the Nash equilibrium of a bimatrix game problem, the market equilibrium problem, the optimal invariant capital stock problem, the optimal stopping problem, etc. [3].

In [4, 5] De Moor introduced the following generalization of the LCP:

Given $Z \in \mathbb{R}^{p \times n}$ and m subsets $\phi_1, \ldots, \phi_m$ of $\{1, \ldots, p\}$, find a non-trivial $u \in \mathbb{R}^n$ such that $\sum_{j=1}^{m} \prod_{i \in \phi_j} u_i = 0$ subject to $u \geq 0$ and $Zu = 0$.

This problem is called the Generalized LCP (GLCP). In Sect. 3 we shall see that the GLCP can be used to determine operating points and transfer characteristics of piecewise-linear resistive electrical circuits.

Another extension of the LCP, the Extended LCP (ELCP), is defined as follows [6, 7]:

Given $A \in \mathbb{R}^{p \times n}$, $B \in \mathbb{R}^{q \times n}$, $c \in \mathbb{R}^p$, $d \in \mathbb{R}^q$ and m subsets $\phi_1, \ldots,$ ϕ_m of $\{1, \ldots, p\}$, find $x \in \mathbb{R}^n$ such that $\sum_{j=1}^{m} \prod_{i \in \phi_j} (Ax - c)_i = 0$ subject to $Ax \geq c$ and $Bx = d$.

In Sect. 4 we shall see that the ELCP can be used to solve many problems that arise in the system theory for max-linear time-invariant DESs.

It can be shown that the ELCP is a generalization of the GLCP and that the homogeneous ELCP and the GLCP are equivalent [6, 7]. In [4] De Moor has developed an algorithm to compute the complete solution set of a GLCP. In [7] we have extended this algorithm in order to compute the complete solution set of an ELCP.

3 The GLCP and Piecewise-Linear Resistive Electrical Circuits

In this section we consider electrical circuits that may contain the following elements: linear resistive elements, piecewise-linear (PWL) resistors (the resistors are not required to be either voltage or current controlled), and PWL controlled sources (all four types) with one controlling variable (the characteristics may be multi-valued). The key idea behind the reformulation of the equations that describe the relations between the voltages and currents in the circuit as a (special case of a) GLCP is an intelligent parameterization of the PWL characteristics.

If x is a vector, then we define $x^+ = \max(x, 0)$ and $x^- = \max(-x, 0)$, where the operations are performed componentwise. An equivalent definition is:

$$x = x^+ - x^-, \qquad x^+, x^- \geq 0, \qquad (x^+)^T x^- = 0 .$$

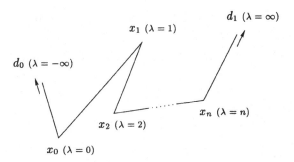

Fig. 1. A one-dimensional PWL curve.

For sake of simplicity we consider only two-terminal resistors since they can be described by a one-dimensional PWL manifold[3]. It is easy to verify that a one-dimensional PWL curve characterized by $n + 1$ breakpoints $x_0, \ldots, x_n$ and two directions d_0 and d_1 (see Fig. 1) can be parameterized as follows [4, 13]:

$$x = x_0 + d_0 \lambda^- + (x_1 - x_0)\lambda^+ + \sum_{k=2}^{n}(x_k - 2x_{k-1} + x_{k-2})(\lambda - k + 1)^+ +$$
$$(d_1 - x_n + x_{n-1})(\lambda - n)^+ . \tag{1}$$

Introducing auxiliary variables $\lambda_i = \lambda - i$ yields a description of the following form:

$$x = x_0 + A y^- + B y^+$$
$$C(y^+ - y^-) = d$$
$$y^+, y^- \geq 0$$
$$(y^+)^T y^- = 0$$

where $y = [\lambda \ \lambda_1 \ \ldots \ \lambda_n]^T$.

If we extract all nonlinear resistors out of the electrical circuit, the resulting N-port contains only linear resistive elements and independent sources. As a consequence, the relation between the branch currents and voltages of this N-port is described by a system of linear equations. If we combine these equations with the PWL descriptions (1) of the nonlinear resistors, we finally get a system of the form:

$$M w^+ + N w^- = q, \qquad w^+, w^- \geq 0, \qquad (w^+)^T(w^-) = 0 , \tag{2}$$

where the vector w contains the parameters λ and λ_i of the PWL descriptions of all the nonlinear resistors. It is easy to verify that after multiplying q by a nonnegative homogenization parameter α and including the extra condition

---

[3] If we also allow multi-terminal nonlinear resistors, which can be modeled by higher-dimensional PWL manifolds, we shall obtain the general GLCP that has been defined in Sect. 2 instead of the special GLCP of (2) (See [4]).

$\alpha \geq 0$, (2) can be considered as a special case of the GLCP. If we solve (2), we get the complete set of operating points of the electrical circuit.

In a similar way we can determine the driving-point characteristic (i_{in} versus v_{in}) and transfer characteristics of the electrical circuit [13].

The behavior of an electrical network consisting of linear resistors, capacitors, inductors, transformers, gyrators and ideal diodes can be described by a model of the form

$$\dot{x}(t) = Ax(t) + Bu(t)$$
$$y(t) = Cx(t) + Du(t)$$

subject to the conditions

$$y(t) \geq 0, \quad u(t) \geq 0, \quad (y(t))^T u(t) = 0 \tag{3}$$

(see e.g., [12]). In order to compute the *stationary points* of such an electrical circuit, we add the condition $\dot{x}(t) = 0$, which leads to an LCP [12]. If we replace (3) by more general conditions of the form $w_i \geq 0$, $z_i \geq 0$, $w_i z_i = 0$, where w_i and z_i are components of u, y or x, then we get (a special case of) an ELCP.

4 The ELCP and Max-Linear Time-Invariant DESs

In general the description of DESs is nonlinear. However, there exists a class of DESs — the so-called *max-linear DESs* — for which the description becomes "linear" when we express it in the max-plus algebra [1, 2]. Loosely speaking we could say that this subclass corresponds to the class of deterministic time-invariant DESs in which only synchronization and no concurrency occurs.

The basic operations of the max-plus algebra are maximization (represented by $\oplus$) and addition (represented by $\otimes$). There exists a remarkable analogy between the basic operations of the max-plus algebra on the one hand, and the basic operations of conventional algebra (addition and multiplication) on the other hand. As a consequence many concepts and properties of conventional algebra (such as Cramer's rule, eigenvectors and eigenvalues, the Cayley-Hamilton theorem, ...) also have a max-algebraic analogue. Furthermore, this analogy also allows us to translate many concepts, properties and techniques from conventional linear system theory to system theory for max-linear time-invariant DESs. However, there are also some major differences that prevent a straightforward translation of properties, concepts and algorithms from conventional linear algebra and linear system theory to max-plus algebra and max-algebraic system theory for DESs.

If we write down a model for a max-linear DES and if we use the symbols $\oplus$ and $\otimes$ to denote maximization and addition, we obtain a description of the following form:

$$x(k+1) = A \otimes x(k) \ \oplus \ B \otimes u(k) \tag{4}$$
$$y(k) = C \otimes x(k) , \tag{5}$$

where x is the state vector, u the input vector and y the output vector. For a manufacturing system $u(k)$ would typically represent the time instants at which raw material is fed to the system for the $(k-1)$st time; $x(k)$ the time instants at which the machines start processing the kth batch of intermediate products; and $y(k)$ the time instants at which the kth batch of finished products leaves the system. In analogy with the state space model for linear time-invariant discrete-time systems, a model of the form $(4)-(5)$ is called a *max-linear time-invariant state space model*.

Let $x, r \in \mathbb{R}$. The rth max-algebraic power of x is denoted by $x^{\otimes^r}$ and corresponds to rx in conventional algebra.

Now consider the following problem:

Given $p_1 + p_2$ positive integers $m_1, \ldots, m_{p_1+p_2}$ and real numbers a_{ki}, b_k and c_{kij} for $k = 1, \ldots, p_1 + p_2$, $i = 1, \ldots, m_k$ and $j = 1, \ldots, n$, find $x \in \mathbb{R}^n$ such that

$$\bigoplus_{i=1}^{m_k} a_{ki} \otimes \bigotimes_{j=1}^{n} x_j^{\otimes^{c_{kij}}} = b_k \qquad \text{for } k = 1, \ldots, p_1 \, , \tag{6}$$

$$\bigoplus_{i=1}^{m_k} a_{ki} \otimes \bigotimes_{j=1}^{n} x_j^{\otimes^{c_{kij}}} \leq b_k \qquad \text{for } k = p_1 + 1, \ldots, p_1 + p_2 \, . \tag{7}$$

We call $(6)-(7)$ a *system of multivariate max-algebraic polynomial equalities and inequalities*. Note that the exponents may be negative or real.

In [6, 10] we have shown that the problem of solving a system of multivariate max-algebraic polynomial equalities and inequalities can be recast as an ELCP. This enables us to solve many important problems that arise in the max-plus algebra and in the system theory for max-linear DESs such as: computing max-algebraic matrix factorizations, performing max-algebraic state space transformations, computing state space realizations of the impulse response of a max-linear time-invariant DES, constructing matrices with a given max-algebraic characteristic polynomial, computing max-algebraic singular value decompositions and QR decompositions, and so on [6, 7, 8, 9, 10].

Although the analogues of these problems in conventional algebra and linear system theory are easy to solve, the max-algebraic problems are not that easy to solve and for almost all of them the ELCP approach is at present the only way to solve the problem.

5   Conclusions and Future Research

We have defined two extensions of the linear complementarity problem (LCP): the Generalized Linear Complementarity Problem (GLCP) and the Extended Linear Complementarity Problem (ELCP). First we have indicated how the GLCP can be used to analyze piecewise-linear resistive electrical circuits, which are examples of continuously variable systems (CVSs). Next we have indicated how the ELCP can be used to solve some problems that arise in the max-algebraic system theory for max-linear discrete event systems (DESs). So generalized LCPs appear in the analysis and modeling of certain classes of CVSs and DESs. Since hybrid systems exhibit characteristics of both CVSs and DESs, it would be interesting to determine whether the GLCP, the ELCP or other — even more general — generalized LCPs also play a role in the modeling and analysis of certain classes of hybrid systems. The results of [11] seem to indicate that this is indeed the case.

Acknowledgment. This research was sponsored by the Concerted Action Project of the Flemish Community, entitled "Model-based Information Processing Systems", by the Belgian program on interuniversity attraction poles (IUAP-50), and by the ALAPEDES project of the European Community Training and Mobility of Researchers Program.

References

1. F. Baccelli, G. Cohen, G.J. Olsder, and J.P. Quadrat, *Synchronization and Linearity.* New York: John Wiley & Sons, 1992.

2. G. Cohen, D. Dubois, J.P. Quadrat, and M. Viot, "A linear-system-theoretic view of discrete-event processes and its use for performance evaluation in manufacturing," *IEEE Trans. on Aut. Control*, vol. 30, no. 3, pp. 210–220, Mar. 1985.

3. R.W. Cottle, J.S. Pang, and R.E. Stone, *The Linear Complementarity Problem.* Boston: Academic Press, 1992.

4. B. De Moor, *Mathematical Concepts and Techniques for Modelling of Static and Dynamic Systems.* PhD thesis, Fac. of Applied Sc., K.U.Leuven, Belgium, 1988.

5. B. De Moor, L. Vandenberghe, and J. Vandewalle, "The generalized linear complementarity problem and an algorithm to find all its solutions," *Math. Prog.*, vol. 57, pp. 415–426, 1992.

6. B. De Schutter, *Max-Algebraic System Theory for Discrete Event Systems.* PhD thesis, Fac. of Applied Sc., K.U.Leuven, Belgium, 1996.

7. B. De Schutter and B. De Moor, "The extended linear complementarity problem," *Math. Prog.*, vol. 71, no. 3, pp. 289–325, Dec. 1995.

8. B. De Schutter and B. De Moor, "Minimal realization in the max algebra is an extended linear complementarity problem," *Syst. & Control Letters*, vol. 25, no. 2, pp. 103–111, May 1995.

9. B. De Schutter and B. De Moor, "Applications of the extended linear complementarity problem in the max-plus algebra," *Proc. of WODES'96 (Internat. Workshop on Discrete Event Syst.)*, Edinburgh, UK, pp. 69–74, Aug. 1996.

10. B. De Schutter and B. De Moor, "A method to find all solutions of a system of multivariate polynomial equalities and inequalities in the max algebra," *Discrete Event Dynamic Systems: Theory and Appl.*, vol. 6, no. 2, pp. 115–138, Mar. 1996.

11. B. De Schutter and B. De Moor, "Optimal traffic signal control for a single intersection," Tech. rep. 96-90, ESAT/SISTA, K.U.Leuven, Belgium, Dec. 1996.

12. J.M. Schumacher, "Some modeling aspects of unilaterally constrained dynamics," *Proc. of the ESA Internat. Workshop on Adv. Math. Methods in the Dynamics of Flexible Bodies*, Noordwijk, The Netherlands, June 1996.

13. L. Vandenberghe, B. De Moor, and J. Vandewalle, "The generalized linear complementarity problem applied to the complete analysis of resistive piecewise-linear circuits," *IEEE Trans. on Circ. and Syst.*, vol. 36, no. 11, pp. 1382–1391, Nov. 1989.

SHIFT:
A Language for Simulating Interconnected Hybrid Systems (Invited Presentation)

Pravin Varaiya

Department of Electrical Engineering and Computer Sciences
University of California
Berkeley, U.S.A.
varaiya@eecs.berkeley.edu

Abstract. SHIFT is a programming language for describing dynamically varying networks of interconnected hybrid automata. In SHIFT, an individual hybrid automaton is an component or instance of a type. New components can be created and interconnected with existing components as the system evolves. The interconnections may change over time. Components may be interconnected through input-output connections or through synchronous composition.

We show that the SHIFT model offers the proper level of abstraction for describing complex applications such as automated highway systems, air traffic control systems, and coordinated submarines for underwater surveillance.

The talk will describe the SHIFT syntax and semantics, the SHIFT runtime system, and some examples.

Author Index

Springer
and the
environment

At Springer we firmly believe that an international science publisher has a special obligation to the environment, and our corporate policies consistently reflect this conviction.
We also expect our business partners – paper mills, printers, packaging manufacturers, etc. – to commit themselves to using materials and production processes that do not harm the environment. The paper in this book is made from low- or no-chlorine pulp and is acid free, in conformance with international standards for paper permanency.

 Springer

Lecture Notes in Computer Science

For information about Vols. 1–1122

please contact your bookseller or Springer-Verlag

Vol. 1160: S. Arikawa, A.K. Sharma (Eds.), Algorithmic Learning Theory. Proceedings, 1996. XVII, 337 pages. 1996. (Subseries LNAI).

Vol. 1161: O. Spaniol, C. Linnhoff-Popien, B. Meyer (Eds.), Trends in Distributed Systems. Proceedings, 1996. VIII, 289 pages. 1996.

Vol. 1162: D.G. Feitelson, L. Rudolph (Eds.), Job Scheduling Strategies for Parallel Processing. Proceedings, 1996. VIII, 291 pages. 1996.

Vol. 1163: K. Kim, T. Matsumoto (Eds.), Advances in Cryptology – ASIACRYPT '96. Proceedings, 1996. XII, 395 pages. 1996.

Vol. 1164: K. Berquist, A. Berquist (Eds.), Managing Information Highways. XIV, 417 pages. 1996.

Vol. 1165: J.-R. Abrial, E. Börger, H. Langmaack (Eds.), Formal Methods for Industrial Applications. VIII, 511 pages. 1996.

Vol. 1166: M. Srivas, A. Camilleri (Eds.), Formal Methods in Computer-Aided Design. Proceedings, 1996. IX, 470 pages. 1996.

Vol. 1167: I. Sommerville (Ed.), Software Configuration Management. VII, 291 pages. 1996.

Vol. 1168: I. Smith, B. Faltings (Eds.), Advances in Case-Based Reasoning. Proceedings, 1996. IX, 531 pages. 1996. (Subseries LNAI).

Vol. 1169: M. Broy, S. Merz, K. Spies (Eds.), Formal Systems Specification. XXIII, 541 pages. 1996.

Vol. 1170: M. Nagl (Ed.), Building Tightly Integrated Software Development Environments: The IPSEN Approach. IX, 709 pages. 1996.

Vol. 1171: A. Franz, Automatic Ambiguity Resolution in Natural Language Processing. XIX, 155 pages. 1996. (Subseries LNAI).

Vol. 1172: J. Pieprzyk, J. Seberry (Eds.), Information Security and Privacy. Proceedings, 1996. IX, 333 pages. 1996.

Vol. 1173: W. Rucklidge, Efficient Visual Recognition Using the Hausdorff Distance. XIII, 178 pages. 1996.

Vol. 1174: R. Anderson (Ed.), Information Hiding. Proceedings, 1996. VIII, 351 pages. 1996.

Vol. 1175: K.G. Jeffery, J. Král, M. Bartošek (Eds.), SOFSEM'96: Theory and Practice of Informatics. Proceedings, 1996. XII, 491 pages. 1996.

Vol. 1176: S. Miguet, A. Montanvert, S. Ubéda (Eds.), Discrete Geometry for Computer Imagery. Proceedings, 1996. XI, 349 pages. 1996.

Vol. 1177: J.P. Müller, The Design of Intelligent Agents. XV, 227 pages. 1996. (Subseries LNAI).

Vol. 1178: T. Asano, Y. Igarashi, H. Nagamochi, S. Miyano, S. Suri (Eds.), Algorithms and Computation. Proceedings, 1996. X, 448 pages. 1996.

Vol. 1179: J. Jaffar, R.H.C. Yap (Eds.), Concurrency and Parallelism, Programming, Networking, and Security. Proceedings, 1996. XIII, 394 pages. 1996.

Vol. 1180: V. Chandru, V. Vinay (Eds.), Foundations of Software Technology and Theoretical Computer Science. Proceedings, 1996. XI, 387 pages. 1996.

Vol. 1181: D. Bjørner, M. Broy, I.V. Pottosin (Eds.), Perspectives of System Informatics. Proceedings, 1996. XVII, 447 pages. 1996.

Vol. 1182: W. Hasan, Optimization of SQL Queries for Parallel Machines. XVIII, 133 pages. 1996.

Vol. 1183: A. Wierse, G.G. Grinstein, U. Lang (Eds.), Database Issues for Data Visualization. Proceedings, 1995. XIV, 219 pages. 1996.

Vol. 1184: J. Waśniewski, J. Dongarra, K. Madsen, D. Olesen (Eds.), Applied Parallel Computing. Proceedings, 1996. XIII, 722 pages. 1996.

Vol. 1185: G. Ventre, J. Domingo-Pascual, A. Danthine (Eds.), Multimedia Telecommunications and Applications. Proceedings, 1996. XII, 267 pages. 1996.

Vol. 1186: F. Afrati, P. Kolaitis (Eds.), Database Theory - ICDT'97. Proceedings, 1997. XIII, 477 pages. 1997.

Vol. 1187: K. Schlechta, Nonmonotonic Logics. IX, 243 pages. 1997. (Subseries LNAI).

Vol. 1188: T. Martin, A.L. Ralescu (Eds.), Fuzzy Logic in Artificial Intelligence. Proceedings, 1995. VIII, 272 pages. 1997. (Subseries LNAI).

Vol. 1189: M. Lomas (Ed.), Security Protocols. Proceedings, 1996. VIII, 203 pages. 1997.

Vol. 1190: S. North (Ed.), Graph Drawing. Proceedings, 1996. XI, 409 pages. 1997.

Vol. 1191: V. Gaede, A. Brodsky, O. Günther, D. Srivastava, V. Vianu, M. Wallace (Eds.), Constraint Databases and Applications. Proceedings, 1996. X, 345 pages. 1996.

Vol. 1192: M. Dam (Ed.), Analysis and Verification of Multiple-Agent Languages. Proceedings, 1996. VIII, 435 pages. 1997.

Vol. 1193: J.P. Müller, M.J. Wooldridge, N.R. Jennings (Eds.), Intelligent Agents III. XV, 401 pages. 1997. (Subseries LNAI).

Vol. 1196: L. Vulkov, J. Waśniewski, P. Yalamov (Eds.), Numerical Analysis and Its Applications. Proceedings, 1996. XIII, 608 pages. 1997.

Vol. 1197: F. d'Amore, P.G. Franciosa, A. Marchetti-Spaccamela (Eds.), Graph-Theoretic Concepts in Computer Science. Proceedings, 1996. XI, 410 pages. 1997.

Vol. 1198: H.S. Nwana, N. Azarmi (Eds.), Software Agents and Soft Computing: Towards Enhancing Machine Intelligence. XIV, 298 pages. 1997. (Subseries LNAI).

Vol. 1199: D.K. Panda, C.B. Stunkel (Eds.), Communication and Architectural Support for Network-Based Parallel Computing. Proceedings, 1997. X, 269 pages. 1997.

Vol. 1200: R. Reischuk, M. Morvan (Eds.), STACS 97. Proceedings, 1997. XIII, 614 pages. 1997.

Vol. 1201: O. Maler (Ed.), Hybrid and Real-Time Systems. Proceedings, 1997. IX, 417 pages. 1997.

Vol. 1202: P. Kandzia, M. Klusch (Eds.), Cooperative Information Agents. Proceedings, 1997. IX, 287 pages. 1997. (Subseries LNAI).

Vol. 1203: G. Bongiovanni, D.P. Bovet, G. Di Battista (Eds.), Algorithms and Complexity. Proceedings, 1997. VIII, 311 pages. 1997.

Vol. 1204: H. Mössenböck (Ed.), Modular Programming Languages. Proceedings, 1997. X, 379 pages. 1997.